Air and Water

Mark W. Denny

The Biology and Physics of Life's Media

Princeton University Press

Published by Princeton University Press, 41 William Street,
Princeton, New Jersey 08540
In the United Kingdom: Princeton University Press,
Chichester, West Sussex

Library of Congress Cataloging-in-Publication Data

Denny, Mark W., 1951–
Air and Water : the biology and physics of life's media / Mark W. Denny.
p. cm.
Based on a symposium lecture given at the annual meeting
of the American Society of Zoologists in 1988.
Includes bibliographical references and index.
ISBN 0-691-08734-2 (CL)
1. Air. 2. Water. 3. Fluid dynamics. I. American Society of
Zoologists. Meeting. II. Title.
QC161.D46 1993
574.19'1—dc20 92-20969

This book has been composed in Linotype Times Roman

Princeton University Press books are printed on
acid-free paper, and meet the guidelines for permanence and durability
of the Committee on Production Guidelines for
Book Longevity of the Council on Library Resources

Printed in the United States of America

10 9 8 7 6 5 4 3

For John Vernon

who continues to remind me why quality ideas are worth having

Contents

Contents

Contents

List of Symbols and Subscripts

Symbol	Definition	Eq. Where First Used
a	Acceleration	3.10
$\mathbf{a}$	Acceleration vector	3.6
A	Area	4.29
$\mathcal{A}$	Amplitude of a sound wave	10.17
b	A coefficient	4.8
B_s	Secant bulk modulus	4.6
c	Speed of a wave (sound, light , or surface)	10.1
c	A coefficient	4.8
C	Concentration	6.31
C_a	Coefficient of added mass	4.31
C_d	Coefficient of drag	4.29
C_ℓ	Coefficient of lift	4.30
C_n	Coefficient of normal force	5.16
C_t	Coefficient of tangential force	5.19
d	Diameter	5.11
	Depth of water column	13.40
$\mathcal{D}$	Diffusion coefficient (various subscripts)	6.19
E	Young's modulus of elasticity	4.14
E	Electric field strength	9.2
f	Frequency	7.26
f'	Shifted frequency	10.45
F	Force	3.10
$\mathbf{F}$	Force vector	3.9
$\mathcal{F}$	Force per length	4.46
Fr	Froude number	13.34
g	Acceleration due to gravity	3.13
Gr	Grashof number	8.11
h	Height	4.12
	Planck's constant	11.8
h_c	Coefficient of heat transfer	8.13
h_m	Coefficient of mass transfer	14.10
H	Wave height	13.10
$\mathcal{H}_r$	Relative humidity	14.3
$\mathcal{H}_s$	Specific humidity	14.32
$\mathbf{i}$	Inertia vector	3.12
I	Current (gas, particles, charges, etc.)	6.40
$\mathcal{I}$	Intensity of sound	10.25
j	Local electrical current flux density (A m^{-2})	9.2
J	Flux (applied to various quantities)	5.46
$\mathcal{J}$	Flux density (various quantities)	6.28
k	Boltzmann's constant	3.15
	Various coefficients	
K_{15}	Conductivity ratio	4.7

List of Symbols and Subscripts

Symbol	Definition	Eq. Where First Used
K	Index of interception	7.77
$\mathcal{K}$	Thermal conductivity	8.2
ℓ	Length	3.11
l	Mean free path	6.52
L	Radiance of a light source	11.16
Le	Lewis number	8.5
m	Mass	3.9
M	Metabolic rate coefficient (W s^{-1} $kg^{-\alpha}$)	5.72
M'	Metabolic rate coefficient (liters O_2 s^{-1} $kg^{-\alpha}$)	8.36
$\mathcal{M}$	Molecular weight	4.3
	Metabolic cost per volume	5.52
n	Number of steps	6.4
	Refractive index	11.12
N	Total number (of molecules, moles, steps, etc.)	4.1
p	Pressure	4.1
P	Power	3.17
Pr	Prandtl number	8.4
$\mathcal{P}$	Power per length	5.51
q	Damping parameter	10.62
Q	Heat	8.1
Q_s	Specific heat capacity	8.2
Q_l	Latent heat of vaporization	14.17
r	Radius	4.19
	Intrinsic rate of increase	7.22
r'	Radius (variable)	5.36
R	Electrical resistance	9.1
Re	Reynolds number (various subscripts)	5.9
$\Re$	Universal gas constant	4.1
S	Salinity	4.8
S_p	Practical salinity	4.8
Sc	Schmidt number	7.62
Sh	Sherwood number	6.25
t	Time	3.3
T	Temperature	3.15
$\mathcal{T}$	Wave period	13.1
u	Velocity along x-axis or speed	3.14
u_*	Shear velocity	7.18
$\mathbf{u}$	Velocity vector along x-axis	3.3
v	Velocity along y-axis	7.3
$\mathbf{v}$	Velocity vector along y-axis	3.4
V	Volume	4.1
$\mathbf{V}$	Overall velocity vector	3.3
$\mathcal{V}$	Voltage	9.1
w	Velocity along z-axis	
$\mathbf{w}$	Velocity vector along z-axis	3.5
W	Energy	3.13
x	Distance along x-axis	3.1

Symbol	Definition	Eq. Where First Used
y	Distance along y-axis	3.6
y_s	Scale height	4.4
z	Distance along z-axis	
	Rouse parameter	7.17
α	Exponent of metabolic scaling	8.19
α_λ	Attenuation coefficient	11.9
β	Thermal expansivity	8.6
γ	Surface tension	12.2
γ_s	Ratio of specific heats (constant p/constant V)	10.10
δ	Step length in random walk	6.4
δ_{bl}	Boundary-layer thickness	7.32
δ_s	Sonic boundary-layer thickness	7.43
ϵ	Eddy diffusivity	7.15
ζ	Fraction of area that is tracheae	6.49
η	Surface elevation	13.19
θ	An angle	3.2
θ_c	Contact angle	12.1
κ	Von Karman's constant	7.18
λ	Wavelength (sound, light, or surface)	10.18
μ	Dynamic viscosity	5.1
ν	Kinematic viscosity	5.10
ξ	Saturation molar fraction	14.12
ρ	Density	4.3
τ	Shear stress	5.3
	Time for one step	6.18
ϕ	An angle	7.26
χ	Conductivity	Table 9.1(b)
ψ	Fraction of shell surface that is pores	6.55
	Resistivity	9.2
Ψ	Index of light transmission	11.27
ω	Angular velocity	7.25

Subscript	
b	Body
e	Effective
max	Maximal
min	Minimal
opt	Optimal
rms	Root mean square
∞	Ambient or mainstream

Preface

This book grew from a symposium lecture I gave at the annual meeting of the American Society of Zoologists in 1988. Margaret McFall-Ngai and Donal Manahan had asked me to provide a brief overview of the physics of air and water as an introduction to talks on how biologists think differently about life in the two media. Because they asked a couple of years in advance, I readily agreed to the task. As the symposium grew closer, however, I began to have second thoughts. What could I possibly say about the biological consequences of fluid physics that hadn't been worked to death in introductory texts?

In a panic, I ran to the library and started to thumb through the *Handbook of Chemistry and Physics*. After a few pages I began to calm down. The variation in viscosity with a change in temperature was much different in air than in water, and that seemed interesting. A few more pages and I discovered that low frequency sounds travel farther through humid air than through dry air. Must be some biological implications in that. Then there was a description of the electrical resistivity of air; it was 20 billion times that of seawater. Wow! That must mean something. And the diffusion coefficient of gases is 10,000 times greater in air than in water. I wonder how that affects metabolism? The more I thumbed, the more fascinated I became, and the chase was on.

The manuscript that resulted gradually grew to the size of a book, and would still be growing to this day if Princeton University Press hadn't imposed a deadline for its receipt. Every day I find a new physical phenomenon I would like to have included, or a new biological example of physics in action. Facts and theories about air and water have become like peanuts to me; once I started consuming them I found it hard to stop. But there are, after all, limits to how many peanuts anyone can eat, and limits to how many facts and theories any reader can be expected to digest. I hope that the text in its present form is not overly filling, and if the readers find find it too short and incomplete, I hope that it has at least whetted some appetites.

A Note about Math

Science texts seem to come in two varieties. In the first, it is assumed that the reader has no mathematical aptitude whatsoever, and math is kept to a bare minimum or left out entirely. In the other, it is assumed that the reader has recently finished a third course in partial differential equations and is hungry for more. In this case, it could appear condescending to include the intermediate steps in an analysis, so these are left out and phrases such as "It can be shown . . ." and "It follows that . . ." are substituted in their place.

It seems to me that this dichotomy in the approach to science writing leaves a large gap into which many interested readers fall. If you are reading this text, for instance, you are likely to have had math at least through introductory calculus. You may have become a bit rusty in the application of this mathematical background, but in the deep recesses of your mind you remember that taking the derivative of a function is equivalent to finding its slope. Algebra and trigonometry, if not second nature to you are at least familiar concepts. To write a text for you that does not include any mathematics would be to deprive you of tools you are capable of using in the application of the ideas presented. On the other hand, you might find it very frustrating to have the mathematics set forth as if you were a practicing mathematician.

In this text I strive for a constructive compromise between these two approaches. I do not spare you the math when I think that the math will prove useful. On the other hand, none of the math is difficult; it is all well within the grasp of a college freshman. As McNeill Alexander once commented, there are a lot of equations in this book, but very little mathematics.

One reason there are so many equations here is because I have tried to include the intermediate steps in each analysis. Thus, with a little perseverance, you should be able to work from the initial concept to the final equation. I try to make it relatively easy for you to follow the mathematical train of thought, and hope that you will thereby gain a more thorough understanding of the final result. Who knows, the practice might even convince you of the utility of a bit of simple math, and thereby serve to knock some of the rust off of your mental gears.

A Note on Nomenclature

There are only twenty-six letters in the Roman alphabet and another twenty-four in the Greek, a shortage that poses a problem for an interdisciplinary book such as this. How can one come up with a unique symbol for each quantity or coefficient with so few symbols from which to choose? I have yet to find a truly satisfactory answer to this problem. Here I have simply done the best I could while trying to follow a few general rules:

- The units of a variable are printed in nonitalic type. Thus newtons (a measure of force) are symbolized by N, which is different from N, a symbol used to denote the number of items. Similarly, the measure of energy, joules, is symbolized by J which is different from J, the symbol for a flux.
- The symbols for variables are printed in italic or calligraphic type, and the different fonts denote different variables. For example, the symbol F is used for force, $\mathcal{F}$ for force per length. A list of symbols follows the table of contents.
- Where possible, the same symbol has not been used for different things. In a few cases, a strong tradition regarding the use of a symbol necessitated its multiple use, but the symbol is carefully redefined at the point where its usage changes.
- Subscripts have been used liberally as a means of differentiating forms of a given variable. The assignment of a subscript is made in the local context, and is meant to serve as a mnemonic. Thus, the subscript a may refer to "air" in one place and "adjusting" in another. Those few subscripts that are consistent throughout the text are given in the list of symbols.
- The symbol k is a special case. As is traditional, k (without any subscript) is used for the Boltzmann constant. However, k with a subscript is used as a general-purpose coefficient, where the subscript is defined locally. Thus k_2 in chapter 6 is different from k_2 in chapter 11.

Acknowledgments

I am indebted to many people for their help in the production of this book. Freya Sommer combined her artistic talent with her biological expertise to create the drawings. The members of Bio. 237 served as guinea pigs for the first draft, with special thanks due to Emily Bell, George Kraemer, George Matsumoto, and Sam Wang for their insight and ideas. John Gosline and Michael LaBarbara read the entire manuscript and provided valuable guidance (and quite a few references) for the final draft. Emily Bell, Kristina Mead, and Brian Gaylord checked my math, corrected my grammar, and laughed at my jokes, for which I am grateful. I hope that the review process has weeded out the worst of the errors in the text; any remaining mistakes are, of course, my responsibility.

The library at Hopkins Marine Station served as both a haven in which to write and a bountiful source of physics facts. Through the labors of Alan Baldridge, Susan Harris, and Mary West, it is simply the best library there is.

My profound thanks to Donald Knuth for creating TeX, without which the typing of equations would have done me in.

It was again a pleasure to interact with Alice Calaprice of Princeton University Press in her continuing quest to teach me how to write. Emily Wilkinson and Eileen Reilly guided this publication project through its many travails, for which I am most grateful.

And finally thanks are due to Sue, my wife, and to my children Katie and Jim for putting up with the many evenings I came home with my mind on air and water when it should have been on family and dinner.

Air and Water

Chapter 1

Introduction

This is a book about air and water, and how the physical differences between them affect life on earth. It should come as no surprise to biologists that the medium in which an organism lives can influence how it functions. After all, many of the basic attributes of life—the size and shape of an organism, how it moves and reproduces, how it captures food, the nature of its sensory capabilities—differ in familiar patterns depending on whether the plant or animal lives in air or water. Fish, shrimp, whales, and kelps live in the sea and no one would mistake them for something that lives on land. By the same token, redwoods, hummingbirds, giraffes, and dragonflies are easily identified as being terrestrial. But most biologists are relatively unfamiliar with how the physics of fluids (in this case, air and water) can help to explain the functional divergence between terrestrial and aquatic life. Therein lies the purpose of this text.

It would be an immense task to enumerate all the known ways in which the physics of air and water have influenced biology, and I will not attempt it here. Instead, this book is limited to two restricted objectives. The first is quite mundane: to provide a compilation of the pertinent physical characteristics of air and water. Although this information can be excavated (with considerable effort) from the literature, there seems to be no single source in which the biologically relevant facts are conveniently assembled. Even when information is obtained, chances are that it is presented in arcane or outdated units that must be transposed before they become useful. Here I provide a brief description of each of the pertinent physical properties of air and water, with values tabulated in standard, Système Internationale (SI) units.

The second objective is to bring attention to how the physics of fluids influences and constrains life. In part, this is an exercise in examining how a terrestrial existence has biased our perception of biology. Some of the physical differences between air and water—for instance, the fact that water is much denser—are so familiar that we often forget to think of them as important. Others—such as the fact that the electrical conductivity of seawater is 20 billion[1] times higher than that of air—are, from our terrestrial perspective, so strange that we fail even to consider the consequences. What would it be like to live in a conductive fluid? The discussion that follows should provide a more objective (or at least different) viewpoint from which to examine life.

This viewpoint can be of particular utility when examining the process of evolution because it provides a rare, time-invariant standard against which life can be measured. There are precious few aspects of biology that have remained constant from the beginning. Perhaps the structure and basic function of DNA are the same now as they were in the Archean, but the body forms and physiological processes for which DNA is the code have been continuously altered through the course of evolution. The physics of air and water, however, were the same for the trilobites and dinosaurs of yesteryear as they are for the lobsters and bears of today.

Exploring the differences between air and water is also a useful exercise in thinking about biology in a physical context. Recent advances in biotechnology can lull even trained biologists into thinking that, through genetic engineering, life

[1] This is a U.S. billion, 1000 million.

can be largely independent of physics, capable of anything. Want to see bacterium that prevents strawberries from freezing? Fiddle for a few weeks in the lab, and voila! How about a tobacco plant that glows in the dark like a firefly? No problem! In these heady days of research triumphs, it is easy to lose sight of the fact that physics still places severe constraints on biology. By exploring the ways in which terrestrial and aquatic organisms have diverged, we are reminded of the importance of physics in understanding how plants and animals work.

This discussion of the physics of air and water takes the form of a series of examples, the selection of which is somewhat biased. Where possible, I avoid the overworked, obvious examples found in introductory texts, and concentrate instead on cases that seem to provide a new perspective. The resulting collection is admittedly quirky. In particular, there is an emphasis on examples that explore how nature might have done things, but for one reason or another has not. In one sense, these "what if ?" examples are a kind of science fiction—the biology of nonexistent organisms—but they can be a convenient vehicle for illustrating a particular point. Furthermore, the examples considered here are presented not so much for their factual content or because they are intended to give a coherent picture of aquatic versus terrestrial life, but rather because they might provide the impetus for creative thought on the biological message of environmental media.

Chapter 2

The Fluid Environment

The choice of subject matter in this text rests on the difference between *fluids*, which we will discuss, and *solids*, which we will not. We concern ourselves primarily with two particular fluids. In one case it is air, a *gas*. As with other gases, air has no defined shape and no defined volume. Our second fluid is water, a *liquid*. It has a defined volume but no defined shape. The physical characteristic that unifies gases and liquids is *viscosity*, their resistance to the rate of deformation. If one deforms a gas or liquid, the force required depends solely on the rate of deformation. For example, a force is required to move a spoon through honey; the faster the fluid is stirred, the more force is required. Note, however, that it does not matter how far the honey is stirred; as long as a constant force is applied, the fluid deforms at a constant rate.

This characteristic of fluids is in contrast to that of solids. A solid is characterized by stiffness—the force required to attain a certain *amount* (rather than rate) of deformation. For example, if one hangs a weight from a rubber band, the rubber, being a solid, deforms a certain amount. This amount is independent of how long the force is applied, and is instead proportional only to force.

Why should we choose to discuss fluids instead of solids? Any complete treatment of the subject would of course deal with both, but would also be unduly long and complex. Here we choose fluids primarily because all living things exist in a fluid medium, either air or water, whereas many organisms are seldom in contact with a solid substratum. In this sense, the properties of air or water as an environment are more basic (or at least more biologically pervasive) than those of solids. At times we will deal with some of the properties of solid objects, but only insofar as they provide insight into the fluid environment.

Before we embark on our study of air and water, let's explore a general context in which to view life's media.

2.1 Size

The earth is a sphere[1] with a radius of 6,371 km, and therefore has a surface area of 5.1×10^8 sq km. Roughly 71% of this area is slightly depressed relative to its surroundings; these depressions form the ocean basins. On average, these basins are 3794 m deep. The volume of water required to fill the oceans is thus 1.37×10^9 km^3, the equivalent of a sphere with a radius of 689 km. In other words, if the water in the earth's oceans were formed into a separate planet, it would be roughly one-tenth the diameter of the earth, about half the diameter of the moon.

The oceans contain virtually all of the earth's free water. For example, the water in lakes and rivers amounts to less than 0.01% of the water available to living things. In contrast, about 2.5% of the earth's water is frozen in ice or sequestered as groundwater (Gross 1990).

The size of the atmosphere is less easily quantified. As we will see in chapter 4, there is no distinct upper edge to the atmosphere; air rarefies with altitude and

[1]Actually, the earth is a spheroid, the distance from its center to a point on the equator being, on average, 21 km greater than the distance from the center to one of the poles. But for present purposes, this small ellipticity is negligible. If the earth were the size of a billiard ball, even the most discerning pool player wouldn't notice that it was out of round.

approaches the vacuum of outer space at an altitude of about 100 km. But no plant or animal strays anywhere near that far from the earth's surface. For instance, the tallest plant ever recorded (a Douglas fir) pushed its uppermost tip a mere 126.5 m above the ground (McFarlan 1990). Birds and insects can fly to relatively great heights; a Ruddell's vulture once met its demise when it collided with an airliner at an altitude of 11 km (McFarlan 1990). But most animal aerialists are physiologically incapable of flying at such altitudes, and we can arbitrarily set 10 km as the biological upper limit of the atmosphere without fear that the limit will often be broken. Using this limit, we can calculate that the biologically useful volume of the atmospheric environment is 5.1×10^9 km^3, about 3.7 times that of the oceans.

2.2 Temperature

Over what range of temperatures do air and water occur on the earth?

For water, the limits are precisely defined. Fresh water freezes at 0° C, and, by becoming a solid, is at that point removed from our consideration. Seawater freezes at about −1.9° C. At sea level, fresh water boils at 100° C, and seawater at 102° C. Both the freezing and boiling points of water are reached some place on earth. On the cold end, we note that about 2% of the earth's water is frozen, mostly in the polar ice caps and in glaciers. Furthermore, much, if not most, of the water in the oceans is very nearly frozen. The deep waters of all the earth's oceans remain at a temperature of about 1° C year-round, and the surface waters in polar regions freeze in winter.

In contrast, surface waters in the tropics are effectively heated by the sun and have an annual average temperature of 26° to 28° C. It is unusual for the oceans to be any warmer than this due to the rate at which they evaporate, a topic we will return to in chapter 14. Shallow lakes and rivers can be warmer than the ocean, reaching temperatures of perhaps 40° C. There are a few places on earth where water is naturally heated to its boiling point (the geysers and hotsprings of Yellowstone National Park and thermal vents in the ocean floor, for instance), but very few plants and animals inhabit these cauldrons. As a general rule, the inhabited waters of earth have a temperature range between 0° C and 40° C.

The absolute temperature range of the atmosphere is not so neatly bounded. The coldest temperature ever recorded on the earth's surface was −89° C at Vostok in the Antarctic on July 22, 1983 (McFarlan 1990). The hottest was 58° C recorded at Al'Aziziyah, Libya, on September 13, 1922 (McFarlan 1990). But there aren't many plants or animals that live on the Antarctic plateau or in the middle of the Sahara, so these absolute extremes are not representative of the practical extremes experienced by most organisms. Few terrestrial animals are active at temperatures below 0° C (the odd polar bird and mammal being the only major exceptions), and even in the desert the air temperature seldom rises above 40° C. In the discussion here, we use this restricted range of atmospheric temperatures (0° to 40° C).

2.3 Speed

Occasions will arise on which we need to know how fast a fluid moves relative to an organism, and it will be useful to explore a few representative examples here. These and other examples are considered in greater depth in later chapters.

First consider the speed of water relative to a stationary plant or animal. The lower limit is zero, but this is a situation that is difficult to find in nature. Even in the deep waters of lakes and oceans, currents of 1 to $2\,\mathrm{cm\,s^{-1}}$ are common. Near the surface of lakes and oceans, and in rivers and streams, organisms have to contend with more rapid water movements. The run-of-the-mill ocean current moves at about $10\,\mathrm{cm\,s^{-1}}$, but in portions of the Gulf Stream (the Florida current) and the Agulhas current off the coast of South Africa speeds of 2 to $3\,\mathrm{m\,s^{-1}}$ are attained.[2] In mountain streams, water may reach speeds of 5 to $10\,\mathrm{m\,s^{-1}}$. The greatest speeds regularly attained by moving water are those found in the surf of wave-swept shores. Velocities as high as 14 to $16\,\mathrm{m\,s^{-1}}$ have been measured (Vogel 1981; Denny et al. 1985), and speeds associated with large storm waves may reach $20\,\mathrm{m\,s^{-1}}$. In summary, naturally occurring water velocities range from 0 to $20\,\mathrm{m\,s^{-1}}$.

Wind speeds have a much broader range. The lower limit is zero, but again this is a situation seldom realized in nature. Even on the calmest of days there is always some breeze, even if only that caused by thermal convection. Except in unusual circumstances, it would be difficult to find air moving slower than 5 to $10\,\mathrm{cm\,s^{-1}}$. The upper extreme for atmospheric speeds near the ground is set by storm winds. Thunderstorms and gales may be accompanied by winds ranging to about $30\,\mathrm{m\,s^{-1}}$. Hurricane winds can exceed $45\,\mathrm{m\,s^{-1}}$, and there are extreme cases where wind speeds have reached $100\,\mathrm{m\,s^{-1}}$. The highest steady wind speed ever recorded was $103\,\mathrm{m\,s^{-1}}$ at Mount Washington, New Hampshire, on April 12, 1934, exceeded only by a brief gust of $125\,\mathrm{m\,s^{-1}}$ recorded in a tornado at Wichita Falls, Texas, on April 2, 1958 (McFarlan 1990). In summary, wind speeds vary from 0 to about $100\,\mathrm{m\,s^{-1}}$.

These are fluid velocities relative to a stationary organism, but speed can also be had by moving an organism relative to a stationary fluid. How fast do animals move through water and air?

The fastest aquatic animals are fishes, probably the yellow-fin tuna (*Thunnus albacares*) and the wahoo (*Acanthocybium solandri*), both of which have been clocked at about $21\,\mathrm{m\,s^{-1}}$ in 10 to 20 s bursts (Walters and Fierstine 1964). The fastest aquatic mammal, the killer whale (*Orcinus orca*), is somewhat slower at about $15\,\mathrm{m\,s^{-1}}$ (McFarlan 1990).

The fastest aerial animal is the peregrine falcon, *Falco peregrinus*, which can reach a speed of $97\,\mathrm{m\,s^{-1}}$ in a stoop. In level flight, ducks and geese can sustain speeds of about $29\,\mathrm{m\,s^{-1}}$, while the fastest insect, the deer bot-fly (*Cephenemyia pratti*), can reach $16\,\mathrm{m\,s^{-1}}$ in short bursts (McFarlan 1990).

It seems, then, that animals can propel themselves through fluids (either air or water) at about the same speeds that winds and currents move relative to the stationary earth. In other words, the maximum *relative* speed is the same for both sedentary and motile organisms.

2.4 History

All of the values cited so far are for the earth as it appears today. But if we are going to use the physics of air and water to try to understand the process through which today's living things evolved, we need to know something about how the

[2]If you are more comfortable thinking in terms of miles per hour than in meters per second, there is an easy conversion: $1\,\mathrm{m\,s^{-1}} \approx 2.2$ mph. Conversely, 1 mph $\approx 0.46\,\mathrm{m\,s^{-1}}$.

Chapter 2

basic characteristics of the fluid environment haved changed over the course of biological history.

The earth is approximately 4.6 billion years old, and the first living things (probably organisms akin to bacteria) appeared about 3.3 billion years ago.[3] The first eukaryotic organisms (cells with a distinct nucleus and chromosomes) arose approximately 1.3 billion years ago, and multicellular organisms made their appearance in the fossil record about 560 million years ago, at the beginning of the Cambrian period. Thus, while life has been present on earth for an astoundingly long time, the bulk of organismal evolution, in terms of the functional radiation of multicellular plants and animals, has a relatively short history.

It is interesting to place the history of life on earth in an absolute context. Current estimates for the age of the observable universe range from 10 to 20 billion years, with the best guess being 12 billion years. Thus, life on earth has been present for fully a quarter of the time the universe has been in existence, and multicellular life for about 5%.

The volume of the oceans has increased over the past 4.6 billion years as water was released by the crust and, perhaps, as icy meteorites impacted the earth. The rate of increase has been a matter of debate, but Schopf (1980) suggests the time course shown in figure 2.1. In this scenario, the volume of the ocean has remained virtually constant for the past 2 billion years. It is likely that the "saltiness" of the oceans has been similarly constant (Schopf 1980; Holland 1984).

The composition of the atmosphere has varied substantially during the 3.3 billion years that life has been present on earth. Until approximately 2.3 billion years ago, the atmosphere was composed primarily of carbon monoxide, ammonia, methane and water vapor, a so-called "reducing" atmosphere. With the evolution of photosynthesis in cyanobacteria about 2.3 billion years ago (Schopf 1978), oxygen was released into the environment (fig. 2.2). The accumulation of oxygen in air was slow at first, because the oxygen released was quickly bound to iron, forming vast deposits of iron oxide (rust). By about 1.8 billion years ago, however, all the free iron had rusted and the concentration of free oxygen increased in the air, resulting in the atmosphere we have today, which is composed primarily of nitrogen and oxygen.

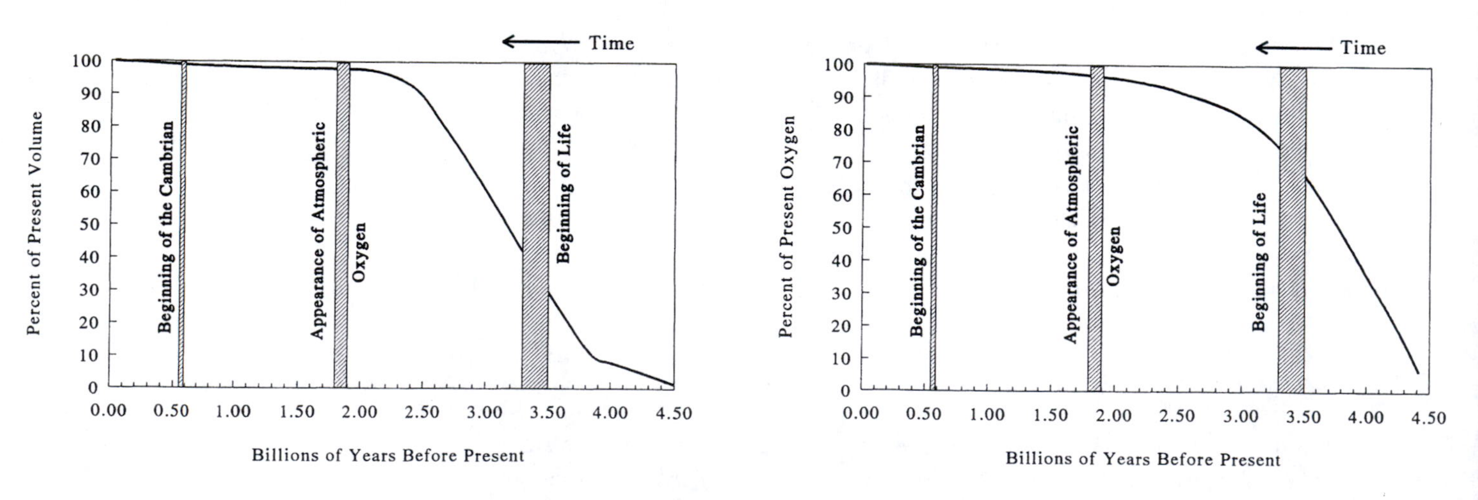

Fig. 2.1 Ocean volume has increased throughout geological time, but the increase has been very slow for the last 2 billion years. (Adapted from Schopf 1980 by permission of Harvard University Press)

Fig. 2.2 Oxygen has slowly been released to the earth's surface throughout geological time. Note that this figure refers to *all* oxygen, not just free, atmospheric oxygen. Until about 1.8 billion years before the present, most of the oxygen released from the crust combined with iron to form rust. Only after the bulk of the iron had rusted could oxygen accumulate in the atmosphere. (Adapted from Schopf 1980 by permission of Harvard University Press)

[3]The history cited here is that according to Schopf (1980). Estimates of the timing of events are likely to change slightly as new evidence becomes available.

In summary, the volume and composition of the ocean and atmosphere have been virtually constant for the last 1.5 to 2.0 billion years. Certainly for the 560 million years in which multicellular organisms have existed, the size and chemistry of the fluid environment has changed very little.

The temperature of the earth's surface is a subject of much current concern and speculation. It is likely, however, that the average temperature of tropical ocean surface waters has varied little through time (less than 5° C, fig. 2.3), while deep-ocean water has varied by about 13° C, from 1° to 14° C. Average air temperature has varied through a range of perhaps 15° C, 10° C warmer than now in the Cretaceous to about 5° C colder during times of worldwide glaciation (Walker et al. 1983; Cloud 1988). Although these variations in temperature have undoubtedly had profound effects on global ecology, in terms of the physics of air and water they are quite small. At no point in the past 560 million years did the average temperature of either air or water get any colder or any hotter than that found somewhere on earth at present. In other words, in the course of evolution, life has had to contend with a variation in fluid temperature similar to that found today.

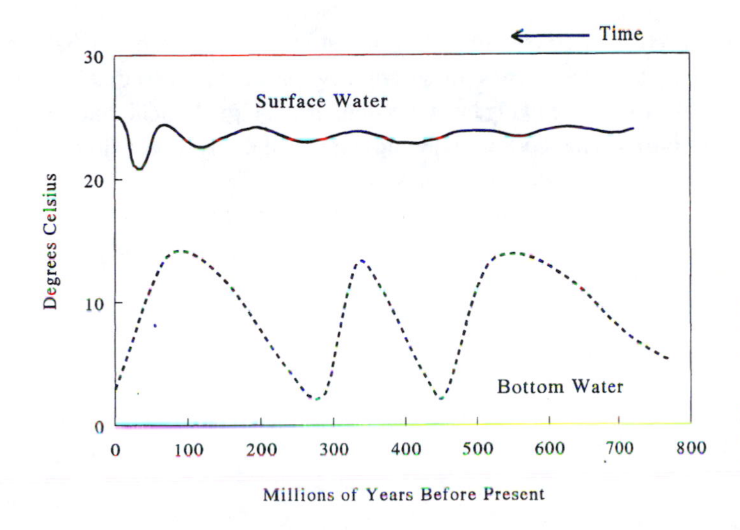

Fig. 2.3 Ocean surface temperature has been virtually constant for the last 800 million years, whereas the temperature of bottom water has varied by as much as 13° C. The major dips in the temperature of bottom water 270 and 450 million years ago coincide with episodes of worldwide glaciation. (Adapted from Schopf 1980 by permission of Harvard University Press)

The extreme speeds at which air and water flowed in ancient times are a matter of speculation, but it seems unlikely that they were much different from those of today. For example, maximal wind speeds are associated with storms, and the most intense storms (hurricanes) are driven by the heat energy stored in surface water of the tropical ocean. Because ocean surface temperatures have never been substantially higher than those now present, we can guess that ancient hurricanes were no more severe than those with which plants and animals must contend today.

2.5 Crossing the Boundary

Life on earth began in the oceans and then extended into fresh water. Only with the advent of a relative high concentration of oxygen in the atmosphere (and consequently, of ozone to shield the earth's surface from much of the sun's ultraviolet light) did plants and animals emerge onto the land. The details of this emergence are still murky, but by the beginning of the Devonian period (about 400 million years ago) two groups of plants—the pteridiophytes (clubmosses, horsetails, and ferns) and bryophytes (liverworts, hornworts, and mosses) had clearly made the transition. These pioneering plants gave rise to the angiosperms and gymnosperms of the present. The first terrestrial animals were scorpionlike (Størmer 1977) and these were followed by subsequent invasions of molluscs (snails and slugs), nemerteans, flatworms, annelids (earthworms), crustaceans (sow bugs), uniramians (the insects, centipedes, and millipedes), and vertebrates.

The transition between air and water has not been one way. Several terrestrial groups have returned to the sea. For example, flowering plants have moved into the oceans in the form of sea grasses, and vertebrates have made this reverse transition at least seven times—cetaceans (whales), seals, sea lions, penguins, ichthyosaurs, plesiosaurs, and pleiosaurs.

These several, separate transitions from water to air and back again provide the basic biological subject matter of this book. How were these transitions accomplished? What changes in shape, size, and physiology were necessary to allow plants and animals to leave the ancestral waters and move into air? What shifts were made in the limits on life?

2.6 Summary

Air and water provide two vast arenas in which life has unfolded. The basic characteristics of these arenas—their size, temperatures, and characteristic speeds—have likely been constant through geological time, and therefore provide a basic framework in which to explore the physics of fluids and its effect on the evolution of plants and animals.

Chapter 3

Thoughts at the Beginning: Basic Principles

Our discussion of the physics of air and water will be assembled with the assistance of a small but important toolbox of basic ideas. Like the hammers, screwdrivers, and wrenches used in the construction of a building, these tools of physics are called upon to perform the host of mundane functions that lead to a finished product. A working understanding of these tools and of the manner in which they can be used safely will allow one to follow along better as various intellectual constructs are assembled, and to appreciate better the beauty and complexity of the final structure. Therefore, it is best to spend a few pages here at the beginning to provide explicit definitions of basic principles and to introduce the terminology applied to them.

If you have a comfortable working knowledge of basic physics, feel free to skip this chapter and move on to the biology in chapter 4.

3.1 The Coordinate System

We begin by establishing a coordinate system through which objects can be located in space. Although polar and spherical coordinate systems will be used sporadically, the familiar, orthogonal, Cartesian coordinate system (fig. 3.1A) serves quite well for most of the situations encountered in this text. To use this coordinate system we need only decide how to align the axes and where to locate the origin. We follow the tradition of most introductory physics texts and specify that the x and z axes lie in a horizontal plane, leaving the y axis to point vertically. The origin is located wherever convenient for a particular situation. For example, when discussing the change in atmospheric pressure with altitude, it is most convenient to locate the origin at sea level, and positive values of y extend upward. The same location serves well for discussing depths in the ocean, but in this case it is more convenient to have positive values of y extend downward. Other locations of the origin will be used as the need arises.

Angles are measured relative to one of the axes (fig. 3.1B). By convention, positive angles increase as the angle increases counterclockwise.

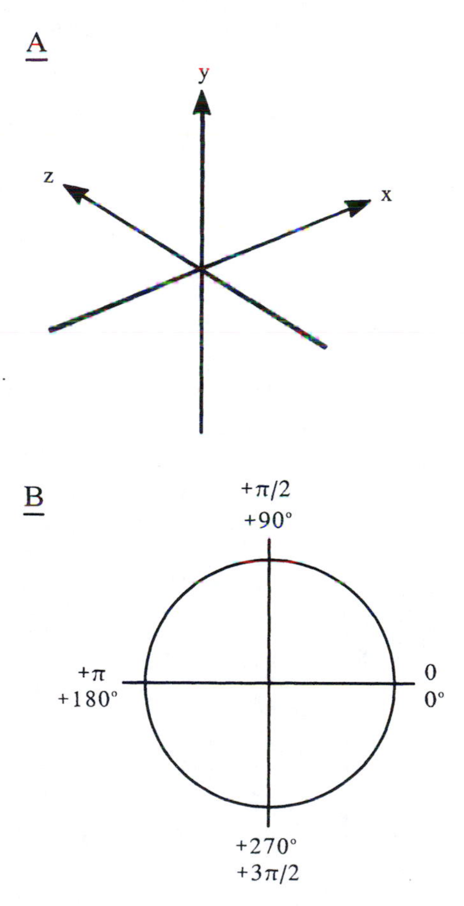

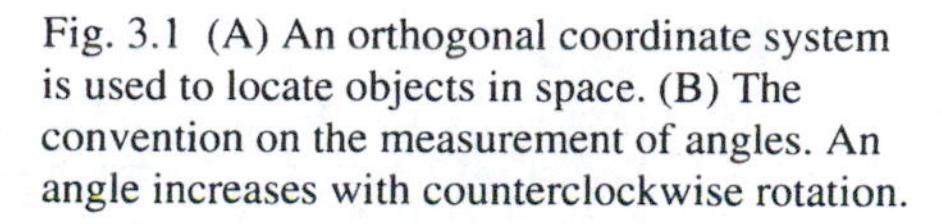

Fig. 3.1 (A) An orthogonal coordinate system is used to locate objects in space. (B) The convention on the measurement of angles. An angle increases with counterclockwise rotation.

3.2 Dimensions and Units of Measure

Having provided ourselves with a coordinate system, we can now proceed to measure things. The basic quantities we need to measure are *size*, *mass*, and *time*. From these, all other quantities can be constructed, as we will see.

Size can be defined in any of three ways. For instance, it is traditional for the size of a fish to be measured by its length, while the size of a plot of land is measured by its area. The size of a glass of beer is measured by its volume. It is fortunate for the simplicity of our discussion that these three measures of size are interrelated. Area has the *dimension* of length squared, L^2, and volume, the dimension of L^3. Thus for any of its three definitions, size can be expressed in terms of a linear dimension, L.

The SI *unit* for the dimension of size is the *meter*, m. Thus, length is expressed in meters, area in m^2, and volume in m^3. The length of a meter was originally defined as one ten-millionth the distance from the earth's north pole to its equator. The best estimate of this length—based on a survey in 1791 of the difference in latitude between Dunkerque and Barcelona—was marked on a platinum bar that for many years served as the standard for length. More recently, the length of a meter has been redefined in more portable fashion. The accepted measure is now 1,650,763.73 times the wavelength in a vacuum of a certain frequency of orange-red light (the unperturbed transition between the level $2p_{10}$ and $5d_5$ of the atom of Krypton 86, to be exact).[1]

For complex reasons of history, the English-speaking countries use a different standard of length, the yard. Originally the yard was defined as the distance from the tip of King Henry I's nose to the end of his outstretched thumb, but on July 1, 1959, the standard yard was redefined to be exactly 0.9144 m, making the conversion from the English measure of length to SI a simple matter of remembering four digits.

The dimension of mass, symbolized as M, has as its SI unit the kilogram, kg. Why this standard unit should be expressed as a thousand of some other unit (the gram) is not readily apparent, but in this case, ours is not to reason why. Kilograms have been decreed, and here they will be used. One kilogram is defined as the mass of one cubic decimeter ($0.001\ m^3$) of pure water at the temperature of its maximum density. It is interesting to note that this definition of mass therefore relies directly on the definition of size. In practice, the standard mass is a cylinder of platinum-iridium alloy residing at the International Bureau of Weights and Measures in Sévres, France. The English unit for mass (the quaintly named *slug*[2]) is equal to 0.06852177 kg, not quite as handy a conversion factor as that for the yard.

The dimension of time is given the symbol T and its standard unit is the second, s. Originally defined as 1/86,400 the length of the mean solar day (based on the tropical year for 1900), the unit of time has, like the unit of length, recently been redefined. The new standard second is equal to the duration of 9,192,631,770 periods of the radiation resulting from the transition between the two hyperfine levels of the fundamental state of the atom of Cesium 133. Here, the English agree with the rest of the world, and no conversion from English to SI seconds is necessary.

3.3 The Measure of Motion

The dimensions of length and time provide us with a means to measure the rate and direction with which an object moves in our coordinate system. Consider an object (a small turtle, for instance) whose position we have just measured. We note that at time t_1 the turtle was located at (x_1, y_1, z_1) relative to the northeast corner of the floor in our laboratory. We then go out for a cup of coffee, and upon returning at time t_2 find our turtle positioned at (x_2, y_2, z_2). During the interval $\Delta t = t_2 - t_1$ the turtle has moved a distance,

$$\Delta\ell = \sqrt{(x_2 - x_1)^2 + (y_2 - y_1)^2 + (z_2 - z_1)^2}. \tag{3.1}$$

The ratio $\Delta\ell/\Delta t$ (a quantity having the dimensions LT^{-1}) is the average *speed* of the turtle during the period of measurement. If we have measured t, x, y, and z in SI units, the resulting speed has the units of $m\ s^{-1}$. Speed, however, is only part of

[1] In this chapter, all physical facts not attributed to a particular source are taken from Weast (1977).

[2] The *pound*, which is commonly thought of as a measure of mass, is actually a force. Pounds are equal to the product of slugs and the acceleration due to gravity.

the information we have regarding the movement of our object. From the initial and final location of the turtle we can deduce its average direction of motion. For instance, in the x-y plane, the turtle has moved at an angle θ_x relative to the x axis (fig. 3.2):

$$\theta_x = \arctan \frac{y_2 - y_1}{x_2 - x_1}. \tag{3.2}$$

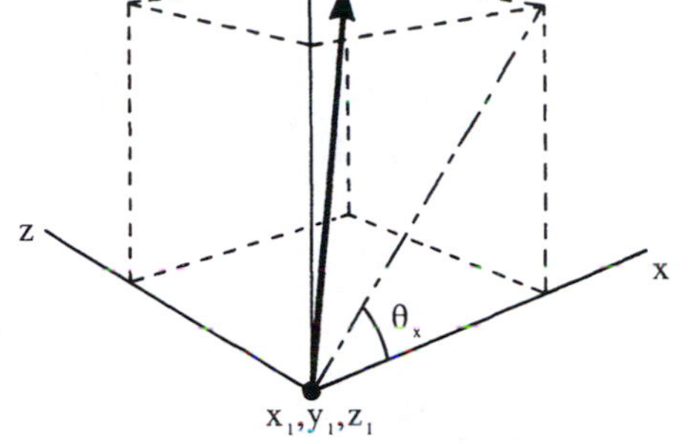

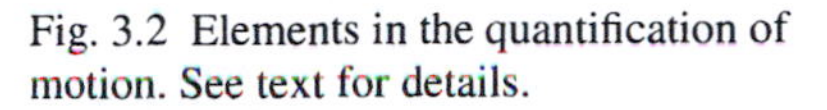
Fig. 3.2 Elements in the quantification of motion. See text for details.

In a similar fashion, the direction of average motion can be measured relative to the other axes. When we take into account both speed and direction we are, in essence, defining a vector quantity, the *velocity*, **V**. Thus, the component of an object's speed parallel to the x axis is the velocity, $\mathbf{V}_x = \mathbf{u}$. Motion along the y and z axes results in the velocity components $\mathbf{V}_y = \mathbf{v}$ and $\mathbf{V}_z = \mathbf{w}$, respectively.

The velocity we have dealt with in this simple example is an average velocity, calculated on the basis of measurements at two discrete points in time. This average suffices for some applications, but it carries within it the possibility of overlooking important information. For example, while we were out drinking coffee between measurements, our turtle could have traveled to Timbuktu by way of Buenos Aires before arriving back in our lab at (x_2, y_2, z_2). If this were the case, its instantaneous velocity would have at times been much higher than the average velocity we calculated. A precise analysis of the time-varying, instantaneous aspects of motion requires the application of a set of mathematical methods collectively referred to as *differential calculus*. In fact, this sort of analysis formed the initial impetus for the development of calculus. A review of this methodolgy is beyond the scope of this book. Instead we proceed on the assumption that the reader has at least a rusty knowledge of what it means to take the derivative of a function. The instantaneous components of the velocity vector **V** are

$$\mathbf{V}_x = \mathbf{u} = \frac{dx}{dt} \tag{3.3}$$

$$\mathbf{V}_y = \mathbf{v} = \frac{dy}{dt} \tag{3.4}$$

$$\mathbf{V}_z = \mathbf{w} = \frac{dz}{dt}. \tag{3.5}$$

Note that **u**, **v**, and **w** are themselves vectors (they have both magnitude and direction), and it is quite possible to work with them in this capacity. A primer on vector calculus (e.g., Schey 1973) will tell you how to add and subtract vectors, plus inform you about several operations (vector products, etc.) that are not possible for simple scalars. But many of these methods, although useful and therefore desirable, aren't absolutely necessary for the physics we explore in this text. With a little extra work, the tricky parts of vector calculus can be avoided. For example, in most instances we can justifiably align our coordinate system so that motion occurs along a single axis. In this case, we can talk about velocity as if it were a simple speed without loss of information. In the few cases where this stratagem won't work, we can avoid at least the notation of vector calculus by treating each component of motion separately, applying the Pythagorean theorem at the end to arrive at a final result.

At times, it will be convenient to treat velocity in a casual fashion in which the directionality of the vector is implicitly understood from the nature of the problem. In these cases, we refer to velocity as speed and give it the symbol u.

The temporal rate of change of velocity is *acceleration*, **a**. Like velocity, acceleration is a vector, and we can easily write its components:

$$\mathbf{a}_x = \frac{d\mathbf{u}}{dt} = \frac{d^2x}{dt^2} \tag{3.6}$$

$$\mathbf{a}_y = \frac{d\mathbf{v}}{dt} = \frac{d^2y}{dt^2} \tag{3.7}$$

$$\mathbf{a}_z = \frac{d\mathbf{w}}{dt} = \frac{d^2z}{dt^2}. \tag{3.8}$$

As with velocity, there are times in which it is convenient to treat casually the directional nature of acceleration. In these cases, we use the symbol a.

To help you keep track of the many symbols used in this text, see the list that immediately follows the table of contents.

3.4 Newton's Laws of Motion

Newton's three laws of motion comprise the basis for parlaying length, mass, and time into the myriad other measures of the physical world:

1. *In the absence of a net applied force, the state of motion of an object remains constant.* In other words, if an object is at rest relative to our coodinate system, it stays at rest unless an unbalanced (or *net*) force is applied. If an object is moving at a constant velocity, its velocity (speed, direction, or both) can be altered only through the application of a net force.
2. *The application of a net force results in the acceleration of a mass.* The net force required to achieve a given acceleration is proportional to the mass, m, of the object.
3. *If object A pushes on object B with a certain force, object B automatically pushes back on object A with a force of equal magnitude acting in the opposite direction.*

The second of these laws is used as a definition of force, **F**:

$$\mathbf{F} = m\mathbf{a}, \tag{3.9}$$

or, in casual form,

$$F = ma. \tag{3.10}$$

Force is that which imparts an acceleration to a mass. As a definition of force, Newton's second law of motion forms the foundation of much of physics. It should be noted, however, that this definition is less straightforward than it might first appear. If we know the acceleration of an object and can measure its mass, we can calculate the net force. We have discussed how to measure acceleration, but how do you measure mass? The devices we normally use to "measure" mass (e.g., balances and spring scales) actually do so by measuring the force exerted by an object when it is accelerated by gravity.

Consider, for instance, a simple balance (fig. 3.3). If you place the mass to be measured, m_1, on one of the pans, it is acted upon by the acceleration of gravity ($g = 9.81\ \mathrm{m\,s^{-2}}$)[3] and pulls down on the balance arm. This force, $m_1 g$, is known

[3]The acceleration of gravity actually varies slightly from one point on the earth to another. In particular, because the earth is an oblate spheroid rather than a sphere, g is a bit lower at the equator than at the poles.

as the object's *weight*. Because it acts at a distance ℓ_1 from the fulcrum, the weight of the unknown mass results in a *moment* or *torque*, $m_1 g \ell_1$, that tends to rotate the balance arm. To measure the unknown mass, one then adds known masses to the other pan. When the torque exerted by the second pan, $m_2 g \ell_2$, equals that exerted by the unknown mass, the beam is balanced. By knowing ℓ_1, ℓ_2, and m_2 we can solve for m_1:

$$m_1 = m_2 \frac{\ell_2}{\ell_1}. \tag{3.11}$$

In other words, if you have an independent measure of force (in this case $m_2 g$), you can measure mass. But to have this independent measure of force, you already need to know m_2. No matter how you work it, you cannot have independent measures of both mass and force—each is defined only in terms of the other. So which should we choose as our "basic" unit? Intuition, rather than logic, dictates that mass be chosen. It seems intuitively "right" to think of mass as an intrinsic property of objects, whereas our perception of everyday events is that force comes into play only when an object moves.

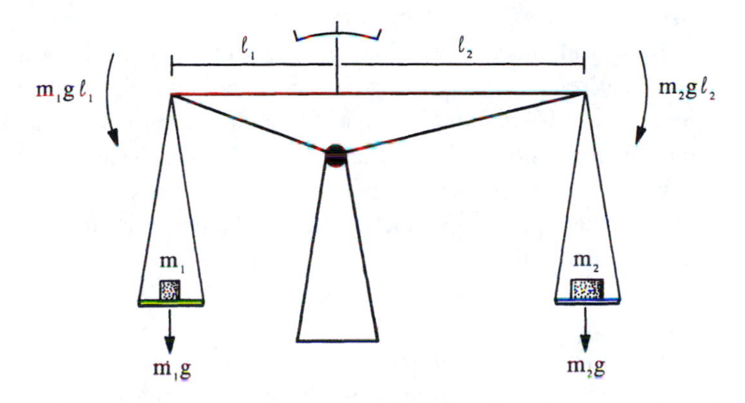

Fig. 3.3 A simple balance compares the moments exerted by two weights.

Given mass as a basic unit, it is easy to arrive at the standard unit for force. Force is the product of mass and acceleration (MLT^{-2}), in SI units, $\mathrm{kg\,m\,s^{-2}}$. In honor of the founder of the modern science of mechanics, the unit of force is called the newton, N.

The *momentum* of an object, **i**, is the product of its mass and its velocity, MLT^{-1}. For example, the x component of momentum is $\mathbf{i}_x = m\mathbf{u}$. Consider two examples: A slowly moving cement truck has a lot of momentum, not because of its velocity, which is low, but because of its great mass. A relatively small shell fired from a cannon may have the same momentum as that of a cement truck, but in this case the momentum is largely due to the shell's great velocity.

The importance of an object's momentum is most tangibly observed when one tries to change it by rapidly decreasing velocity. For instance, a large force must be applied to the brakes of the cement truck to bring it to a sudden halt. Similarly, a large force is required to stop a cannon shell. The explicit connection between momentum and force can be seen by examining the equation for the rate of change of momentum. For an object of constant mass,

$$\frac{d\mathbf{i}_x}{dt} = \frac{dm\mathbf{u}}{dt} = m\frac{d\mathbf{u}}{dt} = m\mathbf{a}_x = \mathbf{F}_x. \tag{3.12}$$

In other words, force and the temporal rate of change of momentum are one and the same. There is no standard SI term for momentum; it is simply expressed as $\mathrm{kg\,m\,s^{-1}}$.

3.5 Pressure and Stress

Often a force is referenced to the area over which it acts. Depending on the situation, this quantity—force per area—is termed *pressure* or *stress*. The two terms are often interchangeable, and the choice of which to use is governed by loose tradition. Pressure and stress have the dimensions $\mathrm{ML^{-1}\,T^{-2}}$, the units $\mathrm{N\,m^{-2}}$, and in SI are expressed as pascals, Pa.

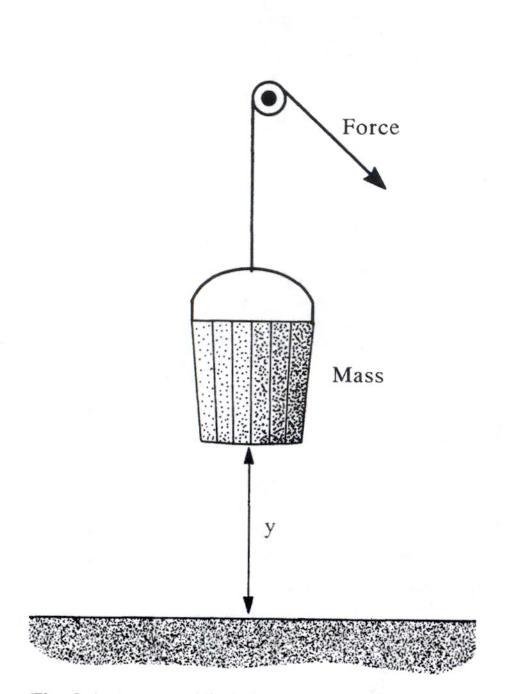

Fig. 3.4 A water-filled bucket on a rope, an example of gravitational potential energy.

3.6 Energy

The concept of *energy* is the next to last of our basic principles. Unfortunately, energy is an accomplished chameleon, changing its appearance to fit the circumstances. We start with its most intuitive form—mechanical energy—and gradually work up to the less intuitive disguises.

Mechanical energy can be thought of as the capacity for doing *work*. When used in this sense, work has a precise definition—it is the product of the net force applied to an object and the distance through which the force moves its point of application. Consider a simple example (fig. 3.4). A bucket of water is attached to a rope, and the rope passes over a pulley. To lift the bucket off the ground, we apply a force to the rope. The product of the applied force and the height to which the bucket is lifted is therefore a measure of the work or energy we have expended in lifting the bucket. Because we have done work on the bucket, the bucket itself now possesses the capacity to do work. For instance, the shaft of the pulley could be attached to a mill; as the bucket falls back to the ground it could provide the energy needed to grind grain. Alternatively, the pulley could be attached to a generator and the fall of the bucket used to produce electricity.

The particular type of energy possessed by the bucket is one form of *potential energy*, in this case, *gravitational potential energy*. It is worth taking a moment to consider this example in greater detail. Assume that the bucket has a mass m, and that the bottom of the bucket lies a distance y above the ground. The force acting on the bucket is simply its weight, that is, its mass times the acceleration due to gravity. As the bucket falls to the ground, it can apply this force, mg, over a total distance, y. Therefore,

$$W = mgy \tag{3.13}$$

is the total amount of work that can be done by the bucket in grinding grain, generating electricity, or in any other energy-requiring activity. In other words, when the bucket is suspended at height y, it has the *potential* for doing this amount of work and is therefore said to have gravitational potential energy equal to mgy.

Because mechanical energy is the product of force and distance, it has the dimensions $\mathrm{ML^2\,T^{-2}}$. In SI, the units are $\mathrm{N\,m}$ (or $\mathrm{kg\,m^2\,s^{-2}}$), a value that is referred to as the joule, J, in honor of James Joule, who conducted pioneering experiments on the relationship between mechanical work and heat.

Gravitational potential energy is only one of many forms of potential energy. For example:

1. A stretched spring has the potential for doing work (e.g., for closing a gate or running a wristwatch), and therefore possesses *elastic potential energy.*
2. When the restraining wires are removed from a bottle of champagne, the compressed gas within the bottle can cause the cork (a mass) to rise to a substantial height against the pull of gravity, in the process demonstrating that a compressed gas has the potential for doing work. This type of energy is termed *pressure-volume potential energy.*
3. The battery in your car has the potential to push electrons through wires, an activity that under normal circumstances requires energy, and the battery therefore possesses *electrical potential energy.*

This list could be extended, but it is better at this point to dwell for a moment on three characteristics that are similar among all types of potential energy.

First, the object possessing the energy may not exhibit any obvious manifestation of being energized. Instead, it merely possesses the *potential* for manifesting its energy in an observable form. For example, one can't tell from the appearance of a champagne bottle or a battery whether or not they contain potential energy.

Second, the amount of work that can be realized from potential energy depends on circumstances outside the object itself. In the above example, we could dig a hole in the ground below the bucket and thereby allow the bucket to fall a greater distance before it came to rest. By thus increasing the effective value of y, we could increase the potential energy of the bucket without ever touching the bucket or the rope by which it is suspended. Similarly, we could reduce the bucket's potential for doing work by placing a stool beneath it, reducing the distance it could fall. The amount of gravitational potential energy clearly depends on how we choose to arrange the environment around the bucket.

The same is true for a champagne cork, which is boosted skyward only because the pressure inside the bottle is greater than the pressure outside. If we were to submerge the bottle sufficiently deep in a lake so that the external pressure was equivalent to the internal pressure, there would be no tendency for the cork to be expelled and no work could be done. Similarly, the potential of a battery for doing work can only be realized when the positive terminal is connected to something with a different electrical potential (e.g., the negative terminal).

And finally, we should make explicit a fact that has been implied throughout this discussion. Potential energy can be transformed into other types of energy—the energy of motion in a ticking clock or a flying champagne cork, the electrical energy of current flowing through a wire. In fact, as long as we don't fiddle with the system by changing our point of reference, the potential energy of a system can *only* be changed by transferring energy to or from another form. Energy, potential or otherwise, cannot appear out of nowhere or simply disappear; it can only be traded back and forth.[4] We will return to this concept—the conservation of energy—later in this chapter.

We turn our attention now to a brief discussion of other forms of energy. The first of these is the energy of objects in motion—*kinetic energy.* The simplest way to understand kinetic energy is to calculate the work required to put an object into motion. Consider an object of mass m, initially at rest. We apply a net force F to the object, which causes the object to accelerate at a constant rate along the x axis

[4] Actually, energy can be traded for mass, but this exchange happens to any appreciable extent only under very extreme circumstances (travel at relativistic speeds, for instance), and need not concern us here.

(fig. 3.5). After a period t, at which time the mass has a speed u, we remove the force and calculate how much work we have done.

During the period in which force was applied, the object's acceleration was equal to the change in speed divided by the time over which speed was changing. This is simply u/t. Thus, the force we must have applied, the product of mass and acceleration, is mu/t. Over what distance was this force applied? The average velocity during the period t is $u/2$, so that the distance traveled during the application of force (the product of speed and time) is $ut/2$ (see fig. 3.5). The work done on the object to bring it to speed u (the product of force and distance) is thus

$$\frac{mu}{t} \times \frac{ut}{2} = \frac{mu^2}{2}. \tag{3.14}$$

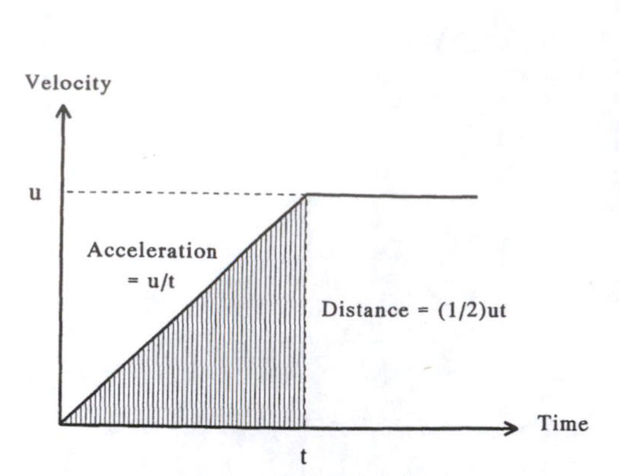

Fig. 3.5 When acceleration is constant, the velocity of an object of known mass increases linearly through time, and the distance traveled can be shown graphically by the area under the velocity curve. Knowing acceleration, mass, and distance, one can calculate the kinetic energy of the object.

This is the object's kinetic energy. Because it has taken energy equal to $mu^2/2$ to accelerate the mass to its present speed, the moving mass now has the capacity to do work. For instance, consider a shell being fired vertically upward from a mortar. As the shell leaves the mortar it has considerable velocity and, therefore, considerable kinetic energy. We could easily arrange things so that the shell exits the mortar exactly at ground level, so that at the instant the shell is fired its gravitational potential energy is zero. As the shell travels upward it continually slows down as a result of the acceleration of gravity. After a time, the shell reaches the top of its flight and for a brief instant hangs motionless. At that instant, the shell's velocity is zero, and so its kinetic energy is zero. But clearly its energy has not disappeared—work has been done by the shell as it moved up against gravity. In fact, in the absence of any friction between the shell and the air, the gravitational potential energy at the apex of flight is exactly equal to the kinetic energy the shell had as it left the mortar.

We previously noted that gravitational potential energy is measured relative to an arbitrarily chosen reference height, and therefore that its value depends on circumstances external to the object possessing the energy. Much the same is true of kinetic energy. In the example discussed above, we measured the velocity of a cannon shell relative to a coordinate system fixed to the earth. If, instead, we measure velocity relative to another object (the shell, for instance), we obtain a different value for kinetic energy. Thus, kinetic energy, like potential energy, is relative. The physical implications of relativity are interesting and complex (as Albert Einstein discovered), and you are urged to consult a standard physics text on the topic. In this text we measure velocities relative to the largest available object—earth's surface—unless otherwise noted. In technical terms, we use an inertially anchored coordinate system.

3.6.1 *Heat*

The concept of kinetic energy serves as a starting point for an examination of another type of energy—*heat*. Consider an example. Pull out the plunger of a bicycle pump, thereby filling the pump with air at room temperature. Then put a plug in the end of the hose so that no air can escape. Push the plunger in. If you now measure the temperature of the air in the pump, you find that it has increased. Somehow the mechanical work you did in pushing in the plunger has been transformed into heat. But what exactly is heat and how is it related to temperature?

The mechanism by which energy is transferred from the movement of the plunger to the gas provides us with an answer to both these questions. At room temperature, molecules of gas move around randomly in the pump, bouncing off the walls and off the face of the plunger. In fact, it is this impact of air molecules that creates the force against which you work as the plunger is pushed in. Now, the inward movement of the plunger gives a little extra boost to the molecules ricocheting from its surface. In other words, the movement of the plunger increases the velocity of the molecules that impact upon it, thereby increasing their kinetic energy. These molecules in turn collide with others, quickly distributing the increase in kinetic energy among the entire population. In this fashion, the mechanical work you do on the plunger is converted into an increase in the average kinetic energy of the gas molecules. This increase in kinetic energy is what we mean when we say that the temperature of the gas has increased. The temperature of a gas (or any other material) is directly proportional to the average kinetic energy of its molecules:

$$T = \frac{m\langle u^2 \rangle}{3k}, \tag{3.15}$$

where $\langle u^2 \rangle$ is the mean square speed of molecules. Thus there is a temperature, *absolute zero*, at which molecular motion ceases.[5]

SI uses the absolute, or Kelvin, scale of temperature in which one kelvin, K, is one-hundredth the temperature difference between the freezing and boiling points of water. Given this scale, the constant, k, in eq. 3.15 (known as *Boltzmann's constant*) is $1.38054 \times 10^{-23}\,\mathrm{J\,K^{-1}}$. Thus, if you know the mass and the mean square speed of molecules, you can calculate the temperature. Conversely, if you know the temperature and the mass you can calculate the average speed. For example, a nitrogen molecule N_2 weighs about 4.65×10^{-26} kg. At room temperature (290 K) its average speed[6] is

$$\sqrt{\frac{3kT}{m}} = 508\,\mathrm{m\,s^{-1}}. \tag{3.16}$$

We will deal with the consequences of this surprisingly high velocity in chapter 6 when we discuss diffusion.

Organisms usually live at temperatures in the range of 273 to 313 K. It can be awkward to work with these large numbers, and it would be more convenient to use a temperature scale for which the zero point had been shifted upward. In practice, this is done by using the Celsius scale, for which 0° is defined as the temperature at which pure water turns to ice, 273.15 K.

We have now seen how temperature is defined in terms of the kinetic energy of gas molecules. But how is this kinetic energy any different from that which we have previously examined? On a molecule-by-molecule basis, it isn't. But when we consider a large number of molecules, each with its own kinetic energy and direction of flight, the net result is an energy that is subtly different from any we have discussed so far.

[5]Actually, this simple statement is slightly misleading. Absolute zero is the temperature at which the molecular motion of an ideal gas (see chapter 4) would cease *if* the gas obeyed the laws of the kinetic theory of gases at that temperature. In reality, gases (and other materials) are not "ideal," and some strange quantum effects govern molecular behavior at very low temperatures. Some motion is possible even at absolute zero.

[6]This average is the square root of the mean square speed, what is referred to as the *root mean square* or *rms* average. We discuss this type of average in more detail in chapter 6.

The crux of the difference lies in the disorder of the system. If somehow all the gas molecules in our bicycle pump were to move toward the face of the plunger, their combined impact would result in a very large force, which could be used to perform mechanical work. But the likelihood of this occurrence is inconceivably small. The frequent collisions between molecules insures that their direction of flight is randomized, and at any time only a small fraction moves directly toward the plunger. As a result, the force of impact, and thereby the plunger's effective capacity for doing work, are reduced below what they might be.

The same is true of the increase in kinetic energy imparted to gas molecules as the plunger is inserted. Because the increased velocity of molecules, which was initially directed away from the plunger, is quickly randomized by intermolecular collisions, it is never possible in practice to retrieve from the gas all of the mechanical energy you put into it. In this respect, the disordered motion of molecules is different from the ordered movement of larger objects. As a practical matter, work that is transformed into the kinetic energy of molecular motion can never be transformed back in its entirety. This is true of solids and liquids as well as of gases. Because of this one-way character of the kinetic energy of molecular motion, we treat the macroscopic effects of this type of kinetic energy differently from others, and call it *heat*.

There are so many important ramifications of heat that an entire branch of physics, *thermodynamics*, is devoted to their study. But the results of these studies, important as they are, are peripheral to our discussion of the biologically relevant physics of air and water, and (with one exception) will not be explored further. If you are interested in pursuing the subject of thermodynamics, you may want to consult the excellent introductory text by Atkins (1984) or the more advanced treatment presented by Reif (1965).

3.6.2 *Friction*

The single exception has to do with the concept of *friction*. The friction of a piston against its walls, the friction between a bearing and its shaft, the friction of one water or air molecule against another—all are ways in which mechanical energy is converted to heat. In fact, friction is a generic term for the many processes by which mechanical work is converted to heat. Because the production of heat invariably acts as a net drain on other forms of mechanical energy, one of the associated effects of friction is the tendency in any physical system for motion eventually to grind to a halt. The dissipative action of friction will be a common theme throughout this text.

3.6.3 *Other Forms of Energy*

One other form of energy will make an occasional appearance in this text. This is *light energy*, or more generally, the *energy of electromagnetic radiation*. Each photon of light (or of any other electromagnetic radiation, such as a gamma ray) has an energy proportional to its frequency. If frequency f is expressed as cycles per second (the SI term is the hertz, Hz), the energy of a photon is hf, where h is Planck's constant, 6.626176×10^{-34} J Hz^{-1}. Under the appropriate circumstances this energy can be transformed into other forms. The most common trade is between light energy and heat, a phenomenon familiar to those who have warmed themselves by basking in the sun. Another trade is between light energy and *chemical energy*, the energy contained in the bonds between atoms. The photosynthetic process in plants, where light energy is used to form carbohydrates from carbon dioxide and water, is a good example of this particular transformation of energy. On close examination, chemical energy is seen to consist partly of electrical energy (accounting for the attraction of one atom to the other) and of the kinetic energy of the electrons in each atom. But these fine points of energy taxonomy need not concern us here.

Energy is also present at the smaller scales of structure. *Nuclear energy* holds together the neutrons and protons of atomic nuclei, and energy is contained in the bonds that hold quarks together within neutrons and protons. However, under "normal" circumstances, these energies are not transformed into potential, kinetic, electrical, or chemical energy, and are therefore irrelevant to our exploration.

3.6.4 *Conservation of Energy*

You may have noticed a certain air of ambiguity in this discussion of energy. We assert that a suspended bucket of water has gravitational potential energy, and use as "proof" the fact that, by falling, the bucket can produce electrical energy. We "prove" that a mortar shell has kinetic energy by showing that the shell's vertical motion results in an increase in the shell's potential energy. Aren't these proofs circular? Can we really demonstrate the existence of one form of energy only in terms of another form? The answer is yes on both counts. But the circularity of our argument is cause for celebration rather than worry! The fact that one form of energy can only be defined in terms of another, is evidence of an important, underlying principle of nature—energy is conserved. In every experiment that has ever been carried out, the disappearance of energy in one form has resulted in the appearance of an equal amount of energy in some other form or forms. It is this conservation—a fact so strongly supported that it is regarded as law—that makes it legitimate to use the gradual appearance of gravitational potential energy in a mortar shell traveling skyward as evidence for the shell's initial kinetic energy. From the law of conservation, we know that the potential energy cannot appear *de novo*, and given the constraints of the system, the only place it can come from is the motion of the shell's mass. Therefore the motion of the shell's mass must represent a form of energy. In this respect, the law of conservation of energy is more basic than the concept of energy itself. If you are interested in the philosophical underpinnings of energy as a concept you should consult the readable discussion by Feynman et al. (1963).

3.7 Power

The temporal rate of change of energy in any system is termed *power*. Its dimensions are ML^2T^{-3}, and its SI unit is the $J\,s^{-1}$, called the watt, W, in honor of James Watt, the inventor of the steam engine. Because energy is the product of force and distance, energy per time (power, P) can be expressed as the product of force and velocity:

$$P = \frac{\text{force} \times \text{distance}}{\text{time}} = \text{force} \times \text{velocity}. \qquad (3.17)$$

This equivalence will prove useful at various points in our exploration, and is worth keeping in mind.

Table 3.1 The basic building blocks of physics, their SI units, and the symbols used when the quantity is a variable.

Name	Dimensions	SI Units	Variable
Length	L	m	ℓ
Time	T	s	t
Mass	M	kg	m
Force	MLT^{-2}	N	F
Pressure	$ML^{-1}T^{-2}$	Pa	p
Energy	ML^2T^{-2}	J	W
Power	ML^2T^{-3}	W	P

3.8 Summary

An orthogonal, inertially anchored coordinate system allows us to locate objects in space and to follow their motions as a function of time. If an object has mass, knowledge of its motion allows us to measure force. In turn, the concept of work—force times distance—provides a practical measure of mechanical energy. The various types of energy—mechanical, electrical, chemical, heat, and light—can be traded among themselves. In most cases these trades can be symmetrical. For instance, gravitational potential energy gained at the expense of kinetic energy can (in the absence of friction) be totally converted back to kinetic energy. But heat is a special case. Other forms of energy converted to heat can never be entirely recovered. It is a basic law of nature that the total amount of energy is conserved. All of these ideas apply to fluids as well as to solids.

The dimensions and SI units for the values discussed in this chapter are summarized in table 3.1.

Chapter 4

Density: Weight, Pressure, and Fluid Dynamics

Perhaps the most familiar physical difference between water and air is the difference in their *density*, in other words, the difference in the mass of each fluid that occupies a given volume. A liter of water has about eight hundred times the mass of a liter of air, and this striking difference results in several important biological consequences. For example, fish may hang motionless above the seabed without flapping a fin, but hovering birds must frantically beat their wings to stay aloft. In this chapter we will see why there are no terrestrial animals that function as hot-air balloons, but that sperm whales can act as their aquatic equivalent; why jet propulsion is a feasible locomotory strategy in water but not in air; and why it might have been difficult for large dinosaurs to drink.

4.1 The Physics

The SI unit of density is $\mathrm{kg\,m^{-3}}$ and is given the symbol ρ. In general, the density of a fluid is a function of both temperature and pressure, and in this section we examine these variables as they apply to air and water.

Some care must be given to exactly what we mean by "air" and "water." For example, biologists often use the term "water" loosely, intending it to apply not only to the specific chemical H_2O, but also to the combination of water and any material dissolved in it. Thus water, in a biological sense, can be fresh or salty and we need to know how its density varies with salinity. The same problem applies to air, which can be dry or humid, clear or polluted.

4.1.1 *Air*

Dry air is primarily a mixture of two gases. Nitrogen makes up 78.08% of the atmosphere by volume, and oxygen, 20.95%. The remaining 0.97% is a mixture of argon, carbon dioxide, and neon, with trace amounts of other gases.

Nitrogen has an average molecular weight of 0.028 kg per mole, where a mole (symbolized by mol) is 6.022×10^{23} molecules. Oxygen has an average molecular weight of $0.032\ \mathrm{kg\,mol^{-1}}$, and the overall average molecular weight of dry air is about $0.0286\ \mathrm{kg\,mol^{-1}}$.

Air often contains an appreciable fraction of water vapor, an amount determined by temperature and local availability. However, even when air is saturated with water vapor (in other words, when the relative humidity is 100%), water contributes only about 2% of the molecules present.

The density of air can be predicted from the kinetic theory of gases. Our understanding of gases is one of the great successes of classical physics and is an interesting story that unfortunately lies beyond the purview of this text. The inquiring reader may wish to consult a standard text on the subject (e.g., Feynman et al. 1963). Here we apply the results of theory with only passing reference to their lineage.

The equation of state for an ideal gas is

$$pV = N\Re T, \tag{4.1}$$

where p is the absolute pressure of the gas, V is the volume in which the gas is

Chapter 4

confined, and T is the absolute temperature. N is the number of moles of gas present in volume V, and $\Re$ is the universal gas constant, 8.3144 J mol^{-1} K^{-1}.

The sense behind eq. 4.1 is this. As noted in chapter 3, the pressure exerted on a container by a gas is due to the impact of molecules on the container's walls. The higher the temperature, the greater the average speed of each molecule, and the larger the force exerted when the molecule collides with the wall. As a result, pressure increases with increasing temperature. By similar reasoning, we can see why pressure depends on N, the number of moles present. The more molecules there are in the container, the more frequent are the collisions with the wall, and the higher the pressure. Note also that for a fixed number of molecules at a given temperature, the product of pressure and volume is constant. As a result, any increase in pressure must be accompanied by a decrease in volume, and vice versa.

Rearranging eq. 4.1 we see that

$$\frac{N}{V} = \frac{p}{\Re T}. \tag{4.2}$$

Multiplying both sides of the equation by the average molecular weight of the gas $\mathcal{M}$ (in kg mol^{-1}) results in an expression that relates density to temperature and pressure,

$$\rho = \frac{m}{V} = \frac{p\mathcal{M}}{\Re T}. \tag{4.3}$$

Note that this equation applies exactly only to an "ideal" gas, one in which the total number of molecules per volume is large but not so large that they take up an appreciable amount of space, and one in which the collisions between molecules and the walls of the container, or between molecules themselves, are elastic. Under sufficiently high or low pressure, air would not behave as an ideal gas, but under the conditions prevailing in the biosphere it fits the assumptions of the kinetic theory quite nicely, and eq. 4.3 can be applied as if it were exact (table 4.1).

Table 4.1 The density at one atmosphere of air, fresh water, and seawater.

T (°C)	Density, ρ (kg m^{-3})		
	Air	Fresh Water	Seawater (S = 35)
0	1.293	999.87	1028.11
3.98	1.274	1000.00	1027.77
10	1.247	999.73	1026.95
20	1.205	998.23	1024.76
30	1.165	995.68	1021.73
40	1.128	992.22	1017.97

Note: The value for air at 0° is taken from Weast (1977), fresh water and seawater values are from UNESCO (1987).

The average pressure at sea level, called the *normal atmosphere*, is 1.01325×10^5 Pa at 15° C, and we use this pressure as a standard when calculating the effects of temperature on the density of air (fig. 4.1). At one normal atmosphere, the density of air varies from 1.292 kg m^{-3} at 0° C to 1.127 kg m^{-3} at 40° C, a variation of about 13% relative to the larger value.

The density of air also varies directly with pressure. What sorts of pressure

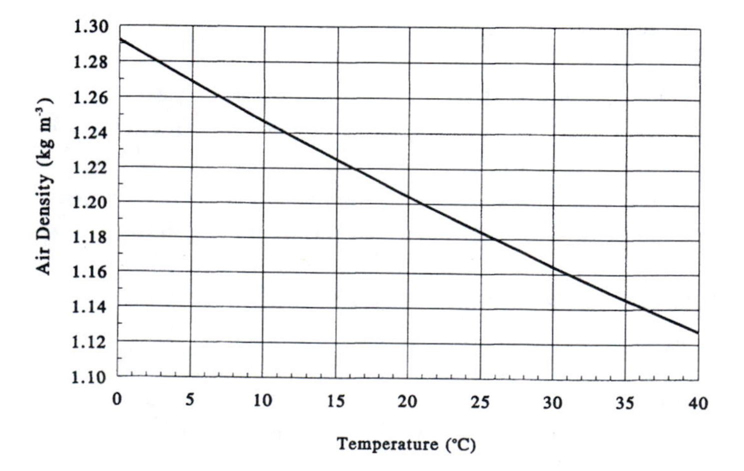

Fig. 4.1 Density decreases as temperature increases for air held at a constant pressure (here assumed to be one normal atmosphere).

variation might we expect to find in the atmosphere? Perhaps the most familiar pressure changes are those caused by day-to-day changes in the weather, the changes in so-called *barometric* pressure. Because these pressures have traditionally been measured using mercury barometers, values are usually quoted in inches of mercury. The pressure exerted by a column of mercury one inch high at 0° C is 3386.39 Pa, and typical barometric pressures vary between 29 and 31 inches of mercury. Thus, on any given day, the pressure at sea level can vary between about 0.98×10^5 and 1.05×10^5 Pa, resulting in a variation in the density of air of about 6.7% of a normal atmosphere.

Atmospheric pressure also varies as a function of altitude, decreasing as one moves upward from sea level. The discomfort one feels in one's middle ear during ascent in an airplane (or even when driving up a mountain) is testament to the fact that altitude can have a major effect on air pressure. The magnitude of the effect is again predicted from the kinetic theory of gases.

The equilibrium distribution of gas molecules as a function of altitude is governed by two factors. First, molecules are subject to the acceleration of gravity—in the absence of other forces, molecules would fall toward the center of the earth, congregating at sea level. However, this trend is offset to a certain extent by the random motions of the gas. As each molecule rattles around with an average kinetic energy corresponding to the local temperature, it occasionally travels upward against the pull of gravity. The higher the altitude, the lower the probability that a molecule will arrive there by chance. Thus, at equilibrium, the number of gas molecules per volume, and thereby the density, is expected to decrease with altitude. An appropriate consideration of the physics leads to the conclusion (expressed in the Boltzmann equation) that at a constant temperature,

$$\frac{\rho}{\rho_0} = \frac{p}{p_0} = e^{-y/y_s}, \tag{4.4}$$

where the subscript 0 denotes a value measured at sea level, y is the altitude above sea level, and y_s is the *scale height*, a value that adjusts the equation to a particular gas and temperature. For air at 15° C, the scale height is 8434.4 m. The implications of this result are graphed in figure 4.2. At a given temperature, the air pressure and density in Denver (altitude 1 mile, or 1609 m) are 17.4% less than in San Francisco (altitude 0). Air at the top of Mount Everest (altitude 8839 m) is 65% less dense than that at sea level.

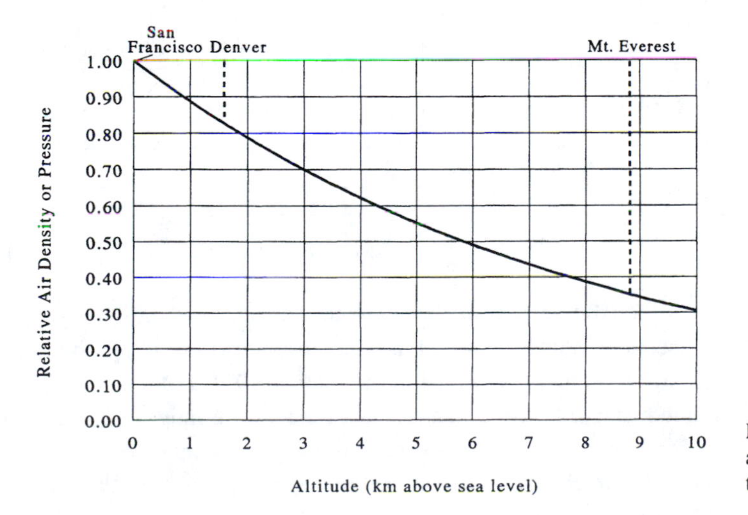

Fig. 4.2 Air density decreases with altitude above sea level. Here we assume that temperature is constant across altitudes.

The density of air at constant temperature and pressure varies slightly with the amount of moisture (water vapor) present. The density of moist air at constant temperature is (Weast 1977)

$$\rho = \rho_d \frac{p - 0.3783 p_v}{p_0}, \tag{4.5}$$

where ρ_d is the density of dry air at that temperature, p_0 is a pressure of one normal atmosphere, and p is the prevailing atmospheric pressure. The term p_v is the ambient vapor pressure of water, or the pressure the available water vapor would exert if it alone were present. As the moisture content of the air increases, the density decreases, but not much. For instance, at 20° C and 100% relative humidity, p_v is 2340 Pa. By applying eq. 4.5 we see that this exceedingly humid air, the sort of "heavy" air one might encounter immediately after a rain, is actually 0.87% less dense than dry air.

The fact that humid air is less dense than dry air may be counterintuitive, but think of it this way. Eq. 4.1 tells us that at constant pressure, a given volume of gas contains the same number of particles, regardless of their composition. Thus, when we add a few water molecules to a volume of air, if the pressure does not change, a few molecules of oxygen or nitrogen must be displaced. Now water, with a molecular weight of 0.018 kg mole^{-1}, weighs less than the molecules it replaces, so the density must decrease.

4.1.2 *Water*

The density of pure water is approximately 1000 kg m^{-3}, 770 to 890 times that of air at sea level (depending on temperature). For purposes of comparison, we use an average ratio between the densities of water and air of 830.

Because water is a liquid rather than a gas, changes in its relative density are small compared to those of air. For example, the density of pure water varies by only about 0.8% over the biological range of temperatures (fig. 4.3, table 4.1) compared to the 13% change in density for air over the same range.

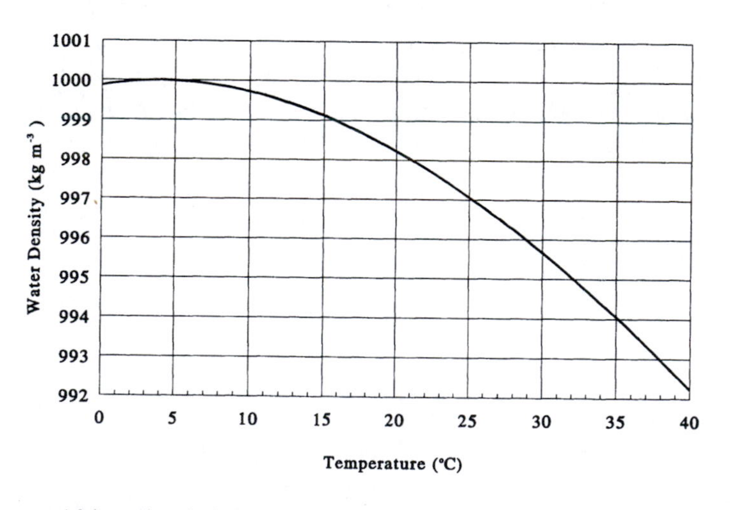

Fig. 4.3 At a pressure of one normal atmosphere, the density of pure water reaches a peak at 3.98°C.

Although relative changes in water density are small, they can have important consequences nonetheless. First, the absolute (rather than the relative) change in density is large. A cubic meter of water at 40° C contains 7.7 kg less mass than a cubic meter at 0° C. Over the same temperature range, a cubic meter of air varies in mass by only about 0.16 kg. Thus, any physical effect that depends solely on the absolute change in density is more variable in water than in air.

Unlike air, the density of pure water is not a monotonic function of temperature—water reaches a maximum density of $1000\,\mathrm{kg\,m^{-3}}$ at a temperature of 3.98° C (fig. 4.3, table 4.1). Thus the densest water is not the coldest, a fact that can have important biological consequences. For example, consider a stably stratified lake in the midst of winter. Because of water's unusual temperature-density relationship, the water at the lake bottom stays at a relatively balmy 3.98° C regardless of the temperature at the surface, thereby facilitating the survival of fish.

Furthermore, water, unlike air, goes through a phase transition within the biological range of temperatures. At 0° C pure liquid water turns into ice, and in the process its density abruptly decreases from $999\,\mathrm{kg\,m^{-3}}$ to about $917\,\mathrm{kg\,m^{-3}}$ (fig. 4.4). Because ice is less dense than water, it floats, a fact that also can have important biological consequences. For example, if ice were denser than water, lakes and ponds would freeze from the bottom up rather than from the top down. Not only would this pose a problem for the ice skaters and hockey players of the world, it would also wreak havoc with the many organisms that seek shelter from winter cold by burrowing into lake bottoms.

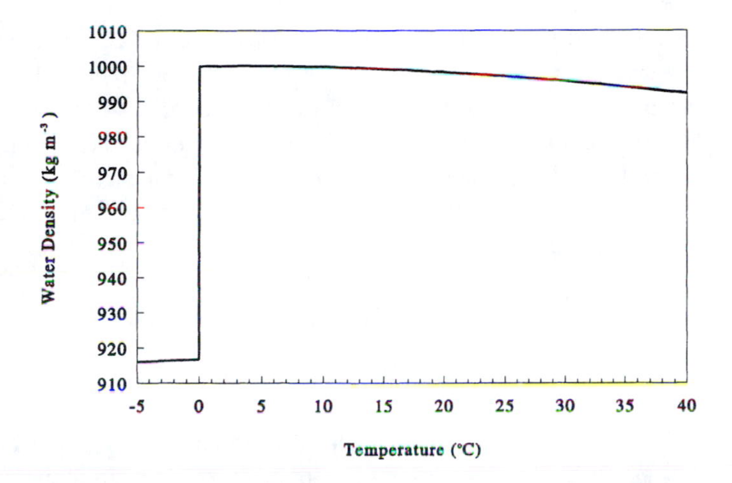

Fig. 4.4 The density of pure water is drastically reduced as the water freezes.

In the matter of density, seawater differs from pure water in two subtle but important respects. First, the dissolved salts in seawater lower its freezing point. At a typical salinity of 35 g of salts per kilogram of liquid, for instance, seawater freezes at −1.9° C. Second, the dissolved salts lower the temperature at which water reaches its maximum density. For salinities above $27.4\,\mathrm{g\,kg^{-1}}$, the temperature of maximum density is lower than the freezing point. Thus, in the ocean, the coldest water *is* the densest water. Surface seawater is cooled to −1.9° C in the polar seas, and this dense water sinks to form the deep waters of the world's oceans. In the process of sinking, however, some mixing occurs with warmer water, with the result that the average temperature of deep water is 0° to 2° C. When seawater freezes, its salt is extruded, so sea ice is essentially the same as freshwater ice, and, like freshwater ice, it floats.

As with temperature, the density of water does not vary much with pressure. The change in pressure required to achieve a relative change in volume (and therefore, in density) of a material is known as the *secant bulk modulus*, B_s:

$$B_s = \frac{p - p_0}{(V - V_0)/V_0}, \tag{4.6}$$

where p_0 is one normal atmosphere, V_0 is the volume at p_0, and p is the ambient pressure.

Chapter 4

Values for fresh water and seawater at various temperatures are given in table 4.2. They are all close to 2×10^9 Pa, implying that a pressure of this magnitude would have to be applied to double the density of water. This is a *very* high pressure indeed. For example, the hydrostatic pressure imposed on water in a lake or ocean increases by about 10^4 Pa for each meter of depth below the surface. Assuming a constant temperature, we calculate that at a depth of 1 km a pressure of 10^7 Pa is applied, but the density of water is a mere 0.5% greater than at the surface. Even in the ocean depths, the change in density is small. The Marianas trench, the deepest point in the earth's oceans, is about 11 km deep, yet, if temperature and salinity are constant, water there is only about 6% denser than water at the surface.

Table 4.2 The isothermal secant bulk modulus of fresh water and seawater at various temperatures (calculated from UNESCO 1987).

T (°C)	Bulk Modulus, B_s (Pa × 10^9)	
	Fresh Water	Seawater (S=35)
0	1.9652	2.1582
10	2.0917	2.2695
20	2.1790	2.3459
30	2.2336	2.3924
40	2.2604	2.4128

The fact that water is nearly incompressible has not always been widely known, leading to some bizarre misunderstandings. For instance, when the Titanic sank in 1912, many relatives of the disaster's victims thought that the ship would descend until the water (which they assumed to be compressible) reached the same density as steel, at which point the wreck would come to rest in midwater (Schlee 1973). The vision of loved ones doomed to float forever in the abyss was cause for many needless nightmares.

As noted previously, the density of water varies as a function of *salinity*, S, the amount (in grams) of dissolved material present in one kilogram of seawater. As a practical matter, the salinity of seawater is now measured by comparing its electrical conductivity to that of a standard aqueous solution of potassium chloride[1] (UNESCO 1983). The ratio of the two conductivities is symbolized by K_{15}. The *practical salinity*, S_p, of a seawater sample is then calculated from the following equation:

$$S_p = 0.0080 - 0.1692K_{15}^{1/2} + 25.3851K_{15} + 14.0941K_{15}^{3/2} - 7.0261K_{15}^2 + 2.7081K_{15}^{5/2}. \tag{4.7}$$

Because the practical salinity is defined in terms of a ratio of conductivities, it has no formal dimensions, which leads to a strange nomenclature. For example, a practical salinity of 35 (no units) should be interpreted as meaning that approximately 35 g of salt are present in each kilogram of solution. To be absolutely precise, it should be noted that the *actual* salinity (in grams of salt per kg of seawater) is (Bearman 1989)

$$S = 1.00510S_p. \tag{4.8}$$

It is evident from these relationships that when the fine details are considered, the physical properties of seawater are somewhat complex. The most recent and accurate description of these complexities can be found in UNESCO (1983, 1987).

At 0° C and one normal atmosphere, the density of seawater (expressed in terms of the practical salinity) is shown in figure 4.5. Seawater with a typical S_p of 35 has a density of 1028 kg m^{-3}, 2.8% greater than that of fresh water.

A general expression for the density of seawater as a function of salinity at other temperatures is explained in table 4.3. For variations due to salinity and temperature at other pressures, consult UNESCO (1987).

As with changes in density due to temperature, changes in density due to salinity have important effects in determining the pattern of circulation in the oceans. These

[1] The standard contains 32.4356 g of KCl per kilogram of solution.

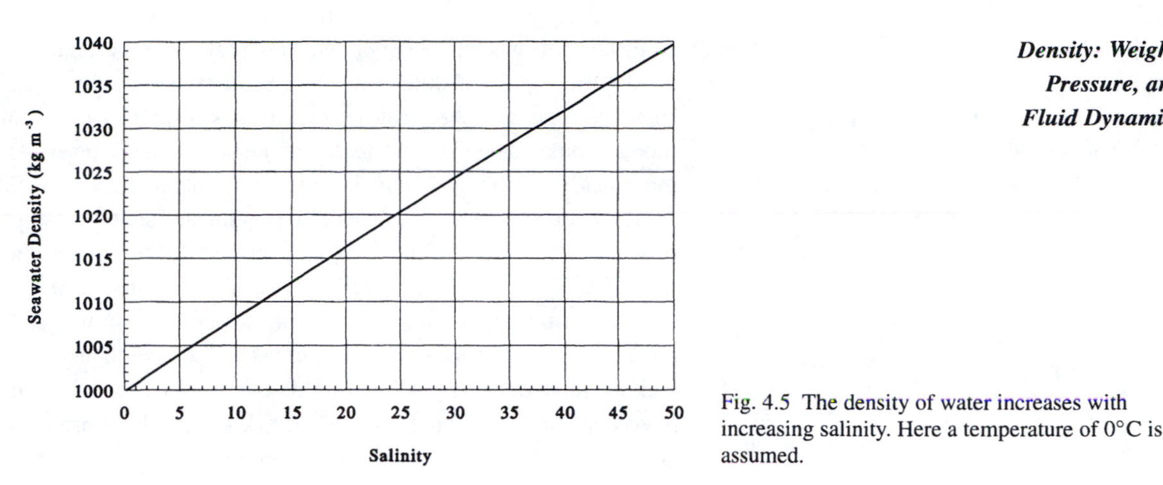

Fig. 4.5 The density of water increases with increasing salinity. Here a temperature of 0°C is assumed.

effects will not be discussed here, but an intriguing introduction to the subject may be found in Mann and Lazier (1991).

Table 4.3 Calculation of the density of seawater at atmospheric pressure (UNESCO 1987).

The density of pure water is calculated as follows:

$$\rho_w = a_0 + a_1 T + a_2 T^2 + a_3 T^3 + a_4 T^4 + a_5 T^5, \quad \text{(a)}$$

where T is in ° C and the constants are

$a_0 = 999.842594$ $\quad a_1 = 6.793952 \times 10^{-2}$

$a_2 = -9.09529 \times 10^{-3}$ $\quad a_3 = 1.001685 \times 10^{-4}$

$a_4 = -1.120083 \times 10^{-6}$ $\quad a_5 = 6.536332 \times 10^{-9}$

This value is then used to calculate the density at salinity S:

$$\rho(S, T) = \rho_w + (b_0 + b_1 T + b_2 T^2 + b_3 T^3 + b_4 T^4)S + (c_0 + c_1 T + c_2 T^2)S^{3/2} + d_0 S^2, \quad \text{(b)}$$

where the constants are

$b_0 = 8.24493 \times 10^{-1}$ $\quad b_1 = -4.0899 \times 10^{-3}$

$b_2 = 7.6438 \times 10^{-5}$ $\quad b_3 = -8.2467 \times 10^{-7}$

$b_4 = 5.3875 \times 10^{-9}$

$c_0 = -5.72466 \times 10^{-3}$ $\quad c_1 = 1.0227 \times 10^{-4}$

$c_2 = -1.6546 \times 10^{-6}$

$d_0 = 4.8314 \times 10^{-4}$

4.2 The Density of Living Things

Before exploring the biological consequences of fluid density, it will be useful to review briefly the density of the materials from which organisms are constructed (table 4.4). All biological materials are very much denser than air, and, with the exception of lipids and some woods, all are even denser than seawater.

Consider the density of animals. The densest part of an animal is generally its skeleton. For instance, skeletons made primarily from calcium carbonate (e.g., the shells of snails, clams, and *Nautilus*) have densities of approximately 2700 kg m^{-3},

Table 4.4 The densities of various inorganic and biological materials.

Material	Density (kg m^{-3})
Inorganic Materials	
Pure water	1000
Seawater	1025
Calcium carbonate	
Calcite	2700
Aragonite	2900
Calcium phosphate	
Apatite	3200
Glass	2400 to 2800
Aluminum	2700
Iron	7870
Organic Materials	
Coral skeleton	2000
Mollusk shell	
Gastropod shell	2700
Nautilus shell	2700
Bivalve shell	2700
Crustacean Shell	1900
Insect exoskeleton	1200 to 1300
Sea urchin spines	2000
Bone	
Femur (cow)	2060
Whale ear bone	2470
Tooth enamel	2900
Muscle	1050 to 1080
Fats and oils	915 to 945
Wood	
Red oak	680
Sweet gum	1000
Lignum vitae	1300

Sources: Data taken from Schmidt-Nielsen (1979), Vogel (1988), Wainwright et al. (1976), and Weast (1977).

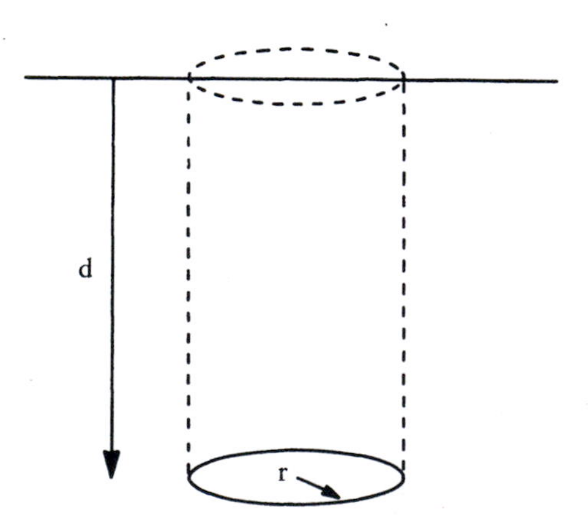

Fig. 4.6 An example used in the calculation of pressure as a function of depth. See text for details.

nearly three times the density of water. Coral skeletons and most sea urchin spines are a bit less dense (2000 kg m^{-3}), but still considerably denser than water. Bone (made of calcium phosphate) typically has a density of about 2000 kg m^{-3}, although some ear bones and teeth can be somewhat denser. Among stiff skeletal materials, the cuticle of flying insects is the least dense (ρ = 1200 to 1300 kg m^{-3}).

The connective tissue in an animal (tendon and ligament) typically has a density of about 1200 kg m^{-3}. Other soft parts are less dense. Muscle has a density of 1050 to 1080 kg m^{-3}, and guts generally have a density of about 1040 kg m^{-3}. Body fluids (blood, hemolymph, etc.) typically have a density of about 1010 kg m^{-3}. Lipids are the only common body constituents less dense than water. Most animal fats and oils have densities of 915 to 945 kg m^{-3}. Taken as a whole, animals typically have densities between 1060 and 1090 kg m^{-3}. There are exceptions, of course, and some of these are discussed later in this chapter.

The density of wood spans a broad range and is very sensitive to the volume fraction of air in the material; the more trapped air, the less dense the wood. As a result, density varies not only from species to species, but it can also vary substantially within an individual depending on the hydration state of the plant. It is common to assume that wood is less dense than water because it floats. This need not be the case, however, and it is a safe generalization only for the wood of temperate-zone trees, where wood densities are typically 500 to 1000 kg m^{-3}. Many tropical trees (e.g., ebony, lignum vitae) are considerably denser than water and sink if immersed. The solid part of wood (cellulose and the various substances that bind it together) has a density of about 1500 kg m^{-3}, and this sets the upper limit to the density of wood.

4.3 Buoyancy

We begin our examination of the biological consequences of fluid density by considering the peculiar force known as *buoyancy*. This is the force that suspends blimps in air and allows boats to float on water. Before we consider the consequences of buoyancy, it will be useful to consider its physics.

We first need to examine how pressure varies with position in a fluid. For simplicity, consider a fluid whose density is ρ_f everywhere. We can place a circle of radius r within the fluid, with its plane perpendicular to the acceleration of gravity (fig. 4.6), and ask what forces act on its area.

Above the circle there is a column of fluid that extends a distance d to the fluid's upper boundary. This column has a volume $\pi r^2 d$ and therefore has a mass $\rho_f \pi r^2 d$. When acted on by the acceleration of gravity, this mass exerts a force $g\rho_f \pi r^2 d$ on the circle. Dividing by the circle's area (πr^2), we see that the pressure (force per area) acting on the circle is $g\rho_f d$. In other words, as long as the density of the fluid is constant, the pressure applied by the fluid increases linearly with distance d from the upper boundary of the fluid and is proportional to the fluid's density.

This thought is easily applied to water, where the density is indeed nearly constant; to a close approximation the hydrostatic pressure in water increases linearly with depth, d. Unfortunately, the same analysis cannot be applied directly to air, for which density changes with altitude (fig. 4.2). The atmosphere gradually thins out as one goes up, changing density in the process, and never reaches a clear upper boundary. Consequently, it is incorrect to think of atmospheric pressure as being exerted by a column of air of constant density and known height. But this does not pose any practical problems. Instead of specifying air pressure by knowing

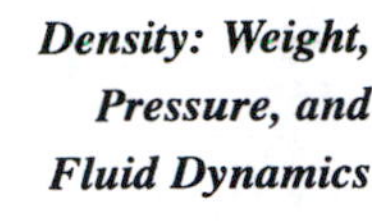

the distance from the upper boundary of the atmosphere, we instead measure the change in pressure from that which is standard for some particular altitude. For small changes in altitude (a few tens of meters), the density of air is unlikely to change appreciably, and therefore (given this restriction) pressure changes almost linearly with altitude.

Now consider the situation depicted in figure 4.7, a solid cylinder of radius r and height h situated in the midst of a large volume of fluid. The cylinder has density ρ_b and the fluid has density ρ_f. What are the forces acting on the cylinder? First we consider the forces imposed by pressure in the fluid. The pressure at the top of the cylinder is p (its absolute value does not matter) and it exerts a downward force on the top:

$$\text{force on the top} = \pi r^2 p. \tag{4.9}$$

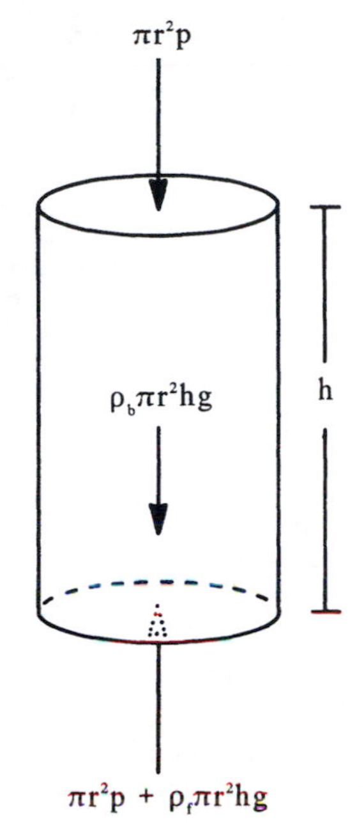

Fig. 4.7 The buoyancy of a solid cylinder results from the interaction between the cylinder's weight and the pressures imposed on its top and bottom. See text for details.

At the bottom of the cylinder (a distance h below the top) the static pressure is greater and it exerts an upward force on the bottom:

$$\text{force on the bottom} = \pi r^2 p + g\rho_f \pi r^2 h. \tag{4.10}$$

The difference between these two forces is the net vertical force,

$$\text{net vertical force} = g\rho_f \pi r^2 h, \tag{4.11}$$

and it acts in an upward direction.

What about the pressure pushing on the sides of the cylinder? Here the situation is simplified by a consideration of symmetry. For example, water pushing on the left side of the cylinder at some depth d exerts a pressure $g\rho_f d$, but the action of this pressure is just offset by the same pressure acting on the right side. For every force acting at one point, there is an equal force acting on the opposite side, and the two cancel. This is true for any value of y. As a result there is no net lateral force due to pressure, and the net force acting on the cylinder is the upward force $g\rho_f \pi r^2 h$. This net force is the *buoyancy* or *buoyant force*.

Buoyancy is resisted by the weight of the cylinder. Now the volume of the cylinder is $\pi r^2 h$, so its mass is $\rho_b \pi r^2 h$ (where ρ_b is the density of the material from which the cylinder is made) and its weight is $g\rho_b \pi r^2 h$. This force acts downward. Thus, the net force on the cylinder (the difference between its actual weight and the buoyant force due to the water pressure) is $g(\rho_b - \rho_f)\pi r^2 h$. This is the effective weight of the object in the fluid and, for obvious reasons, the term $\rho_b - \rho_f$ is called the *effective density*.[2] We give it the symbol ρ_e.

Designating the volume of the cylinder with the symbol V, we see that

$$\text{effective weight} = g\rho_e V. \tag{4.12}$$

Although we have derived this expression using a cylinder, it applies to objects of any shape, and (provided the object isn't more than a few tens of meters in vertical dimension) it applies in air as well as water. If the density of any object is greater than that of the surrounding fluid, the object experiences a downward force proportional to its volume. In this case, the object is said to be *negatively*

[2]The quantity $(\rho_b - \rho_f)$ is often called the "excess" density, but this can lead to confusion when $\rho_b < \rho_f$. The notion of a "negative excess" smacks of double talk, and for this reason the term "effective density" is preferred here.

buoyant. If, instead, the density of an object is less than that of the surrounding fluid, there is a net upward force and the object is *positively buoyant*.

What are the effective densities of plants and animals? Because the density of air is so small, it has little effect on the effective density of terrestrial organisms. For example, a typical density for an animal is 1080 kg m^{-3} and in air its effective density is 1079 kg m^{-3}, a negligible difference. The effective weight in air of a 5000 N cow is 4995 N, for instance. For the same animal immersed in fresh water, however, its effective density is 80 kg m^{-3}, and its effective weight is 370 N, only about 7% of its actual weight. Water obviously has a profound effect on effective density.

Furthermore, because the density of water is so close to the body density of animals, the effective density (and therefore the effective weight) of aqueous organisms is very sensitive to small changes in density of either the body or the surrounding fluid. For example, seawater is only about 2.5% more dense than fresh water, but the effective density of a typical animal (ρ = 1075 kg m^{-3}) is only 50 kg m^{-3} in the ocean compared to 75 kg m^{-3} in a lake. In this case, a 2.5% increase in water density results in a 33% decrease in effective weight. The same holds true if the density change is in the animal. For instance, if an animal reduces its density from 1075 to 1065 kg^{-3}, its effective weight in air changes by only about 1%. In seawater, the same change in body density incurs a 20% change in effective density (from 50 to 40 kg m^{-3}) and a concomitant change in effective weight.

Because the effective density of aquatic organisms is so sensitive to minor changes in body density, it is likely to have important biological consequences, and as a result, the density of aquatic organisms has received much attention. Some of the results are discussed below when we explore balloons and swim bladders. In contrast, because the effective density of terrestrial organisms is very insensitive to minor changes in body density, it has not been important to measure these densities, and information regarding them is scarce. However, a knowledge of effective density allows us to estimate the typical body density of terrestrial mammals.

Most land mammals can float in fresh water and seawater as long as they keep their lungs inflated. If they exhale, however, they sink. Thus the change in body volume during exhalation is enough to change the effective density of the animal from something slightly less than that of water to something greater. If, for the sake of argument, we assume that the animal is just neutrally buoyant with its lungs full, we are then in a position to calculate body density. The lung volume of many mammals has been accurately measured and, regardless of the size of the animal, forms approximately 6% of the body volume (Schmidt-Nielsen 1979). Thus, when the body's mass is confined to the 94% of the body volume that isn't lung, the animals is neutrally buoyant. Expressed as an equation, this thought becomes

$$\frac{\rho_b \times 0.94V}{V} = \rho_f, \tag{4.13}$$

where ρ_b is the average density of the non-lung body tissue, V is the overall body volume (inflated lung included), and ρ_f is the density of the surrounding fluid. Working through the math, we see that in fresh water the body density must be 1064 kg m^{-3} and in seawater 1090 kg m^{-3}. This, then, is an estimate of the density of terrestrial mammals, and it is very close to the range of body densities measured for aquatic organisms. From this exercise we may conclude that the terrestrial organisms in the lineage leading to present-day mammals did not evolve a mechanism for substantially altering their density when they emerged from the sea.

We now turn our attention to the biological consequences of buoyancy.

4.3.1 *The Maximum Height of a Column*

Many plants and animals are shaped like upright columns—trees, grasses, sea anemones, and corals, to name just a few. As with any column, there is a tendency for the these organisms to fall over. If the free end of a vertical column is forced slightly to one side, the column's center of gravity is no longer directly over the column's base, and unless resisted, the resulting *bending moment* causes the column to buckle.[3] Resistance to buckling is provided by the stiffness of the material from which the column is constructed. As the free end of the column is displaced, some of the material in the column is stretched and some compressed. In both cases, the material resists the deformation, tending to force the structure back to its undeflected, upright position. If the column is sufficiently stiff and the mass of the column sufficiently small, minor deflections do not result in sufficient torque to buckle the column, and the structure is *stable*; it can right itself. If the material is insufficiently stiff, or the weight of the column is too great, any deflection of the column's end results in enough bending moment to buckle the structure, and the column is *unstable*; the barest puff of wind pushing the column to one side causes it to topple.

If you were asked to construct from a given material a column with a circular cross section of radius r, how high could you make it before it became unstable? This is a problem in a branch of engineering known as beam theory, one that was solved by A. G. Greenhill in 1881. The critical height h_{max} of the column, the height at which it is just barely stable, is

$$h_{max} = 1.26\left(\frac{Er^2}{g\rho_e}\right)^{1/3}, \tag{4.14}$$

where E is the stiffness of the column's material, a value technically known as the Young's modulus.

The stiffness and density of biological materials vary over a considerable range, but we can solve eq. 4.14 for selected cases to get a "feel" for the maximum heights one might expect. Wood has a Young's modulus on the order of 10^{10} Pa, and we assume that the column has a density of about 900 kg m^{-3}, which, in air, corresponds to an effective density of 899 kg m^{-3}. Given these values, we can calculate the maximum height of a "tree" (a vertical wooden column in air) as a function of its radius (fig. 4.8A). Clearly, trees can get quite high before they become unstable—a wooden column only 10 cm in radius can grow well over 20 m high, and a column 1 m in radius can grow almost 130 m high, higher than the highest tree (126.5 m).

If we try to apply eq. 4.14 to a wooden column in water, we arrive at a curious result—a negative maximum height. Because wood is typically less dense than water, the buoyant force on it more than offsets its weight, and it floats. As a result,

[3] A bending moment is a particular application of a torque, that is, a force applied in such a manner as to cause an object to rotate. In this case, a fraction of the weight of the column acts through a *moment arm* proportional to the distance the center of gravity has been displaced, resulting in the rotation of the column and causing it to buckle.

the taller an upright wooden column is in water, the more stable it is. Conversely, you would need a column of negative height (that is, it would have to be upside down) for it ever to be unstable. For this reason, buoyant water plants, such as water lilies and some kelps, do not have a stability problem in the same sense that trees do.

Some woods and aquatic plants, however, do not have densities as low as we have assumed here. For example, some marine macroalgae have densities of approximately 1080 kg m^{-3}, corresponding to an effective density in seawater of 55 kg m^{-3}. In addition, the material from which they are constructed has a much lower stiffness than that of wood; a typical algal modulus is 10^7 Pa (Denny et al. 1989). Inserting these values into eq. 4.14, we see that an algal column 1 cm in radius (typical of many kelps) could only grow to be a bit over 1.5 m high before it became unstable (fig. 4.8B). The same column in air could grow only about 0.6 m high. Indeed, many intertidal algae that stand upright when the tide is in lie recumbent on the shore when the tide is out.

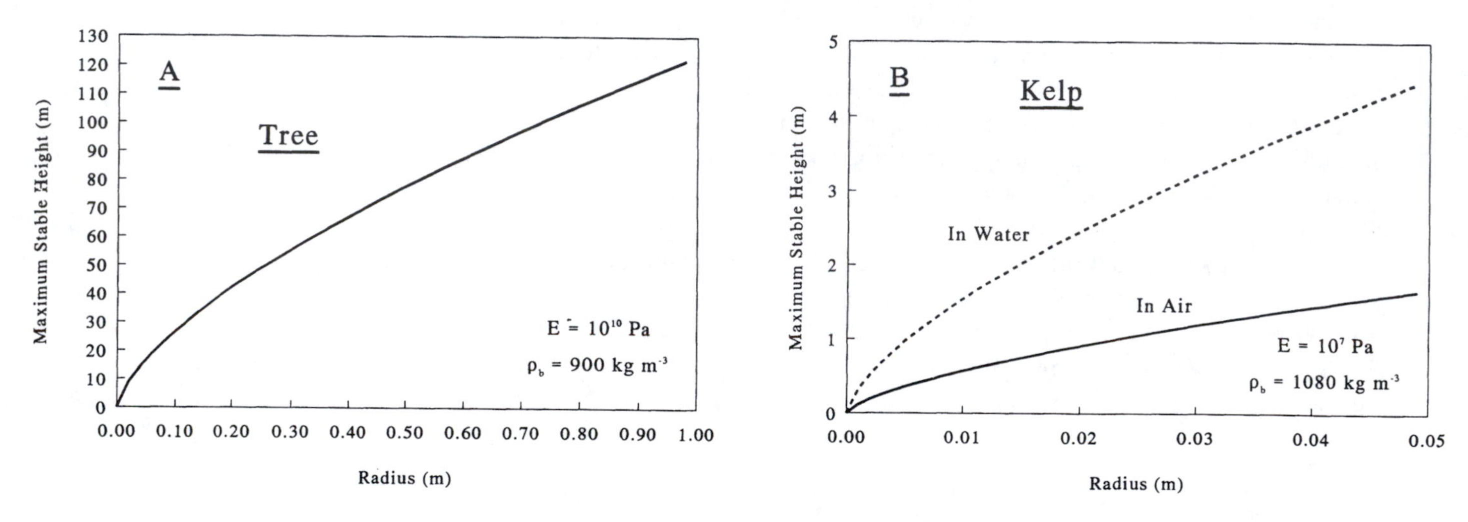

Fig. 4.8 The maximal height of a vertical column in air (a "tree") increases with increasing diameter (A), as does the maximal height of a "kelp" (B) in both air and water. The greater stiffness of wood, however, allows the tree to grow higher than the kelp.

Corals form an interesting counterpoint to trees. The stiffness of coral skeleton (made from a form of calcium carbonate) is of the same order as that of wood. The density of coral skeletal material (about 2000 kg m^{-3}) is such that the effective density of coral in water is about the same as that of wood in air. Thus, one might expect to find the relative dimensions of branching corals to be the same as those of trees. For example, a cylindrical coral 1 cm in radius could grow to a height of about 5 m before becoming unstable. In reality, however, one does not find coral of such slender proportions, suggesting that something other than the stability of the column under its own weight limits the size of corals. We will return to this suggestion when we consider hydrodynamic forces later in this chapter.

As an alternative to the question we have been considering, we can solve Eq. 4.14 for r, and ask the question: What is the minimum radius, r_{min}, of a column if it is to be stable at a given height h? When we note that $1.26^3 = 2$ we see that

$$r_{min} = \sqrt{\frac{h^3 g \rho_e}{2E}}. \tag{4.15}$$

We can take this line of reasoning one step further and ask what minimum volume of a given material is required to construct a column of a certain height. Noting that the volume of a cylinder of constant radius r and height h is $\pi r^2 h$, we calculate that the minimum volume, V_{min}, is

$$V_{min} = \frac{\pi h^4 g \rho_e}{2E}. \tag{4.16}$$

The utility of this rearrangement becomes apparent when one realizes that the metabolic cost of building a structure is likely to be proportional to the volume of material from which the structure is constructed. Thus, eq. 4.16 provides us with a means for comparing the relative costs of producing structurally stable vertical columns in air and in water (fig. 4.9). For any given height of column, independent of the stiffness of the material from which the column is made, about twenty times as much volume is required to construct the column in air as in water. This suggests that it is much more costly in terms of metabolic expenditure to build columns in a terrestrial environment. The fact that vertical columns are nonetheless a much more pervasive theme in terrestrial than in aquatic biology echoes the suggestion made above—that factors other than stability under self-weight have governed the design of columnar organisms.

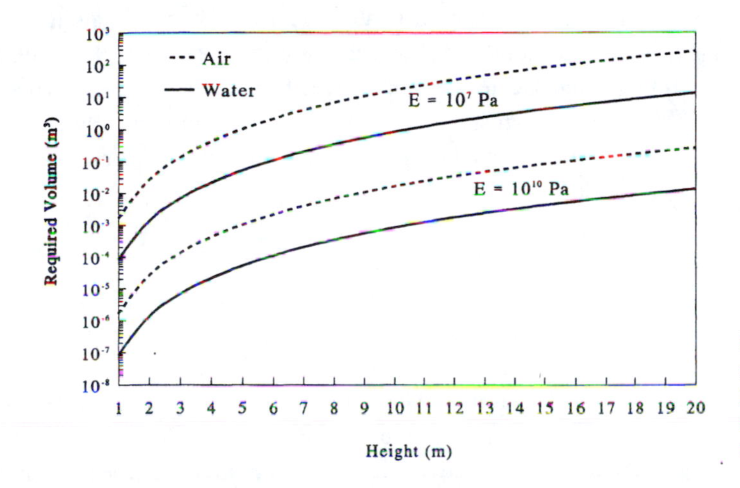

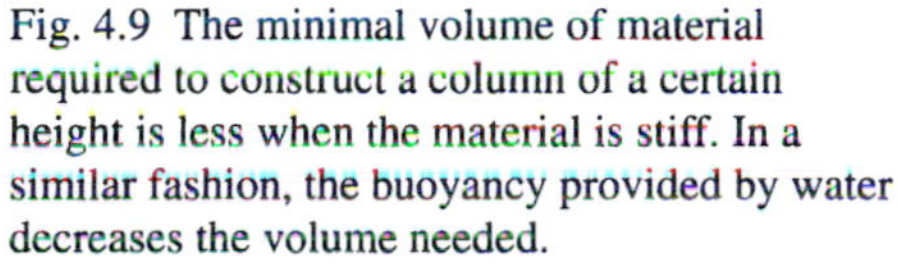
Fig. 4.9 The minimal volume of material required to construct a column of a certain height is less when the material is stiff. In a similar fashion, the buoyancy provided by water decreases the volume needed.

Eq. 4.14 applies only to columns with a circular cross section of constant radius. The equation can be made to apply to columns of varying radii by replacing the value 1.26 with a coefficient, k_c:

$$h_{max} = k_c \left(\frac{Er^2}{g \rho_e} \right)^{1/3}. \tag{4.17}$$

The appropriate value for k_c has been solved for a variety of tapers. For instance, if the column is conical (i.e., it tapers linearly to a point), the appropriate value for k_c is 1.97 (Greenhill 1881). In other words, a conical column could be about 57% higher than a cylindrical column (1.97/1.26 = 1.57). It has been suggested (King and Loucks 1978) that trees grow in a shape such that the radius of the column varies as $\ell^{2/3}$, where ℓ is the distance measured from the free end of the column. With this particular pattern of taper, the appropriate k_c is 2.56, suggesting that real trees might grow twice as high as cylindrical columns with the same basal radius.

Real trees are not simple columns, however; they have branches. To take this complication into account, equations have been derived for columns that are made more realistic by having a mass added to their free end. For example, many trees can be modeled as tapered beams topped by a "crown mass" (King and Loucks 1978).

This model allows us to explore an interesting question: If sea anemones were to invade the land, what form would they take? Let's assume for the sake of argument that as terrestrial anemones evolve, they maintain the basic body plan of a polyp (fig. 4.10) and are constrained to use a beam material similar to their current mesoglea ($E = 5 \times 10^3$ Pa, $\rho_b = 1080$ [Koehl 1977]). Let us further suppose that the weight of the tentacular crown is equal to the weight of its supporting column. In this case, King and Loucks suggest that

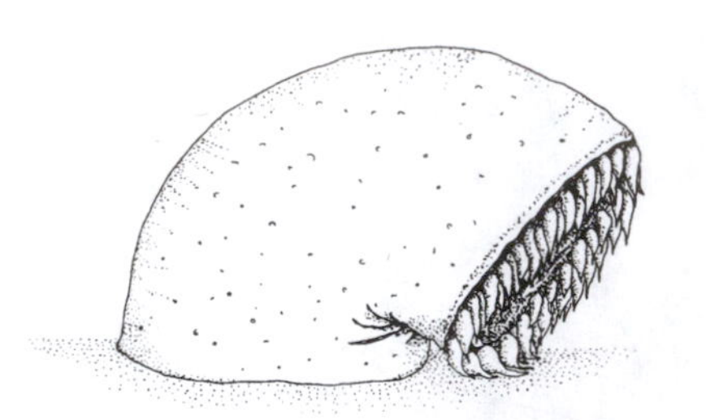

Fig. 4.10 A hypothetical terrestrial anemone 2 cm in diameter buckles when its height is only 3 cm.

$$h_{max} \approx 0.81 \left(\frac{Er^2}{g\rho_e} \right)^{1/3}. \qquad (4.18)$$

Given these assumptions, a terrestrial anemone with a column 2 cm in diameter could grow no more than 3 cm high before it became unstable and buckled. An anemone with ten times the diameter (20 cm) could grow only 4.5 times as high (13.5 cm). Thus, terrestrial anemones would be doomed by gravity and the lack of a stiff structural material to a low-lying life-style.

In this light it is easy to understand why those animals that have been truly successful at invading the terrestrial environment (primarily the insects, arachnids, vertebrates, and gastropods) all evolved from marine organisms that possessed stiff skeletons. Although skeletons may have originally evolved as much in response to a need for defense as for support, they preadapted these lineages for resisting the gravitational exigencies of life out of water.

4.3.2 *Swim Bladders and Balloons*

Like those of terrestrial organisms, the tissues of many aquatic organisms are denser than seawater, and as a consequence these animals have a tendency to sink. Teleost fish have evolved a peculiar mechanism, the swim bladder, as a means of avoiding this problem. The swim bladder is a membrane-enclosed gas pocket within the fish's body cavity. By varying the volume of this low-density organ, the fish can adjust its effective density to that of the surrounding water. The physiology of the adjustment mechanism forms an interesting story—at depth, a fish must somehow pump dissolved gas from its blood, where the partial pressure is low, into the swim bladder, where the partial pressure is very high (Schmidt-Nielsen 1979)—but the details are beyond the scope of this chapter. For present purposes it is sufficient to note that given enough time (usually a matter of hours) a fish can appropriately adjust the volume of its swim bladder. Rapid adjustment of the swim bladder, however, is not possible in most cases.[4]

The slow time-course of the swim-bladder response raises a problem. If a fish cannot instantly adjust its swim-bladder volume, its buoyancy is unstable. Consider the following scenario. A fish at a given depth has adjusted its swim-bladder volume so that the animal is exactly the same density as the surrounding water. All is well until, for example, the fish gets careless and twitches a fin, causing its body to move upward ever so slightly. As the fish rises, the pressure acting on the swim bladder decreases. In response, according to eq. 4.1, the volume of the swim bladder must increase. As a result, the density of the swim bladder, and therefore of the entire

[4] A few fish have a connection between swim bladder and esophagus, allowing them to swim to the surface and gulp air, thereby inflating the swim bladder. The esophogeal connection also allows gas to be released rapidly should the swim bladder become overinflated.

fish, goes down. The fish, which just a moment before was the same density as seawater, is now less dense and the resulting buoyant force causes it to ascend. As it ascends, the pressure decreases further, the fish's density decreases, and it moves upward faster and faster. The system feeds back on itself, and unless quickly checked by some means, the fish rockets to the surface, its swim bladder explodes, or both.

Just the opposite is true if the fish's initial twitch moves it down. In this case, the swim bladder is compressed, the fish becomes more dense, and it sinks. Of the two alternatives, this second one appears to be preferable because it does not involve the potential explosion of the swim bladder, and many fish adjust their effective density so that they are slightly negatively buoyant. In such a case, if the fish loses its concentration for a moment and forgets to adjust its vertical position with its fins, the mistake is reparable. Nonetheless, fish with swim bladders must spend their entire lives being careful of their position in the water column, ever mindful of their buoyancy.

Note that the sensitivity of a swim bladder to pressure depends on depth. Consider, for instance, a swim bladder that has a volume of $10\,cm^3$ at the surface. If a fish containing this swim bladder dives to a depth of 10 m, the pressure is doubled and, as a result, the volume of the swim bladder is decreased by half, a change of $5\,cm^3$. If the fish then dives to a depth of 20 m, the pressure is three times that at the surface, and the volume of the swim bladder is a third of its initial volume, $3.3\,cm^3$. But this is a change of only $1.7\,cm^3$ over the last 10 m change in depth, much less than the $5\,cm^3$ incurred in the initial step of the dive. A subsequent increase in depth to 30 m reduces the swim bladder's volume to $2.5\,cm^3$, a further change in volume of only $0.8\,cm^3$. Thus, because volume is inversely proportional to pressure, the change in volume per change in pressure is less the higher the pressure. As a result, a fish living at great depth need be less mindful of its buoyancy than a fish living near the surface.

A variety of alternative methods have evolved that allow aquatic organisms to avoid the stability problem of swim bladders (Alexander 1990; Smayda 1970). Cuttlefish and *Nautilus* attain neutral buoyancy by varying the gas pressure in a rigid walled structure, the cuttlebone in the case of the cuttlefish, the shell in the case of *Nautilus*. Squids replace some of the sodium ions in their tissues with ammonium, thereby reducing their density. Scyphozoan jellyfish, ctenophores, and some siphonophores reduce their density by actively excluding sulphate from their tissues. Others, such as some fish, crustaceans, and some phytoplankton adjust their buoyancy by accumulating low-density lipids or waxes. As with swim bladders, each of these mechanisms requires considerable time to change the density of an organism substantially. Unlike swim bladders, however, they are relatively insensitive to changes in pressure, and therefore are relatively stable.

Note that these forms of buoyancy control work for aquatic organisms because they have available to them materials (e.g., gases, lipids, etc.) that are less dense than the surrounding medium. Terrestrial organisms aren't so lucky. Their surrounding medium is the lowest-density material in the biosphere, and consequently the mechanisms of buoyancy control we have discussed so far cannot work on land.

Are there mechanisms of buoyancy control of which nature seems not to have taken advantage? One intriguing possibility is suggested by the hot air balloon. If an organism could somehow control the temperature of the fluid in its body, and thereby control its density, it could conceivably regulate its buoyancy.

To explore this possibility, consider a simple example: an organism shaped like a hollow sphere. The "body" of the organism (the wall of the sphere) has a density typical for animals (1080 kg m^{-3}), and it controls the temperature of the fluid in the sphere's lumen. We consider two cases: an organism attempting to become neutrally buoyant in water (ρ_f = 1000 kg m^{-3}) when the lumen is filled with water, and an organism attempting to become neutrally buoyant in air (at sea level, ρ_f = 1.2 kg m^{-3}) when the lumen is filled with air. In both cases we specify that the pressure within the sphere's lumen does not change as the temperature changes; a small orifice somewhere lets fluid in or out as necessary. To what temperature must the organism heat the fluid in its center in order to become neutrally buoyant?

The answer depends on the relative volume of the body and of the fluid in its center. Consider a sphere of radius r. The overall volume of the organism is then

$$\text{overall volume} = \frac{4}{3}\pi r^3. \tag{4.19}$$

If the body wall is thin relative to the radius (as it would need to be in this case), the volume of the body is

$$\text{body volume} = 4\pi r^2 \ell \quad \ell \ll r, \tag{4.20}$$

where ℓ is the thickness of the body wall. It will simplify matters if we express the wall thickness as a fraction k_t of the overall radius. Thus $\ell = k_t r$ and

$$\text{body volume} = 4\pi r^3 k_t, \quad k_t \ll 1. \tag{4.21}$$

The volume of the buoyancy-adjusting fluid is the difference between the overall volume and the body volume.

Knowing the volumes of the body and of the adjusting fluid, we can easily calculate their masses:

$$\text{body mass} = \rho_b 4\pi r^3 k_t; \quad k_t \ll 1 \tag{4.22}$$

$$\text{adjusting fluid mass} = \rho_a \left(\frac{4}{3}\pi r^3 - 4\pi r^3 k_t\right); \quad k_t \ll 1, \tag{4.23}$$

where ρ_a is the density of the buoyancy-adjusting fluid. For the organism to be neutrally buoyant, the sum of these two masses divided by the total volume of the organism must equal the density of the surrounding fluid, ρ_f:

$$\rho_f = \frac{\rho_b 4\pi r^3 k_t + \rho_a(\frac{4}{3}\pi r^3 - 4\pi r^3 k_t)}{\frac{4}{3}\pi r^3}; \quad k_t \ll 1. \tag{4.24}$$

Solving for ρ_a:

$$\rho_a = \frac{\frac{1}{3}\rho_f - k_t \rho_b}{\frac{1}{3} - k_t}; \quad k_t \ll 1. \tag{4.25}$$

In the case of an organism living in air, we can easily relate the density of the adjusting fluid to its temperature. We note from eq. 4.3 that

$$\rho_a = \rho_f \frac{T_f}{T_a}, \tag{4.26}$$

where T_f and T_a are the absolute temperatures of the surrounding fluid and adjusting fluid, respectively. When we couple this fact with eq. 4.25, we see that in air

$$T_a = T_f \frac{\frac{1}{3}\rho_f - k_t\rho_f}{\frac{1}{3}\rho_f - k_t\rho_b}; \quad k_t \ll 1. \tag{4.27}$$

This relationship is shown in figure 4.11. Unless the thickness of the body wall is *very* thin, the temperature of the adjusting fluid must be quite high. For example, if the radius is 100,000 times the body wall's thickness, the gas in the center of the sphere must be heated about 8.5° C above the temperature of the surrounding air. For a body radius 10,000 times the wall thickness, the adjusting fluid would have to be heated to more than 100° C above the temperature of the surroundings.

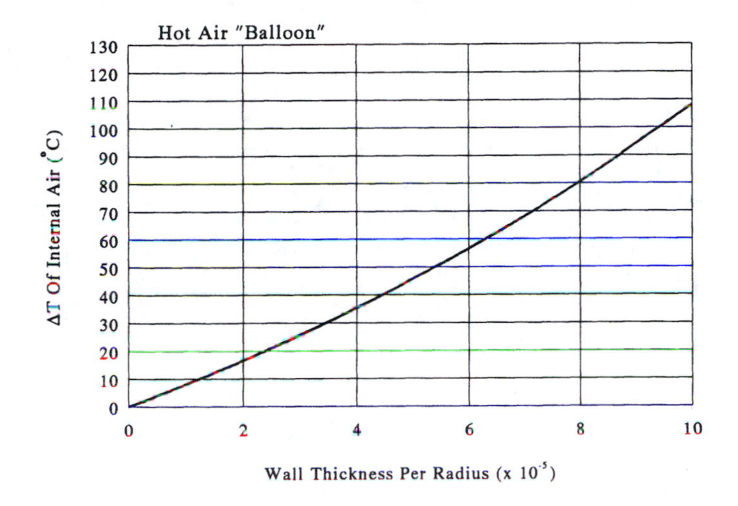

Fig. 4.11 The gas inside a "balloon" organism must be quite hot to keep the organism aloft in air. The thicker the wall of the "balloon," the higher the temperature required.

Perhaps this explains why nature has not utilized this form of aerial buoyancy compensation. An organism with a body wall sufficiently thin to be easily buoyed up would likely be very susceptible to damage and predation. For example, an organism with a radius of 10 cm and an internal temperature 8.5° C above ambient would require a body wall less than 1 μm thick—it would barely be a soap bubble!

Furthermore, unless the organism relied on an external source such as solar radiation to heat its adjusting fluid, the heat necessary to buoy the animal up would have to be provided by the metabolism of the body-wall tissue. The thinner the wall, the lower the temperature to which the adjusting fluid would have to be raised, but the greater the volume of gas that would have to be heated, and these two factors offset each other. If we make the simplifying assumption that all the heat produced by metabolism goes into heating the adjusting fluid, and note that it takes approximately 1000 J to raise the temperature of 1 kg of air one degree celsius at constant pressure, we can calculate that the energy per body mass required to achieve neutral buoyancy is 293 kJ kg^{-1}. This is a lot of energy. For instance, 293 kJ is sufficient energy to lift a metric ton (1000 kg) 29 m against the pull of gravity. And the situation would actually be worse in reality because much of the heat of metabolism would be conducted away by the surrounding air. A more realistic calculation would involve the *rate* at which metabolic energy per body mass would have to be expended in order to maintain the adjusting fluid at the appropriate temperature, but we will defer this calculation until we have examined heat in greater detail in chapter 8.

The possibility of thermal control of buoyancy does not appear quite so remote for organisms in water because their body density is much closer to that of the surrounding fluid. When dealing with water, the relationship between density and temperature is complex, and for each density calculated from eq. 4.25 we must infer the temperature from figure 4.3. The results are shown in figure 4.12. In water, a thin-walled spherical organism must raise the temperature of its adjusting fluid by only a few degrees above that of the surrounding fluid to achieve neutral buoyancy, much less of an increase than is necessary in air. For example, if the ratio of radius to wall thickness is 10,000, an increase of about 0.2° C is needed in water, in contrast to a 108° C increase required in air.

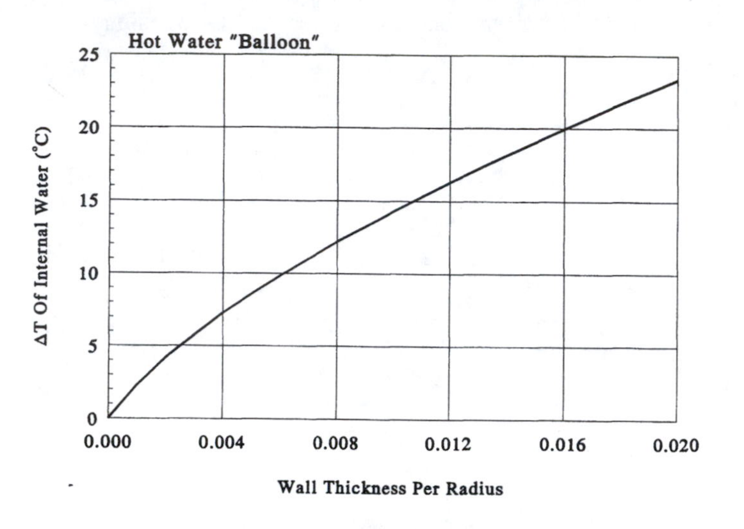

Fig. 4.12 Water inside an aquatic "balloon" organism need not be nearly as hot as the air in a terrestrial ballon to keep the organism neutrally buoyant.

A practical problem arises, however, when we calculate the energy required to heat the water. As we will discuss in chapter 8, water has a very high heat capacity—approximately 4200 J are required to raise the temperature of each kg of water by 1° C. As a result, the energetic cost of buoyancy adjustment by temperature control is actually about ten times higher in water than in air (fig. 4.13). In reality, this high relative cost of heating would be augmented by the fact that water is also a better conductor of heat than is air, and much of the metabolic expenditure would be conducted away by the surrounding water rather than going to heat the adjusting fluid.

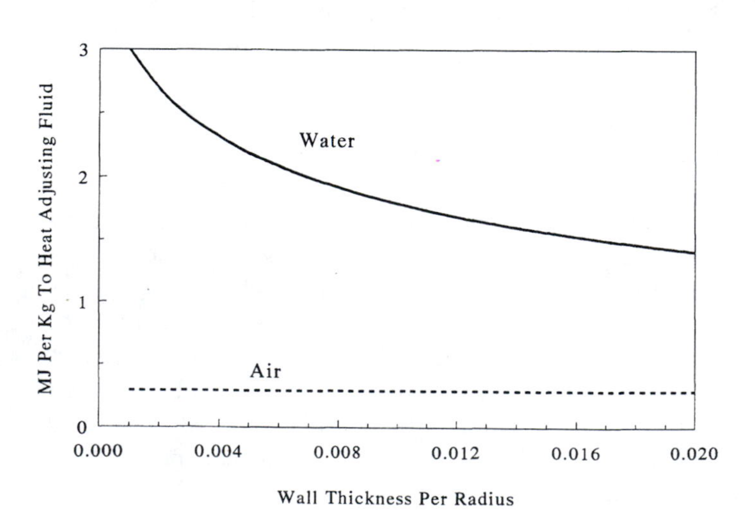

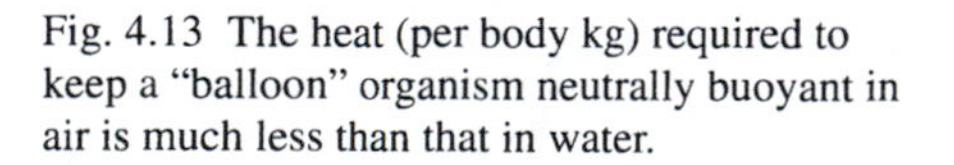
Fig. 4.13 The heat (per body kg) required to keep a "balloon" organism neutrally buoyant in air is much less than that in water.

Despite these problems, there is at least one aquatic organism that has found a way to function as a hot air balloon—the sperm whale (fig. 4.14). The huge "forehead" of this whale (the spermaceti organ) is filled to a large extent with oil, as much as 2.5 tons in a single whale. The particular mixture of lipids in sperm oil exhibits a substantial shift in density over its usual operating range of 33° to 29° C. In effect, the oil gradually "freezes" as the temperature is lowered, becoming more dense in the process. The density variation in the oil over this range (1% to 2%) is much greater than that of water over the same temperature range (about 0.1%), rendering the system much more amenable to control. Through a variety of mechanisms, the whale can rapidly vary the temperature of the oil in the spermaceti organ, and can thereby adjust its buoyancy (Clarke 1979).

We now turn our attention to the effects that density has on the fluid-dynamic forces encountered by plants and animals.

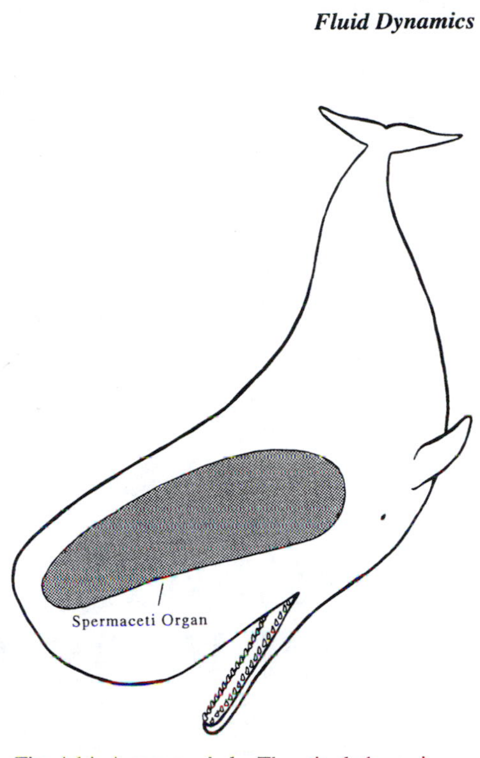

Fig. 4.14 A sperm whale. The stippled area in the head is the spermaceti organ, which allows the animal to act as an aquatic "hot air balloon." (After Clarke 1979. Original figure copyright ©1979 by Scientific American, Inc. All rights reserved)

4.4 Fluid-Dynamic Forces

There are four major hydrodynamic forces, and three of these, pressure drag, lift, and acceleration reaction, are directly proportional to the density of the medium. The fourth, friction drag, will be discussed in chapter 5.

4.4.1 *Drag*

Consider a stationary object suspended in a moving fluid (fig. 4.15). As the fluid moves past the object, its pressure is affected; in general, the pressure is highest on the upstream face and is lower around the sides and downstream face. The reasons for this spatial variation in pressure are somewhat complex and will not be dealt with here (see Vogel 1981 for a readable explanation). The difference in pressure between the upstream and downstream faces of the object results in a force, called *pressure drag*, that tends to push the object downstream.

The magnitude of pressure drag is determined by four factors. The first depends on the magnitude of the pressure difference—the greater the difference, the greater the force. In general, the difference in pressure is proportional to a quantity known as the *dynamic pressure*, the pressure that would be created if the moving fluid were to be brought to a dead stop:

$$\text{dynamic pressure} = \frac{1}{2}\rho_f u^2, \tag{4.28}$$

where u is the velocity of the fluid relative to the object. For a given velocity, the dynamic pressure is directly proportional to the density of the fluid.

Second, because pressure has the units of force per area, the force exerted on an object is proportional to the area over which the net pressure is applied. In the case of drag, this area is usually reckoned as A_f, the area projected in the direction of flow and known as *frontal area*.

Third, drag depends on the shape of the object. Bluff bodies such as flat plates and cylinders, each oriented with their longest dimension perpendicular to flow, substantially affect the pattern of flow in their vicinity, resulting in a large upstream–downstream pressure difference and a large drag. Streamlined objects disturb the pattern of flow to a lesser extent and have a lower pressure drag. Except in a few cases of simple, symmetrical objects at very low velocities, theory is not sufficiently precise to predict the effect of shape. Bowing to the complexity of the

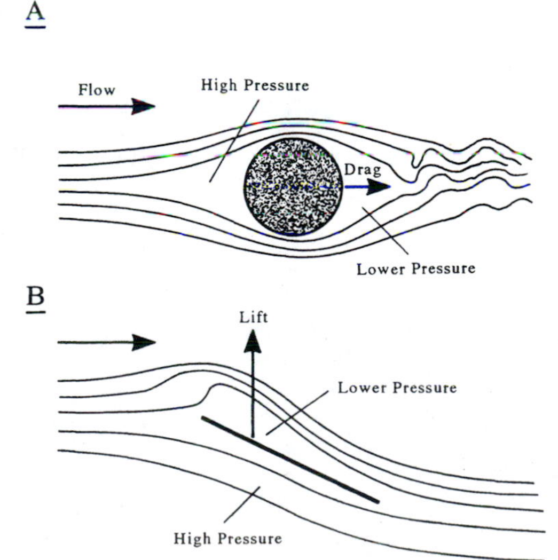

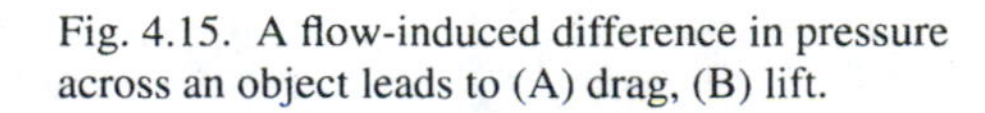

Fig. 4.15. A flow-induced difference in pressure across an object leads to (A) drag, (B) lift.

situation, fluid dynamicists have resorted to describing the effects of shape through the use of an empirically determined (although well-behaved) "fudge factor," the drag coefficient C_d.

Finally, we should note that C_d is not necessarily a constant. It may itself depend on velocity, the size of the object, and the ratio of the fluid's density to its viscosity. These factors are usually grouped together in a dimensionless index for the pattern of flow, the *Reynolds number*. But the effect is typically small, and for the purposes of this chapter we treat C_d as if it were a constant. For the relatively high velocities and large objects dealt with here, this simplification does not lead to any serious problems. The Reynolds number and its implications are discussed in detail in chapters 5 and 7.

Dynamic pressure, frontal area, and drag coefficient combine multiplicatively to give an expression for drag:

$$\text{drag} = \frac{1}{2}\rho_f u^2 A_f C_d. \tag{4.29}$$

Due to the density difference between the two media, drag at a given velocity is about 830 times greater in water than in air for an object of a given C_d and size.

4.4.2 *Lift*

As fluid moves past an object, the pattern of flow may be such that the pressure on one lateral side is greater than that on the opposite, and as a result a force is exerted on the object perpendicular to the direction of flow (fig. 4.15B). This *lift* force behaves in much the same fashion as drag; it is proportional to the dynamic pressure and to the area over which the pressure difference acts. In this case, the relevant area is that projected perpendicular to the direction of flow, and is often referred to as the *planform area*, A_p. As with drag, lift also depends on the shape of the object, a factor that is taken into account through an empirically determined *lift coefficient*, C_ℓ. Circumstances determine what shape gives a large lift. Far from a solid boundary, objects such as appropriately oriented flat plates and the special shapes known as airfoils induce a large pressure difference across their lateral faces. These shapes are commonly seen in the wings of insects, bats and birds, and in the tails of fish and whales. Near a solid boundary, the velocity gradient resulting from the fluid's viscosity can cause almost any shape to experience a lift (e.g., a mussel bed [Denny 1987]).

The overall expression for lift is similar to that for drag:

$$\text{lift} = \frac{1}{2}\rho_f u^2 A_p C_\ell. \tag{4.30}$$

Because of its dependence on the density of the medium, lift at a given velocity is about 830 times greater in water than in air for an object of a given C_ℓ and planform area.

4.4.3 *Acceleration Reaction*

In contrast to lift and drag, the third major hydrodynamic force does not depend on the velocity of the fluid, but rather on how rapidly the velocity changes, in other words on the fluid's acceleration. The reasons why the acceleration of the fluid should impose a force are somewhat complex and will not be explored here (see

Daniel 1984 or Denny 1988 for an explanation). Instead, we treat the phenomenon as a given and explore its biological consequences.

The force an object experiences due to a relative acceleration between itself and the surrounding fluid depends on whether the object is stationary and the fluid is accelerating, or whether the fluid is stationary and the object is accelerating. If the fluid is stationary,

$$\text{acceleration reaction} = \rho_b V a + C_a \rho_f V a, \tag{4.31}$$

where a is the acceleration of the object relative to the stationary fluid and V is the volume of the object. The term $\rho_b V$ is simply the mass of the object, so the first term ($\rho_b V a$) is the force required to accelerate the object's mass. This force would operate if the object were in a fluid or not, and is a straightforward result of Newton's second law of motion. The second term in eq. 4.31 is similar to the first, but the mass that acts as if it is being accelerated ($C_a \rho_f V$) is a mass of fluid rather than the mass of the object. This *added mass* is in one sense a computational fiction. One cannot point at a specific volume of fluid that is dragged along with an accelerating object. Nonetheless, the overall pattern of flow around an accelerating object can be conveniently described as if there were a defined added mass of fluid. The magnitude of this added mass, relative to the mass of the fluid displaced by the object, is described by C_a, the *added mass coefficient*. As with the lift and drag coefficients, C_a depends on the shape of the object. In general, objects that have a high drag coefficient have a high coefficient of added mass.

In the case of a stationary object and an accelerating fluid, the expression for the acceleration reaction is different:

$$\text{acceleration reaction} = (C_a + 1)\rho_f V a. \tag{4.32}$$

Because the object itself is not accelerating, no force is imposed by the object's own mass. However, the object occupies a space that would otherwise be filled with accelerating fluid. As a result, the force due to the motion of the fluid is greater than that in the case of a moving object, hence the term (C_a + 1).

In both cases described here, the acceleration increases with increasing density. For an object with a given C_a and at a given relative acceleration, the acceleration reaction is about 830 times larger in water than in air.

Note that the acceleration reaction is more sensitive to size than is lift or drag. Because acceleration reaction is proportional to the volume of an organism, it scales in proportion to the cube of a linear dimension rather than with the square as for lift and drag.

What does the 830-fold difference in hydrodynamic forces between air and water mean in a biological context? The ramifications are far too numerous to be covered here in any depth, but we can hardly leave the topic without exploring a few examples.

4.5 Density Effects in Locomotion

4.5.1 *Propulsion by Lift and Drag*

First we explore the effect of density on the manner in which animals move. Before we can deal with these we must examine some of the basic physics of locomotion.

We know from Newton's first law that if an animal is initially at rest, a force is required to cause it to move. In the study of locomotion, any force by which an animal tends to move itself is referred to as a *thrust*. If a constant thrust were the only force applied, an organism would continuously accelerate. In general, however, thrust is resisted by some form of drag. For example, when a fish undulates its body and tail, a thrust is created. As the fish begins to move through the water a drag is imposed, the magnitude of which increases with speed. As a result, the fish accelerates to a speed where its drag is just equal to the thrust. At that point there is no *net* force acting on the fish, and it subsequently moves at a constant speed for as long as the thrust is maintained. The same notion applies to birds, bats, and insects as they fly through the air.

For aerial organisms and those aquatic animals that are not neutrally buoyant, there is a third force to be considered in addition to thrust and drag—weight. The acceleration of gravity acting on the animal's mass exerts a force that tends to pull the animal down. As a consequence, to maintain a constant altitude the animal must exert an upward force equal to its weight. This upward force is commonly called lift, but need not be the same as the lift described in eq. 4.30. To avoid any confusion in this section, we refer to the force that resists gravity as *levitation* regardless of the mechanism by which it is produced.

For many animals, thrust is produced either by drag or by the lift created as an appendage is moved through the surrounding fluid. For example, some water beetles (e.g., water boatmen), fish that use their pectoral fins for locomotion, paddling ducks, and swimming humans create thrust by dragging their appendages through the water. This form of locomotion is analogous to a boat being rowed. For others, thrust is created by the lift that acts on an appendage. Flying birds, bats, and insects are obvious examples. Less obvious are swimmers such as tunas, whales, and porpoises, in whom the thrust required to move through the water is provided by the lift created as the tail oscillates back and forth.

How are these forces of locomotion affected by the 830-fold difference in density between air and water? The differences in the magnitude of lift and drag could be offset if the velocities encountered in air were sufficiently higher than those in water. A glance at eqs. 4.29 and 4.30 shows that if the velocity in air is $\sqrt{830}$, roughly twenty-nine times that encountered in water, drag and lift are the same in both media. In the case of locomotion, this is unlikely to occur. For example, birds tend to fly at velocities on the order of $10\,\mathrm{m\,s^{-1}}$, only about one order of magnitude greater than the speeds at which many fish swim. This implies that for animals of the same size, the lift and drag experienced by birds are only about a one-eighth of those experienced by fish.

For many aspects of locomotion, however, it is not the absolute magnitude of lift or drag that is important, but rather their ratio. For example, we have noted that for a fish cruising at a constant velocity, the average thrust provided by lift acting on its tail is just equal to the total drag. Changing the density of the fluid in which the fish swims affects both lift and pressure drag in the same sense and to the same degree, and the swimming ability of the fish is not grossly altered. For this reason, the changes in water density associated with the move from fresh water to seawater and back again should not, in themselves, affect the ability of a salmon as it moves through its life cycle. A similar logic applies to fish and cetaceans that move between water masses of different temperatures.

Some swimming animals have found a way to break this link between thrust and drag. Dolphins, sea lions, and penguins, for example, use the large thrust afforded by water to launch themselves into the air where they have the benefit of a lowered drag (fig. 4.16A). Au and Weihs (1980) have shown that this "porpoising" behavior reduces the cost of locomotion. Flying fish have evolved an even more elegant strategy. When threatened by a predator, they leap through the water's surface, thereby reducing their drag, and they prolong the effect by spreading their large pectoral fins to gain lift (fig. 4.16B). During the initial phase of the flight, the tail remains in the water where, by flapping, it can continue to provide thrust (Fish 1990). In this fashion the fish make maximum use of the difference in density between air and water. After attaining sufficient speed, the fish rises into the air (sometimes to a height of 6 to 7 m) and glides.

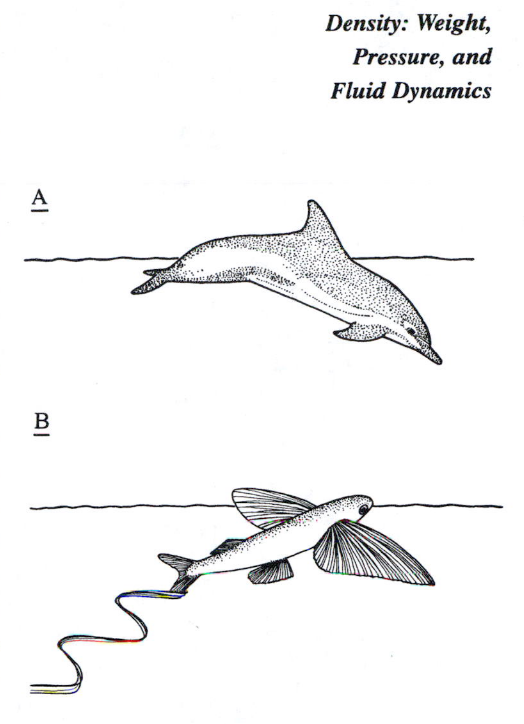

Fig. 4.16 A dolphin can "porpoise" into air to reduce its cost of locomotion (A). A flying fish (here shown in the initial phase of flight) uses a similar strategy, but by keeping its tail in water can retain the advantage of a large thrust.

The importance of the ratio of thrust and drag applies to flying insects, bats, and birds in the same sense it does to fish. In the case of flying animals, however, the absolute magnitude of lift *is* important. If the levitation provided by lift is less than the animal's weight, the animal falls out of the sky. For this reason, variations in air density can affect flight. For instance, Feinsinger et al. (1979) note that the foraging behavior of hummingbirds varies with altitude—at high altitude the birds adopt a feeding mode that minimizes their flight time. The authors suggest that this shift in behavior can be be explained by variations in air density. The higher the altitude, the lower the density, and the faster the birds must flap their wings to stay aloft. The consequent increased metabolic cost of flying at high altitude is thought to have affected foraging behavior.

There are also aquatic animals that require lift to provide levitation. For example, small tunas do not have swim bladders and are negatively buoyant. To maintain their position in the water column, they rely on the lift exerted by their pectoral fins as they move forward—if they stop, they sink (Alexander 1990). If we assume that both tunas and birds have densities of 1080 kg m^{-3} and that birds fly ten times as fast as tunas swim, we can calculate roughly what the relative size of the lifting appendages should be in the two organisms. Consider a bird and a tuna, each with a body volume of 10^{-3} m^3 and appendages having a C_ℓ of 2. The weight of the tuna in seawater is 0.5 N and, at a velocity of 1 m s^{-1}, would require its lifting appendages to have an area of 5×10^{-4} m^2. The weight of the bird is 10.6 N, and at 10 m s^{-1}, requires a lifting area of 9×10^{-2} m^2, 170 times that of the tuna. If the pectoral fins of the fish and the wings of the bird have the same shape, this implies that the bird's wings would have to be $\sqrt{170} \approx 13$ times as long as the tuna's fins. In fact, if you reduce the linear dimensions of a bird's wings by a factor of 13, you end up with an organism that looks amazingly like a tuna (fig. 4.17).

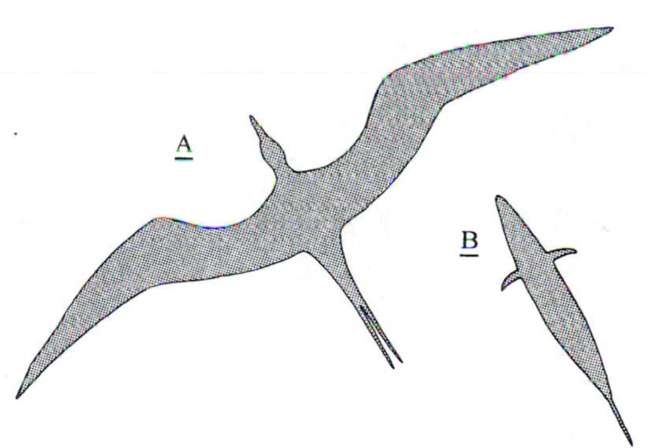

Fig. 4.17 (A) Silhouette approximating that of a frigate bird. (B) The same silhouette, except that the linear dimensions of the wings have been reduced by a factor of thirteen relative to the size of the body. The result looks surprisingly like a tuna. I have taken the liberty of rounding the outline of the bird's beak and combining the forked tail into a single tail, but otherwise the shape of the silhouette has not been changed.

4.5.2 *Jet Propulsion*

Lift and drag are not the only mechanisms by which thrust can be produced. For instance, there are many examples of jet propulsion in water. Squids, octopods, cuttlefish, and *Nautilus* escape from predators by squeezing water from their mantles and rapidly jetting away. Jellyfish propel themselves in a similar fashion, as do scallops. The aquatic larvae of dragonflies also use jet propulsion. But there are no jet-propelled bats, birds, or adult insects. Why? As you might expect, the answer lies (at least in part) with the difference in density between air and water. To see how this works, we briefly examine the physics of jet propulsion.

In jet propulsion, thrust is created by throwing away mass. Consider the simple case shown in figure 4.18. A hollow, spherical organism contains a volume of fluid (either air or water). At the bottom of the sphere is a small orifice so that as the body wall of the animal contracts, fluid is expelled downward as a jet. If the rate at

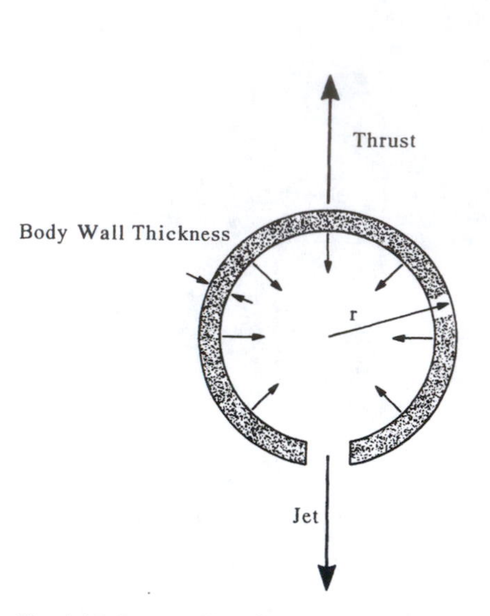

Fig. 4.18 By expelling fluid as a jet, an organism produces thrust.

which the volume of the lumen decreases is dV/dt and if the density of the fluid in the lumen is ρ_f, the rate at which mass is expelled is $\rho_f dV/dt$. If the orifice has area A, the velocity of the fluid as it leaves the sphere must be $(dV/dt)/A$.

Now, the product of the change in mass per time, $\rho_f dV/dt$, and velocity, $(dV/dt)/A$, is the rate at which momentum is being thrown away by the organism. Recalling from chapter 3 that the time rate of change of momentum is the same thing as force, we see that by ejecting fluid in one direction the animal creates a thrust which, by Newton's third law, pushes it in the opposite direction:

$$\text{thrust} = \frac{\rho_f (dV/dt)^2}{A}. \tag{4.33}$$

It is this thrust that propels squids and jellyfish through the water. Note that the thrust is directly proportional to the density of the fluid being ejected. Thus, all other factors being equal, an animal expelling water will have about 830 times the thrust of an animal expelling air.

But that is only part of the story. At any time, the upward thrust from the jet (a levitation) is resisted by the animal's weight and by drag, both of which tend to slow the animal down. The net vertical force available to accelerate an animal is equal to the difference between thrust and the combined effect of weight and drag:

$$\text{net force} = \text{thrust} - (\text{weight} + \text{drag}). \tag{4.34}$$

Noting also that the net force must equal the acceleration of the object times its effective mass (= mass + added mass), we can write

$$\text{acceleration} \times \text{effective mass} = \text{thrust} - (\text{weight} + \text{drag}) \tag{4.35}$$

$$\text{acceleration} = \frac{\text{thrust} - (\text{weight} + \text{drag})}{\text{effective mass}}. \tag{4.36}$$

If at any time we know the thrust, weight, drag, and effective mass, we can calculate the instantaneous acceleration of the animal. If we calculate acceleration as a function of time, we can then calculate velocity and the distance traveled.

It is possible to do this for some simple cases, and the results of one such case are shown in figure 4.19. In making these calculations I have assumed a spherical animal with a radius of 10 cm and a body-wall density of 1080 kg m^{-3}. Fluid is ejected downward from an orifice 1.8 cm in diameter. The organism contracts until all the fluid in its lumen is ejected, the contraction taking 0.5 s to complete. The animal then refills the lumen by drawing fluid back through the orifice.[5] This refilling phase takes 2 seconds. When in water, the animal propels itself by expelling water; when in air, it uses air.

Results are given as a function of the thickness of the body wall. In air, the body wall must be very thin, on the order of one five-hundredth of the sphere's radius, before thrust is sufficient to overcome weight. For the sphere we have used as an example (10 cm in radius), this means that if the body-wall thickness exceeds

[5]When fluid is expelled from the sphere, it emerges as a distinct jet in which virtually all fluid is directed downward. In contrast, when fluid is drawn back into the sphere, it can be drawn from many directions, constrained only by the presence of the "organism." As a consequence, the momentum of the moving bits of fluid during the refilling phase may not all be directed precisely upward as they pass through the orifice, and the calculations here may thereby overestimate the thrust during this period. However, this overestimation is probably minor (Daniel 1982).

200 μm, the animal cannot get off the ground. If the animal is sufficiently thin-walled, however, an aerial jet can perform quite well (fig. 4.19A). For example, when the body wall of a 10 cm radius sphere is 100 μm thick, the animal travels upwards at an average speed of almost 4 m s^{-1}.

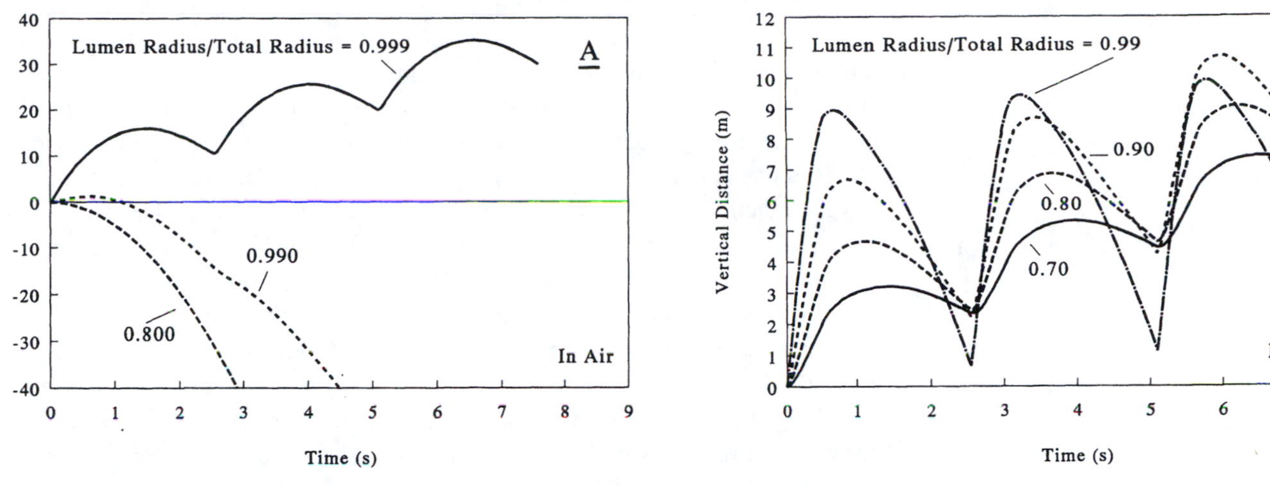

Fig. 4.19 Jet propulsion in air is feasible only when the body wall is very thin (A), but in water a thin wall can result in saltatory motion (B).

There are practical problems, however. A sphere with a muscular or elastic thin body wall might be able to eject air effectively, and thereby propel itself for a while. After all, a blown-up rubber balloon flies quite nicely when released. However, there is likely to be a problem when it comes time to refill. Although muscles or elastic tissue could be arranged radially to effect inflation, in a very thin-walled sphere the structure seems more likely to fold than to inflate.[6]

An aerial jet-propelled organism is also likely to encounter the same sort of problem encountered by a hot-air balloon. The thinner the body wall, the more voluminous the lumen, and the more effective the thrust. However, for a thin-walled sphere the energy required to propel an animal at a given speed is largely independent of the volume of the body wall—it depends primarily on the drag and therefore on the overall size of the animal. Thus, as the body wall is made thinner, the more-or-less constant energy required for locomotion must be provided by a smaller volume of muscle. It seems unlikely that muscle could provide energy at a sufficient rate to power locomotion in an organism thin-walled enough to be effectively jet propelled in air.

A spherical organism can have a much thicker body in water and still be able to move by jet propulsion (fig. 4.19B). In this case, a thin body wall may actually be a hindrance. The relatively large lumen in a thin-walled sphere results in such effective jetting that the animal experiences a large negative thrust during refilling. As a result, the motion of a thin-walled sphere is very saltatory—it may move several meters during contraction, but, during refilling, returns nearly to its starting point. Within certain bounds, the thicker the wall, the more damped the oscillations. For example, a sphere with a body wall 30% as thick as the radius moves with very little backsliding during refilling, and its average speed of 0.8 m s^{-1} is virtually the same as a sphere with a body wall only a third as thick. Furthermore, it is mechanically feasible for a thick-walled organism to refill itself.

[6]Some thin-walled organisms (e.g., squids and the sacklike alga *Halasaccion*) can use the flow-induced low pressure around their bodies to assist in inflation (Vogel 1987; Vogel and Loudon 1985). The effect is unlikely to be sufficient, however, to make aerial jet propulsion feasible.

For example, jellyfish use the elastic properties of their mesoglea (the "jelly") to expand themselves (DeMont and Gosline 1988) and squids use a combination of elastic tissue and radially oriented muscles (Gosline and Shadwick 1983).

4.5.3 *The Cost of Locomotion*

The consideration above of the mechanics of locomotion raises the question of how much metabolic energy it costs for an animal to move. We start by defining the cost of transport. There are two definitions in common use. First, we may define cost as the energy required to move a given mass of animal a given distance, the SI units being $\mathrm{J\,kg^{-1}\,m^{-1}}$. Alternatively, we may define cost as the liters of oxygen required to transport one kilogram of body mass one meter. This may seem like a strange set of units, but it has some practical advantage. The rate at which most animals expend energy can most easily be measured from the rate at which they consume oxygen. The consumption of one liter of O_2 corresponds to the metabolic expenditure of approximately 20.9 kJ, but the exact value depends on what compounds are being metabolized to provide this energy (carbohydrate, protein, or lipid). Because it is not always known what metabolite is used, the conversion from liters of O_2 to joules has not always been attempted, and one often finds values for cost cited only in terms of oxygen consumption.

Schmidt-Nielsen (1972a) has combined the results of many researchers to show that the size-specific cost of transport differs among animals in a clear pattern related to the density of the medium in which an organism moves (fig. 4.20).

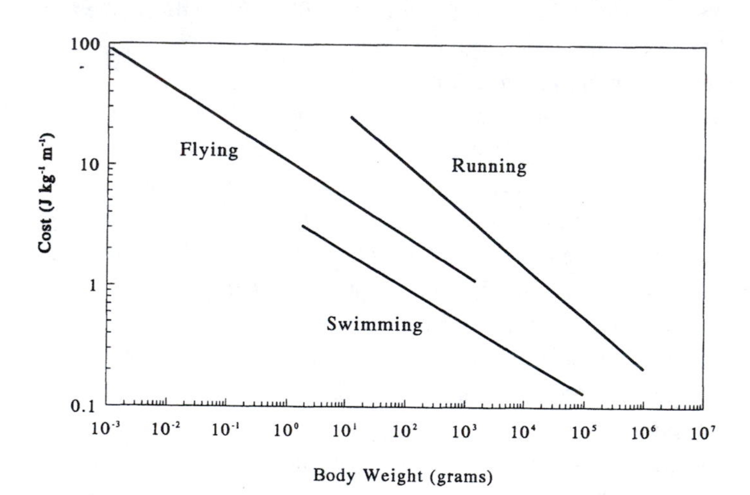

Fig. 4.20 For an animal of given weight, it costs more to fly than to run and more to run than to swim. (Redrawn from Schmidt-Nielsen 1972a. Original figure copyright ©1972 by the American Association for the Advancement of Science)

For an animal weighing 1 kg it costs about 2.7 times more to transport its body mass one meter by walking or crawling than it does by swimming, and it costs 3 times more to transport a given body mass by flying than it does by walking. Although the specific physics that account for this hierarchy of costs are complex, buoyancy (and therefore density) is an important contributing factor. A neutrally buoyant fish expends very little energy in maintaining its position against gravity—its primary transport cost comes in doing work against the fluid as its moves.[7] In contrast, terrestrial animals that walk or crawl must maintain an upright posture. Lacking buoyant support from water, this posture requires an expenditure of energy that increases the cost of transport. Furthermore, as an animal walks, its center of

[7]The additional volume required to become neutrally buoyant (via a swim bladder, for instance) does incur some increased drag when the animal swims (Alexander 1990), but the cost is slight.

mass typically goes through some vertical oscillation, leading to an oscillation in gravitational potential energy. To a certain extent the kinetic energy gained as the center of gravity falls can be used toward returning the center of gravity to its original position, but the reciprocal trading of gravitational potential and kinetic energy is never completely loss-free. As a result there is a net expenditure of energy in overcoming gravitational forces that would not be present if the animal were neutrally buoyant.

Flying is worse than walking. As we have seen, a large part of the energy expended in flight is used simply to maintain the animal's vertical position in the air. We can explore this differential cost by comparing the energy required by birds and tunas to maintain their position in the medium. Recall that tunas, like birds, offset their effective weight by the lift of their pectoral appendages, so it is reasonable to compare the costs entailed.

Alexander (1990) proposes that the power (in joules) expended in remaining aloft is

$$P \propto \frac{m^2 g^2 \rho_e^2}{\rho_f \rho_b^2 u \ell^2}, \tag{4.37}$$

where m is the mass of the organism, u is its speed relative to the surrounding fluid, and ℓ is the animal's wing (or fin) span. This power can be expressed as a cost by dividing eq. 4.37 first by mass to give power per kilogram, and then by velocity to yield energy per meter per kilogram.[8] Thus,

$$\text{cost} \propto mg^2 \frac{\rho_e^2}{\rho_f \rho_b^2} \frac{1}{u^2 \ell^2}. \tag{4.38}$$

If we assume that the body densities of bird and fish are equal, the ratio of cost in air to that in water is

$$\frac{\text{cost in air}}{\text{cost in water}} = \left(\frac{\rho_{e,a}}{\rho_{e,w}}\right)^2 \frac{\rho_w}{\rho_a} \left(\frac{u_w}{u_a}\right)^2 \left(\frac{\ell_w}{\ell_a}\right)^2, \tag{4.39}$$

where the subscripts a and w refer to air and water, respectively.

Now, $\rho_{e,a}/\rho_{e,w}$ is about 20, ρ_w/ρ_a is 830, and we previously showed that it is reasonable to expect ℓ_w/ℓ_a to be 1/13. As before, we may assume that the bird flies ten times as fast as the fish swims. Given these assumptions, we find that the cost of remaing aloft is about nineteen times greater for a bird than for a tuna of the same mass.

We noted above, however, that empirical measurements show that it costs a 1 kg animal roughly $3 \times 2.7 \approx 8$ times as much to fly as to swim. The discrepancy between these two figures (19-fold versus 8-fold) is due to the cost considered. Here we have only considered the cost required to create lift and thereby to stay aloft. Additional costs are associated with moving the body through the medium, and as we have seen these are relatively greater in water than in air. The effect of these additional costs is to narrow the gap between the cost of swimming and that of flying. Nonetheless, without the benefit of being neutrally buoyant, terrestrial organisms spend much more energy in transport than do aquatic organisms.

It should be pointed out that the cost of locomotion as we have used it here refers to the cost of transport relative to a stationary medium. If the medium itself is moving in the direction in which the animal wants to go, the effective cost of

[8] Recall that power equals force times velocity (eq. 3.17).

traveling from one point on the earth to another is less. This is unlikely to affect the net cost for animals that crawl or walk because the ground seldom moves fast enough to be of any consequence. But ocean currents and atmospheric winds could be used by swimming and flying animals, respectively, to substantially lower their net transport cost. Using values available in the literature (Schmidt-Nielsen 1984), it is possible to calculate, for instance, how fast the wind would have to blow in order for the net cost of transport of a bird to be less than that of a fish.

We begin by examining the components of cost. Now, the overall cost of locomotion is equal to the power expended while moving at a constant speed divided by the speed of the animal. For present purposes, we are concerned with the *effective speed*, the speed of the animal relative to the medium plus the speed of the medium relative to the ground. For example, if a bird flies at $10\,\mathrm{m\,s^{-1}}$ relative to the air, but the air itself is blowing at a speed of $10\,\mathrm{m\,s^{-1}}$ relative to the ground in the same direction as the bird flies, the effective speed of the bird is $20\,\mathrm{m\,s^{-1}}$. Conversely, if the bird flies at $10\,\mathrm{m\,s^{-1}}$ into a $10\,\mathrm{m\,s^{-1}}$ wind, its effective speed is zero.

We first calculate the cost of transport for a salmon. The power required to swim at a typical cruising speed (75% of maximum burst speed) has been carefully measured (Brett 1965) and can be expressed as an allometric equation:

$$P = 7.37 \times 10^{-5} m^{-0.084}, \tag{4.40}$$

where P is measured in liters of O_2 $\mathrm{kg^{-1}\,s^{-1}}$, and m is body mass in kilograms. The speed at which the fish swims relative to the water is also a function of its body mass, and from data given by Brett (1965) can be calculated to be

$$u = 1.0 m^{0.17}, \tag{4.41}$$

where velocity is measured in $\mathrm{m\,s^{-1}}$. Thus, the effective cost of transport is

$$\text{effective cost} = \frac{7.37 \times 10^{-5} m^{-0.084}}{m^{0.17} + \text{current speed}}. \tag{4.42}$$

There is a danger in combining allometric equations in this fashion. The data on which these estimates of power and velocity are based show some variation about the general trends, and any correlation between these deviations could affect our estimate of effective cost. For example, if those fish that expend energy at a higher rate than expected have a lower than expected velocity, the effective cost calculated here may be an underestimate. In the absence of more complete data, however, this approach is the only one open to us, and we will cautiously explore its consequences.

By a similar method we can calculate the effective cost of transport for a bird. Again using values cited by Schmidt-Nielsen (1984),

$$P = 3.8 \times 10^{-3} m^{-0.03} \tag{4.43}$$

$$u = 14.6 m^{0.20} \tag{4.44}$$

$$\text{effective cost} = \frac{3.8 \times 10^{-3} m^{-0.03}}{14.6 m^{0.20} + \text{wind speed}}. \tag{4.45}$$

These effective costs are plotted for a range of possible current and wind speeds in figure 4.21. For a body mass of 1 kg, the wind must blow at about 40 m s^{-1} (90 mph) for the effective cost of bird flight to be less than that of a fish swimming through still water. Winds of this velocity can be created in storms, but they are rare, and whether birds can and do take advantage of them is open to question. Perhaps more amenable to transport are the jet stream winds. These high-velocity (30 to 40 m s^{-1}) winds blow in predictable directions over the temperate regions of much of the globe and could potentially be used for transport. Their main disadvantage is the fact that they occur only at altitudes in excess of 10,000 m.

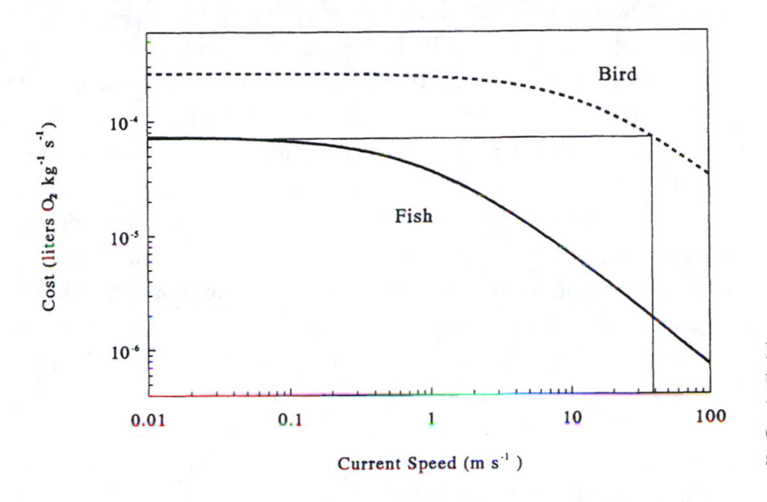

Fig. 4.21 If a bird flies in the same direction as the wind, its cost of transport is reduced. If the wind speed is approximately 40 m s^{-1}, the bird's cost is the same as that of a fish swimming in still water.

Of course, the lowered costs of transport associated with currents and winds can be realized only when the current or wind blows in the direction the animal wishes to travel. Flying or swimming upstream would increase the cost of transport, and as the current or wind approaches the intrinsic speed of the animal, the cost tends toward infinity. For example, we can compare the effective cost of transport between a bird flying in still air and a salmon swimming upstream. For water velocities greater than about 0.7 m s^{-1}, the cost of transport is higher for the fish (fig. 4.22). Thus, the effective cost of transport for a salmon homing upstream may well be as high as that for a bird in flight.

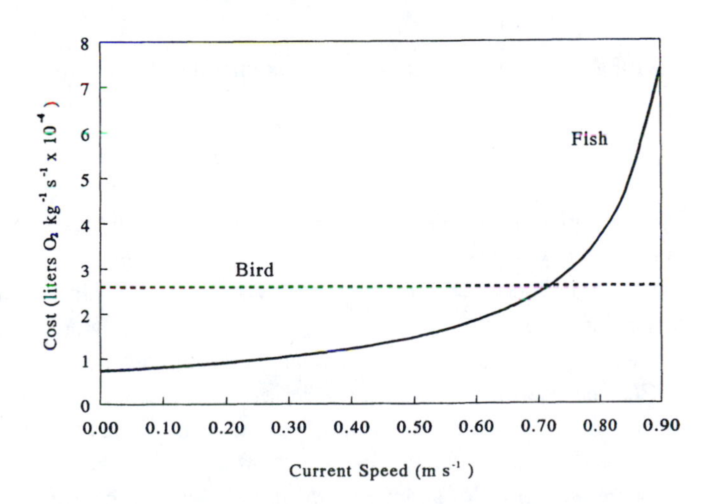

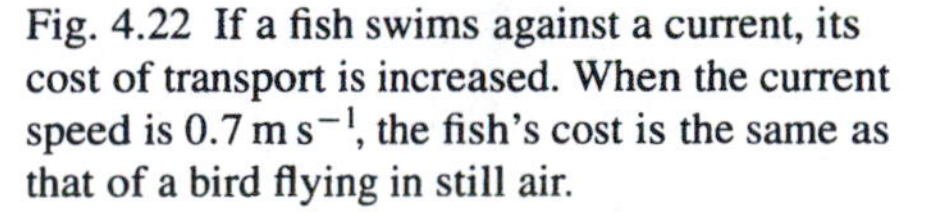

Fig. 4.22 If a fish swims against a current, its cost of transport is increased. When the current speed is 0.7 m s^{-1}, the fish's cost is the same as that of a bird flying in still air.

These examples are not presented as evidence that flying and swimming are, in general, equally costly. In most situations the hierarchy of costs outlined by

Schmidt-Nielsen (1971) is appropriate, and it is only under special circumstances such as those considered here that their order can be reversed.

4.6 Fluid-Dynamic Limits to Size

We now return to the subject of columnar organisms that we left earlier in this chapter. We know from that discussion that there is an upper limit to the height of a stable column of a given radius. From this, one would predict that, due to buoyancy, columnar organisms would reach greater heights in water than in air, all other factors being equal. Having briefly examined fluid-dynamic forces, we are now in a position to show why "other factors" are not equal between air and water, and to provide a mechanical explanation as to why trees are taller than corals.

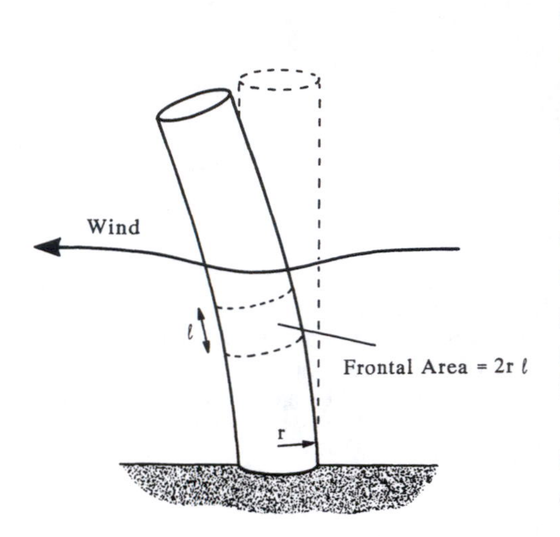

Fig. 4.23 An upright cylinder in flow experiences a drag.

Consider a stationary vertical column subjected to a horizontal fluid flow (fig. 4.23). As fluid moves past the column, a drag is imposed that tends to bend the column to one side. As the column is bent, the material on its upstream side is stretched and material on the downstream side is compressed. An examination of the mechanics of the bending process (Timoshenko and Gere 1972) shows that for a beam of uniform cross section, the bending moment acting on the beam is greatest at the beam's base. As a result, deformation of the beam's material is greatest at the base, and if the beam were to break as a result of the applied drag it would likely break there.

A standard equation from beam theory (Timoshenko and Gere 1972) allows us to calculate the force per area (the tensile or compressive stress) exerted by drag on the material at the beam's base:

$$\text{stress} = \frac{2\mathcal{F}h^2}{\pi r^3}, \tag{4.46}$$

where $\mathcal{F}$ is the force per length imposed on the column by drag, h is the height of the column, and r is its radius.

From eq. 4.29 we can calculate $\mathcal{F}$. The area over which a pressure difference acts as water flows past the column is $2rh$, thus the area per length is $2r$. The drag per length is then

$$\mathcal{F} = \rho_f u^2 C_d r. \tag{4.47}$$

Inserting this value into eq. 4.46, we see that the stress at the base of a column is

$$\text{stress} = \left(\frac{2C_d}{\pi}\right)\rho_f u^2 \left(\frac{h}{r}\right)^2. \tag{4.48}$$

The stress placed on the beam's material increases with the density of the fluid, the square of the velocity, and the square of the ratio of height to radius. If we assume that the drag coefficient in water is the same as that in air, we can then calculate how much higher a beam of a given strength (in other words, a beam capable of withstanding a certain stress) could be in air than in water.

Consider a worst-case scenario. Storm winds occasionally reach velocities of 25 to 30 m s^{-1}, and we take this as a reasonable estimate of the maximum velocity with which a tree might have to cope. Benthic aquatic plants and animals typically encounter velocities of about 5 m s^{-1} in the surf zone of wave-swept shores or in mountain streams. Using these values, we see that the maximal product of ρ_f and u^2 is about 25 times greater in aquatic than in terrestrial habitats. On this basis we would expect that the height-to-radius ratio could be about 5 ($= \sqrt{25}$) times greater in air than in water. Thus, we would expect columns constructed of the same material and of equal radius to grow 5 times as high on land as in water. This result can help to explain why, despite the advantages of buoyancy, columnar organisms are generally smaller in water than in air. In fact, the strength of wood (10^8 Pa) is roughly 2.5 times that of materials such as coral skeleton and algal stipes, which should allow wooden columns in air to grow almost $5 \times \sqrt{2.5} \approx 8$ times as high as algal or coral columns in water, a prediction that is in approximate agreement with reality.

We can take this line of reasoning one step further by noting that drag is not the only force important in constraining the size of aquatic organisms. If the fluid is accelerating, the acceleration reaction must also be taken into account. The importance of the acceleration reaction (the force caused by water's acceleration) becomes apparent when we consider how the shapes of plants and animals change as they grow.

To this point we have considered columnar organisms that "grow" by becoming higher, but which maintain a constant radius. In other words, the shape of the column changes as it grows. What happens if a plant or animal grows isometrically? In this case, the size limitation imposed by drag disappears. This is most easily seen by considering the cubic "organism" shown in figure 4.24. If one side of the cube has length ℓ, there is an area ℓ^2 projecting into flow, and a corresponding drag is imposed on the cube. This force is resisted by the area of the cube attached to the substratum, which likewise has area ℓ^2. What happens if the size of the cube (measured as a linear dimension) increases by a factor of two? There is now four times as much area projecting into the flow and, assuming that C_d is constant, four times as much drag is imposed. By the same token, however, the area resisting drag has increased by a factor of four. As long as the organism grows isometrically, the imposed force and the organism's ability to resist that force increase at the same rate. In such a case, drag (or lift, which scales in the same fashion) does not constrain the size of the organism.

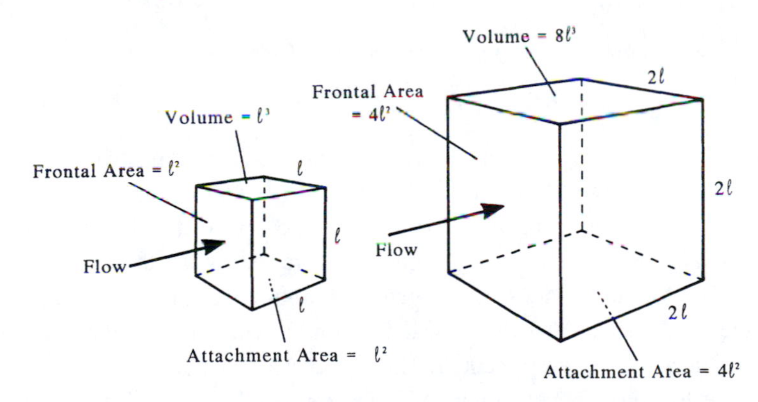

Fig. 4.24 When the size of an object is increased isometrically, the ratio of frontal to attachment area is constant. As a result, the ratio of drag to strength is constant. In contrast, volume increases faster than attachment area, and the ratio of accelerational force to strength increases.

Note, however, that in doubling a linear dimension the volume of the cube increases by a factor of eight instead of four. Because the acceleration reaction is proportional to the volume of fluid displaced (eq. 4.31) rather than to the area projecting into flow, the larger cube experiences a larger acceleration reaction relative to its basal area than does the smaller cube. In this case, stress increases in direct proportion to ℓ. Thus, the fact that acceleration reaction scales with volume rather than area poses a potential limit to the size of organisms, even those that grow isometrically.

Denny et al. (1985) have shown that the rapid water accelerations associated with breaking ocean waves, and the resulting acceleration reaction placed on benthic organisms, can help to explain why wave-swept organisms never get very large. We measured accelerations in the surf zone in excess of 400 m s^{-2}, and calculated that the resulting acceleration reaction might be sufficient to act as a limit to the size of animals such as sea urchins and limpets.

A similar mechanical limitation to size could operate in the terrestrial environment, but because the density of air is only 1/830th that of water, organisms in air could be much larger than those in the surf zone, all other factors being equal. Consider again the cubic organism of figure 4.24. For simplicity, let us assume that the maximal acceleration imposed on this organism occurs when the fluid velocity is zero, at which time the acceleration reaction places a shearing force on the base of the cube:

$$F = (1 + C_a)\rho_f \ell^3 a. \tag{4.49}$$

The shear stress placed on the base of the cube is this force divided by the cube's basal area, ℓ^2:

$$\text{shear stress, } \tau = (1 + C_a)\rho_f \ell a. \tag{4.50}$$

When this stress equals the organism's shear strength, τ_b, the organism is as large as it can get. Solving for this maximal size, we see that

$$\ell_{max} = \frac{\tau_b}{(1 + C_a)\rho_f a}. \tag{4.51}$$

If we assume that the strengths and added mass coefficients of terrestrial and aquatic organisms are equal, we can calculate the relative maximal sizes of organisms in the two environments:

$$\frac{\ell_{max,a}}{\ell_{max,w}} = \frac{\rho_w a_w}{\rho_a a_a}, \tag{4.52}$$

where the subscripts w and a refer to water and air, respectively.

In reality, terrestrial organisms are typically only an order of magnitude larger than their aquatic cousins. If the size of these land-dwelling plants and animals is (like that of some aquatic organisms) limited by the the acceleration reaction, it implies that the product of ρ_f and maximum acceleration must be smaller by a factor of ten in air than that in water. Assuming a maximal acceleration of 400 m s^{-2} for the aquatic environment, this implies that air (which is 1/830th as dense) would have to exhibit accelerations of at least 3.3×10^4 m s^{-2}. This is an acceleration of 3300 gravities! It seems unlikely that such accelerations are ever present in the atmosphere, and we may conclude that the size of terrestrial organisms is unlikely to be limited by the effects of the acceleration reaction.

The examples considered here are just an hors d'oeuvre to the feast of information available on the biological effects of drag, lift, and acceleration reaction. For further information, you should consult sources such as Vogel (1981), Denny (1988), Alexander (1983), Webb (1975), and Grace (1977).

4.7 Blood and Sap Pressures

As our last foray into the biological consequences of density we now explore the physics of blood pressure. Consider a long, upright tube filled with water (fig. 4.25A). From considerations discussed when we examined buoyancy, we know that water at the bottom of the tube has a higher hydrostatic pressure than water at the top. In other words, water in the tube exerts a greater outwardly directed pressure the farther down in the tube one measures it. If the tube is to stay intact, its walls must somehow resist this pressure.

This is no problem if the tube is entirely submerged in water. The hydrostatic pressure inside the tube is just offset by the hydrostatic pressure outside. There is no net pressure across the wall, and thus no force that the wall itself must resist. This scenario (a liquid-filled tube submerged in water) is roughly analogous to the blood vessels in a fish or whale. Because these animals are surrounded by water, any change in the hydrostatic pressure in their blood vessels is counteracted by a commensurate change in pressure in the surrounding water. Actually, in an animal the situation is a bit more complex. In the case of an artery, the blood in the vessel is maintained at a higher pressure than that of the surrounding water by the pumping action of the heart; the increased pressure is needed to drive the blood through the capillaries (see chapter 7). However, in an aquatic organism, the arterial blood pressure is approximately the same everywhere in the body, and the arteries have to resist only the pressure created by the heart.

The situation is drastically different for terrestrial animals. If the tube of figure 4.25A is situated in air rather than water (fig. 4.25B), the pressure exerted by air on the outside of the tube does not vary with vertical position nearly as much as the pressure exerted by blood on the inside of the tube, and near its bottom the walls of the tube must resist a substantial net outward force. As before, this force is increased if in addition to hydrostatic pressure there is a pressure imposed by the pumping action of the heart. This is the situation faced by vessels in your legs and feet, for instance. A few simple calculations serve to illustrate this point.

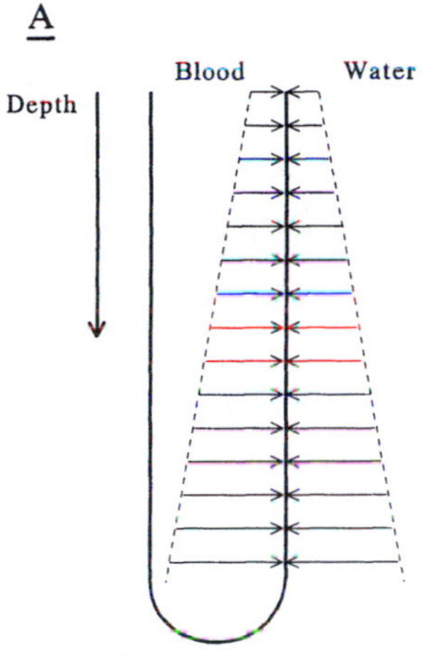

When a physician measures your blood pressure with a sphygmomanometer, the values obtained typically vary between 70 mm of mercury (9.3×10^3 Pa) when the heart is refilling and 120 mm of mercury (1.6×10^4 Pa) when the heart is pumping. These are the values usually cited as *the* blood pressure and are used as a standard. However, the physician is careful to place the cuff on your arm at a level close to that of your heart, and therefore the values measured are those corresponding to the pressure in the arteries as they exit the heart. But what is the blood pressure in your feet? If your feet are 1.5 m below your heart (a reasonable value), the pressure is greater by

$$1.5\rho_{bl}g = 1.5 \times 10^4 \text{ Pa}, \tag{4.53}$$

where ρ_{bl} is the density of blood. This is an increase of 1.6 times the diastolic pressure. The arteries and veins in your feet are, in effect, chronically hypertensive and must be capable of coping with this higher pressure. This, in turn, can lead to problems for astronauts. In the weightless environment of outer space, blood in one's legs and feet is at the same pressure as blood elsewhere. In response, the "extra" elasticity of the vessels in these extremities (normally needed to resist the gravity-induced hypertension) instead tends to force blood into the head.

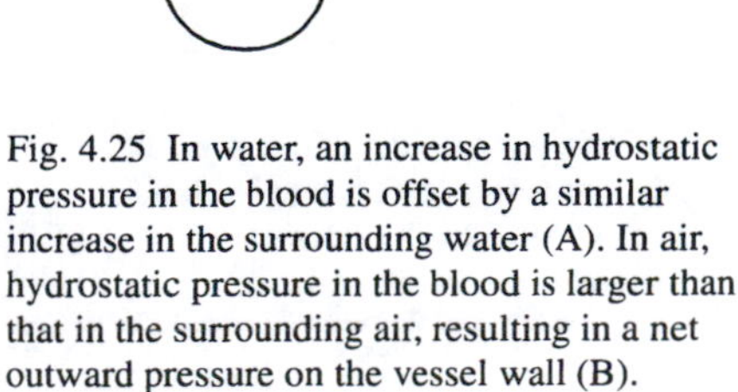

Fig. 4.25 In water, an increase in hydrostatic pressure in the blood is offset by a similar increase in the surrounding water (A). In air, hydrostatic pressure in the blood is larger than that in the surrounding air, resulting in a net outward pressure on the vessel wall (B).

You can convince yourself of the validity of this reasoning by standing on your head, thereby placing your head 0.6 m lower relative to your heart than when you are upright. The pressured feeling you get in your skull, the tendency for your eyes to bulge out and for your face to get red, are all consequences of the increased hydrostatic pressure. If you care to try it, you will note that you do not feel these symptoms standing on your head while submerged in a swimming pool. Another sign that blood pressure varies across your body is the tendency for your ankles to swell when you sit still for a long period. The high pressure in the capillaries of your legs tends to force water out of the circulatory system and into the surrounding tissue. In the absence of muscular activity, this fluid collects in the lymphatic spaces.

If the blood pressure in your head is higher than at your heart when you are upside down, what is the pressure like in your head when you are right side up? Here the situation is just reversed. Because your head is approximately 30 cm above your heart, the pressure in the arteries there is 3×10^3 Pa lower than the standard blood pressure. As a consequence, a minimum standard blood pressure of greater than 3×10^3 Pa (about 25 mm Hg) is needed to provide effective blood flow to your brain. At a lower pressure, you faint.

You might question this conclusion. Why does the blood in your brain have to be above ambient? Why can't blood vessels act like siphons, allowing substantial flow at pressures below atmospheric? The reason is that blood vessels do not have rigid walls. If pressure in a vessel falls below ambient, the vessel collapses and the flow is cut off.

The variations in blood pressure from one part of the body to another are particularly striking in certain species. For instance, consider the problem faced by a giraffe. These curious creatures grow to be 5 m tall, with their heads 2.5 m above their hearts and their feet 2.5 m below. When they lower their heads to drink, the pressure in their cerebral arteries changes by nearly half an atmosphere! In practice, giraffes avoid some of this pressure variation by spreading their front legs as they bend over to drink, thereby lowering the level of the heart at the same time as that of the head (Warren 1974). Giraffes avoid swollen ankles by having tight, elastic skin on their legs that resists the tendency for fluid to be forced out of the capillaries. One wonders how dinosaurs coped with their blood pressures. For instance, large dinosaurs such as *Apatosaurus* were twice as tall as giraffes (fig. 4.26), implying that the blood pressure in their heads might have changed by almost an atmosphere when they bent down to drink. It must have been an interesting sight to see a herd of apatosaurs at a watering hole.

Fig. 4.26 Giraffes spread their front legs to drink, thereby reducing the change in blood pressure in the head. One wonders how dinosaurs such as *Apatosaurus* drank.

One also wonders about snakes, which can grow to lengths of more than 10 m. Do they faint when climbing trees? In fact, climbing snakes are generally small and have evolved several adaptations to a vertical existence (Lillywhite 1987, 1988). Their hearts are placed close to their heads, and they have a tight-fitting skin that resists the pooling of blood in their tails. In contrast, really large snakes such as anacondas do not often climb trees.

The physics of pressure can also be applied to water in plants. For example, water in the xylem of vascular plants forms an unbroken column from the roots to the leaves. In a redwood tree 60 m high, this means that the water in the roots experiences a hydrostatic pressure 6 atmospheres higher than that in the crown. Curiously enough, this does not necessarily mean that water in the roots has a pressure 6 atmospheres greater than that of the surrounding air. The water in a tree is supported both by root pressure (a pressure created by osmosis) and by the fact that the top of the water column does not allow air to enter (see fig. 12.8). Because air cannot flow in, the water column can in effect be "hung" from the leaves. As a result of the mechanical interaction between the roots and the leaves, pressure in the xylem starts out high in the roots, reaches ambient air pressure at a height a few meters above the ground, and continues to decrease with increasing altitude. Much of the water in the xylem is therefore at a *negative* pressure with respect to the surrounding atmosphere. In this respect, xylem has the opposite mechanical problem of an artery—its walls must resist a strong inwardly-directed force. We will return to the mechanics of xylem when we explore surface tension in chapter 12.

4.8 Summary . . .

The 830-fold difference in density between air and water far outstrips the relatively minor changes in density that accompany changes in temperature and pressure. The greater buoyancy provided by water would allow stable aquatic columns to grow higher than terrestrial columns, but the greater fluid-dynamic forces experienced by columns in water more than offset this advantage. A variety of mechanisms have evolved to adjust the buoyancy of aquatic organisms, but no practical mechanism seems to be available for organisms in air. As a result, marine organisms that possesed stiff skeletons were preadapted for the invasion of the terrestrial environment.

The density of the medium has important consequences for the cost of locomotion—because swimming animals have to expend less energy in fighting gravity it costs them less to travel than animals that walk or fly. The low density of air compared to that of water has interesting consequences for blood pressure and the construction of blood vessels in terrestrial animals, and for the support of a water column in plants.

4.9 . . . and a Warning

This chapter provides a picture of how the density of the medium affects biology; a picture, however, which has been painted with broad strokes. In attempting to provide a coherent overview, much of the fine detail has been neglected, and any of the individual cases examined here is, upon close scrutiny, likely to be much more complex than is implied by its cavalier treatment in this chapter. Therein lies a danger. You are hereby warned that the examples explored here, although correct within the limitations of their context, are not meant to be applied as "recipes" for research. Before attempting to apply the information presented here to any specific biological example, you should carefully consult the original literature.

Chapter 5

Viscosity: How Fluid Is the Fluid?

As we have seen, fluids such as air and water are differentiated from solids by their ability to flow. When acted upon by a net force, a solid deforms a bit and then stops. A fluid, on the other hand, responds with a continuous deformation. For example, water in a stream flows continuously downhill, driven by the force resulting from the acceleration of gravity. The steeper the stream bed is, the faster the current. Local differences in atmospheric pressure cause air to flow, producing wind; the greater the gradient of pressure the more brisk the breeze. How rapidly a fluid is deformed by a given driving force is determined by the fluid's *viscosity*, the physical characteristic that is the subject of this chapter.

Water is much more viscous than air, a fact of considerable biological importance. For example, we will see that it takes 2500 times as much pressure to force water rather than air through a pipe at a given rate, and how this difference has guided the evolution of circulatory systems in locusts and water lilies. We will explore the mechanism by which limpets adhere to rocks, see why bacteria might be able to fly faster than they can swim, and explain why it is more costly for fish and insects to breathe when their environment is hot.

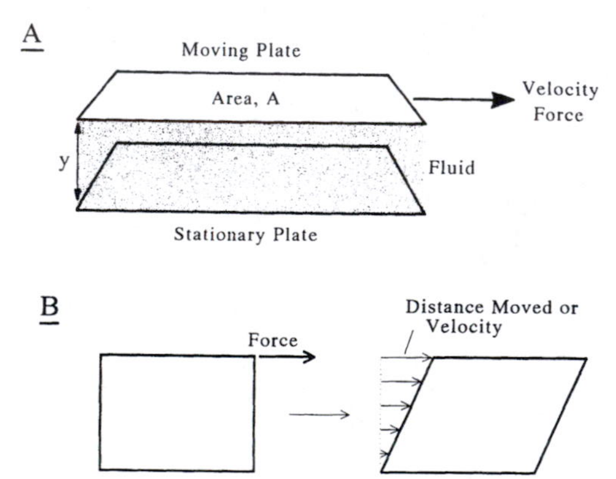

Fig. 5.1 Fluid sandwiched between two plates is deformed by the movement of one of the plates (A). A force is required to create the velocity gradient in the fluid (B).

5.1 The Physics

Consider the following example (fig. 5.1A). A volume of fluid is sandwiched between two horizontal plates that are separated by distance y. Each plate has area A and the top plate slides parallel to the other with velocity u. As a result, the fluid is *sheared*. In other words, each molecule moves slightly faster on average than the molecules directly below it and slightly slower than the molecules directly above, and the fluid as a whole deforms (fig. 5.1B). The rate at which fluid moleclues move past each other (that is, the deformation rate) depends on two factors. First, the faster the moving plate moves, the faster the fluid must deform to accommodate the motion; thus, deformation rate is proportional to u. Second, the rate of deformation is inversely proportional to y; the smaller the spacing between plates, the faster the remaining fluid must deform to keep up.

In turn, the force F required to maintain this rate of deformation depends on two factors. The larger the area of the plates, the larger the volume of fluid that is affected and the greater the force required. In other words, force is proportional to area. And finally, the force required to maintain a given rate of deformation varies from one fluid to another, depending on the "stickiness" of the fluid. This "stickiness" is the fluid's *dynamic viscosity*, symbolized by μ, and is a measure of how difficult it is to persuade one molecule of fluid to slide past its neighbors. The more viscous the fluid, the greater the force per area required to maintain a given rate of deformation.

We can express these ideas mathematically:

$$\frac{F}{A} = \mu \frac{u}{y}, \tag{5.1}$$

and this relationship is used as a definition of dynamic viscosity:

$$\mu = \frac{F/A}{u/y}. \tag{5.2}$$

Dynamic viscosity has the units of N s m^{-2}.

As we have seen, the ratio u/y is a measure of the rate at which the fluid is sheared. This expression for the gradient of velocity is sufficiently precise for the simple geometry used in this example; but for the more general case in which shear can vary from place to place, the gradient is better described as du/dy. Furthermore, the quantity F/A is the *shear stress*, τ, so that one will often see eq. 5.1 and eq. 5.2 expressed in the form

$$\tau = \mu \frac{du}{dy}, \tag{5.3}$$

$$\mu = \frac{\tau}{du/dy}. \tag{5.4}$$

Implicit in this example is the assumption that by moving one solid plate relative to another, the fluid in between is sheared. In essence, this requires the fluid to stick to each of the plates—if fluid could slip along the plates, movement of the plates would not necessarily shear the fluid. It is an empirical fact, however, that fluid directly in contact with a solid surface does not slide along that surface. The physical basis for this "no-slip" condition is complicated and not entirely understood (Khurana 1988), but it has immense practical importance. Because of the no-slip condition, the fluid in contact with a solid object is constrained to move at the same speed as the object itself, and if the object is moving relative to the bulk of the fluid, the fluid must be sheared. From eq. 5.1 or eq. 5.3 we see that this shear is accompanied by a force proportional to surface area and to the viscosity of the fluid. Thus, for an object of a given size, the force it feels in moving through a fluid is (at least in part) proportional to viscosity. Therein lies an important set of differences between water and air.

The dynamic viscosities of air, fresh water, and seawater are given in table 5.1. The first thing to note is that water is very much more viscous than air. For example, at 20° C the viscosity of water is fifty-five times that of air; the liquid is "stickier" than the gas. The second thing to note is that viscosity depends on temperature for both air and water, but in opposite directions. Over the range 0° to 40° C the viscosity of air increases by about 11% (Fig. 5.2), while over the same temperature range the viscosity of water decreases by 64% (fig. 5.3). As a consequence, the ratio of the dynamic viscosity of water to that of air decreases considerably as the temperature rises, from about 100 at 0° C to about 34 at 40° C (fig. 5.4). For the

Table 5.1 Dynamic viscosity of air and water.

T (° C)	Dynamic Viscosity (N s m^{-2})		
	Dry Air	Fresh Water	Seawater (S = 35)
0	1.718 × 10^{-5}	1.79 × 10^{-3}	1.89 × 10^{-3}
10	1.768	1.31	1.39
20	1.818	1.01	1.09
30	1.866	0.80	0.87
40	1.914	0.65	0.71

Sources: Values for air taken from List (1958), those for fresh water and seawater from Sverdrup et al. (1942).

Note: The value for seawater at 40°C is an extrapolation.

sake of simplicity, we use an average ratio of 70 in comparing water to air in this chapter.

Seawater is slightly more viscous than fresh water (table 5.1), but the variation of its viscosity with temperature follows the same pattern as that of fresh water (fig. 5.3).

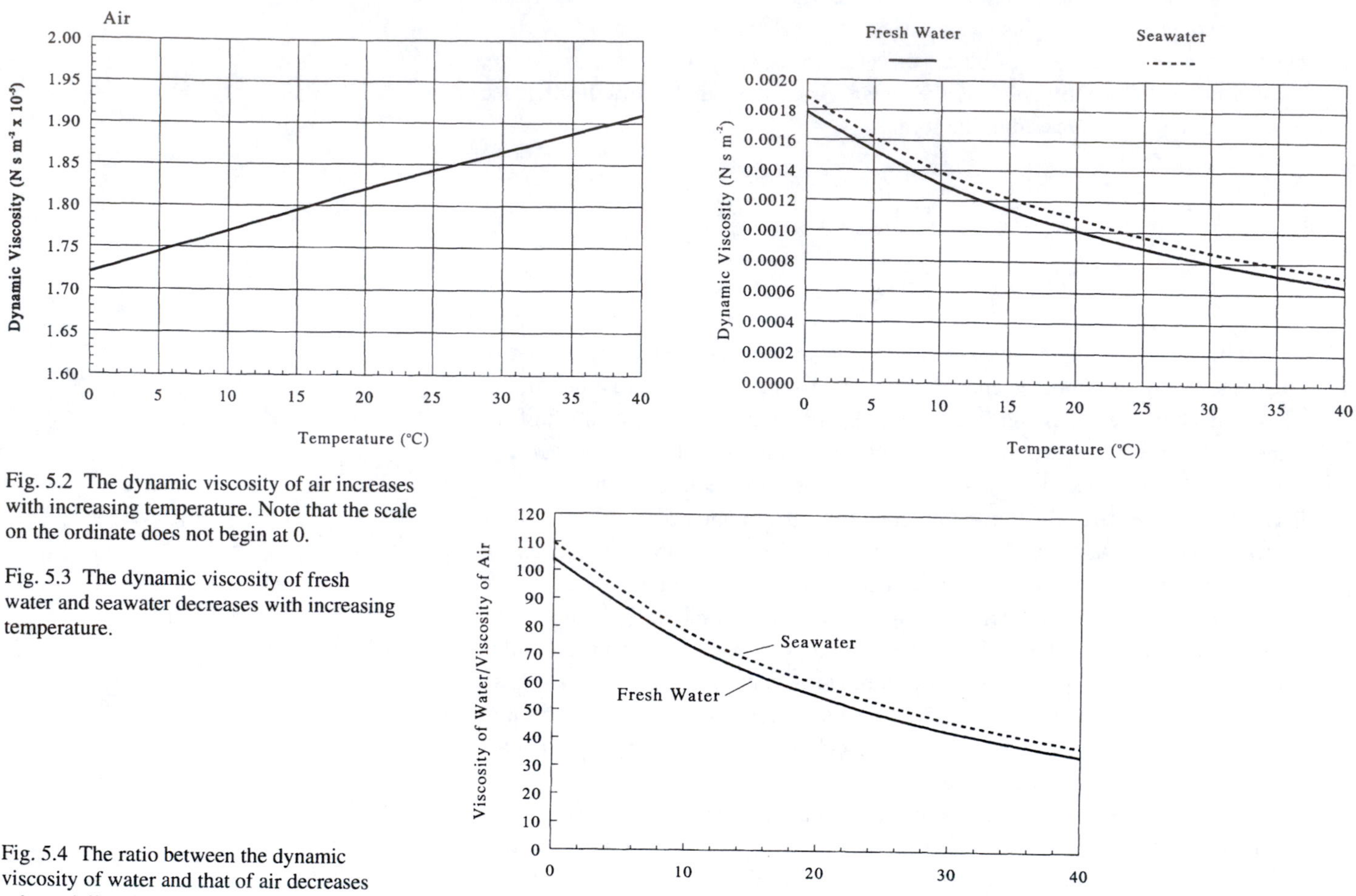

Fig. 5.2 The dynamic viscosity of air increases with increasing temperature. Note that the scale on the ordinate does not begin at 0.

Fig. 5.3 The dynamic viscosity of fresh water and seawater decreases with increasing temperature.

Fig. 5.4 The ratio between the dynamic viscosity of water and that of air decreases substantially with increasing temperature.

5.2 Reynolds Number

In chapter 4 we explored a variety of examples in which fluid mechanics and its biological consequences depended primarily on density. In this chapter we consider cases in which the relevant dynamics depend primarily on viscosity. When viewed in the context of fluid mechanics in general, these two subjects can be seen to form the distant extremes of a broad continuum. In most cases, the dynamics of the situation depend on *both* density and viscosity; the two are intertwined as the yin and yang of fluid mechanics. For example, the ability of many suspension feeding organisms to capture food particles depends as much on the density of the particles as it does on the viscosity of the surrounding fluid. Similarly, the speed at which an organism falls or the thickness of the stagnant layer of fluid at a solid surface depends on both density and viscosity. These cases, which are "typical" in that they involve both viscosity and density, form the subject matter for chapter 7. But

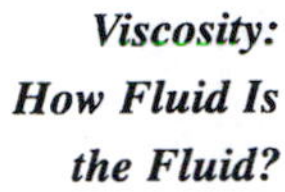

how are we to know *a priori* whether density, viscosity, or both are important? In terms of our present purpose, how are we to know when the effects of viscosity far outweigh those of density? What we need is a rule of thumb.

We begin by carefully defining what we mean when we speak of the "importance" of a particular parameter. In the present context, we are primarily concerned with how density and viscosity affect the pattern of flow in the vicinity of an organism. If we know this pattern, we can specify from it much of the more specific information we might desire, such as forces and the rates of transport of heat and mass.

What determines the pattern of flow? As a small bit of fluid (a so-called *fluid particle*) travels past an object, the path it follows depends on the interplay betweeen the particle's tendency to continue moving (a function of the inertial force acting on the particle) and its tendency to come to a halt (a result of the viscous force imposed on the bit of fluid). Many experiments with fluids have shown that it is the *ratio* of these two forces (inertial and viscous) that determines the pattern of flow around an object of a given shape.

What, then, determines the ratio of inertial and viscous forces? Here we are finally able to grapple with the problem directly. The inertial and viscous forces acting on a fluid particle depend on the size and shape of an object, on the speed of the fluid relative to the object, and (not surprisingly) on the viscosity and density of the fluid. To see how these factors combine, we explore a simple example. Consider a small, imaginary cube located in a flowing fluid of density ρ_f (fig. 5.5). The cube has open sides of length ℓ and it remains stationary as fluid flows freely in its upstream face and out its downstream face. If the fluid has velocity u, the rate at which fluid volume enters the cube is $u\ell^2$ m^3 s^{-1}, and the rate at which mass enters the cube is $\rho_f u\ell^2$ kg s^{-1}. Because momentum is the product of mass and velocity, we see that the rate at which fluid momentum enters the cube is (in kg m s^{-2})

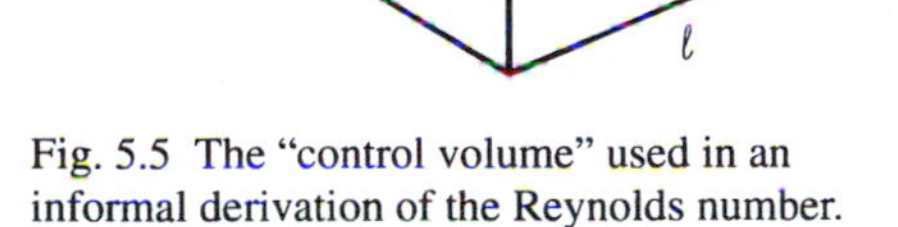

Fig. 5.5 The "control volume" used in an informal derivation of the Reynolds number.

$$\text{momentum per time} = \rho_f u^2 \ell^2. \tag{5.5}$$

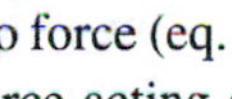

Recalling that the rate of change of momentum is equivalent to force (eq. 3.12), we see that this influx of momentum can be thought of as a force acting on the upstream face of the cube. Of course, fluid flows out of the cube at the same rate as it flows in, imposing an equal and opposite force on the cube's downstream face; but for present purposes, we concern ourselves only with the force on the upstream face. Because it depends on the fluid's momentum, this force (eq. 5.5) is the inertial force we seek.

In contrast, an estimate of the viscous force acting on the cube is arrived at by an examination of flow past one of the cube's lateral faces. If the cube were solid, the no-slip condition would apply at its faces, and a velocity gradient would be established in the fluid. As a result, a shear stress would be applied to the face,

$$\tau = \mu \frac{u}{k_R \ell}, \tag{5.6}$$

where we have approximated the magnitude of the velocity gradient by assuming that velocity u is reached at a distance $k_R\ell$ from the face. Now shear stress, τ, is force per area, so to arrive at a value for the viscous force acting on one face of the cube we multiply eq. 5.6 by the area of the face, ℓ^2:

$$\text{viscous force} = \mu \frac{u\ell}{k_R}. \quad (5.7)$$

Taking the ratio of the inertial force (eq. 5.5) to this viscous force, we arrive at our goal, the dimensionless *Reynolds number*, *Re*:

$$Re = \frac{\rho_f u k_R \ell}{\mu}. \quad (5.8)$$

If we knew the value of k_R we could calculate the Reynolds number for any particular situation. As a general matter, however, it is difficult to predict the precise nature of the velocity gradient around an object, and therefore it is problematic to specify k_R. No matter! The coefficient k_R is important only if we require that the Reynolds number be *exactly* equal to the ratio of inertial and viscous forces. If we lower our sights a bit, and require only that Re be *proportional* to the ratio, we are presented with a solution to our problem. We substitute for the value $k_R\ell$ a "characteristic length" of our object, ℓ_c, and redefine Re as

$$Re = \frac{\rho_f u \ell_c}{\mu}. \quad (5.9)$$

Because we now require only that the Reynolds number be proportional to the ratio of inertial and viscous forces, we are free to choose ℓ_c as is convenient.

As the ratio between inertial and viscous forces, the Reynolds number provides the rule of thumb for which we have been searching. When the Reynolds number is low,[1] viscous forces exceed inertial forces, and if *Re* is sufficiently low, the effects of density can (to a first approximation) be safely neglected altogether. Thus, in this chapter on viscosity we are primarily concerned with situations for which the Reynolds number is very low. A few, quirky cases of high Reynolds number will sneak in, but they are the exception rather than the rule.

Note that the magnitude of the Reynolds number depends on the choice of a characteristic length. We are free to pick any convenient length, but one must realize that by changing the characteristic length we change only the Reynolds number, not the pattern of flow. Unless otherwise stated, ℓ_c is taken to be the length of an object along the direction of flow. In this case, it is usually safe to neglect inertial forces if $Re < 0.1$.

It is apparent from the form of the Reynolds number that it is the ratio of viscosity and density (rather than the magnitude of each separately) that is important in determining the pattern of flow. The ratio μ/ρ_f (units = $m^2\,s^{-1}$) appears so often in fluid dynamics that it is given a name, the *kinematic viscosity*,[2] ν. Thus,

$$Re = \frac{u\ell_c}{\nu}. \quad (5.10)$$

Because water is about 70 times as viscous as air, but roughly 830 times as dense, its kinematic viscosity is smaller than that of air by a factor of eight to fifteen,

[1]I suppose that to be grammatically correct, a Reynolds number should be *small* rather than *low*. But fluid dynamicists traditionally speak of Reynolds numbers being either low or high, and we follow tradition here.

[2]Kinematics is the study of motion without regard to the forces involved. Because the ratio of viscosity to density has dimensions of length and time but not mass, it is a kinematic quantity.

depending on temperature (table 5.2; fig. 5.6). As a consequence, an object of a given characteristic length moving at a given speed has a Reynolds number 8- to 15-fold larger in water than in air. Alternatively, at a given *Re*, an object can be eight to fifteen times as large, or move eight to fifteen times as fast, in air as in water. The maximum product of size and speed for which an organism can qualify as being "low Reynolds number" are shown in figure 5.7.

Table 5.2 Kinematic viscosity of air and water.

T (° C)	Kinematic Viscosity ($m^2\ s^{-1}$)		
	Dry Air	Fresh Water	Seawater (S = 35)
0	1.33×10^{-5}	1.79×10^{-6}	1.84×10^{-6}
10	1.42	1.31	1.35
20	1.51	1.01	1.06
30	1.60	0.80	0.85
40	1.70	0.66	0.70

Sources: Values for air calculated from tables 4.1 and 5.1, those for fresh water and seawater taken from Sverdrup et al. (1942).
Note: The value for seawater at 40° C is an extrapolation.

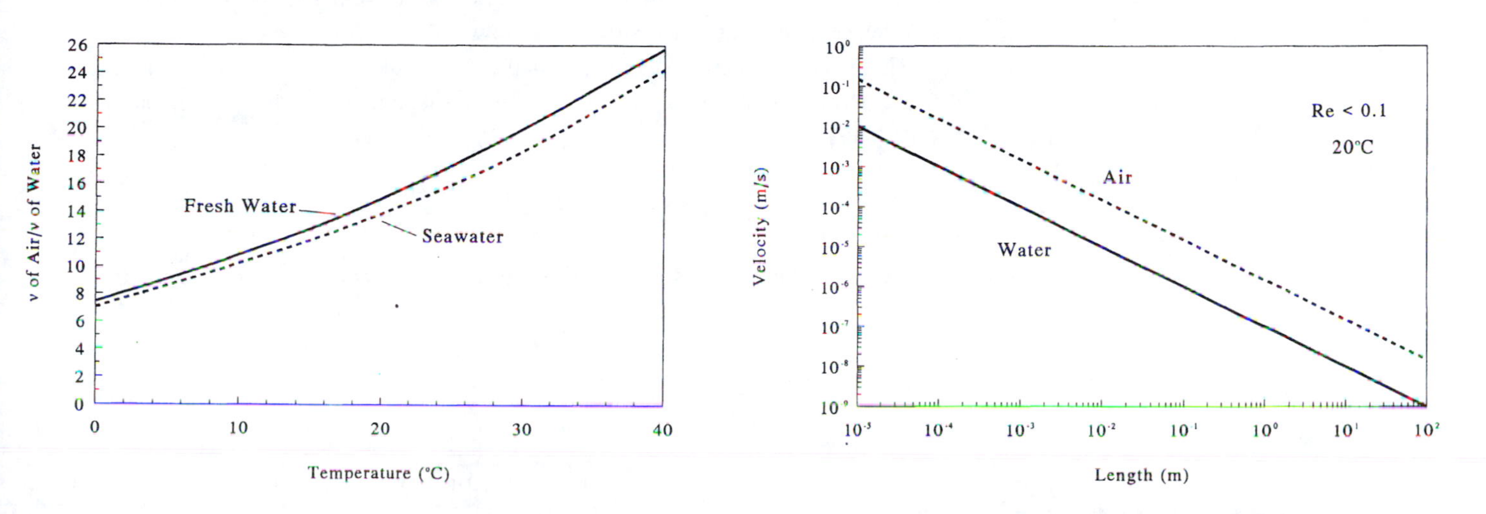

Fig. 5.6 The ratio between the kinematic viscosity of air and that of fresh water and seawater increases with increasing temperature.

Fig. 5.7 The maximum velocity at which an organism can move and still operate at a "low" Reynolds number ($Re < 0.1$) decreases if the length of the organism is increased.

5.3 Locomotion

With the assumption firmly in mind that we are confining ourselves to low Reynolds numbers, we can now turn our attention to the biological consequences of dynamic viscosity, and how they differ between air and water.

Consider first the problem of locomotion. Because of the no-slip condition, any solid object moving through a fluid can do so only by causing the fluid to shear. This shearing is resisted by the "stickiness" of the fluid. Thus, as an organism moves through a fluid (either air or water) it must continuously overcome a viscous resistance or *friction drag* in addition to the pressure drag discussed in chapter 4.

When the size of an object is large or its speed is high, or both—in other words when *Re* is high—pressure drag typically is much larger than friction drag. It was for this reason that in chapter 4 we could at times discuss pressure drag as if it were the only force acting. But for small objects at slow speeds, pressure drag is negligible and friction drag dominates. It is to these examples that we now turn our attention.

In some respects, friction drag is much more "tidy" than pressure drag. When viscous forces dominate, the pattern of flow is orderly and predictable. In the jargon of fluid mechanics, the flow is said to be *laminar*, a term evoking images of fluid layers (or laminae) sliding past one another. These orderly patterns of flow are amenable to mathematical description, and as a result we have a useful set of analytical formulas for the viscous resistance of objects moving at low Reynolds numbers.

Take, for instance, the case of a sphere. Provided the Reynolds number is sufficiently low (less than about 0.1), the force required to propel a sphere through a fluid is (according to an analysis by Stokes)

$$F = 3\pi\mu u d, \tag{5.11}$$

where d is the diameter of the sphere and u is the speed of the sphere relative to the fluid. What could be simpler? The faster the sphere moves or the larger its size, the greater the force required.

More to the point, because the dynamic viscosity of water is seventy times that of air, it takes seventy times as much force to push a sphere through water than it does to push the same sphere through air at the same velocity, provided of course that the Reynolds number is sufficiently low.[3]

This effect is independent of the shape of the object. For example, if, instead of a sphere, we consider a prolate ellipsoid (a shape like a cigar) moving parallel to its long axis,

$$F = \mu \frac{2\pi\ell}{\ln(2\ell/d) - \frac{1}{2}} u, \tag{5.12}$$

where ℓ is the major axis of the ellipsoid, d is its minor axis, and $\ell \gg d$ (Berg 1983). Again, because of viscosity, the force required to propel an object at a given speed is seventy times greater in water than in air. The same is true for a prolate ellipsoid moving sideways through the fluid:

$$F = \mu \frac{4\pi\ell}{\ln(2\ell/d) + \frac{1}{2}} u. \tag{5.13}$$

The force in this case is about twice that required for the ellipsoid moving lengthwise, but it is still seventy times larger in water than in air.

What are the consequences for locomotion? At first one might be tempted to think that it would be seventy times easier for a small organism to propel itself through air than water, but there is a catch. At low Reynolds number, organisms must overcome not only a viscous drag, but they must also rely on viscosity to provide their thrust. The more familiar mechanisms for producing thrust (lift, pressure drag, jet propulsion) all rely on changing the momentum of the fluid. At low Reynolds numbers, where inertial forces are, by definition, relatively small, the thrust produced by these mechanisms is negligible. Instead, organisms operating at low Reynolds numbers typically use the interaction between viscosity and the movements of flagella or cilia to provide thrust. Before we can compare low-Reynolds-number locomotion in air and water, we need to know how these locomotory systems work.

Consider a flagellum (fig. 5.8A). If the flagellum is attached to a bacterium, it is a more-or-less rigid helix that is rotated about its axis (Berg 1983). If the flagellum

[3]In chapter 7 we will reconsider this relationship for high Reynolds numbers.

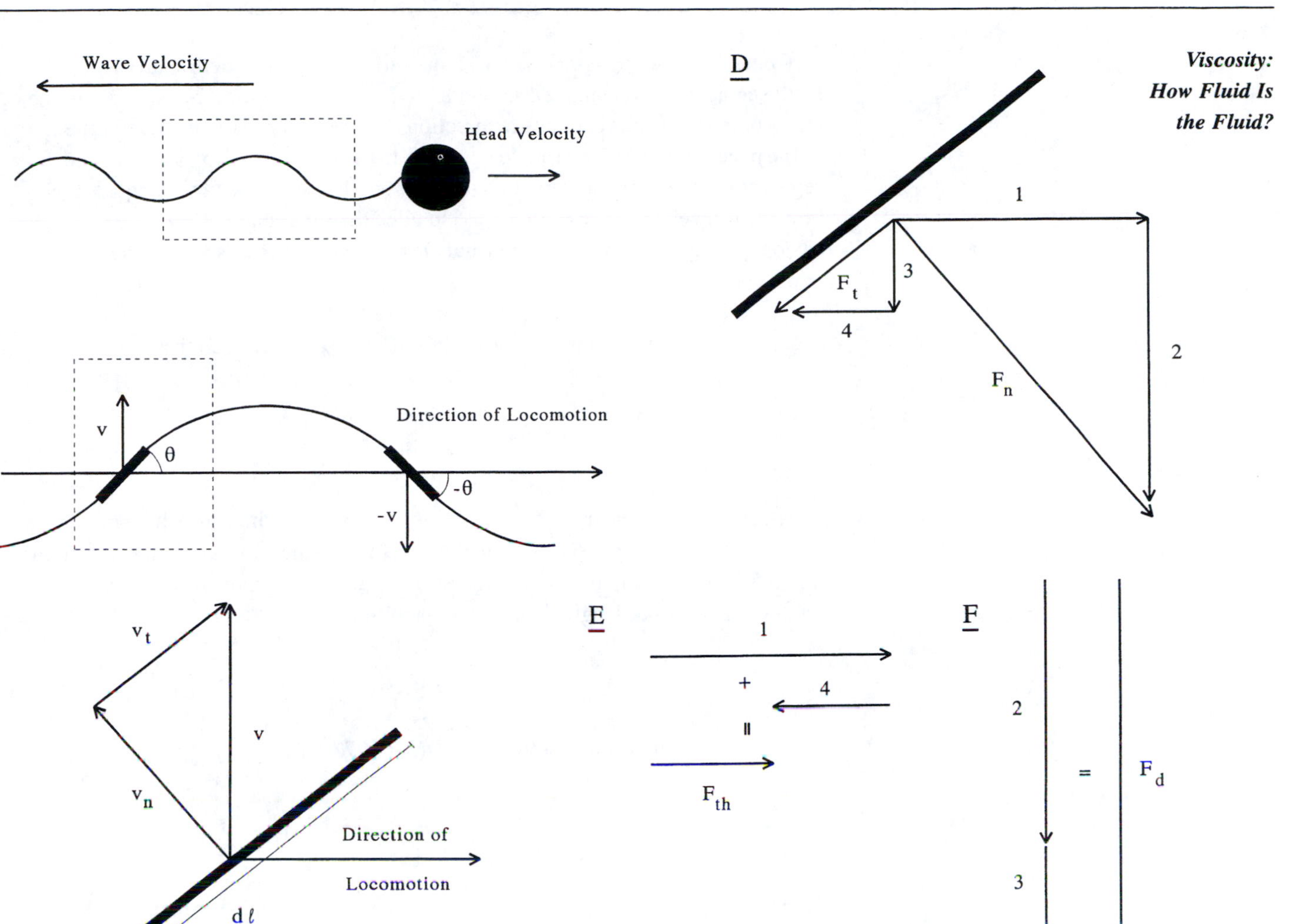

Fig. 5.8 The undulations of a flagellum produce fluid dynamic forces that thrust the flagellum forward. The terms defined here are used in the calculation of forces operating in this type of locomotion. See text for details.

is attached to a eukaryotic cell (such as a sperm), it is a flexible rod, the internal machinery of which causes planar, sinusoidal waves to pass along its length. The mechanics of thrust production in the two types of flagella are similar, but here we examine the sperm's flagellum because, being two-dimensional, its action is somewhat easier to visualize.

5.3.1 *The Mechanics of Flagellar Locomotion*

What follows here is a mechanistic explanation of why a flagellum moves through a fluid. The ideas involved aren't particularly difficult, but the process of understanding them is moderately complex and it is necessary to have some "feel" for how vectors add. If you are willing to stay the course, repeated reference to figure 5.8 will ease the process. The faint of heart may skip to the summation presented in the next section.

During locomotion, each segment of the flagellar rod moves transversely as waves pass down the flagellum and each segment moves forward as the flagellum is propelled through the fluid (fig. 5.8A). In a typical case, the speed of forward movement is an order of magnitude slower than the transverse speed, and for simplicity in understanding the forces generated by flagellar oscillations, we can ignore forward motion without serious repercussions.

Consider now two segments of the flagellum, half a wavelength apart (fig. 5.8B). Both segments have length $d\ell$ and both are moving transversely at speed v in the plane of the page, but in opposite directions. The segment moving toward the top of the page is tilted up by an angle θ relative to the direction of locomotion, and the segment moving toward the bottom of the page is tilted down by an equal angle. For most flagella, θ is less than 45°, smaller than the angle shown here, which has been set artificially large for visual clarity. Because each segment is tilted, its transverse speed v can be decomposed into two components, one perpendicular to the axis of the segment, v_n, and one parallel to the axis of the segment, v_t (fig. 5.8C). From the geometry shown in figure 5.8B we can see that

$$v_n = v \cos \theta \tag{5.14}$$

$$v_t = v \sin \theta. \tag{5.15}$$

Associated with each of these velocities is a viscous drag that, like any drag, acts in a direction opposite that of the velocity of the object through the fluid (fig. 5.8D). For example, the velocity of the segment perpendicular to its axis, v_n, produces a force per length, $\mathcal{F}_n$, on the flagellum:

$$\mathcal{F}_n = \mu C_n v_n, \tag{5.16}$$

where C_n (a form of drag coefficient) is (Wu 1977)

$$C_n = \frac{4\pi}{\ln(2k_f/r) + \frac{1}{2}}. \tag{5.17}$$

Here k_f is a parameter that depends on the "waviness" of the flagellum. Expressed in terms of the wavelength, λ, of the flagellar sinusoid,

$$k_f \approx 0.09\lambda. \tag{5.18}$$

The term r in eq. 5.17 is the radius of the flagellar rod. For most flagella λ/r is 100 to 200, so $2k_f/r$ is 18 to 36. Note that this expression for the "drag coefficient" of a flagellum (eq. 5.17) is similar to that for a prolate ellipsoid moving broadside (eq. 5.13).

There is a similar expression for the component of viscous drag, $\mathcal{F}_t$, due to the velocity v_t along the segment's axis (fig. 8D):

$$\mathcal{F}_t = \mu C_t v_t, \tag{5.19}$$

$$C_t = \frac{2\pi}{\ln(2k_f/r)}, \tag{5.20}$$

where C_t is the "drag coefficient" for tangential motion.

Note that in both cases the force acting on the flagellar segment is directly proportional to dynamic viscosity, μ.

Furthermore, when the radius of the flagellum is small compared to its effective length (which is almost always the case), C_n is 1.4 to 1.7 times C_t. As a result, when the flagellum is waved side to side the force required to move the flagellar segment perpendicular to its axis is considerably greater than that required for motion parallel to the axis.[4] It is this disparity that makes flagellar locomotion possible.

To see how, we need to look at both the magnitude and direction of the forces acting on our flagellar segments (fig. 5.8D). Because $\mathcal{F}_n$ and $\mathcal{F}_t$ are tilted relative to the direction in which the organism moves, each can be decomposed into two components, one parallel to the direction of locomotion (vectors 1 and 4 of fig. 5.8D) and one perpendicular (vectors 2 and 3). For the segment moving toward the top of the page, the net force parallel to the direction of motion is

$$\mathcal{F}_{th} = \mathcal{F}_n \sin\theta - \mathcal{F}_t \cos\theta \tag{5.21}$$

$$= \mu C_n v \sin\theta\cos\theta - \mu C_t v \sin\theta\cos\theta \tag{5.22}$$

$$= \mu(C_n - C_t)v\sin\theta\cos\theta. \tag{5.23}$$

This calculation is shown schematically in figure 5.8E. Because $C_n > C_t$, there is a net axial force $\mathcal{F}_{th}$—a thrust—acting in the direction opposite to that in which the sinusoid moves along the flagellum. This is the force that propels the flagellum through the fluid. Note that thrust is directly proportional to the fluid's viscosity.

We can also add the forces that act perpendicularly to the direction of the organism's locomotion (fig. 5.8F):

$$\mathcal{F}_d = \mathcal{F}_n \cos\theta + \mathcal{F}_t \sin\theta \tag{5.24}$$

$$= \mu(C_n \cos^2\theta + C_t \sin^2\theta)v. \tag{5.25}$$

Obviously, there is a considerable net force resisting the transverse motion of the segment. Again, this force is proportional to viscosity.

The same set of calculations can be carried out for the segment moving toward the bottom of the page. Because both the direction of motion and the tilt of the segment have been switched, the net thrust for the segment still acts in the direction opposite to that of the sinusoidal wave.[5] Thus, as each segment along the sinusoid moves transversely, it provides some thrust tending to propel the flagellum forward.

The net transverse force acting on the segment moving toward the bottom of the page is of equal magnitude to that of the segment moving toward the top, but acts in the opposite direction. As long as there is an integral number of wavelengths on the flagellum, the net transverse force acting on any one segment is just offset by the transverse force acting on some other segment and, taken as a whole, there is no net transverse force on the organism. Of course, the flagellum must still work to move itself back and forth, but this "wasted" work is the price that must be paid to produce thrust.

5.3.2 *The Final Result*

The final result of all this decomposition and shuffling of forces is this: because there is more viscous resistance to moving a rod sideways than along its length, there is a net thrust as the flagellum waves back and forth, a thrust available to propel the flagellum itself and any attached body through the fluid. The amount of thrust depends on the transverse speed of the flagellum, and therefore on the frequency

[4]This assertion is based on the assumption that θ is reasonably small. It would not be true for θ approaching 90°.

[5]Velocity here is $-v$, $\sin(-\theta) = -\sin\theta$ and $\cos(-\theta) = \cos\theta$. As a result, the product of speed and the sine and cosine of the tilt angle are still positive.

and amplitude of the flagellar beat, but in any case it is directly proportional to the dynamic viscosity of the fluid.

This proportionality between thrust and viscosity negates any advantage an organism might have in moving through air rather than water. Because water is seventy times more viscous than air, seventy times more thrust is required to push a body along at a given velocity; but for a given length of flagellum and a given amplitude and frequency of flagellar motion, seventy times as much thrust is available. It all tends to even out.

This, of course, assumes that the size, shape, and velocity of an organism using flagellar locomotion in air would be the same as that in water. Could this indeed be the case? To answer this question we briefly explore the energetics of flagellar locomotion.

Wu (1977), in a detailed examination of the mechanics of flagella, suggests that the power required to propel a flagellum is

$$P = 50\mu u^2 \ell_f, \tag{5.26}$$

where u is the forward speed of the flagellum through the fluid and ℓ_f is the flagellar length. This is the power required for the flagellum alone. If the flagellum pushes a spherical cell of diameter d, the overall power is

$$P = 50\mu u^2 \ell_f + 3\pi\mu u^2 d, \tag{5.27}$$

where we have again made use of the fact that power is equal to the product of force and velocity.

Now, for eukaryotic organisms, the power for flagellar movement is provided by the microfibrillar apparatus within the flagellum itself, and, as a result, the power available for locomotion is probably proportional to the length of the flagellum (Alexander 1971):

$$P = \text{constant} \times \ell_f. \tag{5.28}$$

The value of the proportionality constant is not accurately known, but this poses no problem. We can equate the power required and the power available and see that

$$50\mu u^2 \ell_f + 3\pi\mu u^2 d = \text{constant} \times \ell_f \tag{5.29}$$

$$(50 + 3\pi d/\ell_f)\mu u^2 = \text{constant}. \tag{5.30}$$

As long as d/ℓ_f is small (a reasonable assumption), we conclude that the product of viscosity and the square of velocity is virtually independent of the length of the flagellum. In other words, as long as viscosity is held constant, lengthening the flagellum does not increase the speed of locomotion. Experiments with a variety of aquatic flagellates have shown this conclusion to be valid (Wu 1977).

This means, however, that if the power output of the flagellum is held constant while the viscosity of the medium is reduced, the speed of locomotion should increase (eq. 5.30). Thus, we could expect an organism propelled by a flagellum in air (where the viscosity is only 0.014 times that in water) to move about eight times as fast as the same organism in water as long as both expend energy at the same rate. Given that flagellated organisms in water move at 50 to 150 μm s^{-1}, we could expect an aerial flagellate to zip along at about a millimeter per second.

This line of argument also leads us to the curious conclusion that flagellated organisms might be able to fly. Consider, for instance, an aerial "sperm," the power output and size of which are the same as those of a bull's sperm. From the argument made above, we would expect, in the absence of gravitational effects, that the aerial version could move at a speed u of about $1\ \mathrm{mm\,s^{-1}}$. To do so, the flagellum must be capable of exerting a force (in excess of that needed to move itself) equal to the viscous drag of the sperm's head. If the head of the sperm is a sphere with a diameter d of $2\ \mu$m (a reasonable approximation), this means that the thrust produced by the flagellum in air is (from eq. 5.11)

$$F = 3\pi u \mu d \quad (5.31)$$
$$= 3.4 \times 10^{-13}\ \mathrm{N}. \quad (5.32)$$

What if we now take gravity into account? Most of the mass of the sperm is concentrated in its head, and this mass is accelerated downward by gravity. If the aerial sperm is to stay aloft, the flagellum must be able to apply a net upward thrust to the head equal to the head's weight.[6] For the example here,

$$\text{weight} = \rho_e g (4/3)\pi (d/2)^3 \quad (5.33)$$
$$= 4.4 \times 10^{-14}\ \mathrm{N}, \quad (5.34)$$

where we have assumed that the effective density of the sperm head is 1080 kg $\mathrm{m^{-3}}$. Because its weight is less than its thrust, an aerial sperm should be able to hold itself aloft, and even to make a bit of headway as long as it maintains the same power output as its aquatic cousin. It appears that flagellar locomotion might just be a feasible form of aerial transport.

There may well be practical problems that we have not considered, however. For instance, if a flagellum operating in the low viscosity of air is to maintain the same power output as a flagellum in water, it must beat eight times as fast. Now, flagella in water already beat at a rapid rate, ten to fifteen times per second, and it may be impossible for microfibrillar machinery to function at a sufficiently rapid rate to allow for aerial locomotion. There is also the very real problem of maintaining the flagellar machinery in the desiccating terrestrial environment. The long, thin flagellum of a eukaryote might dry out too fast to be of any use. In this respect, the flagella of bacteria might have an advantage. The driving apparatus of bacterial flagella is contained in the cell membrane rather than in the flagellum; the flagellum itself is little more than a rigid rod that would not necessarily be adversely affected by contact with air. If the cellular machinery that rotates bacterial flagella can maintain its power output in air, bacteria might be able to fly. To my knowledge, however, no one has actually checked to see whether bacteria can swim effectively in air, and in chapter 6 we will consider whether aerial swimming would be advantageous.

Ciliary locomotion obeys the same basic principles as flagellar locomotion, but in a more complex fashion. Each cilium beats with a rowing motion, and the flow around individual cilia is affected by its neighbors. These complexities will not be

[6]Because the sperm's head is so much more massive than the flagellum, there would be a tendency for the organism to assume a head-down posture. To provide upward thrust, then, the flagellar wave would need to move toward, rather than away from, the head. This direction of wave propogation is not unusual; it is found in many flagellates.

explored here; you are urged to consult Wu (1977) for an excellent overview of the subject.

Flagellar and ciliary locomotion are not the only mechanisms by which small particles can remain airborne, and in the next chapter we will explore several alternatives.

5.4 Flow in Pipes

To this point, most of the flows we have considered have been around the outside of organisms. What about flows inside? For instance, all animals larger than flatworms maintain some kind of internal flow to deliver oxygen to their tissues and to remove carbon dioxide, and vascular plants transport water from their roots to their leaves through internal plumbing. How are the mechanics of these flows affected by the difference in dynamic viscosity between air and water?

We first explore flow through pipes, which can serve as a convenient model for flow in blood vessels and other biological tubes. Consider a pipe with a circular cross section (fig. 5.9). The pipe has radius r and length ℓ, and to avoid complications from gravity, we assume that the axis of the pipe is horizontal. There is a pressure difference Δp between the ends of the pipe, and in response, fluid flows from the end with the higher pressure toward the end with the lower pressure. Given this particular *pressure gradient* ($\Delta p/\ell$), our job is to figure out how fast the fluid moves and in what pattern.

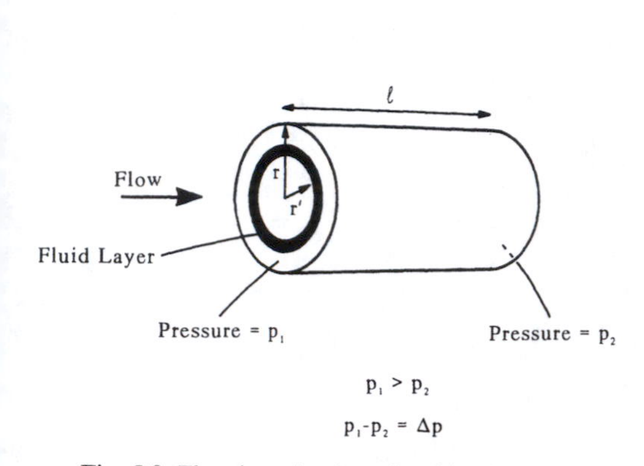

Fig. 5.9 Flow in a pipe is resisted by the viscosity of a fluid. As a result, there must be a difference in pressure between the ends of the pipe before the fluid will flow.

This task is simplified if we deal with orderly, low-Reynolds-number flows. The Reynolds number for a pipe, Re_p, is most conveniently defined not in terms of the pipe's length, but rather in terms of its diameter, d:

$$Re_p = \frac{\rho_f u d}{\mu} = \frac{2\rho_f u r}{\mu}, \tag{5.35}$$

where again r is the pipe's radius. Experiments have shown that flow in a pipe remains orderly (laminar) up to a pipe Reynolds number of about 2000 (Schlichting 1979).

As with any laminar flow, we can think of fluid in the pipe as if it were formed from a series of layers. In this case, the layers take the form of concentric tubes, each telescoping inside another, together filling the pipe. We can give each particular layer an identity by specifying its average radius r' (fig. 5.9). Because the pipe has solid walls, the no-slip condition requires the fluid at the walls to be stationary. Thus we can think of the layer of water in contact with the wall ($r' = r$) as being stuck to the wall like a coat of paint. The next layer in is not touching the walls, so it is free to move, but to do so it must slide over the water held in place by the wall. Because of viscosity, a force is associated with this shearing equal to the product of the shear stress between the layers (τ) and the area over which the layers are in contact ($2\pi r'\ell$). Recalling eq. 5.4, we see that the force F_r resisting the movement of one concentric layer past another is

$$F_r = \tau \times (2\pi r'\ell) = \mu \frac{du(r')}{dr'}(2\pi r'\ell), \tag{5.36}$$

where $u(r')$ is the velocity (along the pipe) of the layer at distance r' from the pipe's center.

As one layer moves past another at a constant rate, a force F_p must be supplied. From Newton's third law (chapter 2) we know that F_p is equal to F_r but acts in the opposite direction. In this example, F_p is provided by the difference in pressure between the ends of the pipe acting over the cross-sectional area of the layers for which the radius is less than or equal to r'. Thus,

$$F_p = \Delta p(\pi r'^2) = -\mu \frac{du(r')}{dr'}(2\pi r'\ell) = -F_r. \tag{5.37}$$

Canceling terms and rearranging, we see that

$$du(r') = -\frac{\Delta p\, r'}{2\ell\mu} dr'. \tag{5.38}$$

Both sides of this equation can be integrated:

$$\int du(r') = -\frac{\Delta p}{2\ell\mu} \int r'\, dr' \tag{5.39}$$

$$u(r') = -\frac{\Delta p r'^2}{4\ell\mu} + \text{constant}. \tag{5.40}$$

The constant of integration can be evaluated by specifying that, because of the no-slip condition, $u(r) = 0$. Thus,

$$\text{constant} = \frac{\Delta p r^2}{4\ell\mu}. \tag{5.41}$$

Substituting this result back into eq. 5.40 we arrive at the final answer:

$$u(r) = \frac{r^2 - r'^2}{4\mu} \frac{\Delta p}{\ell}. \tag{5.42}$$

This expression is graphed in figure 5.10. The velocity is zero at the walls of the pipe (as it must be) and rises parabolically to a maximum at the center, where the velocity is

$$\text{maximum velocity} = \frac{r^2}{4\mu} \frac{\Delta p}{\ell}. \tag{5.43}$$

From eqs. 5.42 and 5.43 we see that at any point in the pipe velocity is proportional to the pressure gradient ($\Delta p/\ell$) and inversely proportional to viscosity.

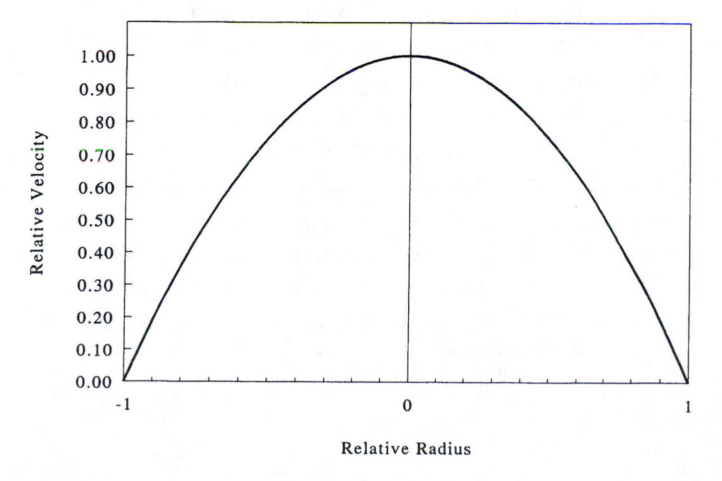

Fig. 5.10 At a Reynolds number less than about 2000, the velocity distribution in a pipe is parabolic.

If we average the velocity across the entire pipe,[7] we find that

$$\text{average velocity, } \langle u \rangle = \frac{1}{\pi r^2} \int_0^r 2\pi r' u(r')\, dr' \tag{5.44}$$

$$= \frac{r^2}{8\mu} \frac{\Delta p}{\ell}. \tag{5.45}$$

In other words, the average is exactly half the maximum.

Now, the rate at which volume moves through the pipe, J, is simply the average velocity times the pipe's cross-sectional area, πr^2, so

$$J = \frac{\pi r^4}{8\mu} \frac{\Delta p}{\ell}. \tag{5.46}$$

As with velocity, J (the *volume flux*) is directly proportional to the pressure gradient and inversely proportional to viscosity. This result is of considerable general utility and has been given its own name—the *Hagen-Poiseuille equation*—in honor of its discoverers.

We can use eq. 5.46 to calculate the rate at which work must be done to push fluid through a pipe. Power (work per time) is equal to force times velocity (see eq. 3.17). Now, the total force applied to the fluid in the pipe is $\Delta p \pi r^2$, and the fluid moves at average velocity $\Delta p r^2/(8\mu\ell)$, so the power required to pump fluid through a pipe is

$$P = J\Delta p = \frac{\pi r^4}{8\mu} \frac{(\Delta p)^2}{\ell}. \tag{5.47}$$

In other words, the power required to pump fluid through a pipe is proportional to the square of the pressure applied.

5.4.1 *Flow in Xylem*

Let us explore some of the biological consequences of these results. Consider, for example, the flow of water through the vascular system of a tree. What gradient in pressure is required to force water to flow from the roots to the leaves?

Rearranging eq. 5.45, we see that

$$\frac{\Delta p}{\ell} = \frac{8\mu\langle u \rangle}{r^2}. \tag{5.48}$$

A typical vessel cell has a radius of 20 μm, and when the tree is actively respiring, water may flow through the cell at an average velocity of 1 mm s^{-1} (Nobel 1983). Thus, at 20° C, a gradient of approximately 2×10^4 Pa m^{-1} is required to keep the water flowing against the action of viscosity. A tree 100 m high would therefore need a pressure difference of 20 atmospheres across the length of its vascular system if water flows at a rate of 1 mm s^{-1}.

Considerable power is required to sustain this flow rate. For example, 200 W per square meter of vessel tubes is expended in our 100-meter-high tree (eq. 5.47). The tree, however, does not have to provide the energy. Unless the surrounding air has a relative humidity of 100%, water evaporates from the leaves, a process driven by the thermal energy of water molecules (see chapter 14).

[7] To do this we multiply the velocity at radius r' by the infinitesimal area at this radius ($2\pi r' dr'$), and sum (i.e., integrate) these area-weighted velocities over the entire tube. We then divide by the overall pipe area to arrive at an average.

Note that the pressure gradient calculated here is in addition to the gradient in hydrostatic pressure (chapter 4), which is 10^4 Pa m^{-1}. Thus water in the leaves of a 100-meter-tall tree must be at a pressure 30 atmospheres lower that that in the roots! In chapter 12 we will explore the mechanism by which such a low pressure can be maintained.

5.4.2 *Viscosity and the Circulatory System*

As an animal respires, its tissues consume oxygen. Lest the animal die, this oxygen must be replenished by the circulatory system at a rate equivalent to that at which it is used. The rate of oxygen replenishment, in turn, depends on two factors.

First, the higher the concentration of oxygen in the circulating fluid ($[O_2]$), the more readily the oxygen in the tissues can be replaced. It is important, then, that the concentration of oxygen in air is twenty to forty-eight times that in water, depending on temperature (table 5.3).

Table 5.3 The ratio of oxygen concentration in air to that in saturated fresh water and seawater.

T(° C)	O_2 Concentration (moles m^{-3})			$[O_2]_a/[O_2]_{fw}$	$[O_2]_a/[O_2]_{sw}$
	Fresh	Sea	Air		
0	0.457	0.359	9.349	20.5	26.1
10	0.352	0.282	9.018	25.6	32.0
20	0.284	0.231	8.711	30.7	37.7
30	0.236	0.194	8.423	35.7	43.3
40	0.200	0.168	8.154	40.8	48.5

Source: Data from Weiss (1970).

Second, the rate at which oxygen can be delivered depends on J, the flux of circulating fluid past a particular site. Eq. 5.46 tells us that if the circulatory fluid is contained in pipes, this rate is proportional to the pressure gradient $\Delta p/\ell$ and to the fourth power of the pipe's diameter, and it is inversely proportional to the circulatory fluid's dynamic viscosity. We can express these ideas as an equation:

$$\text{rate of oxygen delivery} = [O_2] \times J = \frac{[O_2]\pi r^4}{8\mu}\frac{\Delta p}{\ell}. \tag{5.49}$$

This relationship raises some interesting questions. Consider a very simple circulatory system consisting of one pipe of length ℓ. At 20° C, the viscosity of water is fifty-five times that of air, and its concentration of oxygen is 30- to 38-fold less. Thus, for a given pressure gradient, the pipe must have a radius in water that is $\sqrt[4]{2100} \approx 7$ times that in air if the circulatory fluid is to achieve the same rate of oxygen delivery. That is, for a given pressure, an animal must build bigger pipes if it uses water rather than air as its circulating fluid. Alternatively, if the pipe has a fixed radius and the rate of oxygen delivery is to be constant, the pressure gradient must be 1700 to 2100 times larger if water is used rather than air. In either case, it would seem that air would be the fluid of choice for transporting gas through the body.

Why, then, do most animals use an aqueous fluid in their circulatory systems? First, primitive circulatory systems evolved in aquatic animals for whom air was not available. Thus, the constraints of evolutionary history mediate against the use of air. Second, circulatory systems do more than transport gas. They carry dissolved food to the tissues and remove nitrogenous wastes, as well as transporting hormones through the body and serving a variety of other functions. As a result, even those animals which do use air as a circulatory fluid (e.g., insects) must have a secondary aqueous system to serve these other functions.

And finally, animals have evolved tricks for avoiding at least some of the pumping costs associated with circulating water. In many animals, the blood contains molecules that effectively bind oxygen (e.g., hemoglobin and hemocyanin), thereby increasing the oxygen-carrying capacity of the circulatory fluid. For example, the blood of many mammals contains about 8 moles of oxygen for each cubic meter of blood, about the same concentration of O_2 as found in air. In cases such as this, the animal need only do the extra work associated with the viscosity of the medium without the added tax of a decreased oxygen concentration.

5.4.3 *Murray's Law*

A second trick concerns the way in which large vessels are connected to smaller ones. Consider, for instance, the circulation in human beings. Blood exits the left ventricle of the heart through the aorta, an artery with a radius of about 0.5 cm. The aorta then branches to form smaller arteries, which themselves branch again. This pattern of dichotomous branching continues until capillaries are formed. At each level of branching, how large should the arteries be?

A possible answer was proposed by C. D. Murray, who suggested that the circulatory system has been designed in the course of evolution to minimize the overall cost of pumping blood (Murray 1926). This cost has two components.

First, there is the cost associated with forcing blood through arteries against the action of viscosity. To calculate this cost, we first rearrange eq. 5.46 to express the pressure gradient that must exist in a pipe to cause a given volume flux of fluid:

$$\frac{\Delta p}{\ell} = \frac{8\mu J}{\pi r^4}. \tag{5.50}$$

From eq. 5.47 we know that the power required to pump fluid is $J\Delta p$, or, using the value for $\Delta p/\ell$ that we have just calculated, we see that the power of pumping (per length of pipe) is

$$\mathcal{P}_p = \frac{8\mu J^2}{\pi r^4}. \tag{5.51}$$

If the volume flux is held constant, this power increases as the radius of the pipe is decreased.

Then, in addition to the cost of pumping fluid, there is a cost associated with maintaining the pipes and, because blood is a living fluid, with maintaining the fluid itself. Murray proposed that this maintenance cost is proportional to the volume of the system (pipes plus blood). Thus, the cost per length of pipe is

$$\mathcal{P}_m = \mathcal{M}\pi r^2, \tag{5.52}$$

where $\mathcal{M}$ is the metabolic cost per volume used in maintenance. In this case, the smaller the vessel, the smaller the cost.

The overall cost of a length of circulatory system is the sum of these two costs:

$$\mathcal{P} = \mathcal{P}_p + \mathcal{P}_m = \frac{8\mu J^2}{\pi r^4} + \mathcal{M}\pi r^2. \tag{5.53}$$

We may suppose that in the course of evolution, the design of the system has been adjusted to minimize this cost. By taking the derivative of $\mathcal{P}$ with respect to r and setting it equal to zero, we can relate the size of vessels and the volume flux of fluid that they carry in a system that minimizes the overall cost:

$$\frac{d\mathcal{P}}{dr} = \frac{-32\mu J^2}{\pi r^5} + 2\mathcal{M}r = 0 \tag{5.54}$$

Working through the algebra, we find that

$$r_{opt} = J^{1/3}\left(\frac{16\mu}{\pi\mathcal{M}}\right)^{1/6}. \tag{5.55}$$

In other words, the optimal radius is proportional to the cube root of the volume flux, and the actual value is set by the ratio of viscosity to metabolic cost. If we assume that μ and $\mathcal{M}$ are the same throughout a given animal, we may simplify the form of this equation by replacing $\sqrt[6]{16\mu/\pi\mathcal{M}}$ by a constant, k_p. Thus,

$$r_{opt} = k_p J^{1/3}, \tag{5.56}$$

or, rearranging,

$$J = k_p r_{opt}^3. \tag{5.57}$$

This is an interesting conclusion. Consider the case of an artery splitting into two arterioles. Lest blood accumulate in the artery, the volume flux through the arterioles must sum to that in the artery. Thus,

$$J = J_1 + J_2, \tag{5.58}$$

where J is the volume flux in the artery and J_1 and J_2 are the volume fluxes in the two arterioles. But we can replace each volume flux by its equivalent expressed in terms of radius (eq. 5.57), with the result that

$$r^3 = r_1^3 + r_2^3. \tag{5.59}$$

In other words, in a circulatory system where the overall cost is minimized, the sum of the cubes of radii after a bifurcation should equal the cube of the pipe radius before the bifurcation. Thus, if one knows the radius of the aorta, one can predict the radii of the two vessels into which it splits. This prediction is known as *Murray's law*.

Murray's law has been tested against a wide variety of circulatory systems (LaBarbera 1990). In general, systems that pump water fit the prediction quite well, while those that pump air (such as insect tracheae) do not. To see why there should be this disparity, we make use of two intermediate steps in the derivation of Murray's law. Inserting r_{opt} (eq. 5.55) into the expression for the total cost (eq. 5.53) gives us a means to see how the overall cost of circulation depends on the physical properties of the circulating fluid. With appropriate manipulation we find that

$$\mathcal{P} \propto J^{2/3}\mu^{1/3}\mathcal{M}^{2/3}. \tag{5.60}$$

Now, the viscosity of air is 70-fold less than that of water, and we may suppose that because an insect does not expend energy maintaining the air that it breathes while other animals must maintain their blood, $\mathcal{M}$ is likely to be smaller in insects than in other animals.

We cannot at present assign values to $\mathcal{M}$ either for animals circulating water or for those circulating air, so we cannot calculate the exact relative costs of the two systems. It can be proposed, however, that the cost is so much lower in air than in water that it has not served as an effective selective factor guiding the design of insect tracheal systems.

One should also recall that only the larger vessels in the insect tracheal system have air flowing through them. The smaller vessels transport gas by diffusion alone, and therefore would not be expected to obey Murray's law.

How does an aqueous circulatory system know to set the radii at a bifurcation according to Murray's law? Sherman (1981) and LaBarbera (1990) propose that the developing system responds to the shear stress acting on the walls of the pipe.

To see how this works, we return to eq. 5.42. Taking the derivative of u with respect to r', we see that the velocity gradient at the walls of the tube ($r = r'$) is

$$\frac{du(r')}{dr'} = -\frac{r\Delta p}{2\mu\ell}. \tag{5.61}$$

The shear stress acting on the walls is the product of viscosity and this velocity gradient:

$$\tau = \mu\frac{du(r')}{dr'} = -\frac{r\Delta p}{2\ell}. \tag{5.62}$$

In light of this result, we can rewrite eq. 5.46:

$$J = \frac{\pi r^4 \Delta p}{8\mu\ell} = \frac{\pi r^3}{4\mu}\frac{r\Delta p}{2\ell} = \frac{\pi r^3 \tau}{4\mu}. \tag{5.63}$$

Rearranging one more time, we see that

$$\tau = \frac{4\mu J}{\pi r^3}. \tag{5.64}$$

In other words, the shear stress acting at the wall of a pipe is set by the ratio of volume flux to the cube of radius. But we showed above that in a system that follows Murray's Law, volume flux is itself proportional to r^3:

$$\tau = \frac{4\mu k_p r^3}{\pi r^3} = \frac{4\mu k_p}{\pi}. \tag{5.65}$$

Thus in a Murray's law system, the shear stress acting on the walls of the pipes is constant. If the epthelial cells of the developing system can detect the shear stress acting on them (and there is evidence that they can [LaBarbara 1990]), they thereby have a signal as to whether they are growing in the optimal form.

Fig. 5.11 Flow between two plates is similar to that through a pipe. The terms shown here are used in the calculation of laminar flow between parallel plates. Note that the width, $k_w h$, is much larger than the vertical spacing between plates, h. See text for details.

5.4.4 *Rectangular Pipes*

The relationships between viscosity and flow that we have derived for pipes with circular cross section can, with minor modification, be applied to flow through channels of other shapes. For example, an analysis similar to that of eq. 5.36 through eq. 5.47 can be applied to the flow between two horizontal parallel plates (fig. 5.11). Here we assume that the plates are separated by a distance h and that the width of the plates, $k_w h$, is much larger than h. Vertical position between the plates is denoted by y. Given these conditions,

$$u(y) = \frac{hy - y^2}{2\mu} \times \frac{\Delta p}{\ell} \tag{5.66}$$

$$\text{maximum velocity} = \frac{h^2}{8\mu} \times \frac{\Delta p}{\ell} \tag{5.67}$$

$$\text{average velocity} = \frac{h^2}{12\mu} \times \frac{\Delta p}{\ell} \tag{5.68}$$

$$J = \frac{k_w h^4}{12\mu} \times \frac{\Delta p}{\ell} \tag{5.69}$$

$$P = \frac{k_w h^4}{12\mu} \times \frac{(\Delta p)^2}{\ell}. \tag{5.70}$$

The similarity to flow through a pipe with circular cross section is apparent. The maximum velocity occurs in the middle of the channel, although in this case it is three halves (rather than twice) the average velocity. Both velocity and volume flux are directly proportional to the pressure gradient and inversely proportional to viscosity. And finally, volume flux and power are proportional to the fourth power of the spacing between plates.

Eqs. 5.66 to 5.70 could be used, for instance, to examine the flow between filaments in the gills of fish, and the results would be similar to those for flow of blood through pipes. We can use this similarity to examine the effects of temperature on the ability to acquire oxygen. In particular, what effect does temperature have on the cost of pumping air through the tracheae of insects and the cost of pumping water through the gills of fish?

In both cases, the rate at which oxygen is delivered to the body (in moles per second) is equal to the product of the volume flux of fluid through the system and the concentration of oxygen in the fluid:

$$\text{rate of oxygen delivery} = J \times [O_2]. \tag{5.71}$$

As the temperature varies, we desire to keep this value equal to the metabolic rate of the organism, M.

In the case of a circular pipe such as a trachea, J is calculated from eq. 5.46, from which we may deduce that

$$M = \frac{\pi r^4 \Delta p [O_2]}{8\mu\ell}. \tag{5.72}$$

Rearranging, we find that the pressure required to deliver oxygen at this rate is

$$\Delta p = \frac{8\mu\ell M}{\pi r^4 [O_2]}. \tag{5.73}$$

Now, we can insert this value of Δp into eq. 5.47 to calculate the power required to pump fluid at the required rate, with the result that

$$P = \frac{8\mu\ell}{\pi r^4}\left(\frac{M}{[O_2]}\right)^2. \tag{5.74}$$

Power increases with an increase in viscosity, and decreases with an increase in oxygen concentration.

As temperature rises, the viscosity of air increases (table 5.1) and the concentration of oxygen decreases (table 5.3). Inserting appropriate values for μ and $[O^2]$ into eq. 5.74 we find that it costs an insect 46% more power to deliver oxygen to its tissues at 40° C than it does at 0° C.

This analysis can be repeated for the rectangular pipes of a fish's gills, with the result that

$$P = \frac{12\mu\ell}{k_w h^4}\left(\frac{M}{[O_2]}\right)^2. \tag{5.75}$$

Now, the viscosity of warm water is less than that of cold, and one might initially expect the power of oxygen delivery to decrease with an increase in temperature. However, the oxygen content of water decreases drastically as temperature rises (table 5.3), with the result that the cost of oxygen delivery at 40° C is 1.9 times that at 0° C in fresh water. Thus, both insects and fish must expend more power to breathe when the fluid around them is warm.

In this analysis we have assumed that metabolic rate is independent of temperature. In most cases, however, a rise in ambient fluid temperature would be accompanied by a rise in metabolic rate, an effect that would amplify the increase in cost calculated here. For example, the metabolic rate of many "cold-blooded" animals approximately doubles with each 10° increase. In this case, the metabolic rate at 40° C is sixteen times that at 0° C, and the cost to an insect of circulating air would be 24-fold higher at 40° C than at 0°. The cost to a fish would be about 30-fold higher.

5.5 Flow in Porous Materials

The same basic physics can be applied to flows through porous materials such as dirt and sand, and to flow through the interstices of aggregated animals or plants. In these cases, the haphazard contact of adjacent objects forms a series of small "pipes" of nonuniform size and shape. The heterogeneous nature of these pipes makes it difficult to predict precisely the resistance to flow, and in these cases it is standard practice to invoke *Darcy's law* for flow through porous media:[8]

$$\frac{J}{A} = u = \frac{k_p}{\mu}\frac{\Delta p}{\ell}, \tag{5.76}$$

where u is the velocity of the fluid through the medium averaged over an area large compared to the pore size, and k_p is the *permeability* of the medium, the units of which are m^2. Permeability depends on the size, shape, and volume fraction of pores in the medium, and in most cases must be determined experimentally. For example, the permeability of loosely packed sand is 2 to 18×10^{-7} m^2 (Scheidegger 1971).

Despite its reliance on an empirically determined coefficient, Darcy's law is expressive of the same message conveyed by the more elegant formulas previously examined—the pressure gradient required to achieve a given rate of flow is directly proportional to the viscosity of the fluid. As a result, a larger force and a greater power are required to force water than to force air through porous media.

[8]Named for H. Darcy and a classical experiment he described in 1856 regarding the public fountains of Dijon, France.

Again, this fact is reflected in the design of plants and animals. For example, many aquatic or semiaquatic plants face a problem: their roots, which, like animals, require oxygen to live, are buried in water-logged soil. The flow of water, and therefore of oxygen, through the soil is impeded by water's high viscosity, and delivery of oxygen to the roots from the surrounding dirt is sluggish at best. In fact, many such soils are virtually devoid of oxygen. When the oxygen supply is not sufficient, aerobic metabolism in roots is blocked and ethanol (an end product of glycolysis) builds up in the tissues.[9] If anaerobic conditions persist, the alcohol concentration eventually reaches toxic levels and the roots die. How, then, do the roots of aquatic plants survive?

Plants such as water lilies and rice use essentially the same strategy as that used by insects. Interconnected spaces (lacunae) within the stem form a continuous (though not necessarily pipelike) path between the roots and the air above water level. A pressure gradient is maintained such that the air in the roots is at lower pressure than air higher in the stem, and gas flows to the roots. Because plants do not have muscles, the pressure gradient that can be maintained is very small (on the order of 200 Pa in water lilies), requiring that a low viscosity fluid be used if a sufficient flow of oxygen is to be maintained. In this respect air, rather than water, is the obvious choice. In water lilies, flow rates of 1.5 cm s^{-1} have been observed.

The methods by which plants maintain a pressure gradient between root and stem is an interesting story, but one well outside the scope of this chapter. If you are interested you should consult Dacey (1981) or Raskin and Kende (1985) for the physiological details.

5.6 Stefan Adhesion and Repulsion

Our consideration of the flow of fluid in confined spaces raises questions regarding another consequence of dynamic viscosity. Consider the situation shown in figure 5.12. Two parallel circular plates of radius r are immersed in a large volume of fluid and are separated by a distance y. If we pull the plates away from each other, fluid must flow into the gap between. We have already seen that a force is required to persuade fluid to flow between solid walls in this fashion, so we can guess that a force is required to separate the plates. In other words, we predict that the presence of a viscous fluid between two plates acts as an adhesive, tending to keep the plates together. How good an adhesive might this be? Stated another way, what force **F** is required to pull the plates away from each other at rate dy/dt? The answer to this question was calculated by J. Stefan in 1874, and this form of adhesion (*Stefan adhesion*) is named in his honor:

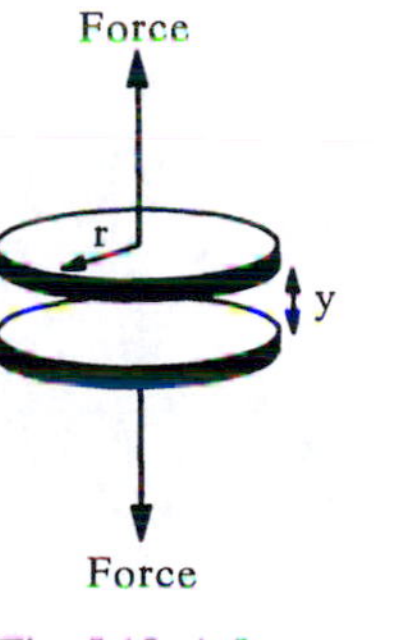

Fig. 5.12 A force must be applied to separate two disks in a viscous fluid, an effect known as Stefan adhesion. The terms shown here are used in the calculation of this form of adhesion. See text for details.

$$\mathbf{F} = \mu \frac{3\pi r^4 \, dy/dt}{2y^3}. \tag{5.77}$$

where **F** (a vector) is positive when it acts to increase y and negative when it acts to decrease y.

As one might have suspected, the "stickiness" of the fluid, in the sense of its ability to act as an adhesive, is directly related to its "stickiness" on the molecular level, that is, to its dynamic viscosity. For plates of a given radius, it takes seventy

[9]The thought of roots soaked in their own ethanol gives a whole new meaning to the term "potted plant."

times more force to separate the plates at a given rate if the plates are immersed in water rather than in air.

Consider an example. Limpets, abalones, and a few other gastropods have a relatively rigid foot that is closely applied to the rock on which the animal sits (fig. 5.13). If such an animal has a circular foot with a radius of 1 cm and the layer of water between the foot and the rock is 10 μm thick, an initial force of about 47 N would be required to pull the animal from the rock at a rate of 1 mm s^{-1}. This is a substantial force, similar to the actual force required to pull a limpet of this size from a rock.

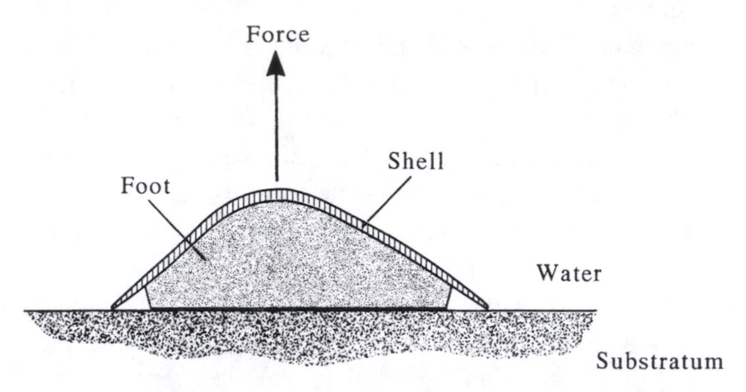

Fig. 5.13 A limpet adheres to a rock by creating a low pressure beneath its foot.

There are problems with this simple argument, however, and they serve to highlight the difficulty of using Stefan adhesion as a practical means of sticking to a surface. The forces calculated in the above example depend on an arbitrary specification of the rate at which the animal is pulled from the substratum. If instead of dislodging the aquatic animal at 1 mm s^{-1} we pull at a rate of only 1 μm s^{-1}, a force of only 0.047 N (rather than 47 N) would be required eventually to separate the animal from its substratum. Of course the process would take much longer, but if we were sufficiently patient we wouldn't need to apply much force at all.

Therein lies one difficulty with the use of viscous fluids as adhesives. In water, Stefan adhesion can be useful in holding two objects together, but only if the force tending to separate the objects is applied for a short period. This is a particular problem when it is gravity—which never takes any time off—that applies the force. Indeed, it appears likely that limpets and other gastropods do not rely on Stefan adhesion as their primary means of resisting dislodgment. Instead, they use a form of suction, in principle much like the mechanism that holds a toy dart to the surface of a refrigerator (A. Smith, pers. com.).

Alternatively, we could pull on our limpet at a rate of 1 cm s^{-1}, in which case the force of dislodgment is predicted to be 470 N. Acting over the area of attachment, this force translates to a negative pressure of 1.5×10^6 Pa, 15 atmospheres. This may well be more tension than water can stand, and the water beneath the foot may cavitate before the full force of 470 N is realized. Thus, cavitation may place an upper limit on the strength of a Stefan adhesive that uses water. The tensile strength of water is discussed in greater detail in chapter 12.

If, instead of water, our limpet had air under its foot, only 0.67 N (rather than 47 N) would be required to dislodge the animal at a speed of 1 mm s^{-1}. Stefan adhesion is much less effective in air than in water.

There is another aspect of Stefan adhesion that is worthy of comment. For more typical adhesives such as suction and glue, the force required to separate two adherent objects is proportional to the area of the adhesive. In other words, the force per area (stress) required to break the bond is constant. Not so for Stefan adhesion. Dividing both sides of eq. 5.77 by the area of the adhering disks (πr^2), we see that the force per area is itself a function of size:

$$\text{force per area} = \frac{\mathbf{F}}{\pi r^2} = \mu \frac{3r^2\, dy/dt}{2y^3}. \tag{5.78}$$

In fact, the force per area is proportional to the square of radius; large disks are therefore much more strongly adherent than small disks. As a consequence, if a disk in air were $\sqrt{70} = 8.4$ times the radius of a disk in water, the two would have the same Stefan adhesive capabilities. This strong size dependence does not, however, seem to have been sufficient to render Stefan adhesion advantageous to terrestrial organisms—I know of no terrestrial animal or plant that relies on the viscosity of air as an adhesive.

A second look at eq. 5.77 raises another question about the role of viscosity in relation to objects near solid surfaces. What if, instead of being pulled away from each other, the plates are being pushed together? In other words, what happens when dy/dt is negative? Evidently, the closer the plates approach each other, the greater the force required to keep them moving at a constant rate. Alternatively, if a constant force is applied to one of the plates, the rate at which it approaches the other continuously decreases. Taken to its logical limit, this implies that an infinite period is required to bring the plates into actual contact. In reality, we cannot expect eq. 5.77 to apply when the spacing between plates approaches the size of individual molecules of gas or liquid. We can, however, use eq. 5.77 to calculate the time required for an organism to approach within a distance which, for all practical purposes, is the equivalent of actual contact. Rearranging eq. 5.77, we see that

$$dt = \frac{3\pi\mu r^4}{2\mathbf{F}y^3} dy. \tag{5.79}$$

Integrating both sides of eq. 5.79, we get

$$t = -\frac{3\pi\mu r^4}{4\mathbf{F}y^2} + \text{constant}. \tag{5.80}$$

We can evaluate the constant of integration by specifying when we start our clock. For example, if we want to know how long it takes to get from an initial distance y_0 to some closer distance y, we set $t = 0$ at $y = y_0$ and

$$t = -\frac{3\pi\mu r^4}{4\mathbf{F}}\left(\frac{1}{y_0^2} - \frac{1}{y^2}\right). \tag{5.81}$$

Note that in this case **F** is negative because it acts to decrease y.

We can apply this equation to situations of biological interest. Consider first a small insect trying to land on a flower, or a planktonic larva attempting to attach to a rock. In either case, once the animal is within a certain distance of its objective it can extend a leg and grab ahold. But to get within grasping distance, the fluid between the body and the substratum must be squeezed out of the way. Neither object is a circular disk, so we cannot expect eq. 5.81 to apply in any exact sense, but it can provide qualitative predictions. For an object of a given size, acted upon by a given force, it takes seventy times as long for the object to get within grasping distance in water as in air. Alternatively, the aquatic organism would have to apply seventy times the force to get itself down in a given period. Note, however, the factor of r^4 in the numerator of eq. 5.81. If the terrestrial organism is $\sqrt[4]{70} = 2.9$ times the size of its aquatic counterpart, equal forces imply equal times.

We can get some feeling for the magnitude of these effects by examining an admittedly contrived example. Consider an organism shaped like a disk with a thickness, ℓ, equal to one-tenth its radius, r. The body has a density of 1080 kg m^{-3}, so that its effective density in air is 1079 kg m^{-3} and in fresh water, 80 kg m^{-3}. Acted upon by its weight alone, the organism falls toward a horizontal substratum. How long does it take the organism to travel from a distance equal to its radius above the substratum to a distance equal to $0.1\,r$, at which point it can extend a leg and attach? The results (calculated using eq. 5.81 at 20° C) are shown in figure 5.14. Because, in this case, the driving force increases as r^3 while the distance that must be traveled increases only as r, it takes less time for a large organism to arrive within effective grasping distance of the substratum. Arrival, of course, is much quicker in air than in water. In neither case is the time very long.

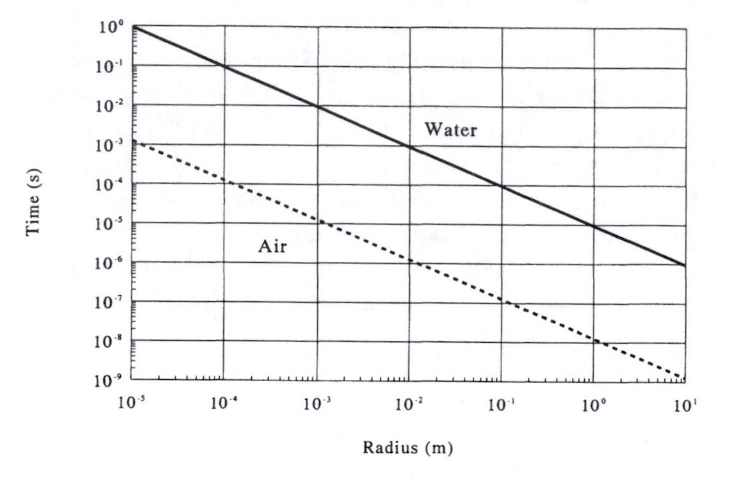

Fig. 5.14 The time required for a disk to approach a solid wall is less in air than in water due to air's lower dynamic viscosity. In the case shown here, the distance traveled is a fixed fraction of the disk's radius and the force propelling the disk is its own weight. The larger the disk, the shorter the time required.

What if, rather than traveling a distance proportional to its size, an object must travel some fixed distance? This is a situation that might occur during certain kinds of locomotion. For example, sea stars and sea urchins walk using tube feet. At each step, the disklike end of the tube foot is lifted from the substratum, moved forward, and then placed back down, a motion very similar to a human step but lacking the pronounced heel-to-toe rotation of the foot. Typically the tube foot is lifted about a millimeter and, for the sake of argument, let us assume that it must return to within 10 μm of the substratum to regain effective contact. We further assume that the force pushing down on the tube foot is proportional to its area (100 Pa, say). How is the time to contact the substratum affected by the size of the tube foot?

The answer (calculated using eq. 5.81) is shown in figure 5.15. In this case, the smaller the radius of the tube foot, the shorter the time needed to contact the substratum. In water, an object of radius 1 mm (typical of tube feet) requires about 0.08 s to travel the distance; an object 1 cm in radius requires about 8 s, and an object 1 m in radius requires over 80, 000 s. To maintain a reasonable stepping rate, sea stars, sea urchins, and other aquatic pedestrians need to have small feet.

The same physics can affect other organisms. For instance, a vertical approach to the substratum might be quite time consuming for large, disklike organisms such as skates, rays, and flatfishes. However, these animals tend to glide down to the bottom at a shallow angle, thereby avoiding the problem. In air, the time to contact is less by a factor of seventy, and therefore poses much less of a problem.

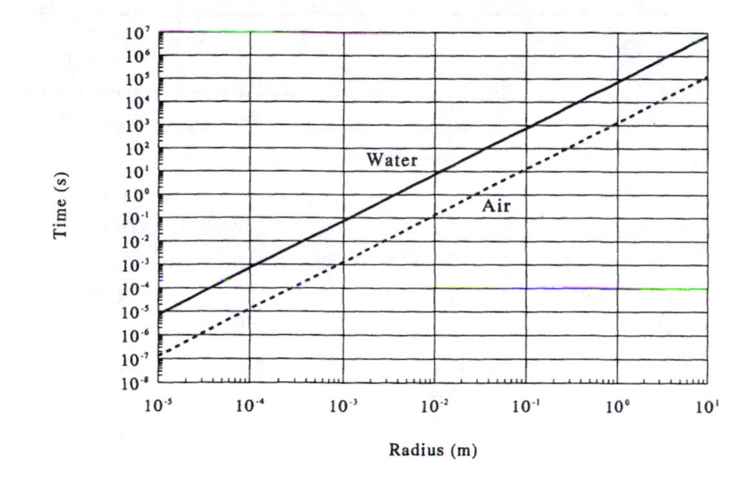

Fig. 5.15 When a disk must travel a fixed absolute (rather than relative) distance toward a solid wall, the time required increases with the size of the disk. Here the disk is propelled by the force resulting when a pressure of 100 Pa is applied to the disk's area.

5.7 Summary . . .

Viscosity results in an unavoidable resistance to the motion of a fluid relative to a solid object. For small objects moving at slow speeds (i.e., for low Reynolds numbers) viscous resistance may be a dominant force.

Because water is forty to one-hundred times as viscous as air (depending on temperature), a given pressure gradient can move the same amount of fluid through a smaller pipe in air than in water. Insects have taken advantage of this fact in their tracheal circulatory systems. Conversely, for a given flux of fluid and size of pipe, a much smaller pressure gradient is required in air than in water, a factor relied upon by some aquatic plants.

In many cases the design of branching circulatory systems has evolved to minimize the cost of the system according to Murray's law. Circulatory systems that pump air, however, seem to be exempt.

For small, flagellated organisms, both the viscous drag of the body and the production of thrust depend on viscosity. As a result, locomotion using flagella is feasible (at least in a mechanical sense) in both air and water, leading to the thought that perhaps bacteria can fly.

The viscosity of fluids can make it difficult for organisms to make either a rapid exit from or a rapid approach to a substratum, but the effect should be much less noticeable in air.

5.8 . . . and a Warning

Once again, you are warned to view the results of this discussion in their proper context. The goal of this chapter has been to provide a broad overview of the biological consequences of the difference in dynamic viscosity between air and water. In the process, many of the finer points have been swept under the rug. In particular, circulatory physiology is much more complex than one might surmise from the treatment presented here. If you wish to examine in detail any particular example, consult the original references.

Chapter 6

Diffusion: Random Walks in Air and Water

This chapter deals with the process of random motion and the way in which it can act as a mechanism of transport. The focus here is on the process of molecular diffusion and its biological consequences. For instance, we will see how the diffusion of gases can limit the size of plants and animals and explain why an ostrich egg is more porous than that of a hummingbird. We will explore the role of diffusive transport in foraging strategy: Is it better to go looking for food or to wait for the food to come to you? And we will see how the mean free path of molecules in air sets the minimal size of insect tracheoles.

The principles outlined in this chapter are quite general, however, and we will make considerable use of them in later chapters.

6.1 The Physics

6.1.1 *Molecular Velocity*

In chapter 3 we defined temperature in terms of the average kinetic energy of molecules using the relationship

$$\frac{3kT}{2} = \frac{m\langle u^2 \rangle}{2}, \tag{6.1}$$

where m is the mass of the molecule, u its speed, and k is Boltzmann's constant, 1.38×10^{-23} J K^{-1}. The brackets, $\langle\rangle$, denote that velocity is averaged over time for a single molecule, or averaged over many molecules at any one time. This simple expression has some intriguing consequences.

Consider, for instance, a typical molecule of air (nitrogen) with a weight per molecule of about 4.7×10^{-26} kg. At room temperature (290 K), we have seen (chapter 3) that the average speed of a nitrogen molecule is

$$\sqrt{\langle u^2 \rangle} = \sqrt{\frac{3kT}{m}} \approx 508 \text{ m s}^{-1}. \tag{6.2}$$

This is a sizable speed.[1] For example, an unimpeded nitrogen molecule would finish the hundred-yard dash in about a fifth of a second. And this speed is not peculiar just to nitrogen. At 290 K oxygen moves at an average speed of 475 m s^{-1}, carbon dioxide at 405 m s^{-1}, and water vapor at a brisk 634 m s^{-1}.

The high velocity of molecules at room temperature raises visions of extremely effective transport systems. If you want to move oxygen from point A to point B, you simply take the gas one molecule at a time, point it in the right direction,[2] and

[1] Of the total kinetic energy of a molecule, 1/3 (on average) is associated with motion along each particular axis. Thus, at 290 K the average velocity of a nitrogen molecule along the x axis is

$$\sqrt{\langle \mathbf{u}^2 \rangle} = \sqrt{\frac{kT}{m}} \approx 292 \text{ m s}^{-1}. \tag{6.3}$$

The overall speed of 508 m s^{-1} is $\sqrt{\langle \mathbf{u}^2 \rangle + \langle \mathbf{v}^2 \rangle + \langle \mathbf{w}^2 \rangle}$.

let it go. Thermal kinetic energy does the rest! But these visions seem at odds with everyday experience. If you put a drop of dye in a glass of still water, it may take hours to spread throughout the container. Similarly, it can take several minutes for the scent from a bottle of perfume opened in one corner of a room of still air to reach the opposite corner. How can this be if molecules are moving at several hundred m s^{-1}?

Even worse, these everyday experiences are at odds with careful experiments. In an ordinary glass of water, convection currents created as the water evaporates account for most of the movement of dye. If one makes sure that no convection is present, a drop of dye may take months (rather than hours) to spread throughout the container. Similarly, convection currents in air account for most of the transport of perfume. Clear evidence of these currents are the motes of dust that dance in a sunbeam. In the absence of convection, scent takes hours (rather than minutes) to spread across a room.

Our task in this chapter is to reconcile these experimental results regarding the macroscopic transport of molecules in fluids with the microscopic behavior of individual molecules. This reconciliation is found in the process of random motion. Each molecule moves at considerable speed, but it cannot move very far in a straight line without colliding with another fluid molecule. Like a billiard ball on a crowded table, at each collision the molecule flies off in a new direction only to collide again. The direction taken by a molecule after a collision is purely a matter of chance. If it hits another molecule squarely, it may rebound back from where it came. A glancing blow results in a less drastic change in direction. As a result of its collisions, each molecule performs a *random walk* through space. It is to the characteristics of such walks that we now turn our attention.

6.1.2 *Random Walks*

The concept of a random walk is best understood through the use of a simple example. Consider motion along the x axis. We start by placing a hypothetical particle at the origin, $x = 0$, and every τ seconds we allow the particle to step a distance δ along the axis. Thus the average speed of the particle is δ/τ. The direction of each step, however, is purely a matter of chance. Half the time the particle steps to the right, half the time to the left. In practice, we can simulate this process by tossing a coin before each step. If the coin comes up heads, the particle moves to the right; if tails, it moves to the left.

After each step in this random walk we note the position of the particle. In this manner we can track the particle along the axis as a function of the number of steps taken, or, equivalently, as a function of time. Because each step is chosen at random, the precise path taken by a particle is impossible to predict (fig. 6.1). However, if we repeat the experiment a number of times, we can perceive a pattern in the *average* manner in which particles move. It is this predictability of the average that renders the concept of a random walk so useful.

To explore the properties of the average, we examine the statistics of particle motion. To do so, we track the random walks of N particles. Let $x_i(n)$ be the

[2] You may object to the idea of "pointing" a thermally agitated molecule. Because thermal motion is disordered, the act of aiming a molecule in essence cools it. It would be possible in theory, however, to let each molecule rattle around in a box until by chance it is moving in the proper direction, and then suddenly release it.

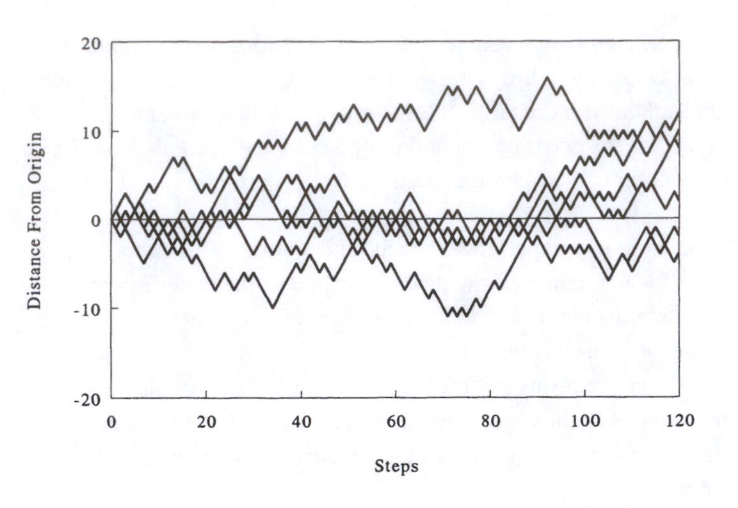

Fig. 6.1 Six particles, all starting at the same spot, take quite different paths in a random walk. On average, however, the particles go nowhere at all.

location on the x axis of the ith particle after the particle has taken n steps. We know from our assumptions that

$$x_i(n) = x_i(n-1) \pm \delta. \tag{6.4}$$

In other words, the position of the particle after n steps is either one step length to the right or one step length to left of the position of the particle after $n-1$ steps. After we have repeated our experiment N times, we can compute the average position of particles after n steps:

$$\langle x(n) \rangle = \frac{x_1(n) + x_2(n) \ldots + x_N(n)}{N} \tag{6.5}$$

$$= \frac{1}{N} \sum_{i=1}^{N} x_i(n), \tag{6.6}$$

where the symbol Σ means that we add the position after n steps of all particles numbered 1 to N.

We can expand this expression by replacing $x_i(n)$ with $x_i(n-1) \pm \delta$, which we know to be equivalent (Eq. 6.4). Thus,

$$\langle x(n) \rangle = \frac{1}{N} \sum_{i=1}^{N} [x_i(n-1) \pm \delta] \tag{6.7}$$

$$= \frac{1}{N} \sum_{i=1}^{N} x_i(n-1) + \frac{1}{N} \sum_{i=1}^{N} \pm\delta. \tag{6.8}$$

But, since half the time δ is to the right and half the time to the left, the average of $\pm\delta$ must be 0. As a consequence,

$$\langle x(n) \rangle = \langle x(n-1) \rangle. \tag{6.9}$$

In other words, the average position of a particle after n steps is the same as the position after $n-1$ steps. Taken to its logical conclusion, this means that, on average, particles starting at the origin remain at the origin. This is just what you might expect given equal probability of stepping right or left.

The fact that particles go nowhere on average does *not* imply that all particles remain at the origin, however. All that the average tells us is that for every particle that ends up at $x = +3$, for instance, there is likely to be a particle that ends up at $x = -3$. As long as an equal number of particles have net movements left and right, the average can still be 0, although there is considerable spread of particles.[3] To see how far away from the origin particles are likely to travel, we need to calculate our average in a different way.

It would be possible to avoid the problem of averaging positive with negative values by taking the absolute value of each particle's location after n steps and averaging these values as before. Instead, it has become traditional to circumvent the problem by taking the square of the location on the x axis. Because the squares of both positive and negative numbers are always positive, we can, by averaging the squares of locations, gain information as to the average distance traveled from the origin. Again we begin by expressing the location after n steps in terms of the location after $n - 1$ steps. Expanding the expression, we see that[4]

$$x_i^2(n) = [x_i(n-1) \pm \delta]^2 \tag{6.10}$$

$$= x_i^2(n-1) \pm 2\delta x_i(n-1) + \delta^2. \tag{6.11}$$

When we average this value for N particles, we see that

$$\langle x^2(n) \rangle = \frac{1}{N} \sum_{i=1}^{N} [x_i^2(n-1) \pm 2\delta x_i(n-1) + \delta^2] \tag{6.12}$$

$$= \frac{1}{N} \sum_{i=1}^{N} x_i^2(n-1) + \frac{1}{N} \sum_{i=1}^{N} \pm 2\delta x_i(n-1) + \frac{1}{N} \sum_{i=1}^{N} \delta^2 \tag{6.13}$$

$$= \frac{1}{N} \sum_{i=1}^{N} x_i^2(n-1) + \frac{1}{N} \sum_{i=1}^{N} \delta^2 \tag{6.14}$$

$$= \langle x^2(n-1) \rangle + \delta^2. \tag{6.15}$$

In this computation we have again made use of the fact that the average of $\pm\delta = 0$.

This result tells us that the average (or *mean*) square of position after n steps is greater by δ^2 than the average square of position after $n - 1$ steps. Now, we know from our initial assumptions that the mean square position after 0 steps is 0 [i.e., $\langle x(0) \rangle = 0$], so $\langle x^2(1) \rangle = \delta^2$. After two steps, the average square of position is $2\delta^2$, and so forth. Thus,

$$\langle x^2(n) \rangle = n\delta^2. \tag{6.16}$$

In other words, the square of the displacement from the starting point increases directly with the number of steps taken. We can convert from the square of displacement to displacement itself simply by taking the square root of eq. 6.16:

$$\sqrt{\langle x^2(n) \rangle} = x_{rms} = \sqrt{n}\delta. \tag{6.17}$$

[3]The distinction between average and individual movement reminds me of the story of two statisticians who went duck hunting. A bird flew past, and both hunters fired. One shot passed 2 m in front of the duck, the other 2 m behind. The statisticians proceeded to congratulate themselves because on average they had hit the bird dead center.

[4]Recall that $x_i(n-1)$ is the x position of the ith particle after $(n-1)$ steps. The square of this value is $x_i^2(n-1)$, *not* $x_i^2 \times (n-1)^2$.

The value $\sqrt{\langle x^2(n)\rangle}$ is called the *root mean square* or *rms* displacement, and it is a measure of how far, on average, particles have moved from their starting point after n steps. Readers familiar with statistics may note that the mean square displacement is the same as the variance of displacement, and the root mean square displacement is the standard deviation.

We now return to our original assumptions regarding the random walk and recall that the particle takes a step every τ seconds. Thus, the number of steps taken is equal to t/τ, where t is the time since the particle started its walk. Inserting this expression for n in eq. 6.16, we see that

$$\langle x^2(t)\rangle = \frac{\delta^2}{\tau}t. \tag{6.18}$$

The mean square distance traveled by a particle (in other words, the *probable* square of distance) increases directly with time at a rate governed by the ratio of δ^2 to τ.

6.2 The Diffusion Coefficient

We now are in a position to relate this theoretical consideration of a random walk to the process of diffusion. To do so, we define a *diffusion coefficient*, $\mathcal{D}$:

$$\mathcal{D} \equiv \frac{\delta^2}{2\tau}. \tag{6.19}$$

The reason for the factor of $1/2$ will become clear later in this chapter; it eventually makes the math simpler. The diffusion coefficient has the units $m^2\,s^{-1}$.

Using this definition, we can restate our conclusion regarding the random walk:

$$\langle x^2(t)\rangle = 2\mathcal{D}t \tag{6.20}$$

$$x_{rms} = \sqrt{\langle x^2(t)\rangle} = \sqrt{2\mathcal{D}t}. \tag{6.21}$$

Eq. 6.21 is of special importance, and is therefore worth dwelling upon for a moment. It says that the distance a particle is likely to travel from its point of origin while performing a random walk (the rms distance) increases not with time, as one might guess, but rather with the square root of time. In other words, if a particle travels an average of 1 cm in 1 s, it takes an average of 100 s to travel 10 cm, and 10,000 s to travel 100 cm.

As a result, if we try to express the rate of travel as a "diffusion velocity" we come to a curious conclusion. Setting velocity equal to rms distance per time,

$$\text{“diffusion velocity”} = \sqrt{\frac{2\mathcal{D}}{t}}. \tag{6.22}$$

The shorter the period over which we measure velocity, the greater the velocity we measure. Conversely, the longer the period of measurement, the slower the velocity measured. The fact that the rate of transport in diffusion is a function of time is one of the principal characteristics of diffusive processes.

Note that eq. 6.22 can be applied only when $t > \tau$. It was calculated assuming that the particle changes its direction at random. For periods shorter than τ, we have assumed that the particle is in the midst of a step and not subject to directional change. Therefore, at $t < \tau$, our assumption is violated and eq. 6.22 leads to spurious results.

We now examine the diffusion coefficients for three molecules with biological importance: oxygen, carbon dioxide, and water vapor.

6.2.1 *Diffusion Coefficients in Air*

The diffusion coefficients for oxygen and water vapor in air can be described by the equation,

$$\mathcal{D}(T,p) = k_1 T^{k_2} \frac{p_0}{p}, \tag{6.23}$$

where $\mathcal{D}(T,p)$ is the diffusion coefficient at absolute temperature T and pressure p and p_0 is one normal atmosphere. The coefficients k_1 and k_2 vary from one type of molecule to the next. For oxygen, $k_1 = 1.13 \times 10^{-9}\,\mathrm{m^2\,s^{-1}}$ and $k_2 = 1.724$ (Marrero and Mason 1972). For water vapor, $k_1 = 0.187 \times 10^{-9}\,\mathrm{m^2\,s^{-1}}$ and $k_2 = 2.072$. The form of eq. 6.23 tells us that the diffusion coefficient rises with any increase in temperature, but decreases with any increase in pressure.

The diffusion coefficient for carbon dioxide in air is described by a slightly more complex equation (Marrero and Mason 1972),

$$\mathcal{D}(T,p) = k_1 T^{k_2} \exp(-k_3/T) \frac{p_0}{p}, \tag{6.24}$$

where k_3 is an additional parameter. In this case $k_1 = 2.70 \times 10^{-9}\mathrm{m^2\,s^{-1}}$, $k_2 = 1.590$, and $k_3 = 102.1$ K.

These results are summarized in table 6.1 and figure 6.2. The diffusivity of water vapor is about 50% greater, and the diffusivity of O_2 about 30% greater, than that of carbon dioxide. For all three gases, the diffusion coefficient is about 30% higher at 40° C than at 0° C. At the top of Mount Everest, where the pressure is only a third of that at sea level, the diffusion coefficients of these gases would be three times those shown here.

The distance (x_{rms}) that an oxygen molecule is expected to travel in air is shown as a function of time in figure 6.3. In one second a molecule will, on average, move about a centimeter from its starting point. Several hours are required, however, before the molecule reliably travels one meter.

Table 6.1 The diffusion coefficients of various gases in air at one atmosphere.

T(° C)	Diffusion Coefficient, $\mathcal{D}_m$ ($\mathrm{m^2\,s^{-1} \times 10^{-6}}$)		
	O_2	CO_2	H_2O
0	17.9	13.9	20.9
5	18.5	14.4	21.7
10	19.1	14.9	22.5
15	19.7	15.4	23.3
20	20.3	16.0	24.2
25	20.9	16.5	25.1
30	21.5	17.0	26.0
35	22.1	17.6	26.8
40	22.7	18.1	27.7

Source: Calculated from data presented by Marrero and Mason (1972).

Notes: Note that these values represent the best fit to empirical measurements, but that actual measured values may vary by ± 5% from those shown here. The symbol $\mathcal{D}_m$ is used here to distinguish molecular diffusion from other types of diffusion (e.g., the diffusion of heat).

6.2.2 *Diffusion Coefficients in Water*

The diffusion coefficients of oxygen and carbon dioxide are about 10,000 times smaller in water than they are in air (table 6.2). This tremendous difference in diffusivity will dominate our discussion of the biological consequences of diffusion, but there are two further contrasts to note. First, the variation in diffusion coefficient with temperature is much greater in water than in air. For example, at 40° C the diffusion coefficient of O_2 in water is 3.2 times that at 0° C, whereas in air it is increased by a factor of only about 1.3. The variation is somewhat less for CO_2; $\mathcal{D}$ in water is about 2.1 times higher at 40° C than at 0° C.

Second, the dependence on temperature affects the rank order of the diffusivities in water (table 6.2; fig. 6.4). At low temperature, the diffusion coefficient of CO_2 is larger than that of O_2, but above about 6° C, the diffusivity of oxygen is greater than that of carbon dioxide.

The distance (x_{rms}) that an oxygen molecule is expected to travel in water is shown as a function of time in figure 6.3. These distances are surprisingly small.

For example, relying on diffusion alone, a molecule moves only about 5 cm in a million seconds.

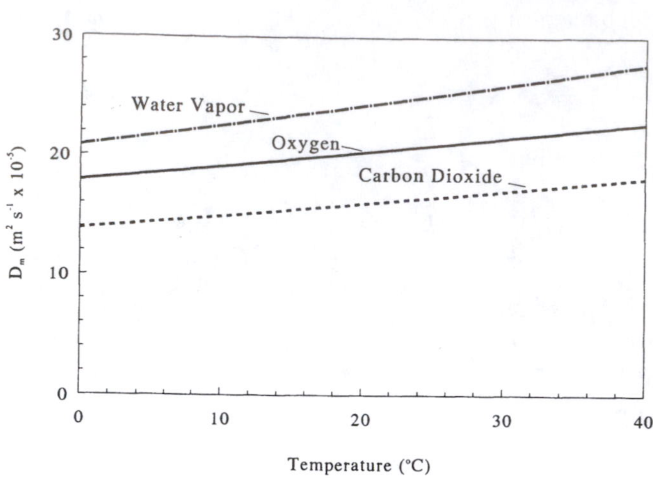

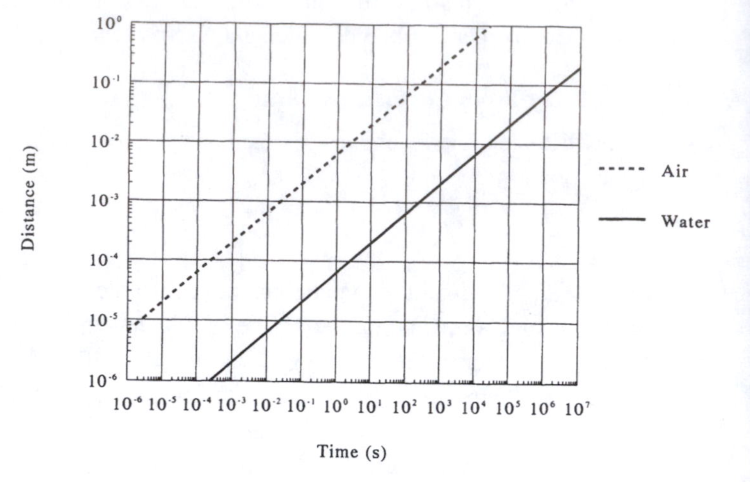

Fig. 6.2 The diffusion coefficients of gases in air increase slightly with an increase in temperature. (Data from table 6.1)

Fig. 6.3 The root-mean-square distance traveled by an oxygen molecule is much greater in air than in water. Note, however, that in either medium it takes a very long time to travel a meter by diffusion alone.

Table 6.2 The diffusivity of O_2 and CO_2 in water at one atmosphere (from Armstrong 1979).

T(° C)	Diffusion Coefficient, $\mathcal{D}_m$ ($m^2\ s^{-1} \times 10^{-9}$)	
	O_2	CO_2
0	0.99	1.15
5	1.27	1.30
10	1.54	1.46
15	1.82	1.63
20	2.10	1.77
25	2.38	1.92
30	2.67	2.08

Note: The symbol $\mathcal{D}_m$ is used here to distinguish molecular diffusion from other types of diffusion (e.g., the diffusion of heat).

6.2.3 *Mean Free Path*

Let us now examine the concept of a diffusion coefficient in greater detail. We first note that because the coefficient has dimensions of length squared per time, it can be treated as the product of velocity and a distance. Thus, the rate at which a particle is diffusively transported depends both on how fast it moves while in free flight between collisions ($u = \delta/\tau$) and on the average distance it moves before again colliding ($\delta/2$), a distance called the *mean free path, l*. We have already calculated how fast fluid molecules move at room temperature ($\delta/\tau \approx 500$ m s^{-1} [eq. 6.2]), and are now in a position to estimate their mean free path.

To do so, we rely on the diffusion coefficients as presented in tables 6.1 and 6.2. We will discuss later in this chapter how these empirical measurements are made, but for now accept them as given. The diffusion coefficient for an oxygen molecule in air is about 2×10^{-5} $m^2\ s^{-1}$. Noting that $\delta = 2\mathcal{D}/u$, we calculate that the mean free path in air is 8×10^{-8} m. In other words, every 0.08 μm an oxygen molecule collides with another air molecule and careens off in a new direction. Traveling at its velocity of 500 m s^{-1}, the O_2 molecule travels this distance in about 0.00016 μs. To put it another way, an air molecule experiences about 6.25 billion collisions every second, each one capable of changing the direction of its flight. With so many collisions going on, it is no wonder that air molecules behave differently in bulk than might be expected from the behavior of a single molecule on its own.

In water, the diffusion coefficient of an oxygen molecule is 10,000 times smaller than that in air, 2×10^{-9} $m^2\ s^{-1}$. We know, however, that at a given temperature, an oxygen molecule dissolved in water has the same kinetic energy as one in air, and therefore must have the same average velocity. As a consequence, the reduced diffusion coefficient in water must be due to a reduction in the mean free path. Again equating δ and $2\mathcal{D}/u$, we calculate that the mean free path of an oxygen molecule in water is about 10^{-11} m, only a twentieth of the diameter of a hydrogen atom! Because water molecules are packed so tightly together, there is virtually no "free space" through which an oxygen molecule can fly, and the randomly walking molecule undergoes about 60,000 billion collisions per second. The net result is a reduction in the rate at which the molecule is transported from one spot to another.

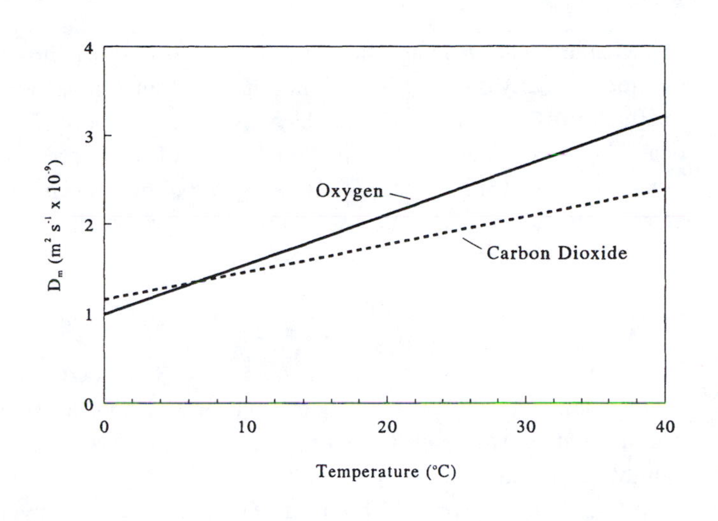

Fig. 6.4 The diffusion coefficients of both oxygen and carbon dioxide gases in water increase with an increase in temperature, but the increase in $\mathcal{D}_m$ is greater for oxygen. (Data from table 6.2)

6.3 The Sherwood Number

Before we consider the biological consequences of diffusion and the 10,000-fold difference in diffusion coefficients between air and water, we need to take a moment to consider under what circumstances diffusion is the primary means by which substances are transported through the fluid environment to or from an organism. In particular, we need to compare transport by diffusion with the rate at which a substance can be transported by *convection*, where convection includes transport by either the bulk movement of a fluid relative to an organism or by movement of an organism relative to the fluid. The ratio of transport by convection to that by diffusion is quantified by a dimensionless quantity, the *Sherwood number*, Sh:

$$Sh = \frac{u\ell_c}{\mathcal{D}}, \tag{6.25}$$

where u is the relative velocity between the organism and the bulk of the fluid, ℓ_c, is a characteristic length of the organism (here taken to be the length along the direction of motion), and $\mathcal{D}$ is the diffusion coefficient. We derive the Sherwood number later in this chapter.

When Sh is large, convective transport to an organism is much greater than that by diffusion. But $\mathcal{D}$ is so small in either air or water (10^{-9} to $10^{-5}\,\mathrm{m^2\,s^{-1}}$) that the Sherwood number is bound to be large unless the organism in question is exceedingly small, the relative velocity is exceedingly slow, or both.

To give this concept some tangibility, we calculate the Sherwood number for representative organisms. Consider first a sessile plant or animal. If a stationary terrestrial organism is 1 cm long, the Sherwood number is 500 when the wind blows at just $1\,\mathrm{m\,s^{-1}}$. The breeze would have to fall below $2\,\mathrm{mm\,s^{-1}}$ before Sh would be less than 1, at which point diffusion would begin to outweigh convective transport. A velocity this slow is unusual. Even in the complete absence of wind, the convective currents set up by hot or cold objects are virtually guaranteed to exceed $2\,\mathrm{mm\,s^{-1}}$, and, as a result, a terrestrial organism 1 cm in length almost always operates at a Sherwood number greater than 1.

In water, the convective flow must be slower still before diffusive transport outweighs convective transport. For example, a stationary organism 1 cm long would be serviced primarily by convection unless the current speed dropped below $0.2\,\mu\mathrm{m\,s^{-1}}$!

Mobile organisms can effect the Sherwood number through the increase in relative velocity due to their locomotion. How small must a mobile organism be before diffusive transport outweighs convective transport? We may make a rough calculation by noting that the maximal speed of many organisms is about ten of their own body lengths per second (see fig. 6.5). Thus, we can estimate the velocity term in the Sherwood number as at most 10 ℓ_c per second and

$$Sh \approx \frac{10\ell_c^2}{\mathcal{D}}. \tag{6.26}$$

This relationship is shown in figure 6.6. The Sherwood number is less than 1 (indicating that diffusion is the dominant form of transport) when ℓ_c is less than about 1.4 mm in air. Thus aerial organisms up to the size of gnats and no-see-ums might live in a diffusion-controlled world. In this chapter we will generally limit ourselves to the discussion of terrestrial organisms smaller than this size.

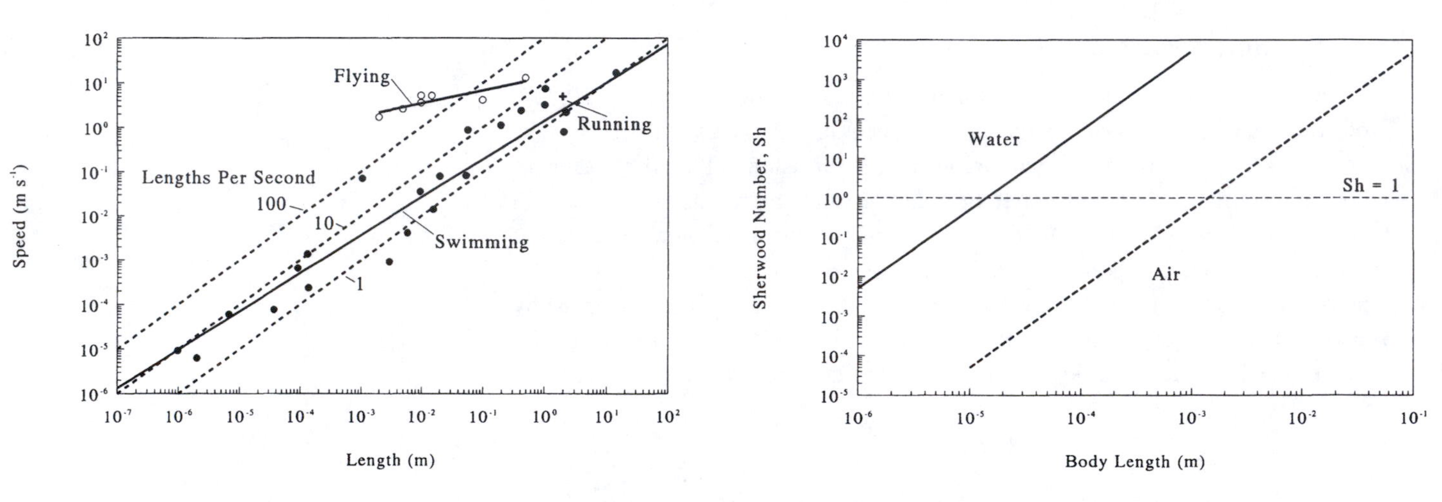

Fig. 6.5 The speed of locomotion increases in an orderly fashion as the size of an animal increases. Swimming data (solid dots) span sizes from bacteria to whales (Okubo 1987). The best-fit line for these data is speed ($\mathrm{m\,s^{-1}}$) = $1.4 \times \ell^{0.86}$. Flying data (open dots) span sizes from fruit flies to ducks (insect data from Denny 1976, duck datum from McFarlan 1990). The best-fit line for these data is speed ($\mathrm{m\,s^{-1}}$) = $13.8 \times \ell^{0.30}$, but given the small number of points, this line should not be taken too seriously. The point for running (cross) is for a human being (2 m tall) running at 10 $\mathrm{m\,s^{-1}}$ (fast sprint pace). (Adapted and expanded from a figure in Mann and Lazier 1991 by permission of Blackwell Scientific Publications, Inc.)

Fig. 6.6 The Sherwood number (Sh) for organisms moving at a speed equal to ten times their body length per second is much larger in water than in air.

In water, ℓ_c must be less than about 14 μm before diffusion is the dominant form of transport. As a consequence, we will limit our discussion of aquatic organisms in this chapter to small phytoplankton and bacteria.

In summary, the message of the Sherwood number is this: for most large organisms and most real-world flows, transport by diffusion is small compared to transport by convection.

I do not mean to imply that diffusion is unimportant for other organisms. We will explore several examples in chapter 7 in which diffusive transport forms an integral and important part of the system by which gases and nutrients are delivered. However, in these later examples, it is primarily convection that controls the rate at which diffusion transports material (by a mechanism we have not yet discussed), and it is for this reason that these examples are best left to a later chapter. Here we focus on examples where diffusion alone governs the rate of transport.

6.4 Fick's Equation

Before we begin this exploration, we need one further mathematical tool. To this point we have treated diffusion from a microscopic perspective, following individual molecules as they undergo random walks. For many practical problems, it is easier to calculate transport when diffusion is examined from a more macroscopic point of view.

Consider the situation shown in figure 6.7. Again we deal with random motion in a single direction, but do so in a three-dimensional container. For example, at time t we may assume that $N(x)$ particles are located at x, a distance $\delta/2$ to the left of the plane shown in the figure. Similarly, $N(x+\delta)$ particles are located at $x+\delta$, a distance $\delta/2$ to the right of the plane. All of these particles undergo the same type of motion as prescribed in the previous example, which is to say that at time $t+\tau$ half the particles at x have by chance moved to the right through the plane and end up at $x+\delta$. Similarly, half the particles at $x+\delta$ move to the left. The *net* number of particles that cross the plane in the positive x direction in time τ is

$$\text{net number crossing} = \frac{N(x)}{2} - \frac{N(x+\delta)}{2}. \tag{6.27}$$

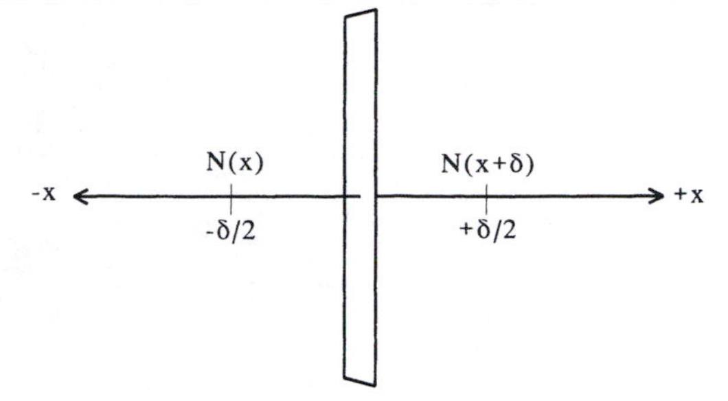

Fig. 6.7 In a derivation of Fick's law, we keep track of the number of particles (N) at two points on the x-axis. Particles taking a step to the left from $(x+\delta)$ cross through the plane shown, as do particles taking a step to the right from x.

If we assume that A is the area of the plane through which particles pass, we can express this process as a *flux density*, $\mathcal{J}_x$, i.e., the net number of particles crossing our plane per time per area. Dividing eq. 6.27 by A and rearranging, we see that

$$\mathcal{J}_x = \frac{-[N(x+\delta) - N(x)]}{2A\tau}. \tag{6.28}$$

We now perform a bit of mathematical legerdemain. Multiplying the righthand side of this equation by δ^2/δ^2 and again rearranging, we conclude that

$$\mathcal{J}_x = -\frac{\delta^2}{2\tau}\frac{1}{\delta}\left[\frac{N(x+\delta)}{A\delta} - \frac{N(x)}{A\delta}\right]. \tag{6.29}$$

This expression can be simplified. First, we recall that by our definition $\delta^2/(2\tau)$ is the diffusion coefficient, $\mathcal{D}$. The factor of 1/2 that appears in this equation as a result of the averaging process is the reason we included a factor of 1/2 in our definition of $\mathcal{D}$ (eq. 6.19). Next we note that the product $A\delta$ is a volume, so that the terms in brackets each represent the number of particles per volume occurring at a given position. That is, these terms are measures of concentration, C. Thus,

$$\mathcal{J}_x = -\mathcal{D}\frac{C(x+\delta) - C(x)}{\delta}. \tag{6.30}$$

The fraction in this equation is expressed in the form used to define a derivative; in other words, it represents the change in concentration per change in distance along the x axis. As $\delta \to 0$, we may express this concentration gradient[5] as $\partial C/\partial x$, with the final result,

$$\mathcal{J}_x = -\mathcal{D}\frac{\partial C}{\partial x}. \tag{6.31}$$

The flux of particles in the x direction is proportional to the gradient in concentration in the x direction and to the diffusion coefficient. The negative sign tells us that transport proceeds from a position of higher concentration to one of lower concentration.

This differential equation, known as *Fick's first equation of diffusion*, is the basis for much of the analysis of diffusive transport that follows.

Fick's equation as expressed here refers to the flux through a plane. At times it will be more convenient to deal with the flux through a spherical surface. In this

[5]Because the concentration may also vary along the y and z axes, we express this gradient as a partial derivative. In the rare circumstances where C varies only with x, we could use dC/dx.

case, the net movement of molecules is directed radially, either into or out of the sphere, and the flux density, $\mathcal{J}_r$, is

$$\mathcal{J}_r = -\mathcal{D}\frac{\partial C}{\partial r}, \tag{6.32}$$

where r is radial distance.

Note that concentration, as used in these equations, can be expressed in a variety of ways. If one is concerned with the transport of mass, concentration expressed as mass per volume is appropriate, and flux density has the units of $\text{kg m}^{-2}\,\text{s}^{-1}$. Alternatively, if one is concerned primarily with how many particles are transported regardless of their mass, one should express concentration as moles per volume. For ease of computation, it is convenient to express concentration as moles per cubic meter rather than the more traditional moles per liter (1 liter = $10^{-3}\,\text{m}^3$). In this case, $\mathcal{J}$ has units of $\text{mol m}^{-2}\,\text{s}^{-1}$. In this chapter we deal primarily with the diffusive flux of metabolic gases into and out of organisms, and therefore care more about the number of particles than their mass.

6.5 Deriving the Sherwood Number

Before leaving Fick's equation, we use it to derive the Sherwood number. Recall that Sh is the ratio of the rate at which mass is transported by convection to that at which mass is transported by diffusion. Consider these transports in the following simple situation (fig. 6.8). A planar surface lies next to a gradient in concentration such that the concentration rises from 0 at the surface to C_∞ over a distance ℓ_c perpendicular to the surface. A substance dissolved in the fluid can reach the surface either by moving diffusively down the concentration gradient or by moving convectively toward the surface.

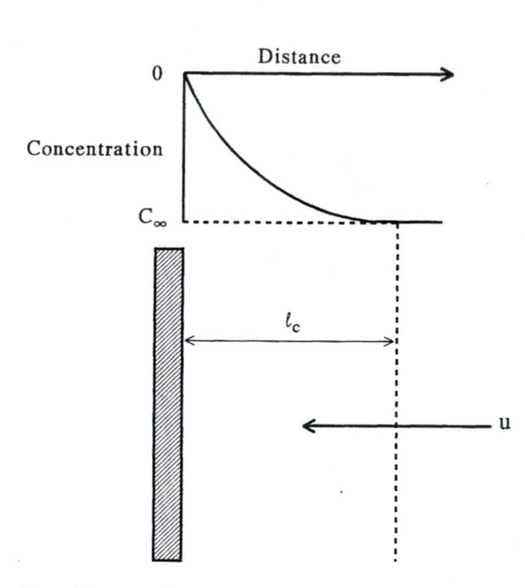

Fig. 6.8 The situation shown here is used in an informal derivation of the Sherwood number.

Consider convection first. The amount of dissolved substance delivered to the surface per time is equal to the product of moles per volume in the fluid (concentration) and the volume per time delivered to the surface. Now, volume is equal to the product of area and length, so volume per time is equivalent to area times length per time, or area times velocity. Thus the rate at which a substance is delivered by convection is

$$\frac{\text{moles}}{\text{time}} = C_\infty A u, \tag{6.33}$$

where u is the convective velocity of the fluid toward the surface. The rate of delivery per area (a flux density) is thus

$$\text{convective flux density} = C_\infty u. \tag{6.34}$$

Now, the rate at which moles are delivered by diffusion (per area) is simply the diffusive flux density as expressed by Fick's equation. In this particular case,

$$\text{diffusive flux density} = \mathcal{D}\frac{C_\infty}{\ell_c}, \tag{6.35}$$

where the gradient in concentration has been expressed as the total change in concentration (C_∞) over the entire thickness of the gradient ℓ_c.

Dividing convective flux density by that due to diffusion, we see that

$$Sh = \frac{u\ell_c}{\mathcal{D}}, \tag{6.36}$$

as proposed in eq. 6.25. Note, however, that ℓ_c here is a measure of the thickness of the concentration gradient, whereas earlier we defined it as a characteristic length of an object. As we will see in chapter 7, the thickness of a concentration gradient is often a function of the size of an organism, and therefore this subtle switch in nomenclature does not present a problem. It does, however, mean that unless ℓ_c is chosen specifically to be the thickness of the concentration gradient (i.e., the *boundary layer* as discussed in chapter 7), Sh is not *equal* to the ratio of transport by convection to that by diffusion, but rather is *proportional* to the ratio. As a result, the Sherwood number is best thought of as an index (rather than a direct measure) of the ratio of convective to diffusive transport.

6.6 Other Forms of Diffusion

In this chapter we deal with the transport of molecules as they move diffusively due to thermal agitation. However, there is nothing in our analysis so far that is intrinsically tied to molecules. Anything that is transported as particles undergo a random walk can be described by the concepts presented here and modeled as a diffusive process. For example, in chapter 7 we use an analogue of Fick's equation to describe the way in which momentum is transported, and in chapter 8 we use the concept of diffusion to quantify the transport of heat. The notion that "things" are carried down a gradient through the action of random motion is one of the grand, unifying concepts of physics, and there are a number of different diffusion coefficients, each tailored to what it is that is diffusing. To avoid confusion when dealing with the diffusion of molecules we refer to the *molecular diffusion coefficient*, $\mathcal{D}_m$.

6.7 Diffusion Velocity vs. the Speed of Locomotion

We are now in a position to explore the biological consequences of diffusion, and we begin by comparing the time it takes a mobile organism to move a certain distance with the time required for a molecule to diffusively move that same distance. Consider first an organism that releases a chemical signal into its environment. For instance, the carbon dioxide excreted by an aerobic bacterium might act as a "come hither" signal to nearby bacteria-eating flagellates. Which is likely to arrive at the predator first, the chemical signal or the bacterium? Let us suppose that the bacterium is 10 μm away from the flagellate and swims toward it. A bacterium with a body length of 2 μm moves at a speed of about 20 μm s^{-1} (Berg 1983). Thus it takes the bacterium half a second to swim to its captor. In contrast, CO_2 molecules travel 10 μm in only about 0.03 s. In other words, at the small scale of this predator-prey interaction, the chemical signal announcing the presence of prey is diffusively transported more than ten times faster than the prey itself.

The same logic applies to the delivery of food or oxygen. On the large scale at which terrestrial mammals live, one is often conscious of depleting the local resources and having to move elsewhere. For instance, a cow must move to find food as it consumes what is available locally. At the small size of a bacterium, however, diffusion replaces the local supply faster than the organism can move, as if the grass grew faster than the cow could eat it. As Berg (1983) points out, the only reason for a bacterium to move is if it finds itself in a particularly poor pasture.

Consider an example. The "diffusion velocity" (eq. 6.22) of an oxygen molecule in water is equal to the swimming speed of a bacterium (20 μm s^{-1}) when the period under consideration is about 10 s. At shorter times, diffusive transport is faster than swimming; at longer times swimming is faster than diffusion. In 10 s a bacterium can swim 200 μm. Thus, if the local concentration of oxygen is favorable at a distance greater than 200 μm away, it is advantageous for the bacterium to swim there rather than wait for delivery by diffusion. On the other hand, if the area of favorable oxygen concentration is only 100 μm away, the bacterium is better off staying put.

In air, these arguments apply over larger scales. For example, in the last chapter we proposed that bacteria might be capable of swimming in air. The argument just made, however, makes it unlikely that locomotion would ever be of any substantial advantage in terms of aquiring food or oxygen because the swimming speed of an aerial bacterium (which we calculated to be about 1 mm s^{-1}) exceeds the diffusion velocity only if the time considered is greater than about 40 s. Thus, only if the bacterium needed to move to a greener pasture 4 cm away (a huge distance relative to the size of the organism) would it be better to swim to the pasture rather than to wait for the pasture to come to it. Oxygen would be rapidly delivered and carbon dioxide effectively removed without the bacterium ever moving a flagellum.

6.8 Diffusion and Metabolism

Having seen that diffusion can effectively deliver oxygen to small organisms, we now explore the limits of the process: at what rate can an organism metabolize if its only means of receiving gaseous fuel is through diffusion? For instance, how fast can an animal or plant respire if it must rely on diffusion for the delivery of oxygen? Or, how fast can a plant photosynthesize if carbon dioxide is transported by diffusion alone?

We first carefully define the situation. To keep matters simple, we assume that our organism is a sphere of radius r immersed motionless in a stationary fluid where, in the absence of the organism, the gas in question is present at concentration C_∞ (expressed as mol m^{-3}). We assume that at a great distance from the sphere, this concentration is maintained even when the organism consumes gas. Furthermore, we may assume that any gas molecule that contacts the surface of the sphere is immediately bound and made available to metabolic processes inside the cell; as a result, the concentration at the surface is maintained at 0. This is a feat beyond the capability of any real organism, but it allows us to calculate the maximum rate at which O_2 or CO_2 can be delivered in theory if not in practice.

Note that the approach outlined here treats the inside of the organism as a "black box" that consumes oxygen or carbon dioxide at a rate determined solely by the rate at which gas is delivered to the periphery of the sphere. The specifics of internal transport and the mechanism of metabolism are left intentionally vague.

We now proceed to solve Fick's equation to calculate the flux of gas to the surface. To do so we need to know the diffusion coefficient and the concentration gradient. For the diffusion coefficients we rely on the empirical values cited in tables 6.1 and 6.2. Specifying the concentration gradient is less straightforward, but we can employ one of the classical methods of solving such problems: we look it up. Books such as Berg (1983), Crank (1975), and Carslaw and Jaeger (1959) are filled with the analytical solutions to a variety of diffusion problems, of which this is one.

Under these circumstances, the concentration at a distance r' from the center of the sphere is (Berg 1983)

$$C(r) = C_\infty\left(1 - \frac{r}{r'}\right), \quad r' > r. \tag{6.37}$$

Taking the derivative of this expression with respect to r', we find that the concentration gradient is

$$\frac{dC}{dr'} = C_\infty \frac{r}{r'^2}, \tag{6.38}$$

leading to the conclusion that the flux density of gas at the sphere's surface (radius $= r$) is

$$\mathcal{J}_r(r) = -\frac{\mathcal{D}_m C_\infty}{r}. \tag{6.39}$$

Now, this flux density is the number of moles of gas passing through unit area of the surface per second. To calculate the total rate at which moles enter the sphere we need to multiply by the sphere's surface area, $4\pi r^2$. Thus, the inward flux of gas, J, is

$$J = -4\pi \mathcal{D}_m C_\infty r, \tag{6.40}$$

where the negative sign denotes that gas moves into the sphere.

This influx must supply the metabolism of the sphere, which has a volume of $4\pi r^3/3$. As a result, the maximum metabolic rate, M, i.e., the moles of gas supplied by diffusion per body volume per time, is

$$M = \frac{3\mathcal{D}_m C_\infty}{r^2}, \tag{6.41}$$

a relationship shown in figure 6.9. This value is proportional to the maximal rate at which metabolism can be sustained. The larger the sphere, the lower its metabolic rate must be if gas is provided by diffusion alone. Because the diffusion coefficient of both oxygen and carbon dioxide in air are 10,000 times that in water, a sphere

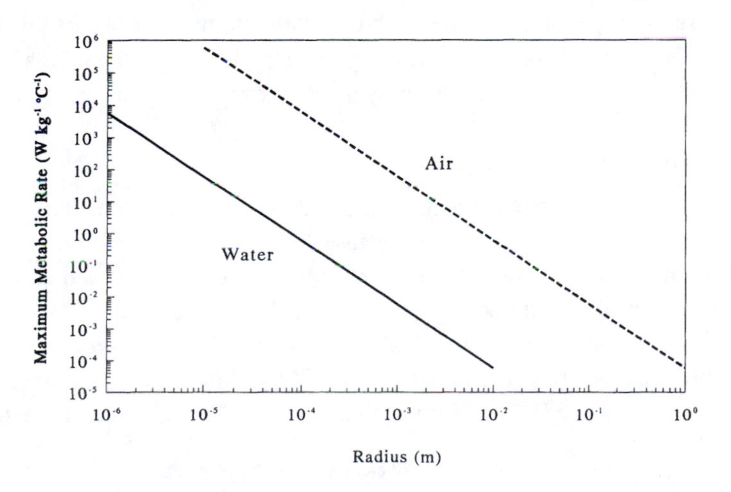

Fig. 6.9 As the size of a spherical organism increases, its metabolic rate must decrease if oxygen is delivered by diffusion alone. This constraint is much more severe in water.

in air can be $\sqrt{10,000} = 100$ times larger than a sphere in water and maintain the same rate of respiration or photosynthesis.

This assumes, of course, that the ambient concentration of gas, C_∞, is the same in both environments. Therein lies an interesting story, to be explored in the following section.

6.8.1 *Metabolism in Air*

Oxygen forms 20.95% of the volume of air. Thus, at standard temperature and pressure (0° C, one atmosphere), 1 m^3 of air contains 0.2095 m^3 of O_2. Now, at this temperature and pressure 1 mole of gas occupies 0.02241 m^3, so the concentration of O_2 in air is 9.35 mol m^{-3} (table 6.3). This concentration decreases with increasing temperature to 8.15 mol m^{-3} at 40° C. In contrast, CO_2 forms only 0.033% of the volume of air, and its concentration varies from 0.0147 mol m^{-3} at 0° C to 0.0128 mol m^{-3} at 40° C at one normal atmosphere. In other words, C_∞ is about 635 times higher for oxygen than for carbon dioxide, and therefore the rate of delivery of oxygen by diffusion in air is 635 times that of CO_2.

Under what circumstances do these rates limit metabolism? Consider first aerobic respiration in which oxygen is consumed. Hughes and Wimpenny (1969) report that typical protozoa consume O_2 at a rate of about 0.1 moles per cubic meter of cell per second, but that some bacteria (*Azotobacter vinelandi*, for instance) may consume O_2 at up to 20 moles m^{-3} s^{-1}. We can see what consequences these metabolic needs entail by rearranging eq. 6.41

$$r_{max} = \sqrt{\frac{3C_\infty \mathcal{D}_m}{M}}, \tag{6.42}$$

where r_{max} is the maximum radius to which the organism can grow given that it has a metabolic rate M.

Even at the tremendous metabolic rate of a bacterium, (20 mol m^{-3} s^{-1}) a spherical organism in air can be adequately supplied by oxygen if it has a radius less than about 5.3 mm. At the metabolic rate of the typical protozoan (0.1 mol m^{-3} s^{-1}), the cell could be almost 7.5 cm in radius before it begins to be deprived of oxygen, well into the range where convection would normally begin to augment transport. We can conclude, therefore, that diffusive transport of oxygen in air is unlikely to limit the metabolic rate of small organisms.

A typical single-celled plant consumes about 1 mole of CO_2 per cubic meter of organism per second,[6] at which rate the plant must have a radius less than about 0.76 mm if carbon dioxide is transported by diffusion alone. This is 7- to 100–fold smaller than the size limit set by the delivery of oxygen.

Table 6.3 The concentration of oxygen and carbon dioxide in air.

T (°C)	Gas Concentration (mol m^{-3})	
	Oxygen	Carbon Dioxide
0	9.353	0.0147
10	9.022	0.0142
20	8.714	0.0137
30	8.427	0.0133
40	8.154	0.0128

6.8.2 *Metabolism in Water*

The rates of metabolism in water are affected not only by the diffusion constant, but also by the solubility of the gases involved, and carbon dioxide is very much more soluble in water than is oxygen (table 6.4). If oxygen and carbon dioxide were present at the same pressure in air, 28 times as much CO_2 would dissolve in a given volume of water at 20° C. For water in contact with air (where the partial pressure of O_2 is 635 times that of CO_2), this means that the concentration of O_2 is only $635/28 \approx 23$ times that of CO_2. Thus, in water the disparity between

[6]This value is based on the rate of carbon uptake by phytoplankton.

the rates at which oxygen and carbon dioxide are delivered is much smaller than that in air.

In terms of the absolute amount of gas present, there is still another contrast. The molar concentration of carbon dioxide in saturated water is similar to that in air. Thus, it is only the decreased diffusion coefficient (rather than a decrease in concentration as well) that reduces the rate of transport of CO_2 in water. The concentration of oxygen in water is only about 5% of that in air, and the transport of O_2 in water is thus slower than that in air due both to a decreased diffusion coefficient and to a drastically decreased concentration.

Table 6.4 Concentration of molecules in water in equilibrium with air at 1 atmosphere and at a pH of 8.0.

T °C	Concentration (mol m^{-3})					
	O_2		CO_2		HCO_3^-	
	Fresh	Sea	Fresh	Sea	Fresh	Sea
0	0.457	0.359	0.0233	0.0189	1.804	1.462
10	0.352	0.282	0.0161	0.0132	1.674	1.369
20	0.284	0.231	0.0117	0.0097	1.545	1.279
30	0.236	0.194	0.0089	0.0075	1.425	1.198
40	0.200	0.168	0.0071	0.0061	1.319	1.129

Source: Data calculated from Weiss (1974) and Mehrbach et al. (1973).

When we apply these facts to eq. 6.42, we find that a rapidly respiring bacterium can have a radius no larger than about 9 μm before it becomes limited by the rate at which oxygen can be supplied. A rapidly photosynthesizing phytoplankter that uses CO_2 as its only carbon source can have a radius no larger than about 7 μm before it is limited by the diffusive delivery of CO_2. Thus organisms that utilize O_2 and CO_2 in water must be much smaller than those in air, and the disparity in size between plants and animals is virtually erased.

This result is somewhat unrealistic, however. When carbon dioxide dissolves in water, much of it chemically combines with its surroundings to produce carbonic acid (H_2CO_3), which then dissociates into a hydrogen ion and a bicarbonate ion (HCO_3^-). Now many, if not most, aquatic plants have evolved the capability of using bicarbonate as a source of carbon for photosynthesis. The accrued advantage becomes apparent when one realizes that at the typical pH of seawater (8.0) the bicarbonate concentration is one hundred to two hundred times that of CO_2 (table 6.4). Thus, at 20° C, a marine phytoplankter can have a radius of about 80 μm before the diffusive delivery of bicarbonate limits its metabolism. Although this is ten times the size of an aquatic cell limited by the delivery of dissolved carbon dioxide, it is still an order of magnitude smaller than single-celled plants can grow in air.

These calculations should be taken with a large grain of salt. First, we have assumed that the concentrations of O_2, CO_2, and HCO_3 go to 0 at the surface of the organism. This is tantamount to assuming that the organism would not leak any oxygen, carbon dioxide, or bicarbonate into the medium if the ambient partial pressure of these species were lowered to zero. This, clearly, is a rash assumption. For real organisms there is some finite lower limit to the concentration of oxygen, carbon dioxide, or bicarbonate that they can maintain in their vicinity. As a result, real concentration gradients will be lower than those used here, and the rate of diffusive delivery will likewise be lower. In other words, the calculations made here probably overestimate maximum size.

Similarly, we have neglected the role of diffusion within the organism itself. This may be particularly important in air, where delivery of O_2 or CO_2 to the cell's surface is unlikely to pose a real limit to size. Instead, the internal transport of these gases may be the real limiting factor: if the organism is too large, gas may not be delivered to the interior at sufficient rate. Any calculation of this limitation, however, involves the complicated topic of diffusion in living tissue, and lies well outside the purview of this text. The interested reader may wish to consult Weis-Fogh (1964), Alexander (1966), or Dejours (1975) on this subject.

One should also note that the concentration of bicarbonate in water is quite sensitive to pH—the lower the pH, the lower the concentration (fig. 6.10). Thus, at low pH plants may be limited to smaller sizes, a fact that may have some importance given the pervasive reduction in the pH of lakes and rivers due to acid rain.

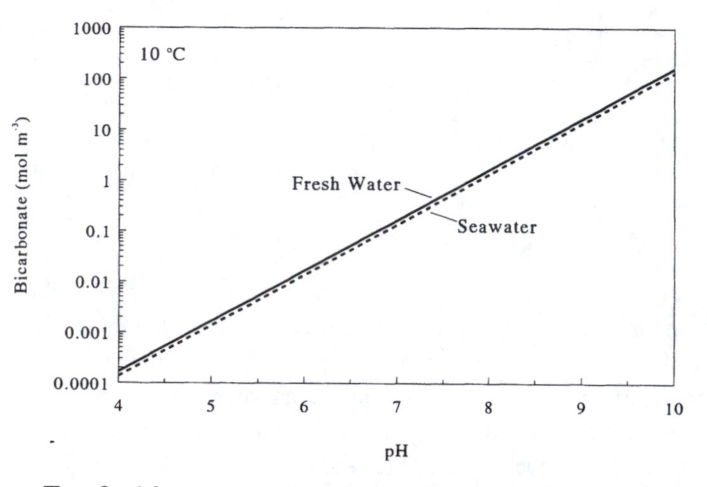

Fig. 6.10 The concentration of bicarbonate ion increases with increasing pH.

Two final facts are worthy of note. Carbon dioxide and oxygen are both slightly less soluble in seawater than in fresh water. Thus, diffusive limits to size may be slightly more stringent in the sea. Second, although the dissolved concentration of both gases decreases with increased temperature, the concomitant increase in the diffusion coefficient tends to offset any effect on size. In fact, the diffusive delivery of oxygen should be slightly higher at 40° C than at 0° C.

6.9 Following the Scent: Flux from a Sphere

In the last section, we dealt with the pattern of concentration around an absorbing sphere. It is interesting to turn the problem around and calculate the concentration gradient around a sphere, when it is the sphere itself that produces the molecules in question. This particular situation arises in a variety of situations. For example, the sperm of some marine invertebrates are chemotactic; they turn to move up a concentration gradient of some (as yet unknown) substance that leads them to the egg (e.g., Miller 1982, 1985a,b). The presence of a concentration gradient is thus a signal that may augment the probability of fertilization. The mobile spores of some kelps are attracted toward increasing concentrations of nutrients such as nitrate and ammonium (Amsler and Neushul 1989). Bacteria (e.g., *Escherichia coli*, the common gut bacterium) also move up concentration gradients, in this case as a means of moving to greener pastures (Berg 1983). In another example, copepods and other small zooplankters can apparently sense the "smell" of phytoplankton when the plant cells are still some distance away (Koehl and Strickler 1981). Given the small size of these phytoplankton cells, the concentration of the odor molecules around them may be set by diffusion. In all these cases it is the pattern of concentration that is biologically important. What is this pattern like?

To analyze this situation, we again rely on a published solution to the radial form of Fick's equation. Carslaw and Jaeger (1959) show that if a spherical organism exudes a substance at a continuous rate of J_{prey} moles per second, the concentration in the surrounding medium is

$$C(r) = \frac{J_{prey}}{4\pi \mathcal{D}_m r}, \tag{6.43}$$

where r is distance from the center of the source (fig. 6.11).

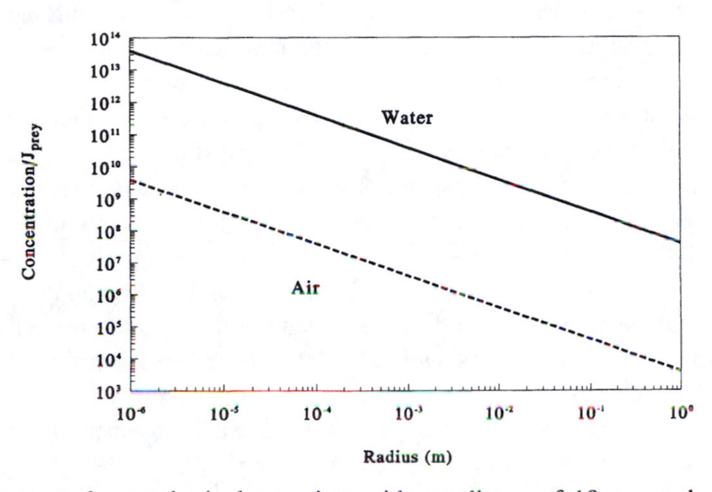

Fig. 6.11 The concentration of "scent" (normalized to the scent flux from the prey) decreases rapidly with radial distance from the prey.

Consider, for example, a spherical organism with a radius r of $10\,\mu$m, and therefore with a volume of 4.19×10^{-15} m^3. Let us assume that the organism produces its "scent" at the rate of 1 mole m^{-3} s^{-1}, a high rate, roughly equivalent to that at which oxygen is consumed. Multiplying this rate by volume we see that J_{prey} is 4.19×10^{-15} moles per second. If the diffusion coefficient of the scent is typical of small molecules in air (10^{-5} m^2 s^{-1}), the concentration of scent at the surface of the organism is only 3×10^{-6} moles per cubic meter. Expressed in more conventional terms, this is 3×10^{-9} moles per liter, i.e., a 0.003 micromolar concentration. In other words, even if the rate of production of a substance is very high, diffusion in air is so effective at this small scale that the concentration of scent in the vicinity of the organism is always low.

In contrast, if the scent has a diffusion coefficient typical of small molecules in water (10^{-9} m^2 s^{-1}), the concentration at the organism's surface is 3×10^{-2} moles m^{-3}, or 30 micromolar, a much more respectable concentration. Thus the lower diffusion coefficient in water results in a very substantial increase in local concentration.

What do these results mean in terms of locating an organism from its scent? We view the situation from the perspective of an organism (the *hunter*) attempting to locate and "capture" the sphere (the *prey*). We examine the question in two steps.

6.9.1 *Detection*

The first step in locating the prey involves the initial detection of scent. There are two ways in which this process can be viewed. First, we might assume that before detection is possible, scent must be present in some minimum concentration. But we know that the concentration of scent for a given J_{prey} is inversely proportional to the product of $\mathcal{D}_m$ and r (eq. 6.43). In other words, because diffusivities in air are 10,000 times those in water, an aerial hunter must be 10,000 times closer to its prey before reaching a detectable concentration of scent than does a hunter in water. For example, if scent is produced at the high rate described above and can be sensed at a concentration of 0.001 μM, an aerial hunter must approach within 33 μm of the center of the prey (a mere 23 μm from its surface for the size prey used here) before detecting the organism. In water, the same level of scent is reached at a distance of 33 *centimeters* from the prey!

This conclusion may be misleading, however. Let us briefly consider the physics by which a cell actually detects molecules in the surrounding medium. In order for a cell to detect a scent, it must be capable of physically interacting with molecules of that chemical at points on the cell's surface. In general, this interaction involves the binding of the scent molecule to a particular receptor embedded in the cell membrane. Changes in the receptor's configuration then signal to the cell that a scent molecule has been detected, in other words, binding turns the receptor "on." To function as a practical detector, there must also be a mechanism to turn the receptors off. If, for instance, scent molecules bind permanently to receptors, a single, brief encounter with a cloud of scent permanently turns the receptors on and the cell cannot sense when the scent is no longer present. As a result, practical receptors bind signal molecules reversibly, and the fraction of a cell's receptors that are on at any time (i.e., the strength of the signal to the cell) is governed by the rate at which signal molecules are delivered to the cell membrane rather than by their concentration. As before, we assume that delivery is via diffusion.

One can model each receptor as a small, reactive patch in a surface that otherwise does not interact with scent molecules. For simplicity, we assume that after a scent molecule is bound by a receptor it is immediately transferred into the cell or otherwise removed from the medium, and the receptor may then bind another molecule. Berg (1983) has shown that under these conditions the rate at which molecules are diffusively delivered to a circular patch is

$$J_{patch} = 4\pi \mathcal{D}_m r_p C_\infty, \tag{6.44}$$

where r_p is the radius of the patch and C_∞ is the concentration of signal molecules in the adjacent fluid.

This leads to a curious conclusion regarding the ability of hunters to detect the diffused signal from prey. In eq. 6.43 we showed that the concentration of scent at distance r from the prey is $J_{prey}/(4\pi\mathcal{D}_m r)$. Using this value for C_∞ in eq. 6.44,[7] we see that the rate of delivery of scent molecules to a receptor is

$$J_{patch} = \frac{r_p J_{prey}}{r}, \tag{6.45}$$

an expression that is independent of the diffusion coefficient! In other words, at a given distance from a prey cell in air, the concentration of scent is small, but because the diffusion coefficient is high, the few molecules present are effectively delivered to receptors. In water, the local scent concentration is high, but because the diffusion coefficient is small, scent molecules are not effectively delivered. The two trends just offset each other. Thus, if it is the *rate of delivery* of scent

[7] C_∞ is not constant in this case; it varies with r. However, at the small scale of a receptor (about 1×10^{-9} m), the local gradient in C_∞ is small compared to the gradient induced by the receptor's absorption of molecules.

molecules (rather than *concentration*) that governs detection, aerial hunters should be just as adept at detecting prey as are aquatic predators. Furthermore, within a single medium, the detectability of a scent should be independent of the scent's diffusion coefficient. For instance, large molecules, which have small diffusion coefficients, will be detected as readily as small molecules.

It remains to be seen whether it is concentration or rate of delivery that governs detectability. I know of no empirical test of these ideas.

In the calculations made here, we have assumed implicitly that receptor patches were scattered sparsely over the cell membrane. As the spatial density of patches increases, the pattern of concentration around each individual patch begins to interact with that of its neighbors. As a result, the flux of scent molecules reaching each patch decreases. This effect is negligible, however, unless patches are quite tightly packed. An interesting discussion of this effect is given in Berg (1983).

6.9.2 *Following the Gradient*

Detection of the scent is only the first step in the process of capture. To effect consummation, a hunter must take a second step and move toward the prey. This may not be easy, however. Consider the strategy that you or I might employ in moving toward a source of scent. We would move in one direction, sniffing as we went, until we sensed a change in the concentration of scent. If we perceived the concentration to be increasing, we would continue moving in the same direction. If the concentration were decreasing, we would change direction and try again. How feasible is it for a microorganism to use this strategy? To evaluate this question, we explore the properties of the concentration gradient.

Taking the derivative of eq. 6.43 with respect to r, we see that radial gradient in concentration is

$$\frac{dC}{dr} = -\frac{J_{prey}}{4\pi\mathcal{D}_m r^2}. \qquad (6.46)$$

The farther the hunter is from its prey, the more gradual the change in concentration.

In many cases the concentration gradient is so gentle that it would be well nigh impossible for a sperm or bacterium to detect the minuscule change in concentration along its body. For example, if dC/dr is 0.33 moles m^{-4} (as would be the case at a distance of 1 mm from the aquatic cell used in the example above), the change in concentration along the 2 μm length of a bacterium is only 0.67 micromoles per cubic meter. This is a difference of only 0.00067 μM between the head and tail of the bacterium, a difference that may well be difficult to detect.

The problem inherent in detecting this small spatial change in concentration can perhaps be circumvented if the hunter can detect a change in concentration as it swims. In other words, if the hunter can remember what the concentration was at some time in the past and can compare this remembered value to that of the present, it can substitute time for distance to detect a spatial gradient. To see this, we note that $dr = u_r dt$, where u_r is the radial component of the hunter's velocity and t is time. When we substitute this expression into eq. 6.46, we see that

$$\frac{dC}{dt} = -\frac{u_r J_{prey}}{4\pi\mathcal{D}_m r^2}. \qquad (6.47)$$

The faster the organism swims, the faster concentration changes with time and the easier the gradient is to detect.

This strategy is effective, however, only at high speeds. For example, if the bacterium moves at a maximal speed of $20\,\mu\text{m}\,\text{s}^{-1}$, dC/dt in the example used above is only $1.34 \times 10^{-8}\,\mu\text{M}$ per second. In other words, for the speeds with which microorganisms move, the change in concentration with time at one point on the body is small compared to the change in concentration along the body at one instant in time. Thus, memory is of little use to an aquatic microorganism in orienting to a gradient.

The situation is even worse in air, where the the concentration gradient (with respect either to time or distance) is 10,000 times smaller than in water. As a consequence, even if a hunter can detect its prey in air, it will be considerably more difficult for it to orient toward it. Only if the aerial hunter can move through the fluid 10,000 times faster than an aquatic hunter will it receive the same temporal stimulus. For instance, an aerial bacterium would have to move at $20\,\text{cm}\,\text{s}^{-1}$ to have the same sensitivity to a concentration gradient as its aquatic cousin, a speed well outside the realm of possibility. Thus, even if bacteria can indeed "swim" in air, it seems unlikely that they can home toward greener pastures by orienting to a chemical gradient.

Microorganisms have evolved at least two strategies to circumvent the problems inherent in detecting a gradient in scent. In one, used by bacteria such as *E. coli*, the chemotactic organism performs a random walk that is biased not by the chemical gradient, but rather by the instantaneous concentration (Berg and Brown 1972). In essence, the hunter takes long steps when at high concentration and short steps at low concentration. The net result is movement toward the source of scent. In the second method, the organism swims in a helical pattern, which requires it to rotate its body simultaneously about two axes. If the organism responds to instantaneous concentration by an appropriate change in the relative rates of these rotations, it moves up the gradient (Crenshaw 1990).

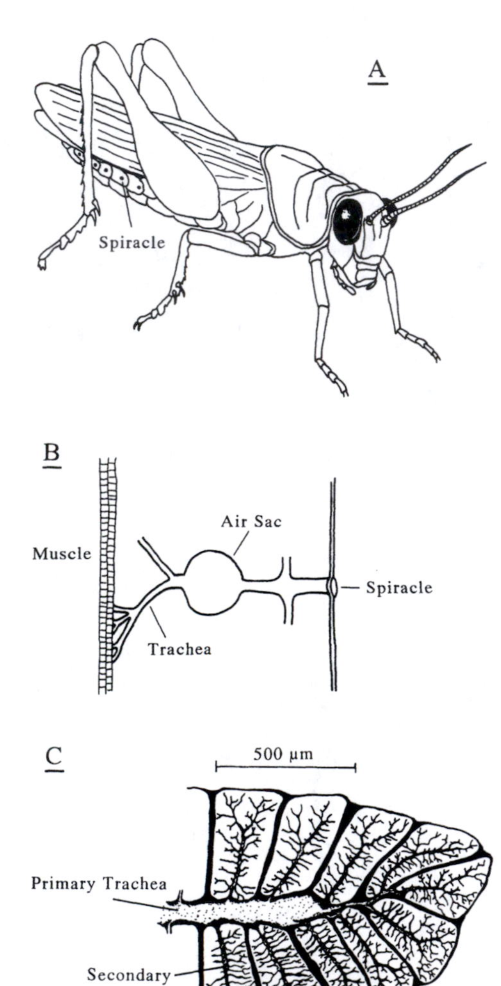

Fig. 6.12 Spiracles (located on the sides of the abdomen) are the external openings to the locust's tracheae. Internally, the tracheae go through various bifurcations (B, after Meglitsch 1972) before they arrive at a muscle. Within the muscle, the tracheae undergo further ramification (C, redrawn from Weis-Fogh, 1964, courtesy of the Company of Biologists, Ltd.).

6.10 Diffusion in Tubes

To this point we have obeyed the dictum of the Sherwood number and have confined our exploration to small, slow-moving organisms. There are, however, a few special cases of large organisms in which pure diffusion is of overriding importance. We begin with an exploration of gas exchange in insects.

6.10.1 *Insect Tracheae*

In chapter 5 we explored the consequences of viscosity for the design of respiratory systems, and showed that it was feasible for insects to pump air through the larger of the pipes in their tracheal tree. We noted at that time, however, that the small pipes that actually deliver air to cells in an insect's body are blind-ended and therefore cannot be actively ventilated. These small pipes—the secondary and tertiary tracheae and tracheoles—may be only about $0.2\,\mu\text{m}$ in diameter at their finest and up to about 1 mm long (fig. 6.12). In the case of flight muscle, these tiny, unventilated tubes must be able to supply oxygen to the surrounding cells at a prodigious rate. For example, during flight the muscles of some insects consume oxygen at the rate of 6.5 moles per m^3 of tissue per second. In other words, each cubic meter of muscle consumes the oxygen contained in a cubic meter of air every 1.3 seconds. Can diffusion actually deliver oxygen at this rate? Could such a system work in water?

To answer these questions we consider a simplified model of tracheae and tracheoles and their surrounding muscle tissue (fig. 6.13). The muscle is modeled as a block of cross-sectional area A and length ℓ, and the tracheae and tracheoles as a single hollow tube with a cross-sectional area A_t. The ratio A_t/A we call ζ; a typical value is 0.1 (Weis-Fogh 1964). We use the variable x to measure length along the tracheolar tubes; x is 0 at the opening of the tracheae to the air and $x = \ell$ at their distal, blind ends.

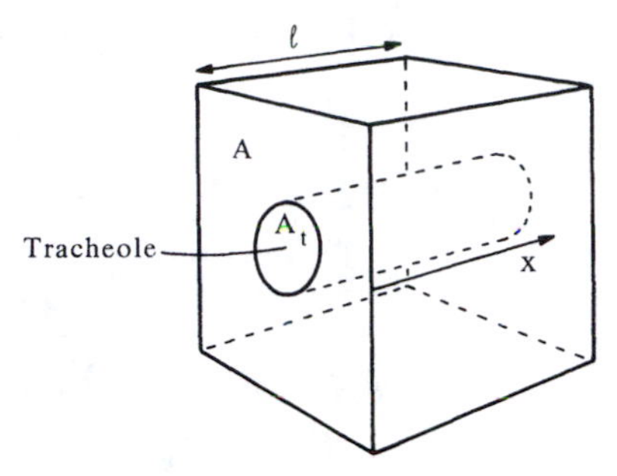

Fig. 6.13 Each tertiary tracheole supplies oxygen to a portion of muscle. The dimensions of the schematic tracheole-muscle system shown here allow us to calculate the maximal length of tracheae.

We assume that the muscle fiber consumes oxygen at a rate of M mol m^{-3} s^{-1}, where for simplicity volume is taken to include both muscle and tracheae. This allows us to calculate the rate at which oxygen must flow through the tracheae at any x. For example, the volume of muscle lying distal to a point x is $A(\ell - x)$, and this volume consumes oxygen at a rate $A(\ell - x)M$. All this oxygen must flow diffusively through the tracheae at x.

From Fick's equation, we know that the flux density of oxygen along the tracheae (moles per combined tracheal area per second) is proportional to the diffusion coefficient and the gradient of concentration. If we set the rate at which O_2 is consumed by the muscle equal to the rate at which it is supplied by the tracheae we see that

$$A(\ell - x)M = -\mathcal{D}_m A_t \frac{dC}{dx}. \tag{6.48}$$

We can rearrange this equation to separate dC from dx with the result,

$$\frac{(\ell - x)M}{\zeta \mathcal{D}_m} dx = -dC. \tag{6.49}$$

Each side of this equation may now be integrated between 0 and ℓ, to give the final answer:

$$\Delta C = \frac{M\ell^2}{2\mathcal{D}_m \zeta}, \tag{6.50}$$

where ΔC is the difference in oxygen concentration between the opening of the tracheole and its blind end.

We may rearrange this equation to tell us how long tracheae may be for a given ΔC:

$$\ell_{max} = \sqrt{\frac{2\Delta C \mathcal{D}_m \zeta}{M}}. \tag{6.51}$$

Now, Weis-Fogh (1964) reports that insect flight muscle can remove at most 25% of the oxygen present in air. Thus, at 20° C the maximal ΔC is about 2.3 moles m^{-3}. Using typical values for the other parameters in the equation (ζ = 0.1 and M = 6.5 moles m^{-3} s^{-1}), we find that the maximal length of tracheae is about 1.2 mm. If insects are to have a longer path from air to muscle, diffusion alone will not suffice. It is for this reason that large insects need actively to ventilate their primary tracheae.

The same type of diffusion-based respiratory system clearly cannot work in water. Because the diffusion constant of O_2 in water is 10,000 times smaller than that in air, tracheole length would need to be less than about 12 μm to supply oxygen to tissue respiring at the same rate as that in insect flight muscle. If aquatic organisms are to rely on diffusion for delivery of oxygen, they must either drastically increase the fraction of the tissue given over to respiratory tubes (ζ), maintain a drastically reduced metabolic rate, or minimize the distance between muscle and oxygenated water.

Increasing ζ sufficiently is impractical. Given the small diffusion coefficient of O_2 in water, a muscle would be virtually all tracheae and no fiber if diffusion were to suffice as a means of delivering oxygen.

The metabolic rates of many aquatic invertebrates are considerably lower than those of flying insects (see chapter 8), but generally not 10,000-fold lower. Thus, the lower metabolic rate of aquatic animals is not sufficient to allow a tracheal system to function effectively.

Some aquatic organisms (flatworms, for instance) have indeed opted for the third strategy; their flat shapes keep all parts of the body close to the surrounding fluid. Most aquatic organisms, however, have, in an evolutionary sense, given up on diffusion as the sole source of oxygen transport and have evolved convective respiratory systems.

It is interesting to note that the high diffusivity of gases in air has been an exceptionally potent selective factor. A tracheal respiratory system has evolved at least three separate times in terrestrial arthropods—in the uniramians (e.g., insects, centipedes, and millipedes); in the chelicerates (e.g., scorpions and spiders); and in isopods (e.g., sow bugs)—as well as in the vascular plants. In this respect, the potential for aerial diffusion as a transport system has been a unifying factor among terrestrial organisms.

6.10.2 *Size Limits to Tracheae*

The fact that tracheoles in insects can have diameters as small as 0.2 μm raises an interesting question. We calculated earlier in this chapter that the mean free path of molecules in air, l, is about 0.08 μm. Thus, the size of these small tubes approaches the average distance a molecule travels in air before colliding with another molecule. What would the effect be if tracheoles were even smaller than they are so that their diameters were actually less then the mean free path?

Pickard (1974) addressed this question and showed that any reduction in the size of tracheoles beyond that now existing would reduce the effective diffusion coefficient, and thereby reduce the rate at which oxygen can be delivered to the muscle. To be precise, Pickard proposed that

$$\mathcal{D}_e = \mathcal{D}_m \frac{1}{1 + (9\pi/16)(l/d)}, \tag{6.52}$$

where $\mathcal{D}_e$ is the effective diffusion coefficient in the tracheole, l is the mean free path (about 0.08 μm in air), and d is the diameter of the tracheole. This expression is graphed in figure 6.14. At diameters below about $3l$ (roughly the size of tracheoles), the effective diffusion coefficient decreases rapidly, suggesting that the minimum size of tracheoles may have been set by the mean free path of molecules in air. One wonders about insects at high altitudes where the mean free path is relatively long. Do they have larger tracheoles? To my knowledge, this question has not been explored.

Because the mean free path of molecules in water is much smaller than the diameter of a hydrogen atom, no physical tube can ever restrict the diffusion coefficient in water in the same way that tracheoles potentially restrict it in air.

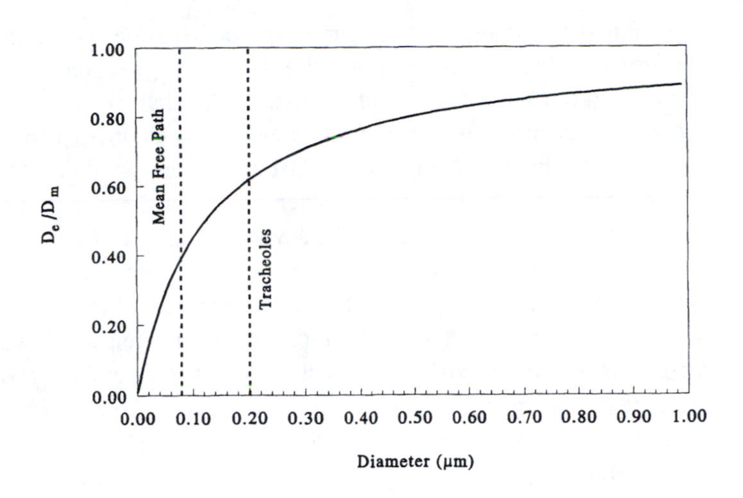

Fig. 6.14 The effective diffusion coefficient ($\mathcal{D}_e$) is decreased in very small tracheoles (eq. 6.52).

6.10.3 *Birds' Eggs*

There is at least one other class of large organisms in which diffusion (separate from convection) is of paramount importance—birds' eggs. A typical egg is shown schematically in figure 6.15. The eggshell is formed of calcium carbonate, providing a rigid protective housing for the developing chick. Although the shell appears solid, it is in fact pierced by thousands of small pores, each forming a small tube that connects the inner and outer surfaces of the shell. The inner surface of the shell is lined by two membranes, and within these the embryonic bird floats in a liquid. Like virtually all animals, the developing bird requires a source of oxygen and a means of dissipating the carbon dioxide formed during metabolism. The proximal site of exchange for these gases is the chorioallantois, the avian equivalent of the placenta in mammals. This well-vascularized organ is closely applied to the inner shell membrane so that gases in the blood can be exchanged with gases in the pores of the shell. However, the ultimate exchange of gases with the atmosphere must occur across the shell via the pores. Thus, gas exchange across the shell of a bird's egg occurs in much the same fashion as gas exchange in insect tracheae, with a variety of consequences.

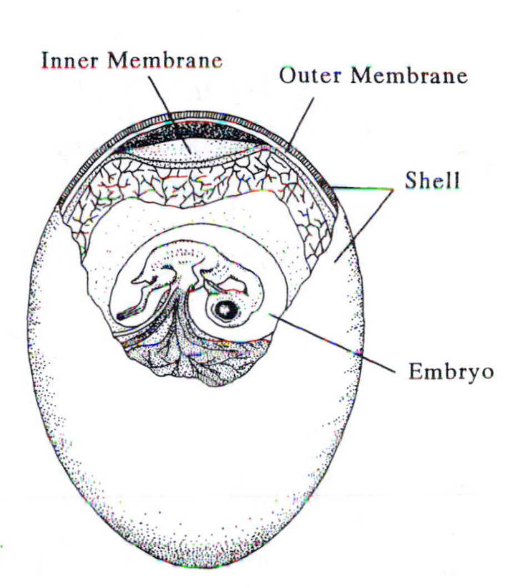

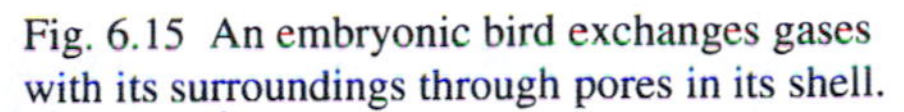
Fig. 6.15 An embryonic bird exchanges gases with its surroundings through pores in its shell.

For example, Rahn and Paganelli (1979) have shown that the embryonic bird can maintain an oxygen concentration of about 6.55 moles m^{-3} at the outer surface of the egg membranes. Recalling that the concentration of oxygen in air is 9.35 moles m^{-3}, we see that there is a concentration gradient of 2.8 moles m^{-3} across the thickness of the shell, ℓ. As a result, the rate at which oxygen can be delivered to the chick is

$$\text{moles } O_2 \text{ per second} = \mathcal{D}_m A \frac{2.8}{\ell}, \tag{6.53}$$

where A is the combined area of the pores. Any increase in thickness of the shell must be accompanied by an increase in the area of pores if oxygen is to be delivered at a constant rate. As a result, there is a trade-off between the metabolic needs of the embryo and the structural strength of the egg. A thicker shell might better protect the developing chick but, without an increase in pore area, would reduce the rate at which O_2 could be delivered. This, along with the fact that the chick must eventually be able to escape from the shell, can help explain why eggs are so fragile.

In terms of the transport of oxygen and carbon dioxide, more pores may be better because any increase in number increases the rate of gas exchange. However, there is a limit to the allowable area of pores set by the rate at which water is lost. The growing chick must make do with the water initially enclosed in the egg. If pore area exceeds some limit, water is lost too fast and the embryo dies from dehydration. We will consider this problem in chapter 14.

The shape of pores is quite variable. With respect to diffusion, a few large pores have the same effect as a large number of small pores; it is only the total pore area that matters. Note, however, that the surface area of the egg increases with the square of the egg's length, while the rate at which oxygen is utilized by the embryo increases (to a first approximation) with the mass of the embryo and therefore with the cube of egg length. As a result, ostrich eggs (which may weigh 1.5 kg) must have more pore area per shell area than do the eggs of hummingbirds, which may weigh less than a gram.

We can calculate the fraction of shell surface area that must be present as pores. First we assume that the developing organism has a metabolic rate M (again expressed in mol O_2 m^{-3} s^{-1}). For simplicity, assume that the egg is spherical, in which case the overall demand for oxygen by the chick is

$$\text{oxygen demand} = \frac{4\pi r^3 M}{3}, \tag{6.54}$$

where r is the egg's radius.

The rate at which oxygen is delivered can be calculated from eq. 6.53,

$$\text{oxygen delivery} = 4\mathcal{D}_m \pi r^2 \psi \frac{\Delta C}{k_s r}, \tag{6.55}$$

where ψ is the fraction of the shell surface area ($4\pi r^2$) that is pores, ΔC is the difference in oxygen concentration between the inner and outer ends of the pores, and k_s is the thickness of the shell expressed as a fraction of r (i.e., $k_s = \ell/r$). Equating eqs. 6.55 and 6.54 and solving for ψ, we see that

$$\psi = \frac{M k_s r^2}{3\Delta C \mathcal{D}_m}. \tag{6.56}$$

In a chicken egg, $k_s \approx 0.02$ and the chick consumes oxygen at a maximal rate of about 0.006 mol m^{-3} s^{-1}. The radius of a sphere equivalent in volume to the egg is about 2.3 cm, and ΔC is 2.8 moles^{-3}, as before. Under these circumstances, pores need form only about 0.04% of the shell surface.

Rahn and Paganelli have shown that ΔC is roughly the same for eggs of all sizes. If we assume that k_s and M are also constant, we can calculate that a hummingbird shell (r = 3 mm) needs only about 0.0006% of its area as pores. An ostrich egg (r = 7 cm) needs about 0.37% of its area as pores. In reality, k_s (the relative thickness of the shell) is smaller for large eggs than small eggs (Rahn and Paganelli 1979), probably in response the necessity of having a shell fragile enough for the chick to be able to escape. As a result, our estimate here of pore fraction for an ostrich is probably an overestimate. Nonetheless, it is obvious that because of the high diffusion coefficient of oxygen in air, only a small fraction of the shell area need be pores for the chick to survive.

This conclusion can be applied to other aerial organisms as well. For example, it can help to explain how insects can respire so actively when only a small fraction of their exoskeleton is pierced by tracheae, and how leaves can actively photosynthesize when only a minor fraction of their surface area is open (via stomata) to the diffusion of carbon dioxide.

Solid shells with pores are not feasible in water. For example, a developing chicken in water would need pores totaling eight times the shell's surface area to receive sufficient influx of oxygen! Given the 10,000-fold lower diffusion coefficient of oxygen in water, aquatic eggs can only survive if they are very much smaller than birds' eggs, have metabolic rates much lower than those of birds, and have shells that are virtually all pore. This indeed seems to be the case. For example, the egg cases of skates and rays (mermaid's purses) may be 10 cm in their largest dimension, but are flattened so that their surface area is large relative to the enclosed volume. Furthermore, the leathery material from which the cases are constructed is permeable to oxygen, and the entire case therefore acts as a "pore." Many other aquatic eggs do not have a solid shell of any sort, and even then may not receive enough oxygen by diffusion alone. For example, fish often must "fan" their egg masses, thereby using convection to augment transport.

6.11 Measuring Diffusion Coefficients

Much of our discussion in this chapter has relied on the empirically measured diffusion coefficients listed in tables 6.1 and 6.2. How were these values determined? Our exploration of diffusion across eggshells provides us with the idea for a practical measurement scheme.

Consider the apparatus shown in figure 6.16. A wall of thickness ℓ separates two chambers and is pierced by N tubes, each of radius r. We now place in one of the chambers the substance whose diffusion coefficient is to be measured and provide a means by which the concentration of this substance is held at a known, constant level, C_1. For example, if we are measuring the diffusivity of water vapor, we could put a wet sponge in the chamber, thereby ensuring that the air remained saturated with water. We then maintain the second chamber at a known, but different, concentration (C_2) and measure the rate J (in moles per second) at which the substance enters the second chamber. In the case of water vapor, we could place a desiccant in the second chamber to maintain C_2 at 0. To then measure the rate at which the diffusing substance moves from one chamber to the other, we record the rate at which the weight of the desiccant increases.

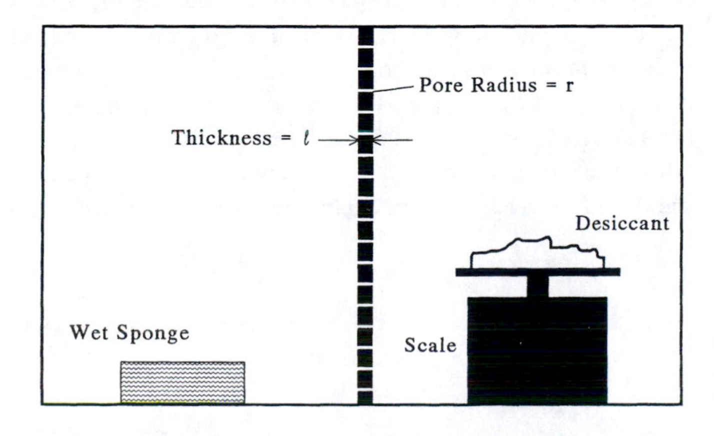

Fig. 6.16 An apparatus such as that shown here can be used to measure the diffusion coefficient of water vapor in air.

From Fick's equation we know that

$$J = -\mathcal{D}_m \pi r^2 N \frac{C_2 - C_1}{\ell}, \tag{6.57}$$

and, rearranging, we see that

$$\mathcal{D}_m = \frac{J\ell}{\pi r^2 N(C_1 - C_2)}. \tag{6.58}$$

Thus, an apparatus very much like a bird's egg provides the means by which diffusion coefficients can be measured.

Rahn and Paganelli (1979) actually used another rearrangement of this equation to measure the area of pores in birds' eggs. They knew the diffusion coefficient of water from previous empirical measurements and could easily measure the thickness of the shell ($= \ell$). By placing each egg in a desiccator and measuring the rate at which it lost mass, they could then solve for the effective pore area.

Many other methods have been devised for measuring diffusion coefficients. To explore these consult Marrero and Mason (1972).

6.12 Summary . . .

The diffusion coefficients of gases are 10,000 times larger in air than in water. As a result, terrestrial organisms that rely on the diffusive delivery of oxygen and carbon dioxide can be larger and can metabolize faster than their aquatic counterparts.

The effective transport of gases by diffusion in air makes it possible for terrestrial animals and plants to respire using a tracheal respiratory system, and this possibility has been realized repeatedly in the course of evolution. The high diffusivity of gases in air also makes it possible for birds' eggs, insects, and leaves to metabolize even though only a small fraction of their surface area is permeable to gases. The mean free path of a molecule in air sets a lower limit to the effective size of pores, and small insect tracheoles approach this limit.

6.13 . . . and a Warning

The process of diffusion can lead to nonintuitive results, and a complex geometry or the presence of convection can have a drastic effect on the rate of diffusive transport. As a consequence, you are cautioned against the blind application of the conclusions reached in this chapter. When faced with the complexities of the real world, you are urged to consult a standard text on diffusion (e.g., Crank 1975).

By no means does this brief discussion exhaust the subject of diffusion and biology. In this chapter we have limited ourselves to cases where the Sherwood number is less than 1. But many important and interesting effects of the disparity in diffusion coefficients between air and water are seen at higher Sherwood numbers, and these form a basis for the explorations of the next two chapters.

Density and Viscosity Together: The Many Guises of Reynolds Number

Much of fluid dynamics involves the tug-of-war between inertia, with its tendency to support continued motion, and viscosity, which tends to bring objects to a halt. As we have seen in chapter 5, the relative contributions of these tendencies in governing patterns of flow can be described conveniently by the Reynolds number, and in this chapter we explore the ways in which Reynolds numbers can be used to highlight relevant differences between air and water. We will see why the high terminal velocities of terrestrial organisms make for a sparse aerial plankton but efficient aerial "filter feeders." We will show how turbulence and a rapid reproductive rate allow diatoms to avoid the abyss. We calculate the maximal speed at which animals can walk in air and water, see how crickets use boundary layers as an aid in discriminating the frequency of sounds, and explore the mechanics of fish olfaction.

7.1 *Re* Revisited

Before embarking on this exploration, let us briefly review the concept of a Reynolds number.

Recall that Re is a dimensionless value proportional to the ratio of inertial and viscous forces. Its mathematical expression involves four parameters: the density of the fluid, ρ_f; the fluid's dynamic viscosity, μ; the relative velocity between object and fluid, u; and a characteristic length, ℓ_c:

$$Re = \frac{\rho_f u \ell_c}{\mu}. \tag{7.1}$$

In this chapter we are primarily concerned with the combined effects of density and viscosity. It is appropriate, then, to use an alternative expression in which ρ_f and μ are combined into the kinematic viscosity, $\nu = \mu/\rho_f$:

$$Re = \frac{u\ell}{\nu}. \tag{7.2}$$

The kinematic viscosity of air is about fifteen times that of water, varying slightly with temperature (table 7.1; fig. 7.1). As a result, the Reynolds number for a given ℓ_c and u is about fifteen times larger in water than in air. Therein hangs the tale of this chapter.

Now, eq. 7.2 is only one particular example of a class of expressions known generically as Reynolds numbers, all having this same general form. The precise manifestation of the Reynolds number applicable to a particular case depends typically on the choice of velocity, characteristic length, and density, and these can be chosen in creative fashions that give the Reynolds number a variety of different "looks." The Re we have dealt with so far involves the length of an organism along the direction of flow, the velocity of the organism relative to the bulk of the surrounding fluid, and the density of the fluid itself. In this chapter we encounter several alternative guises for Re in which length, velocity, and density are chosen so as to scale particular phenomena to certain characteristics of the flow. In each

Table 7.1 The kinematic viscosity of air and water.

T (°C)	Kinematic Viscosity ($m^2\ s^{-1} \times 10^{-6}$)		
	Dry Air	Fresh Water	Seawater (S = 35)
0	13.3	1.79	1.84
10	14.2	1.31	1.35
20	15.1	1.01	1.06
30	16.0	0.80	0.85
40	17.0	0.66	0.70

case, however, it should be remembered that *Re* is an index for the pattern of flow as governed by the ratio of inertial to viscous forces.

We begin our exploration with an examination of the rates at which things fall.

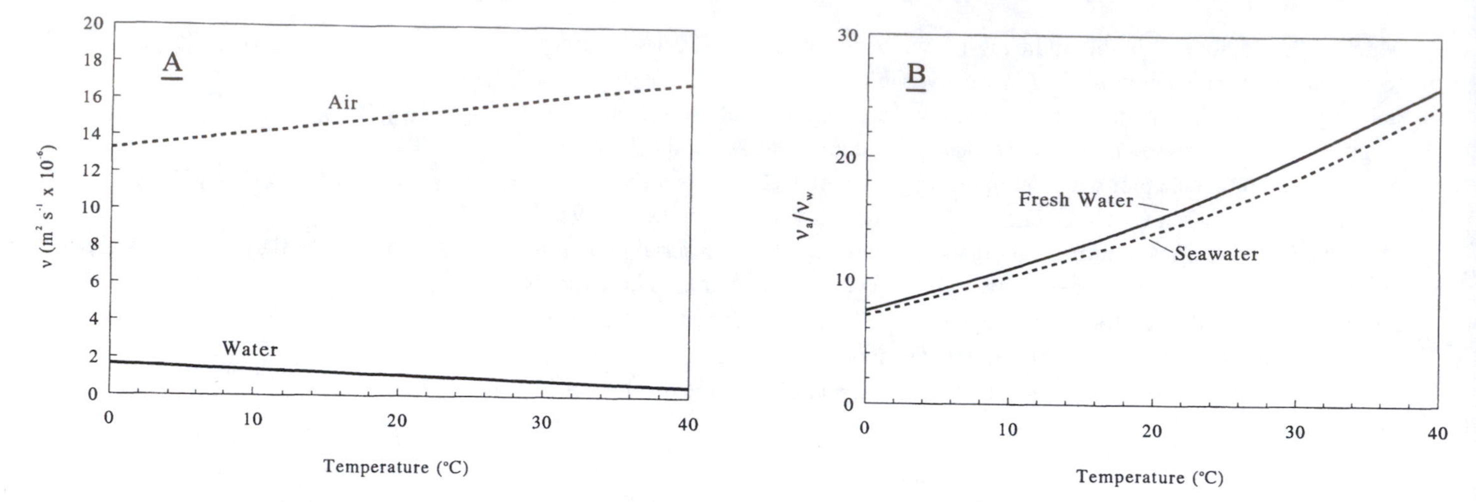

Fig. 7.1 (A) The kinematic viscosity of air is relatively high and increases with increasing temperature while that of water is relatively low and decreases. (B) The ratio of air's kinematic viscosity to that of fresh water and seawater increases with increasing temperature

7.2 Terminal Velocity

7.2.1 *The Descent of Man*

If you step off a cliff or jump out of an airplane, your mass is acted upon by the acceleration of gravity, and your vertical velocity rapidly increases. However, it should be clear from the discussion of pressure drag in chapter 4 that velocity cannot increase without limit. The faster you plummet, the greater is the drag acting on your body, and at some particular velocity the upward force of drag just equals the downward force of your weight (fig. 7.2). At this velocity there is no *net* force acting, your acceleration ceases, and subsequent motion does not result in any further increase in velocity.

Fig. 7.2 The descent of man, showing the relevant forces.

The velocity at which drag equals weight is known as the *terminal velocity*, not because of the effect it has when you eventually strike the ground, but rather because it is the highest velocity an object reaches in free fall.

How fast is the terminal velocity? Recall from chapter 4 that

$$\text{drag} = 0.5\rho_f v^2 A_f C_d, \tag{7.3}$$

where A_f is the frontal area measured perpendicular to the direction of flow, C_d is the drag coefficient, and v is the speed relative to still air. Note that for a falling object movement is vertical, and the symbol v is used to distinguish this motion from the horizontal speed of wind, for instance, which is symbolized by u.

We know that at terminal velocity, drag equals weight. Thus, the terminal velocity, v_t, of a human being can be calculated if the drag coefficient, size, and weight are known:

$$\text{weight} = \frac{\rho_f v_t^2 A C_d}{2} \tag{7.4}$$

$$v_t = \sqrt{\frac{2 \times \text{weight}}{\rho_f A C_d}}. \tag{7.5}$$

For example, Hoerner (1965) reports that $C_d \approx 1$ for a human falling headfirst and the corresponding projected area is about $0.12\,\mathrm{m}^2$. We may suppose that our reckless human weighs about 700 N. Noting that air has a density of $1.2\,\mathrm{kg\,m}^{-3}$ at 20° C, we calculate a terminal velocity of about $100\,\mathrm{m\,s}^{-1}$. It is this tremendous velocity that gives skydiving such a morbid thrill and renders one so leary of stepping off cliffs.

Note that terminal velocity is likely to be different for organisms of different sizes. For instance, let us assume for the moment that C_d is constant. In that case, the drag on an organism increases with profile area, and therefore with the square of some characteristic dimension of the animal. On the other hand, weight, which is proportional to volume, increases as the cube of the characteristic dimension. Because weight increases faster than area, large animals have higher terminal velocities, with obvious effect on the end result. It is indeed true that the bigger you are, the harder you fall. As J. B. S. Haldane (1985) noted, "You can drop a mouse down a thousand-yard mine shaft; and arriving at the bottom, it gets a slight shock and walks away. A rat is killed, a man is broken, a horse splashes."

Actually, a rat probably survives; certainly cats do. Diamond (1989) cites evidence that only a small fraction of cats are killed in falls from New York skyscrapers. Strangely enough, cats survived falls from eight stories more frequently than from four stories, perhaps because they have time to assume a drag-increasing posture.

In contrast to the well-founded trepidation with which you might step into a mineshaft, few people worry about the speed they will attain when they step into the deep end of a swimming pool. What is the terminal velocity in water? Recall from chapter 4 that the effective density of a terrestrial mammal is near that of water, and we sink only when we exhale. Assuming that the average human is neutrally buoyant with lungs full and that the lungs form 6% of the body volume, we can calculate that a 70 kg human being weighs at most 42 N in water. This decrease in weight relative to that in air, accompanied by the much higher density of water (about $1000\,\mathrm{kg\,m}^{-3}$) results in a benign terminal velocity of only about $1\,\mathrm{m\,s}^{-1}$.

The low terminal velocities of objects in water can have practical biological consequences. For example, shorebirds commonly break the shells of clams and mussels by dropping the molluscs from a height of 20 to 30 m onto rocks. The streamlined shape of clams and mussels insures that they have a high terminal velocity, and thereby attain sufficient kinetic energy in their fall to crack the shell. The same strategem does not work for shells dropped in water.

These examples have been presented primarily to provide some tangibility for the notion of terminal velocity. In other respects they can be misleading. For example, in calculating v_t, we have assumed that the drag coefficient is unaffected by the size of the animal and the speed at which it moves or by the density and viscosity of the fluid. In other words, we have assumed that C_d is independent of Re. When C_d is indeed independent of Re, the effect of the fluid's properties on terminal velocity is confined solely to the presence of the density term in eq. 7.5; there is no role for μ.

As we will see, the assumption that C_d is independent of Re (and therefore that drag depends only on ρ_f) is reasonable only for certain cases at high Reynolds numbers. Given the definition of Re, this makes sense. At high Reynolds numbers, inertial forces, which are proportional to fluid density, outweigh viscous forces to the extent that viscosity can sometimes be neglected without serious repercussion.

7.2.2 *Terminal Velocity at Low Reynolds Number*

What happens at low Reynolds numbers where viscous forces dominate flow? In this case it is best if we take leave of our free-falling human, who due to his large size is doomed to a high-Re existence, and turn to a simpler, more universal shape, the ever-handy sphere. At Reynolds numbers less than 1, the drag on a sphere is described accurately by Stokes's equation:

$$\text{drag} = 6\pi\mu v r, \tag{7.6}$$

where r is the radius of the sphere. As expected for an object at low Reynolds number, force is independent of density and depends on dynamic viscosity alone.

Now the weight of a sphere of radius r is $4g\rho_e\pi r^3/3$, where ρ_e is the effective density of the material from which the sphere is constructed (chapter 4). Equating drag and weight, we see that the terminal velocity of spheres at low Re is

$$v_t = \frac{2r^2\rho_e g}{9\mu}. \tag{7.7}$$

What size spheres fall slowly enough so that Stokes's equation can be used to describe their descent? Substituting v_t as defined by eq. 7.7 for u and $2r$ for ℓ_c into the expression for Re (eq. 7.2) and setting the result equal to 1, we see that

$$r_{crit} = \sqrt[3]{\frac{9\mu^2}{4\rho_e g\rho_f}}, \tag{7.8}$$

where r_{crit} is the critical radius below which Stokes's equation applies.

For spheres with a typical biological density of 1080 kg m^{-3}, r_{crit} is about 40 μm in air, 140 μm in fresh water, and 160 μm in seawater. In other words, an object must be exceedingly small for its free fall in air to be governed by viscosity alone, whereas in water (especially seawater) the requirements are somewhat less stringent.

To give these results some tangibility, we can calculate the terminal velocity for biological particles 40 μm in diameter, a size roughly that of pollen grains in air and algal spores in water. We predict that at 20° C pollen grains fall at about 5.1 cm per second,[1] whereas algal spores of the same size and density fall at only 70 μm s^{-1} in fresh water and a mere 50 μm s^{-1} in seawater. Again, the higher viscosity of water serves to slow the fall of particles, as does the reduction in particle weight due to the higher density of the liquid.

We can tie these results to the theme of the Reynolds number by returning to eq. 7.6. Setting this expression for drag equal to the more familiar expression given by eq. 7.3 and rearranging, we see that at low Reynolds number,

$$C_d = \frac{24}{Re}, \tag{7.9}$$

where we have used the diameter of the sphere (= $2r$) as the characteristic length in the Reynolds number. Thus, at low Re we see quite explicitly that the drag coefficient is a function of Reynolds number.

[1] This is somewhat faster than the sinking velocity of many pollen grains, which are about 2 to 2.5 cm s^{-1} (Niklas 1982a). The reasons for this discrepancy are discussed later in this chapter.

7.2.3 *A General Expression*

We now have two visions of the drag coefficient for a sphere: at high Reynolds number it is constant, at low Reynolds number it is inversely proportional to Re. These disparate visions can be smoothly joined, however. For any Re less than about 100,000, the drag coefficient of a sphere can be calculated from an expression given by Vogel (1981):

$$C_d = 0.4 + \frac{24}{Re} + \frac{6}{1+\sqrt{Re}}. \tag{7.10}$$

This expression is shown in figure 7.3. At low Reynolds number, the terms 0.4 and $6/(1+\sqrt{Re})$ are small compared to the term $24/Re$, and the expression approximates that derived in eq. 7.9 from Stokes's equation. At high Reynolds numbers, the terms containing Re are small compared to 0.4, and C_d is relatively constant, a situation analogous to that we assumed for a free-falling human being. At intermediate Reynolds numbers, all terms in the equation need to be taken into account.

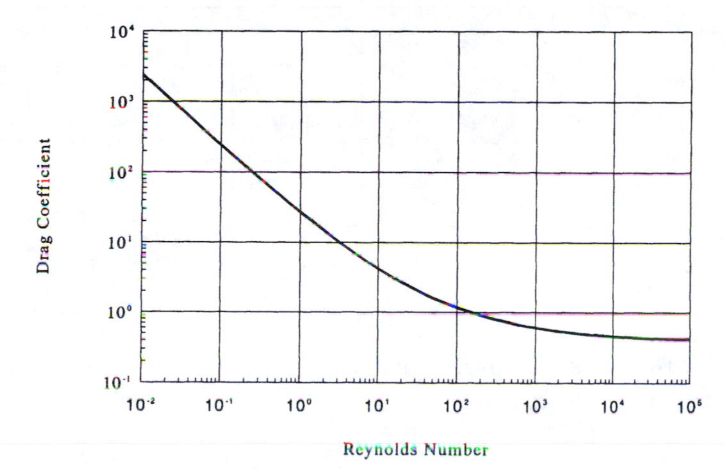

Fig. 7.3 At a Reynolds number less than 100,000, the drag coefficient of a sphere decreases with increasing Reynolds number (eq. 7.10).

This variation in C_d can be used to give a more general expression for terminal velocity. For a sphere of radius r we again use $2r$, the diameter, as the characteristic length in the Reynolds number. In this case,

$$C_d = 0.4 + \frac{12\nu}{rv} + \frac{6}{1+\sqrt{2rv/\nu}}. \tag{7.11}$$

Inserting this value for C_d into eq. 7.3, we see that the drag on a sphere is

$$\text{drag} = \frac{\rho_f \pi r^2 \left(0.4v^2 + \frac{12\nu v}{r} + \frac{6v^2}{1+\sqrt{2rv/\nu}}\right)}{2}, \tag{7.12}$$

where we use πr^2 as the frontal area of the sphere.

Equating eqs. 7.12 with the weight of the sphere and canceling terms, we arrive at the conclusion that

$$0 = 0.2v_t^2 + \frac{6\nu v_t}{r} + \frac{3v_t^2}{1+\sqrt{2rv_t/\nu}} - \frac{4\rho_e r g}{3\rho_f}. \tag{7.13}$$

Solving this equation for terminal velocity is a job for a computer, the results of which are shown in figure 7.4 for spheres with a density of 1080 kg m^{-3}. At all sizes, the terminal velocity in air is considerably greater than that in water. For instance, a sphere 100 μm in diameter falls 750 times faster in air than in water; a sphere 5 m in diameter falls 130 times faster.

At Reynolds numbers above 100,000, the relationship between drag coefficient and Re becomes complex (fig. 7.5). First, at a Reynolds number of about 500,000, the drag coefficient of a sphere abruptly decreases, a phenomenon associated with the transition from a laminar to a turbulent boundary layer around the object. The concept of a boundary layer is discussed later in this chapter. At still higher Reynolds numbers, the drag coefficient gradually rises. Due to these complications, the expression for C_d given in eq. 7.11 cannot be used above $Re = 100,000$.

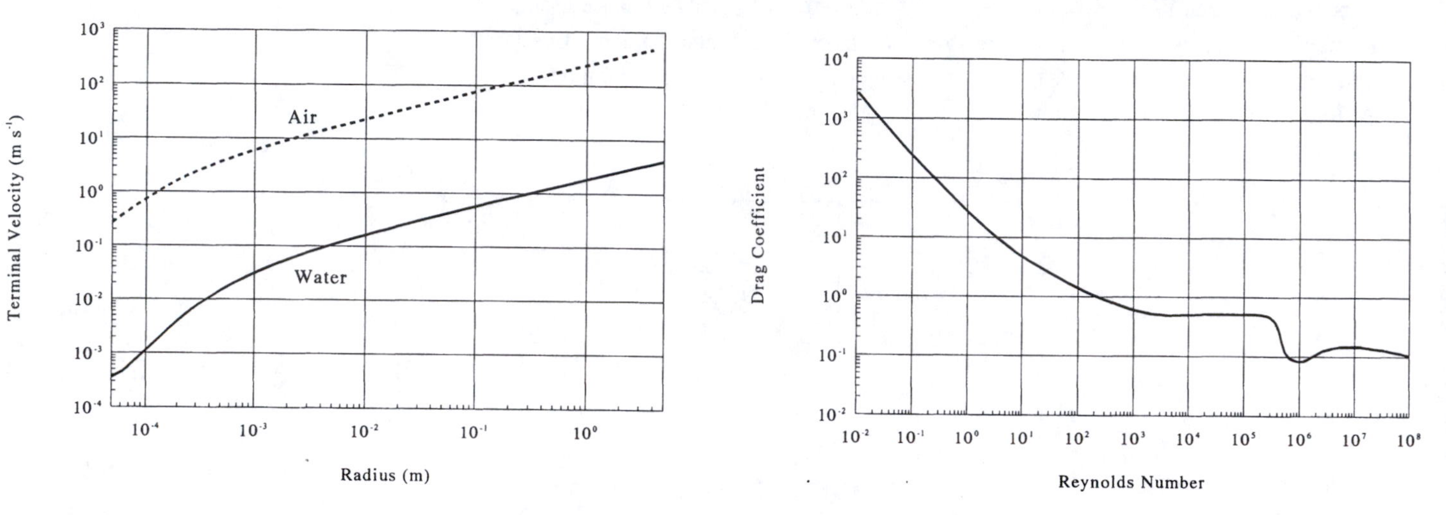

Fig. 7.4 The terminal velocity of a sphere increases with the size of the sphere. Terminal velocities are about 1000-fold larger in air than in water.

Fig. 7.5 The drag coefficient of a sphere abruptly decreases when the boundary layer becomes turbulent. (Redrawn from Vogel 1981)

The fact that terminal velocities are higher in air than in water has had profound biological consequences, several of which we now explore.

7.3 Why Are There So Few Aerial Plankton?

A general characteristic of aquatic (especially marine) environments is the presence of planktonic life. A cubic meter of water taken from virtually anywhere in a stream, lake, or ocean is teeming with small, suspended organisms. In fact, the concentration of these plants and animals is such that many kinds of invertebrates, including clams, mussels, anemones, polychaete worms, and bryozoans, can reliably use planktonic particles as their sole source of food. In contrast, air is relatively devoid of suspended matter. A cubic meter of air might contain a few bacteria, a pollen grain or two, and very occasionally a flying insect or wind-borne seed. Air is so depauperate compared to the aquatic "soup" that few terrestrial animals manage to make a living by straining their food from the surrounding fluid. Web-building spiders are the only example that comes to mind.

The scarcity of planktonic particles in air and their abundance in water is directly due to the relative terminal velocities in the two media. To see why, we need to digress briefly to consider the nature of turbulence.

7.3.1 *Turbulent Mixing*

The motion of fluids is seldom orderly. One need only watch cream as it is mixed into a cup of coffee or a plume of smoke as it is wafted by the wind, to realize that many, indeed most, flows are accompanied by chaotic swirling motions that act to scatter suspended particles. These random fluctuations in fluid velocity are present even in situations where the average flow is quite regular and easily described. Smoke issuing from a chimney is an apt example. It is an easy matter to mount an anemometer on a chimney, and thereby measure the speed of the wind. However, even on a day when our anemometer tells us that the air is moving at a more or less steady rate for which we can easily calculate an average, the inevitable gustiness of the wind serves to mix smoke particles through the surrounding air, and the diameter of the plume expands as smoke moves downwind (fig. 7.6).

Fig. 7.6 Smoke exiting a stack forms a turbulent plume.

The chaotic motions of a fluid which can cause the mixing of particles are known, generically, as *turbulence*, and their study forms one of the important disciplines in fluid dynamics. Turbulence is a difficult subject to explain with any precision, one best left for another forum. Here we raise the issue only because the presence of turbulence in flow allows us to make a useful analogy to the process of molecular diffusion discussed in chapter 6.

Just as random thermal motion drives the process of molecular diffusion, the random motion of turbulent fluids can lead to the transport of macroscopic particles. As particles suspended in either air or water are carried by turbulent eddies, their motion can be characterized as a random walk. As a result, particles carried by turbulence have a tendency to move down a concentration gradient in much the same fashion that random-walking molecules do (Chapter 6), and the rate at which particles are transported can be described by a turbulence equivalent of the molecular diffusion coefficient, the turbulent diffusivity, ϵ.

Turbulent diffusivity is a property of how the fluid moves. It is large when the fluid is actively being mixed and is small when it is not. To a first approximation, ϵ is independent of the type of particle, so there is no need to measure ϵ_{O_2} as distinct from ϵ_{CO_2}.

Just as we related the flux density of molecules, $\mathcal{J}$, to the concentration gradient via a molecular diffusion coefficient,

$$\mathcal{J} = -\mathcal{D}_m \frac{\partial C}{\partial y}, \tag{7.14}$$

we can relate the flux density of particles to the concentration gradient via the turbulent diffusivity:

$$\mathcal{J} = \frac{dC/dt}{A} = -\epsilon \frac{dC}{dy}, \tag{7.15}$$

where C is the concentration of particles (number per volume), A is the area over which particles are transported, and y is the axis along which transport occurs. For present purposes, we assume that y is vertical.

7.3.2 *The Distribution of Plankton*

We are now in a position to combine turbulent mixing and terminal velocity in a calculation of the vertical distribution of particles. Consider a fluid in which particles are suspended at a concentration C and a horizontal area A within that fluid (fig. 7.7). If these particles sink at a rate v_t, what is the flux of particles through our area? In a period of 1 s, particles within v_t meters of the plane of the area pass through it. Thus, in every second the particles in a volume Av_t are transported

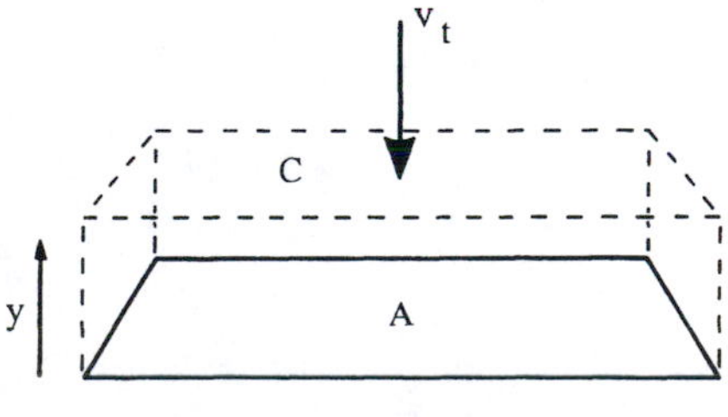

Fig. 7.7 Definition of terms used in calculating the vertical distribution of plankton (see text).

through the area. This volume contains Av_tC particles. As a result, the downward flux density (in particles per area per second) is v_tC.

This is the flux of particles in the absence of any turbulent mixing, and it tends to concentrate particles near the bottom of the fluid's container. In other words, in the absence of turbulent mixing, particles in air tend to concentrate on the ground and particles in water tend to concentrate on the bottom of the lake, stream, or ocean. We can state this idea in mathematical terms. If we locate the origin of our coordinate system at the level of the ground or seabed and take y as being positive upward, the sinking of particles creates a concentration gradient such that dC/dy is negative; that is, concentration decreases the farther away from the bottom one samples.

Now let us introduce turbulence into the system. We know that the action of turbulent mixing is to cause particles to move down the gradient in concentration, which in this case is the same as saying that turbulent mixing acts to move particles upward against gravity. Viewed on a mechanistic level, the presence of turbulent fluctuations in fluid velocity causes particles to undergo a random walk, and because the presence of the ground or seabed limits downward movement, the net result of the random walk is to move particles up.

We therefore have two transport processes (sinking and turbulent mixing) which tend to offset each other. For some particular vertical gradient of particle concentration, the downward flux due to particle sinking just equals the upward flux from turbulent mixing. This gradient represents the equilibrium status of the system. In other words, the system is in a dynamic equilibrium when

$$v_tC + \epsilon\frac{dC}{dy} = 0. \tag{7.16}$$

To calculate the equilibrium gradient in concentration, all we need do is solve this simple equation. Although straightforward, this process involves a lengthy definition of the properties of ϵ near a substratum, the details of which are not relevant to this chapter. Rather than burden the discussion with these details, we simply state the result here and direct the interested reader to Middleton and Southard (1984) for a complete derivation. The concentration at distance y from the ground or seabed is

$$C(y) = C(h)\left[\left(\frac{d-y}{y}\right)\left(\frac{h}{d-h}\right)\right]^z, \tag{7.17}$$

where $C(h)$ is the concentration at a reference height h very near (but not directly at) the substratum, d is the total depth of fluid affected by the turbulence, and z, known as the *Rouse parameter*, is

$$z = \frac{-v_t}{\kappa u_*}. \tag{7.18}$$

In other words, the concentration at height y depends on the relative magnitude of the terminal velocity and the mysterious product κu_* (fig. 7.8). When $-v_t/(\kappa u_*)$ is large, most particles are confined in a narrow band adjacent to the substratum. When $-v_t/(\kappa u_*)$ is small, particles are spread throughout the fluid.

What, then, is this parameter κu_*? The term κ is a dimensionless, empirically determined coefficient known as *von Karman's constant*, which is generally accepted to have a value of approximately 0.4 (Middleton & Southard 1984). The term u_* (pronounced "u-star") is the *friction* or *shear velocity*, and it has the units $\mathrm{m\,s^{-1}}$.

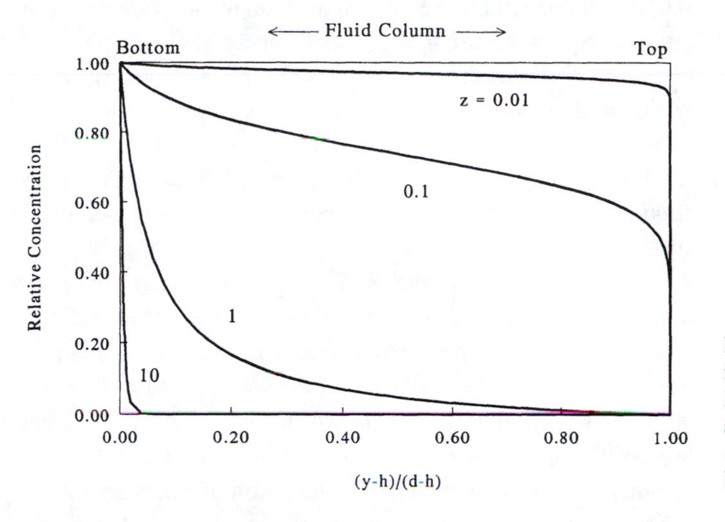

Fig. 7.8 The relative concentration of particles in a turbulent fluid column is a function of the Rouse parameter, z. At low z, particles are spread throughout the water column. At high z they are confined to a layer adjacent to the bottom.

Friction velocity is a peculiar measure of the intensity of turbulence. The larger u_* is, the more turbulent the flow and the larger are the fluctuations in velocity relative to the average velocity. For a thorough explanation of u_*, you should consult Schlichting (1979) or Middleton and Southard (1984).

In practice, u_* is determined from a knowledge of the shear stress, τ_b, acting on the substratum:

$$u_* = \sqrt{\frac{\tau_b}{\rho_f}}. \tag{7.19}$$

As a matter of convenience, however, we may assume that we are dealing with a "typical" intensity of turbulence, in which case the fluctuations in velocity (characterized by u_*) are about 5% of the mean velocity of the fluid over the substratum, u_∞. In other words,

$$u_* \approx 0.05 u_\infty \tag{7.20}$$

(Middleton and Southard 1984). This approximation should be taken with a large grain of salt; the value of u_* can be affected substantially by factors such as the roughness of the substratum and the turbulence of the mainstream, and can be as high as $0.15u_\infty$.

Using this rough estimate for u_*, we can rewrite z as

$$z = \frac{-50 v_t}{u_\infty}. \tag{7.21}$$

What, then, are values for z in air and water? Let us use as an example the terminal velocities calculated earlier for biological particles 40 μm in diameter and a wind or current speed of $1\ \mathrm{m\,s^{-1}}$. In air, where v_t is $-5.1\ \mathrm{cm\,s^{-1}}$, $z \approx 2.6$, and a glance at figure 7.8 shows that most particles of this size (e.g., pollen and dust) are likely to be concentrated very close to the substratum. Particles any larger than this have higher sinking rates, and are even more closely confined to the vicinity of the substratum.[2] This prediction matches reality: a breeze of $1\ \mathrm{m\,s^{-1}}$ (2 mph) does not kick up much dust.

In contrast, the v_t of 40 μm particles in seawater is only about $-50\ \mu\mathrm{m\,s^{-1}}$, $z \approx 0.0025$, and particles are expected to be mixed thoroughly through the entire water column. Again our prediction matches reality. For instance, Amsler and Searles (1980) showed that the spores of red algae were commonly found throughout the water column when the water's depth was 20 m.

In summary, the sinking rates of particles in air overwhelm the ability of turbulence to keep them aloft, with the result that aerial plankton are scarce. The slow sinking rates of particles in water makes it easy for turbulent mixing to keep them suspended.

Although the analysis presented here is valid, there are several reasons it should be used with caution. First, we have yet to give a precise definition to the value d used in eq. 7.17. In some cases, a precise definition is easily given. For instance, in a stream or river the entire depth of the water column is influenced by turbulence and d is simply the water's depth. In the ocean or atmosphere, however, d is less well defined. The value for ϵ used in the derivation of eq. 7.17 assumes that turbulence is associated with the interaction of a flowing fluid and the substratum. In this case, d extends away from the substratum to a distance where the flow is no longer affected by the presence of the ground or seabed. This distance is roughly equal to the thickness of the turbulent *boundary layer*, a concept discussed later in this chapter. In air, this distance is typically on the order of 100 m (Monteith 1973), in the ocean it is on the order of 10 m (Grant and Madsen 1986). As a result, figure 7.8, which is scaled to d, overemphasizes the differences between the two media. In practice, however, this should not pose a problem. The differences in sinking rates between air and water are so great that they can offset any reasonable difference in the scale of the turbulent boundary layer.

The comparison made above is also biased by the choice of mainstream velocity. A u_∞ of $1\ \mathrm{m\ s^{-1}}$ is a gentle breeze in air, but a notable torrent in water. What is the particle distribution like during a wind storm where u_∞ may reach 20 to 30 $\mathrm{m\ s^{-1}}$? In this case, z is approximately 0.1 for a 40 μm particle, and these particles can be expected to mix through much of the depth of air affected by turbulence. The dust storms encountered in deserts are evidence to support this conclusion. Note, however, that 40 μm particles are really quite small. What about particles 1 mm in diameter? In air, these particles have a sinking rate of $-3.7\ \mathrm{m\,s^{-1}}$, and even in a wind of $30\ \mathrm{m\,s^{-1}}$ have a z of 6 and are therefore confined to a thin layer next to the substratum. In other words, even in strong winds, small insects cannot rely on turbulence to keep themselves aloft.

In general, then, our conclusion seems justified. The sinking rates of particles in air are just too high to allow them to remain passively suspended and as a result, aerial plankton are sparse. In water, slow sinking speeds insure that many particles are suspended, and the plankton is plentiful. The abundance of aquatic suspension feeders and the scarcity of terrestrial ones, can therefore be thought of as a direct consequence of the differences in density and viscosity between air and water.

[2]There is one prominent exception to this rule: sand. Because sand particles are made from hard, resilient materials, they bounce when they hit the ground. As a result, the average height of sand particles in a sand storm is set more by the resilience of the sand than by the nature of the aerial turbulence. Even in a severe storm, however, sand seldom rises above 2 m from the ground. For a lucid, interesting discussion of the behavior of windblown sand, consult Bagnold (1942).

Note that eq. 7.17 implies that particles with different sinking rates are distributed differently in the water column. For instance, dense, inorganic sediments are likely to be confined more closely to the substratum than are less-dense organic particles. This fact is apparently used by at least one suspension-feeding organism to separate organic "wheat" from the inorganic "chaff." Muschenheim (1987) has shown that the passive suspension-feeding polychaete *Spio setosa* positions its feeding apparatus 4 to 5 cm above the substratum, thereby increasing the ratio of organic to inorganic matter in its diet.

The conclusions regarding the roles of terminal velocity have been reached using spherical particles as an example, but many organisms have evolved strategies to reduce their terminal velocities below those predicted here. The most widespread strategy is to assume a nonspherical shape. For instance, pollen grains and radiolarians often have highly sculpted extensions to their more-or-less spherical central cores (fig. 7.9). The small diameters of these extensions put them in the very low Reynolds number range where C_d is quite high, and as a result the overall drag on the particle is increased without increasing its weight.

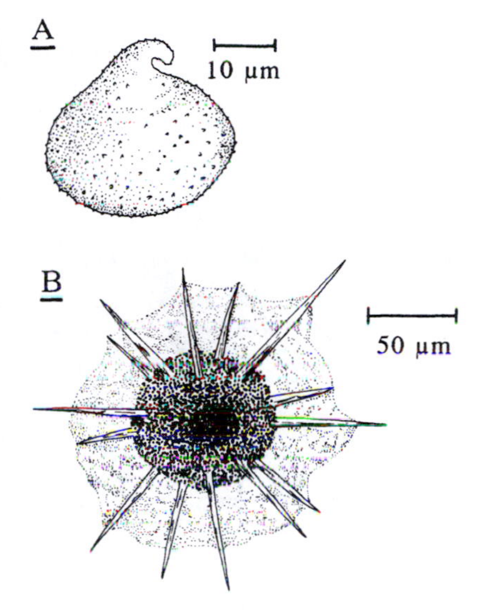

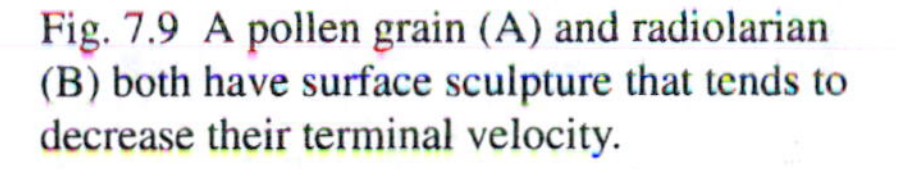
Fig. 7.9 A pollen grain (A) and radiolarian (B) both have surface sculpture that tends to decrease their terminal velocity.

Employing a similar strategy, the young of some spiders exhibit a remarkable behavior in which they climb to the apex of a blade of grass, extend their abdomen into the wind, and pull from their spinnerets a skein of very fine silk fibers. The drag on the fibers is sufficient to carry the young aloft, and Darwin reported having these "ballooning" spiders land on the *Beagle* while still many miles at sea. Humphrey (1987) has carried out a precise analysis of this ballooning behavior.

Other common examples involve the seeds of plants that are dispersed by the wind. By decreasing the rate at which they sink through the air, these seeds increase the distance they can be carried by the wind. The various fascinating strategies used by plant seeds are reviewed by van der Pijl (1982), Vogel (1981), and Augspurger and Franson (1987).

7.3.3 *Sinking and Reproduction*

In the presence of a solid substratum, an individual particle may rest on the bottom for a while, but it never disappears from the system and there is always the chance that it will eventually be resuspended by a passing eddy. Thus, all particles play a role in the maintenance of the type of concentration gradient we used above to explain the equilibrium suspension of plankton. In the surface layers of the ocean, however, there is in effect no bottom, and the ocean depths can be viewed as an endless sink for phytoplankton. As each individual performs its turbulent random walk, it eventually sinks out of the surface layer, and, in the absence of sunlight, dies. How, then, are near-surface plankton populations maintained?

The answer lies with reproduction. Phytoplankters not only sink, they also grow and divide. If a population reproduces fast enough, the odds are that a few individuals will divide during the time that turbulent mixing has by chance carried them close to the surface. If enough individuals reproduce high in the water column, they can just replace those that sink out of the system. In an elegant application of the theory of turbulent mixing, Riley et al. (1949) showed that a stable plankton population can be maintained if

$$\frac{v_t^2}{r} < 4\epsilon. \tag{7.22}$$

Here r is the *intrinsic rate of increase* in the population, a measure of the rapidity with which organisms reproduce, defined by

$$\frac{dN}{dt} = rN, \tag{7.23}$$

where N is the number of individuals in the population. Viewed another way, if we assume that plankton reproduce by binary fission,

$$r = \frac{0.693}{t_d}, \tag{7.24}$$

where t_d is the period between the "birth" of a phytoplankter and the time it splits into two. For example, if a phytoplankter divides once per day (86,400 s), $r \approx 8 \times 10^{-6}$ s^{-1}.

The value of ϵ in the open ocean varies greatly with surface conditions and depth, but we may use 0.01 m^2 s^{-1} as a typical ϵ value for vertical mixing (Bowden 1964). Inserting these values into eq. 7.22, we see that for a phytoplankter that divides once per day, terminal velocity must be less than 566 μm s^{-1} if a stable population is to be maintained. For a spherical organism with a density of 1080 kg m^{-3}, this sinking velocity corresponds to a diameter of about 150 μm, well within the range of existing phytoplankton. Thus, it is feasible for small aquatic organisms to maintain a stable population near the water's surface if there is turbulence and if they reproduce at a rapid rate.

This is only one possible solution to the problem of sinking. As we have discussed in chapter 4, it is possible for phytoplankton (and other small organisms) to adjust their buoyancy and thereby slow their rate of descent. The energetic trade-offs between rapid reproduction and the maintenance of neutral density in planktonic organisms are discussed by Alexander (1990). His conclusion is that rapid reproduction requires less overall energy expenditure than does buoyancy compensation if organisms are smaller than about 100 μm in diameter.

In contrast, it seems unlikely that many aerial organisms can maintain a stable population by rapid reproduction. For example, an aerial phytoplankter 40 μm in diameter could maintain a stable population in the upper atmosphere only if the product of its intrinsic rate of increase and the turbulent diffusivity were greater than about 0.00065. If the organism splits once per day, an ϵ of 81 m^2 s^{-1} would be required. This is a huge diffusivity, one unlikely to be maintained in the atmosphere.

Bacteria or blue-green algae, however, could make the strategy work. At their diminutive size (a radius on the order of 1 μm), sinking rates in air are similar to those for phytoplankton in water (about 0.1 mm s^{-1}). Furthermore, under ideal conditions, bacteria and blue-green algae can divide several times per hour. In such a case, very little turbulence would be needed to maintain a stable aerial population. It seems unlikely, however, that air is a sufficiently commodious medium for bacteria or algae to reproduce rapidly. In fact, they are likely to dry up before they reproduce at all (chapter 14).

7.4 Limits to the Speed of Walking

Terminal velocity can have an important effect on at least one other aspect of biology. Consider the physics of walking. At the simplest level, a walking organism can be depicted as shown in figure 7.10. The body (or one segment of the body) is held above the substratum by two legs. With each step, the foot of one leg is planted on the ground and the body vaults over that leg. The other foot is then planted and the process is repeated. In walking, then, there is always at least one foot in contact with the ground. This need not be so for running, but that is a different matter.

Alexander (1982) has shown that this simple representation of walking is useful in analyzing the gaits of terrestrial animals. As the body vaults over a leg, it behaves much like a weight on the end of a string in that it describes a circular path. Associated with this motion is a centrifugal acceleration equal to the product of leg length, ℓ, and the square of the angular velocity of the leg as it rotates around the foot, ω^2 (fig. 7.10). At the top of the vault, this centrifugal acceleration is directed upward and the only thing keeping the animal on the ground is the downward acceleration of gravity, g. If $\omega^2\ell > g$, the animal has in effect jumped upward, its feet must leave the ground, and it is, by definition, no longer walking. Thus, for the animal to walk ,

$$\frac{\omega^2\ell}{g} < 1. \tag{7.25}$$

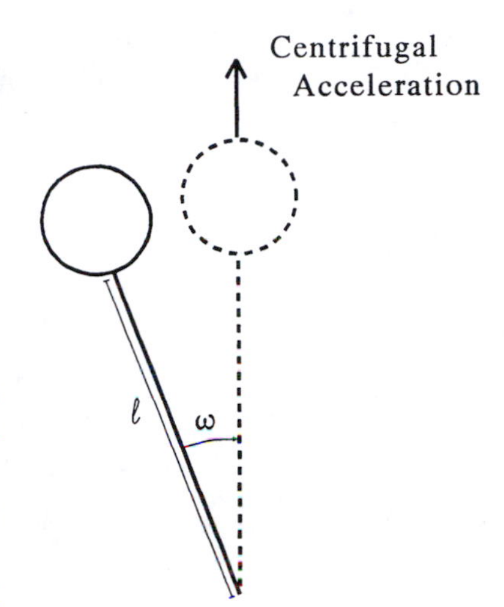

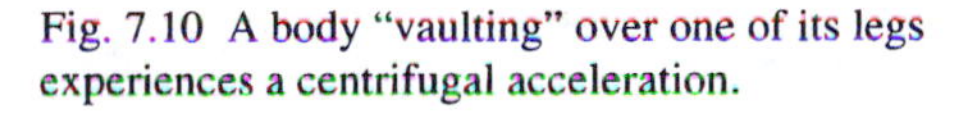
Fig. 7.10 A body "vaulting" over one of its legs experiences a centrifugal acceleration.

For a given length of leg, an upper limit exists to ω and therefore to the speed at which the animals can walk. This simple analysis can accurately predict the speed at which a wide variety of animals switch from walking to running.

The ratio $\omega^2\ell/g$ is one example of a *Froude number*, Fr, a concept we will encounter again in chapter 13.

The analysis above assumes, however, that there is no hindrance to the vertical motion of the body. In particular, it assumes that after the body has reached the top of its vault, it freely accelerates downward to finish the step. The presence of drag can set a limit to the validity of this assumption. In particular, if the stepping rate requires the body to fall faster than its terminal velocity, the legs lift from the ground, and the animal can no longer walk. Is this indeed a problem?

To pursue this thought, we consider the example shown in figure 7.11. A spherical body of diameter d walks on a pair of legs each of length ℓ. We will want to examine organisms of different sizes but all of the same shape, and for this purpose it is convenient to define leg length in units of body diameter. Thus $\ell = k_\ell d$, where k_ℓ is a constant.

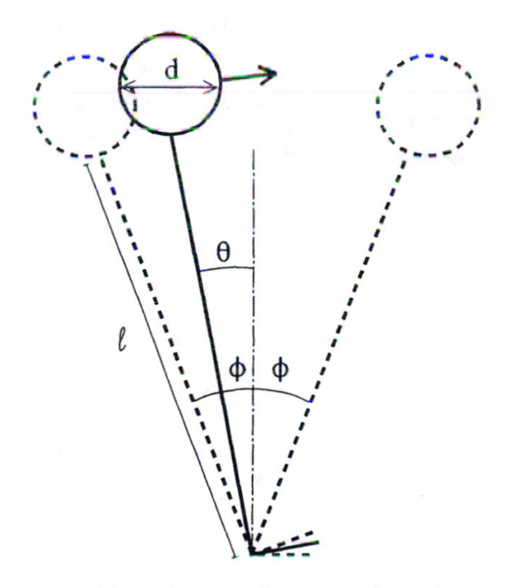

Fig. 7.11 Definition of terms used to calculate the maximal speed of walking.

During a step, each leg swings through an angle 2ϕ as shown. At any time, the angle between the leg and the vertical is measured by θ. The frequency of stepping (steps per s) is f. Given these assumptions, we see that the angular velocity of the leg during a vault is $\omega = 2\phi f$, and the tangential velocity of the body (the velocity perpendicular to the long axis of the leg) is $\omega\ell = 2\phi f k_\ell d$.

At any point during the vault, the tangential velocity of the body can be decomposed into a horizontal velocity u in the direction of walking and a vertical velocity v. A consideration of the geometry shows that

$$u = 2\phi f k_\ell d \cos(\theta), \tag{7.26}$$

$$v = 2\phi f k_\ell d \sin(\theta). \tag{7.27}$$

For a given frequency and angle of leg swing, the forward component of velocity can be averaged over a step to give an average speed of walking:

$$\langle u \rangle = 2 f k_\ell d \sin(\phi). \tag{7.28}$$

The maximum vertical velocity occurs at the end of the vault when $\theta = \phi$, in which case

$$v_{max} = 2\phi f k_\ell d \sin(\phi) = \phi\langle u\rangle. \quad (7.29)$$

It is interesting to note that the maximum vertical velocity at a given forward velocity depends on the angle through which the legs swing, but is independent of the frequency of stepping or the ratio of leg length to body diameter.

We now have a relationship between the average forward speed and the maximum velocity with which the body must fall. If we set this fall velocity equal to the terminal velocity as defined by eq. 7.13, we can calculate the maximum average speed of walking, u_{max}. When this is done, we see that

$$0 = 0.4u_{max}^2 + \frac{24\nu u_{max}}{\phi d} + \frac{6u_{max}^2}{1+\sqrt{\frac{\phi d u_{max}}{\nu}}} - \frac{4\rho_e d g}{3\rho_f \phi^2}, \quad (7.30)$$

an equation that my computer can solve for u_{max} as a function of d. The results for $\phi = 0.5$ radians are shown in figure 7.12.

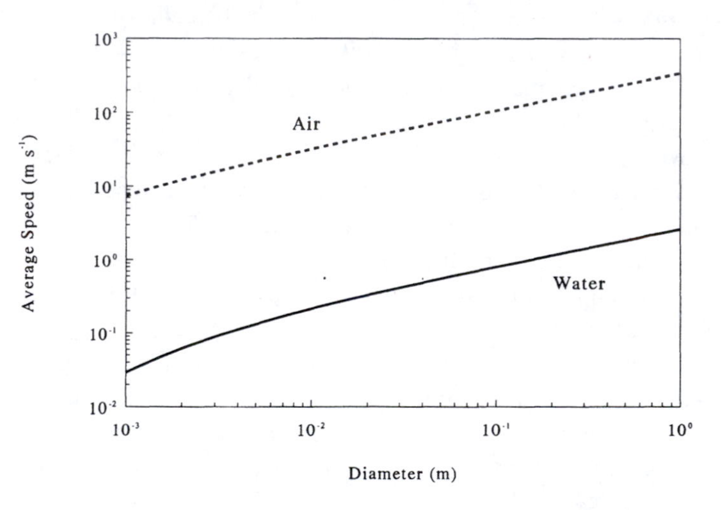

Fig. 7.12 The maximal speed of walking is greater in air than in water, and increases with the size of the organism.

In air, kinematic viscosity does not pose a practical limit to the speed of walking. For instance, a body 1 mm in diameter (an ant, for instance) could walk at a speed of nearly 7.4 m s^{-1} before it couldn't fall fast enough to keep its legs on the ground. This would be a shockingly fast pace.

In water, however, kinematic viscosity can severely restrict the speed of walking. For example, a body 10 cm in diameter could walk at a rate of only about 81 cm s^{-1} (8.1 body lengths s^{-1}) before it became waterborne. This is less than the ten body lengths per second we view as standard. The limitations posed by viscosity get relatively worse as the size of the organism increases. For example, a body 1 m in diameter could walk at a rate of only about 2.6 m s^{-1} (2.6 body lengths s^{-1}) before its feet no longer remained on the ground. It seems, then, that kinematic viscosity may limit aquatic pedestrians to a pedestrian rate of locomotion.

This conclusion should not be taken too seriously. This analysis is based on a very simplified view of the motion of the body during a step. Aquatic organisms may well flex their knees while walking so as to keep the body at a constant height above the ground, and thereby circumvent the problem. If they do, however, it is in contrast to most terrestrial organisms, even those with multiple legs. Anyone who has ever ridden a horse can attest to the fact that the body of a walking quadruped goes up and down.

Aquatic pedestrians could also increase their walking speed by increasing their effective density. For instance, the heavy shell of a lobster may be an adaptation to an increased speed of walking.

7.5 Boundary Layers

Recall from chapter 5 that fluid in contact with the surface of a solid does not move relative to that solid, a fact termed the no-slip condition. This condition holds for all fluids, be they liquid or gas, and has profound implications for fluid motion. Here we address one of the most basic consequences of the no-slip condition, the formation of a velocity gradient around objects.

Consider the situation of figure 7.13A, a body of fluid flowing in a uniform fashion to the right with a velocity known as the *mainstream velocity*, u_∞. Because all of the fluid has the same velocity, there is, by definition, no velocity gradient. We now follow the fluid to the right, to the point where it makes contact with a stationary, thin, flat plate lying parallel to the flow (fig. 7.13B). Here the status quo must change. The inertia of the fluid tends to keep it moving, but the no-slip condition tells us that fluid in contact with the plate must be stationary. In response to these conflicting tendencies, a velocity gradient *must* be established. The character of this gradient is such that the velocity increases in a direction perpendicular to the plate. It is zero directly at the plate's surface (as dictated by the no-slip condition) and rises to equal that of the mainstream at some distance away from the plate (as dictated by the momentum of the fluid).

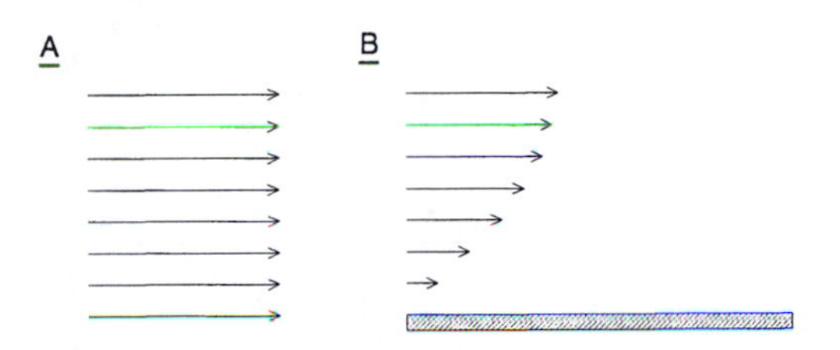

Fig. 7.13 As mainstream flow (A) encounters a solid substratum (B), flow is retarded, forming a boundary layer.

Fluid in the region of this velocity gradient is acted upon by a viscous force. Recall from chapter 5 that the shear stress (the force per area) acting in a fluid is equal to the product of dynamic viscosity and the velocity gradient:

$$\tau = \mu \frac{du}{dy}, \tag{7.31}$$

where in this case we take y as directed perpendicular to the plate. In other words, the presence of a solid object in flow is inevitably accompanied by the imposition of a shear stress.

The fluid affected by these viscous forces, that is, the fluid in the velocity gradient around the object, is known as the *boundary layer*, and it is the properties of this boundary layer that form the subject matter of this section.

The first thing we must note about the boundary layer is that it has a distinct inner edge—the surface of the solid—but no distinct outer edge. As distance increases from the solid surface, the velocity of the fluid rises asymptotically to match that of the free stream. As such, it is virtually impossible to point to a particular spot where velocity first exactly equals u_∞. In this respect, the thickness of the boundary layer is undefinable.

As a practical matter, however, one can pick an arbitrary level within the velocity gradient and dub it "the edge." There are a variety of ways in which this can be done, but it has been traditional among biologists to define the outer edge of the boundary layer as that point where the velocity is equal to 99% that of the mainstream. Admittedly, by this definition there is still some velocity gradient left outside of the boundary layer, but this leftover gradient is so slight that it exerts negligible shear stress. Given this definition, we can define the *boundary-layer thickness*, δ_{bl}, as the perpendicular distance from the solid surface to the boundary layer's outer edge.

Boundary layers come in two basic forms: *laminar* and *turbulent*. In a laminar boundary layer, fluid flows in an orderly fashion and it is reasonable to think of the velocity gradient as composed of a stack of fluid layers (the laminae mentioned in chapter 5) sliding relative to each other, the bottom-most layer being held in place by the no-slip condition. In a turbulent boundary layer, flow is chaotic. Turbulent eddies continually mix fluid perpendicular to the surface of the plate, and, as a result it is unlikely that at any time there is a distinct gradient in velocity. If, however, velocities are averaged over a long period, there is a gradient in the *average* velocity near the plate, and it is this time-averaged gradient that defines the turbulent boundary layer.

Turbulent boundary layers are complex, and their characteristics are in some important ways independent of the density and viscosity of the fluid. For example, the velocity gradient at one point within a turbulent boundary layer can be determined by how rugose the substratum is upstream, a factor that has little to do with the properties of air or water. Because of their relative independence from density and viscosity, we will not discuss turbulent boundary layers here in any depth. Instead, we briefly examine the rules that determine when a boundary layer becomes turbulent, and then leave the industrious reader to explore the nature of turbulent boundary layers by consulting texts such as Middleton and Southard (1984) and Schlichting (1979).

With these definitions and provisos in hand, our task now is to examine the properties of laminar boundary layers and to see how they differ between air and water.

7.5.1 *Boundary-Layer Thickness*

The thickness of a boundary layer increases with downstream distance from the point at which flow makes contact with a solid surface. To see why, consider the course of events in the fluid as it arrives at the leading edge of a flat plate. Fluid at the exact level of the plate comes to a screeching halt the instant it contacts the solid surface. As this layer slows, the layer of fluid immediately above it slides by. But due to the action of viscosity, the velocity of this second layer is also decreased. The third layer then must slide by the second, and so on. Thus, as time goes by the effects of viscosity extend out into the fluid by reducing the velocity of laminae. Now, during the time that these viscous effects are extending out from the substratum, the fluid has moved along the plate. The farther downstream we look from the leading edge of a plate, the more time the effects of viscosity have had to extend into the fluid, and the thicker is the boundary layer. To see how thick, we need to digress for a moment to draw an analogy between the growth of a boundary layer and the process of molecular diffusion.

We begin by noting that adjacent layers in a fluid are akin to two trains traveling in the same direction on parallel tracks, but at different speeds. If there is no exchange of mass between trains, there is no tendency for their velocities to equalize. However, if workers on each train avidly throw bags of sand onto the other train, momentum is exchanged. Each sack thrown from the faster moving train adds to the momentum (and thereby the speed) of the slower train, and each sack thrown from the slow train tends to impede the movement of the faster train. Now, the rate at which momentum is exchanged between fluid layers (analogous to the rate at which sand bags are exchanged between trains) is governed by the momentum equivalent of the molecular diffusion coefficient, which turns out to be nothing other than the kinematic viscosity, ν. In other words, ν is the diffusion coefficient for fluid momentum. Like all diffusion coeficients, ν has the units $m^2\ s^{-1}$.

Given that the spread of viscous effects through a fluid behaves diffusively, we can now venture a guess as to the rate at which a boundary layer grows. In the last chapter we noted that it is a characteristic of diffusive processes that they spread as the square root of time. For instance, the average distance traveled during a random walk is proportional to the square root of the number of steps taken, and if steps are taken at a constant rate, distance traveled increases with the square root of time. The longer each step is (i.e., the larger the diffusion coefficient) the farther the distance traveled. By analogy, we predict that the thickness of a boundary layer (which is set by the diffusion of momentum) should increase as the square root of the time since fluid first encountered the plate, and that the actual thickness should be proportional to the kinematic viscosity.

This indeed turns out to be the case. Provided one doesn't look too close to the leading edge of a plate, boundary-layer thickness, δ_{bl} , is

$$\delta_{bl} \approx 5\sqrt{\nu t}. \tag{7.32}$$

"Too close" in this case is a distance x from the leading edge such that $\delta_{bl} < 0.2x$. At positions closer to the leading edge, the rapid decrease in horizontal velocity causes water in the oncoming mainstream to be diverted away from the plate with an appreciable vertical velocity, which complicates the local velocity gradient (Schlichting 1979; Vogel 1981).The constant 5 in this equation is an approximation; various authors cite values ranging from 4.65 to 5.47 (Vogel 1981).

Eq. 7.32 allows us to compare the thickness of laminar boundary layers in air and water. Because the kinematic viscosity of air is roughly 15 times that of water, the thickness of a boundary layer is about $\sqrt{15} \approx 3.9$ times greater in air than in water, given equal mainstream flows.

It is useful at this point to re-examine briefly eq. 7.32. Time after fluid first encounters the plate can be expressed as

$$t = \frac{x}{u_\infty}. \tag{7.33}$$

Substituting this value for x in eq. 7.32 tells us that

$$\delta_{bl} \approx 5\sqrt{\frac{x\nu}{u_\infty}}. \tag{7.34}$$

Thus the thickness of the boundary layer increases with the square root of distance from the leading edge. The faster the mainstream flow, the less time viscosity has had to act when fluid reaches a distance x, and the thinner is the boundary layer.

It is also useful to tie the concept of the boundary layer to that of the Reynolds number. If we divide both sides of eq. 7.34 by x and rearrange, we see that

$$\frac{\delta_{bl}}{x} \approx 5\sqrt{\frac{\nu}{xu_\infty}}. \tag{7.35}$$

Now, the fraction xu_∞/ν has the form of a Reynolds number in which the characteristic length is the distance x from the leading edge. This new Reynolds number is called the *local Reynolds number* and is given the symbol, Re_x. Thus, we may rewrite eq. 7.35 as

$$\frac{\delta_{bl}}{x} \approx 5Re_x^{-1/2} \tag{7.36}$$

The larger the local Reynolds number, the smaller the boundary layer thickness is as a fraction of x. Because Reynolds numbers are larger in water for a given velocity, boundary layers there are thinner.

The local Reynolds number can be used as an index to predict whether a boundary layer will be laminar or turbulent. Experimental studies have shown that the boundary layer on a flat plate is usually turbulent if Re_x is greater than 3.5×10^5 to 10^6. For example, at a distance 1 m downstream from a leading edge, mainstream velocity in air must exceed 5.3 to 15 m s^{-1} before the boundary layer becomes turbulent. In water, u_∞ need only exceed 0.34 to 1.0 m s^{-1} for the boundary layer to be turbulent.

A more precise prediction than this is not possible because the onset of turbulence can be affected by the roughness of the substratum. The rougher the substratum, the lower the velocity that is needed to maintain a turbulent boundary layer. This effect can be quantified through the use of yet another Reynolds number. The *roughness Reynolds number*, Re_* is defined as

$$Re_* = \frac{u_* d}{\nu}, \tag{7.37}$$

where u_* is again the shear velocity and d in this case is the height of the "roughness elements" on the substratum. These can be anything from the hairs on a caterpillar to the barnacles on a rock. Again using $0.05u_\infty$ as a typical value for u_*, we can restate the roughness Reynolds number as

$$Re_* \approx \frac{u_\infty d}{20\nu}. \tag{7.38}$$

Experiments have shown that when Re_* exceeds 6, the outer reaches of the boundary layer begin to become turbulent. If $Re_* > 75$, turbulence is present throughout the boundary layer (Nowell and Jumars 1985). Using the approximation of eq. 7.38, we see, for example, that with roughness elements 1 cm high, the boundary layer is turbulent if u_∞ exceeds 2.3 m s^{-1} in air, but is turbulent in water if mainstream velocity exceeds only 15 cm s^{-1}.

It is possible to write an equation similar to eq. 7.36 for the growth of a turbulent boundary layer (Schlichting 1979):

$$\frac{\delta_{bl}}{x} = 0.376Re_x^{-1/5}. \tag{7.39}$$

In this case, boundary layers in air are thicker than those in water by a factor of $\sqrt[5]{15} \approx 1.7$. You are reminded, however, that turbulent boundary layers are fundamentally different from laminar ones. The thickness of a turbulent boundary layer is calculated using time-averaged values of velocity, and at any instant velocities equal to or even exceeding that of the mainstream may be present virtually anywhere within the boundary layer.

The thickness of laminar boundary layers has many biological consequences, three of which are explored here.

7.5.2 *Hiding in the Boundary Layer*

Because a boundary layer is a region of reduced velocity, it can serve as a refuge from the mainstream flow. What are the velocities actually like within the boundary layer, and how do they compare between air and water?

The shape of the velocity gradient within a boundary layer is difficult to describe with precision, but for the region nearest the solid surface (y less than about $0.4\delta_{bl}$) the following expression corresponds closely to reality:

$$\frac{u}{u_\infty} = 0.32y\sqrt{\frac{u_\infty}{x\nu}}. \quad (7.40)$$

In other words, in the lower reaches of a laminar boundary layer, velocity increases linearly with distance from the substratum, and the rate at which velocity increases depends on both distance from the leading edge and kinematic viscosity. The greater x is and the larger ν, the more slowly velocity increases.

What does this mean for an organism living in the boundary layer? We might ask, for example, how close to a leading edge an organism of a given size must venture before it encounters flow of a certain velocity. Because the kinematic viscosity of air is fifteen times that of water, an organism in air must move fifteen times closer to the leading edge before it experiences the same fraction of the mainstream velocity as an organism in water.

This has led to interesting behavioral adaptations in small insects and mites. For example, scale insects are exceedingly small (only about 300 μm long) and all but the short-lived, fragile males lack wings. How can these insects manage to disperse? The terminal velocity for the small creature is relatively slow (about 26 cm s^{-1} [Washburn and Washburn 1984]), so they may be transported with some effectiveness by aerial turbulence if they can get above the laminar boundary layer of the leaf on which they are born. Therein lies the problem. A wind velocity of about 3.7 m s^{-1} is required to impose sufficient drag on the insect to pry it free from the substratum. At a typical mainstream velocity of 4 m s^{-1} (Washburn and Washburn 1984), this local velocity is reached a mere 300 μm from the leading edge at the height of a standing scale insect (100 μm). In other words, as long as the insect is standing on all sixes, it must be virtually at the edge of a leaf before it can effectively enter the wind stream. If, however, the insect stands on its hind legs and extends its forelegs into the breeze, its center of area is about 300 μm above the substratum (as opposed to 100 μm). At this height, wind speed sufficient for takeoff is reached approximately 3 mm from the edge of a leaf. Thus, by assuming an upright posture, an insect considerably increases the chance that it will experience sufficient force to be borne aloft. Upright takeoff postures have been observed in a variety of scale insects and mites (Washburn and Washburn 1984).

Viewed from a different perspective, eq. 7.40 tells us that there is more space in which an organism can hide from the mainstream in an aerial boundary layer than in an aquatic one. For example, consider a mainstream velocity of 10 cm s^{-1} and an organism located 10 cm downstream of a leading edge. If the organism requires a velocity of less than 1 cm s^{-1} in order to feed, for instance, it can be 1.3 mm high in air, but only 0.3 mm high in water.

Eq. 7.40 applies only in the lower reaches of the boundary layer. Farther from the substratum the velocity no longer increases linearly, and a more complicated equation is needed to describe the velocity gradient. Vogel (1981) purports that the following fits the observed velocity gradient within about 5%:

$$\frac{u}{u_\infty} = 0.39y\sqrt{\frac{u_\infty}{x\nu}} - 0.038\frac{u_\infty y^2}{x\nu}. \tag{7.41}$$

Many fascinating examples of the biological utility of boundary layers are discussed in Vogel (1981), and we will not repeat that discussion here. Instead we will explore one curious example of hiding in the boundary layer.

7.5.3 *Listening to the Boundary Layer*

This example concerns the nature of the boundary layer set up by sound waves. We will explore the subject of sound in great detail in Chapter 10. Here we simply note that sound waves are accompanied by a displacement of the fluid through which they pass, and as a consequence, when sound passes a solid object, some sort of boundary layer must be established. In this section we examine the nature of this boundary layer.

Consider the situation shown in figure 7.14. A sound wave passes parallel to a flat plate, and in the process imposes an oscillating motion to the fluid.[3] Fluid moves in one direction for a short period, comes to a halt, and then moves back. Accompanying each cycle of displacement is the formation of a boundary layer.

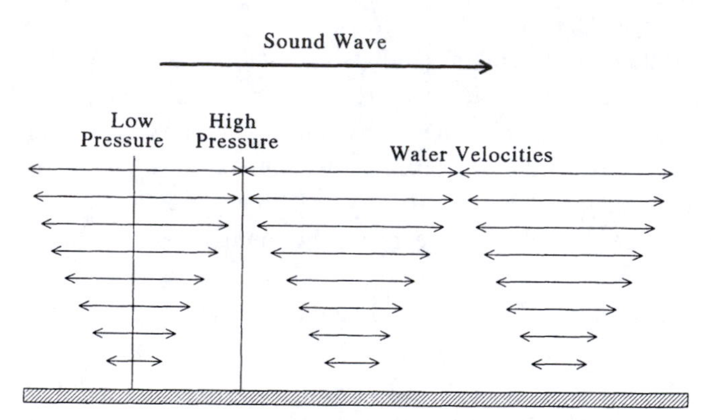

Fig. 7.14 A sonic boundary layer. The direction of fluid velocity varies along a plate, leading to the variation in pressure that forms the sound wave. Velocities near the plate are reduced by the action of viscosity.

We concluded earlier that the thickness of a boundary layer was proportional to the square root of the time allowed for the layer to develop. In the case of the boundary layer associated with a sound wave, this means that the thickness of the layer is governed by f, the frequency of sound. The higher the frequency, the less time fluid moves during each cycle, and the thinner is the boundary layer.

An analysis of oscillating flow next to a solid substratum shows that velocity is virtually the same as that in the mainstream at a distance

$$\delta_s = \frac{\pi}{2}\sqrt{\frac{\nu}{\pi f}}. \tag{7.43}$$

[3]If the fluid is still except for the sound wave, its motion relative to the plate can be described by an equation first derived by G. G. Stokes:

$$u(y,t) = u_0 e^{-(y\sqrt{\pi f/\nu})} \cos\left(2\pi f t - y\sqrt{\frac{\pi f}{\nu}}\right) - u_0 \cos(2\pi f t), \tag{7.42}$$

where $u(y,t)$ describes the instantaneous fluid velocity at time t at a height y above the substratum, and f is the frequency of the sound (in cycles per second). Of particular interest in this equation is the factor $e^{-(y\sqrt{\pi f/\nu})}$, which specifies that the velocity of the fluid approaches that of the mainstream ($-u_0 \cos[2\pi f]$) as distance from the substratum increases.

Thus δ_s can be thought of as the boundary-layer thickness in oscillating flow. Note that δ_s is indeed inversely proportional to frequency.

The thickness of these acoustic boundary layers is shown in figure 7.15 as a function of frequency. At a frequency of one cycle per second, the acoustic boundary layer in air is about 4 mm thick, while at 1000 Hz, it is only 0.1 mm thick. The thickness of acoustic boundary layers in water is smaller still by a factor of about $\sqrt{15} \approx 3.9$.

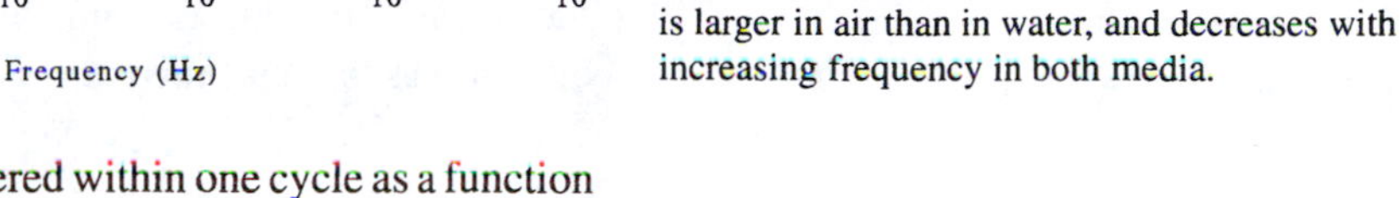

Fig. 7.15 The thickness of sonic boundary layers is larger in air than in water, and decreases with increasing frequency in both media.

The profile of the maximum velocity encountered within one cycle as a function of distance from the substratum in the boundary layer of an oscillating flow is shown in figure 7.16. The profile is quite similar to that obtained for boundary layers in unidirectional flow.

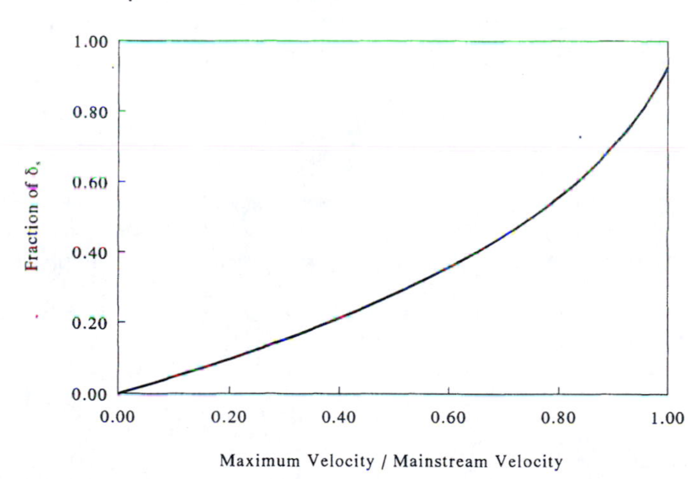

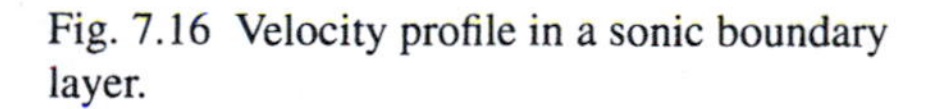
Fig. 7.16 Velocity profile in a sonic boundary layer.

At least one organism makes use of the acoustic boundary layer in the way it senses sound. The posterior end of the abdomen in the common cricket *Gryllus bimaculatus* is covered with a carpet of small sensory hairs of varying lengths. The longer of these cercal hairs (about 1500 μm) are sensitive to low frequency sounds, while the shorter hairs (as short as 30 μm) are sensitive to high frequencies. In a mechanistic sense, a hair is sensitive to a sound wave if the velocity associated with that wave imposes sufficient drag on the hair to push it a critical distance to the side. Once displaced, the hair sets off a reaction from an attached nerve cell, and information regarding sound is transmitted to the central nervous system.

Shimozawa and Kanou (1984) have shown that the length of hairs is precisely tuned to the thickness of the acoustic boundary layer. For example, short hairs, which otherwise would be sensitive to both low and high frequencies, are so short that they are embedded deep within the thick acoustic boundary layer of low frequency sounds and are therefore hidden from this potential interfering factor. At higher frequencies, however, they effectively extend into the mainstream and can be displaced by the moving air. In this fashion, the properties of the boundary layer are used as one part of a filtering system to discriminate sounds of different frequencies.

This is far from the full story, however. For example, why are the long hairs not sensitive to high frequency sounds? If they extend out of the thick, low-frequency boundary layer, they certainly extend out of the thin high-frequency layer. The answer lies in the differential stiffness of long and short hairs and the distance that each must be deflected before it triggers a nerve. These details, although fascinating, are beyond the scope of this chapter, and you are invited to consult Shimozawa and Kanou (1984).

As noted above, acoustic boundary layers in water are about 3.9 times thinner than those in air. If any aquatic animals use an auditory system similar to that of the cricket, all of the hairs will be scaled down by this factor. The small setal hairs on the antennae of copepods may fit the bill (Yen and Nicoll 1990), but neurological evidence as to their function is lacking at present.

7.6 Viscous Drag

Eq. 7.40 not only tells us what the velocity is at a given position in a boundary layer, but also allows us to quantify the velocity gradient at the substratum. By taking the derivative with respect to y, we see that

$$\frac{du}{dy} \approx 0.32 u_\infty \sqrt{\frac{u_\infty}{x\nu}}. \tag{7.44}$$

If we multiply this gradient by the dynamic viscosity of the fluid, we have a measure of the viscous drag per area acting at a given position on the substratum:

$$\frac{\text{drag}}{\text{area}} \approx 0.32 \mu u_\infty \sqrt{\frac{u_\infty}{x\nu}}. \tag{7.45}$$

Rearranging, we see that

$$\frac{\text{drag}}{\text{area}} \approx 0.32 \frac{\mu u_\infty Re_x^{1/2}}{x}. \tag{7.46}$$

This is the shear stress for each particular position on the plate. We can average this value for all x's from the leading to the trailing edge of the plate to give the overall drag per plate area:

$$\frac{\text{average drag}}{\text{area}} \approx \frac{0.64 \mu u_\infty Re^{1/2}}{x}. \tag{7.47}$$

Note that this overall drag is described in terms of Re rather than Re_x.

Now, if our flat plate is thin enough, it has effectively zero frontal area and therefore no pressure drag. In other words, all the drag on a flat plate is skin friction or viscous drag. In this situation we can define a drag coefficient

$$C_{d,p} = \frac{2}{\rho_f u_\infty^2} \frac{\text{average drag}}{\text{plate area}}, \tag{7.48}$$

where the plate area is the product of average length and average width. Combining eq. 7.47 with eq. 7.48, we see that

$$C_{d,p} \approx 1.28 Re^{-1/2}. \tag{7.49}$$

In other words, the drag coefficient of a flat plate is a function of the Reynolds number: the smaller the Reynolds number, the larger the drag coefficient.

In deriving this value, we have used an approximate expression for the velocity gradient. Using the exact expression derived by Blasius, Schlichting (1979) gives the relationship as

$$C_{d,p} = 1.338 Re^{-1/2}. \tag{7.50}$$

This drag coefficient is for the force acting on one surface of the plate. If both surfaces are in contact with fluid, C_d must be doubled:

$$C_{d,p} = 2.676 Re^{-1/2} \quad \text{(both sides exposed to flow).} \tag{7.51}$$

Let us use this expression for the viscous drag coefficient to calculate the drag on plates of width b and length ℓ in both air and water at a given velocity:

$$\text{drag} = \frac{\rho_f u_\infty^2 b\ell C_{d,p}}{2} \tag{7.52}$$

$$= \frac{1.338 \rho_f u_\infty^2 b\ell \sqrt{\mu}}{\sqrt{\rho_f u_\infty \ell}} \tag{7.53}$$

$$= 1.338 \sqrt{\mu \rho_f \ell}\, u_\infty^{3/2} b. \tag{7.54}$$

At a given velocity, the ratio of viscous drag in water to that in air is

$$\frac{\text{drag in water}}{\text{drag in air}} = \frac{\sqrt{\rho_w \mu_w}}{\sqrt{\rho_a \mu_a}}, \tag{7.55}$$

where the subscripts a and w refer to air and water, respectively. At 20°C, this ratio equals 215. Thus, the viscous drag in water is very much higher than that in air.

This is in contrast to the situation at very low Reynolds number (flagellar locomotion, for instance) where the drag force required to push an object through a fluid is directly proportional to the dynamic viscosity and therefore about sixty times larger in water than in air. The difference in scaling is due to the fact that at Reynolds numbers above about 1, the shape of the velocity gradient (which determines the viscous shear stress) is strongly affected by the fluid's density. The greater ρ_f, the thinner the boundary layer, and the greater is the velocity gradient next to the substratum. At very low Reynolds numbers, the velocity gradient is virtually independent of ρ_f.

Because viscous drag at high Re scales as the square root of μ, the effects of temperature on the drag of large animals are minimized. For instance, the drag on a large, streamlined fish (which is primarily skin friction), increases by about 65% as the temperature of the surrounding water decreases from 40° to 0°C. In contrast, over the same temperature range, drag on a slow-moving microorganism in water nearly triples, in direct proportion to the increase in dynamic viscosity.

The drag ratios calculated here have been for organisms in air and water but moving at the same speed. What if organisms in the two media move at the same Reynolds number? To explore this question, we note that for a flat plate of length ℓ,

$$u_\infty = \frac{Re\mu}{\rho_f \ell}. \tag{7.56}$$

Inserting this expression for u_∞ into eq. 7.52, we see that

$$\text{drag} = \frac{1.338 Re^{3/2} \mu^2 b}{\rho_f \ell}. \tag{7.57}$$

As a consequence, the ratio of viscous drag in water to that in air at a given Reynolds number is

$$\frac{\text{drag in water}}{\text{drag in air}} = \frac{\mu_w^2 \rho_a}{\mu_a^2 \rho_w}. \tag{7.58}$$

At 20°C, this ratio is about 3.7. Thus, even if organisms in water and air move at the same Reynolds number, viscous drag is higher in water.

7.7 Mass Transport

In chapter 6 we explored the process of diffusion in a variety of cases in which the concentration gradient was well defined. We noted, however, that fluid flow can at times affect concentration gradients and can thereby affect the diffusive flux of molecules. Flow in a boundary layer is one such time, and we now pick up where we left off.

In exploring the effects of boundary layers on the diffusive flux of molecules, we rely on an analogy between momentum and molecular concentration. We noted earlier that momentum (which is proportional to velocity) is transported through a boundary layer in a fashion similar to that of molecules themselves. Here, we turn this argument around and propose that our knowledge of the velocity gradient in a boundary layer provides us with information about the corresponding chemical gradient.

Consider, for instance, a situation in which fluid containing carbon dioxide at concentration C_∞ passes over a flat leaf that (in the process of photosynthesis) absorbs CO_2. More specifically, we assume that the leaf's affinity for CO_2 is so great that every CO_2 molecule that contacts the leaf is immediately absorbed, in which case the concentration of carbon dioxide at the leaf surface is 0. How fast can CO_2 be delivered to the leaf?

We know from Fick's equation that the flux of CO_2 to the leaf is equal to the product of the diffusion coefficient of carbon dioxide and the concentration gradient of carbon dioxide near the leaf. But the concentration gradient is affected by the flow. While CO_2 is absorbed at a point, new CO_2 is brought into the system by convection. Thus, the concentration gradient is set by the interplay of diffusion and convection.

To see how this interaction works, we rely on an analogy between the velocity gradient and the concentration gradient. As fluid moves over the leaf, momentum "diffuses" out of the mainstream into the solid surface. In other words, the mainstream velocity of the fluid is decreased by the presence of the solid, and the rate at which momentum is transported through the fluid determines the gradient of velocity in the boundary layer. It is reasonable to suppose that CO_2 diffuses from the mainstream fluid by a process similar to that of momentum. Thus, we may guess that dC/dy, the concentration gradient, is somehow proportional to du/dy, the velocity gradient.

We begin the analysis of this connection by multiplying du/dy by C_∞/u_∞, thereby converting the units of the velocity gradient (s^{-1}) to those of the concentration gradient ($\mathrm{mol\ m}^{-4}$). This leaves us with the conclusion that

$$\frac{dC}{dy} \propto \frac{C_\infty}{u_\infty}\frac{du}{dy}. \tag{7.59}$$

Next, we can surmise from eq. 7.44 that $du/dy = 0.32 u_\infty Re^{1/2}/x$. Thus,

$$\frac{dC}{dy} \propto 0.32\frac{C_\infty Re_x^{1/2}}{x}. \tag{7.60}$$

Multiplying both sides of this proportionality by the diffusion coefficient, we see that

$$\mathcal{J} \propto 0.32\frac{\mathcal{D}_m C_\infty Re_x^{1/2}}{x}, \tag{7.61}$$

where $\mathcal{J}$ is the flux density of molecules into the leaf at a distance x from the leading edge.

This proportionality can be made into an equation through the use of an appropriate constant. As one might suspect, this constant is set by the ratio of the rate at which momentum diffuses (the kinematic viscosity, ν) to the rate at which molecules diffuse ($\mathcal{D}_m$), a dimensionless value known as the *Schmidt number*, Sc:

$$Sc = \frac{\nu}{\mathcal{D}_m}. \tag{7.62}$$

A careful analysis of the situation (Bird et al. 1960) shows that flux density scales with the cube root of the Schmidt number. Thus,

$$\mathcal{J} = 0.32\frac{\mathcal{D}_m C_\infty Re_x^{1/2}}{x} Sc^{1/3}. \tag{7.63}$$

If we expand the Reynolds number and the Schmidt number and rearrange the equation, we arrive at our final answer:

$$\mathcal{J} = 0.32\frac{C_\infty u_\infty^{1/2} \mathcal{D}_m^{2/3}}{x^{1/2}\nu^{1/6}}. \tag{7.64}$$

In other words, the flux density of carbon dioxide to the leaf increases with increased mainstream velocity and decreases with increased distance from the leading edge. As a result, plants grow best if there is at least a slight breeze, and eq. 7.64

suggests that it is more effective to have many small leaves (where each point is near a leading edge) than a few large leaves (where points in the center may receive little influx of CO_2). Any increase in the concentration of CO_2 in the atmosphere results in a proportional increase in the rate of delivery to the leaf.

Eq. 7.64 provides us with a useful tool with which to compare the transport properties of air and water. Consider a point on a flat plate (a leaf) a given distance from the leading edge. What are the relative rates of CO_2 delivery to this point if the leaf is in water as opposed to air? If the mainstream velocity is the same in each case,

$$\frac{\mathcal{J}_a}{\mathcal{J}_w} = \frac{C_a}{C_w}\left(\frac{\nu_w}{\nu_a}\right)^{1/6}\left(\frac{\mathcal{D}_{m,a}}{\mathcal{D}_{m,w}}\right)^{2/3}. \tag{7.65}$$

Now, the kinematic viscosity of air is about fifteen times that of water, so $(\nu_w/\nu_a)^{1/6} = 0.64$. The diffusivity of carbon dioxide in air is 10,000 times that in water, so $(\mathcal{D}_{m,a}/\mathcal{D}_{m,w})^{2/3} = 464$. Thus, if the concentration of CO_2 is the same in water as in air (and at 10° to 20°C, it is) , a terrestrial leaf receives CO_2 at a rate about 300 times that of an aquatic leaf. This is in distinct contrast to the 10,000-fold difference one would expect based on a knowledge of the diffusion coefficient alone.

This analysis suggests that aquatic plants would have to live where water flowed swiftly or have smaller "leaves" than their terrestrial cousins to maintain an adequate delivery rate of carbon dioxide. However, most aquatic plants can use bicarbonate as their source of carbon. At pH = 8.0 and 20°C, the HCO_3^- concentration in water is about 130 times higher than the CO_2 concentration in air (tables 6.3, 6.4). Thus, the delivery of carbon to a leaf in air (for a given mainstream velocity and at a given distance from a leading edge) is only about $300/130 \approx 2.3$ times that in water.

In chapter 6 we concluded that diffusion alone was much more effective at delivering molecules to small organisms than any movement they might muster. But the analysis we have just finished seems to suggest that for objects the size of leaves, movement of the fluid relative to the object can substantially increase the rate of molecular transport. Is there a specific size or velocity at which convection first becomes important? To answer this question, we consider the rate at which molecules are diffusively delivered to a sphere.

Bird et al. (1960) show that for a sphere of diameter d,

$$\mathcal{J} = \frac{2\mathcal{D}_m C_\infty}{d} + \frac{0.6 C_\infty}{d} Re^{1/2} Sc^{1/3}, \tag{7.66}$$

where the term $2\mathcal{D}_m C_\infty/d$ is the flux density in the absence of convection. Expanding the terms for Re and Sc, we find that

$$\mathcal{J} = \frac{2\mathcal{D}_m C_\infty}{d} + \frac{0.6\mathcal{D}_m^{2/3} C_\infty u_\infty^{1/2}}{d^{1/2}\nu^{1/6}}. \tag{7.67}$$

If we assume that the maximum velocity that an organism can attain by its own exertion is about ten body lengths per second, we can replace u_∞ in eq. 7.67 with $10d$. Thus,

$$\frac{\mathcal{J}}{C_\infty} = \frac{2\mathcal{D}_m}{d} + \frac{1.9\mathcal{D}_m^{2/3}}{\nu^{1/6}}. \tag{7.68}$$

And finally, we can compare the flux density (per mainstream concentration) at a given speed to the flux density in the absence of convection:

$$\frac{\mathcal{J}_{rel}}{C_\infty} = 1 + \frac{0.95d}{\mathcal{D}_m^{1/3}\nu^{1/6}}. \tag{7.69}$$

This expression is graphed in figure 7.17.

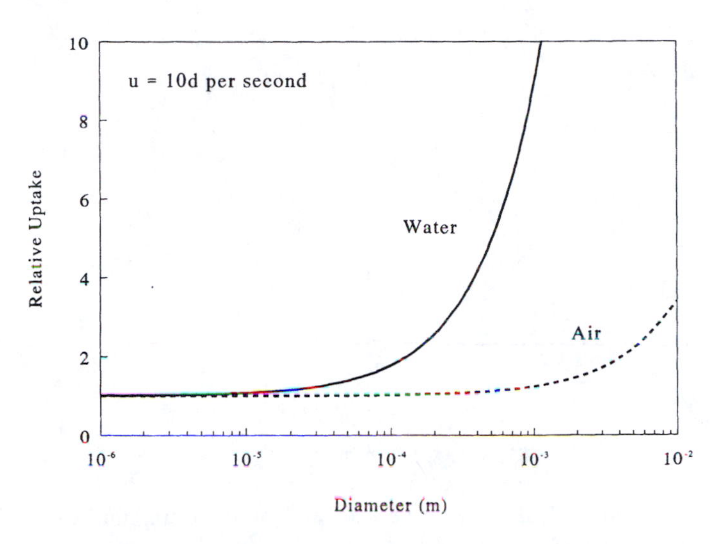

Fig. 7.17 The relative rate at which a sphere can take up molecules is lower in air than in water when the sphere travels at a speed of 10 diameters per second.

In water, locomotion begins substantially to affect the delivery rate of molecules when a spherical organism is about 100 μm in diameter. Thus, by swimming, aquatic organisms larger than the size of small larvae or copepods can actively increase the rate at which they gain oxygen or lose carbon dioxide. Smaller organisms are just as well off relying on diffusion. For a more thorough discussion of the advantages of locomotion in aquatic organisms, consult Berg and Purcell (1977) or Mann and Lazier (1991).

In air, an organism must be larger than about 5 mm in diameter before locomotion enhances diffusive transport of molecules.

To this point we have assumed that the organism increases convection by swimming, but many organisms, plants in particular, are incapable of locomotion. An alternative strategy is available: the organism can sink. Thus, phytoplankton might be able to enhance their uptake of dissolved nutrients if they can sink fast enough. The rate at which a phytoplankter can sink, however, is limited by its terminal velocity. How big must a spherical aquatic plant be before it is better to sink than to swim?

If we assume that the organism is small enough so that it sinks at a low Reynolds number, we can use Stokes's equation (eq. 7.6) to predict its maximal sinking speed:

$$v_{max} = \frac{d^2 \rho_e g}{18\mu}. \tag{7.70}$$

Substituting this expression for u_∞ in eq. 7.67 and solving as before, we find that

$$\frac{\mathcal{J}_{rel}}{C_\infty} = 1 + \frac{0.07d^{3/2}\rho_e^{1/2}g^{1/2}\rho_f^{1/6}}{\mathcal{D}_m^{1/3}\mu^{2/3}}. \tag{7.71}$$

This expression is graphed in figure 7.18. As with motile organisms, a sinking phytoplankter must be about 100 μm in diameter before it moves fast enough relative to its surroundings substantially to affect its ability to acquire nutrients. Curiously enough, the same size limit applies in air. Unfortunately, an aerial phytoplankter 100 μm in radius would have a terminal velocity well in excess of 1 m s^{-1}, and therefore would not be able to sink for long before it hit the ground.

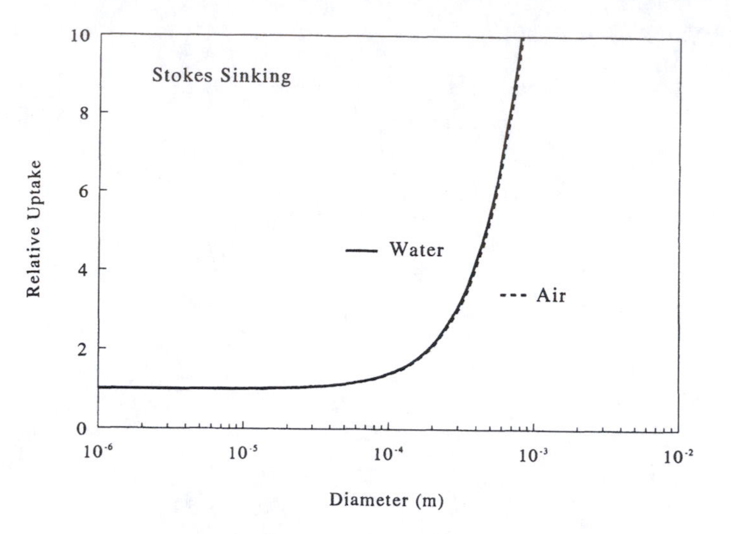

Fig. 7.18 The relative rate at which a sphere can take up particles when sinking at its terminal velocity is similar between air and water.

The conclusions we have reached here regarding the advantages of movement have tacitly assumed that the fluid through which the plant or animal moves is stationary. If instead the fluid is turbulently mixed, the rate of exchange between organism and environment can be increased. For a discussion of this effect, consult Lazier and Mann (1989).

In this section, we have explored two examples in which the presence of a boundary layer affects the exchange of molecules between an organism and its environment. The boundary layers can also affect the rate at which heat is transported, a topic that will be discussed in chapter 8.

7.8 Smelling the Boundary Layer

Before leaving the subject of mass transport in boundary layers, we explore one more aspect. As we have noted, the action of viscosity serves to decrease the speed of fluid in the boundary layer below that of the adjacent mainstream. This can be a problem for those animals that orient themselves through the use of smell. For example, many fish hunt by homing in on the smell of their prey. If the scent reaching the nose must travel in a boundary layer, it is quite possible for a fish to spend a considerable period waiting for the smell to reach its nose. This delay means that the fish smells the water where it has been rather than where it is. This effect could make tracking prey extremely difficult.

To see how bad the problem is, we return to eq. 7.40 in which the velocity of fluid in the boundary layer is specified as a function of y for each distance x downstream of the leading edge. We assume that the fish swims at a speed of u_∞. If water enters the boundary layer at the leading edge of the fish's head, we can then calculate how long it takes to reach the position of the nose if the nose is located at height y above the fish's body surface. Noting that velocity is dx/dt, we see that the time it takes to travel a small distance dx at one level in the inner boundary layer is

$$dt = \frac{\sqrt{\frac{x\nu}{u_\infty}}}{0.32u_\infty y}dx. \tag{7.72}$$

Integrating both sides of this equation, and setting $t = 0$ at $x = 0$, we find that the time needed to travel a distance x is

$$t = \frac{2.08\sqrt{\nu}}{y}\left(\frac{x}{u_\infty}\right)^{3/2}. \tag{7.73}$$

Now, it takes water in the mainstream a time x/u_∞ to travel the distance x. The difference in time between the smell arriving through the boundary layer and its arrival at a position parallel to x, but outside the boundary layer, is thus

$$\Delta t = \frac{2.08\sqrt{\nu}}{y}\left(\frac{x}{u_\infty}\right)^{3/2} - \frac{x}{u_\infty}. \tag{7.74}$$

During this lag time, the fish (which swims at velocity u_∞) has traveled a distance

$$\Delta x = \frac{2.08\sqrt{\nu}}{y}\frac{x^{3/2}}{\sqrt{u_\infty}} - x \tag{7.75}$$

$$\approx x\left(\frac{0.4\delta_{bl}}{y} - 1\right), \tag{7.76}$$

for $y < 0.4\delta_{bl}$.

This expression is shown in figure 7.19. When y is small (i.e., when the opening to the nostrils lies more or less flush with the body surface), the fish may swim a considerable distance before the smell finally arrives at its nose. Even a slight elevation of the entrance to the nostril is sufficient, however, to alleviate the problem. In fact, many fish place the upstream opening to their nostrils on short stalks (fig. 7.20), sufficiently high to avoid most of the lag time between swimming through a volume of water and smelling it.

Note that it would also be advantageous to place the nostrils as far forward

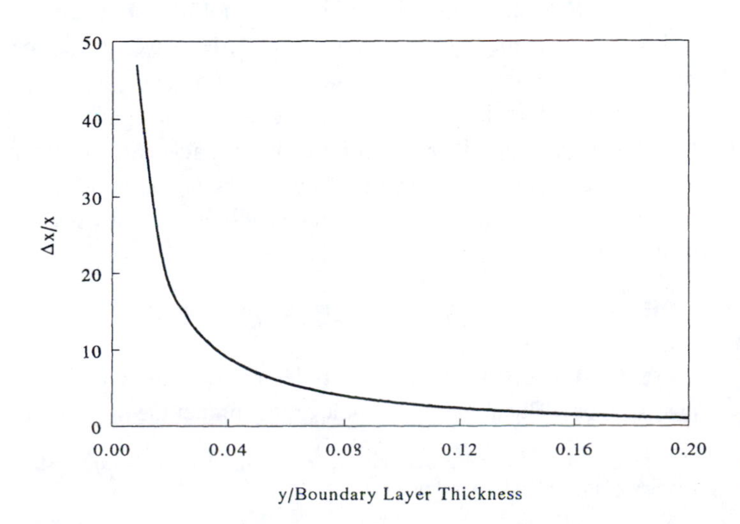

Fig. 7.19 The distance a fish must swim before scent travels from the leading edge of its head to its nose, expressed as a multiple of the distance from leading edge to nose. If the opening to the nose is flush with the surrounding body surface, the fish smells where it has been rather than where it is.

on the head as possible (thereby minimizing x) and to swim fast. The relative importance of these considerations has not, to my knowledge, been examined.[4]

Because the lag time before a smell is sensed increases as the square root of ν, smelling in air (where ν is about fifteen times that in water) might pose a larger problem than smelling in water. As a practical matter, it probably does not. Smell-orienting animals such as deer and dogs actively draw air into their noses, thereby avoiding the kind of lags described above. Insects that orient to scents (moths, for example) generally perch their smelling apparatus on antennae carried well in advance of the body and thereby minimize the effects of the boundary layer. The problems that moths might have with the smaller boundary layers around individual filaments in their antennae are examined by Vogel (1983).

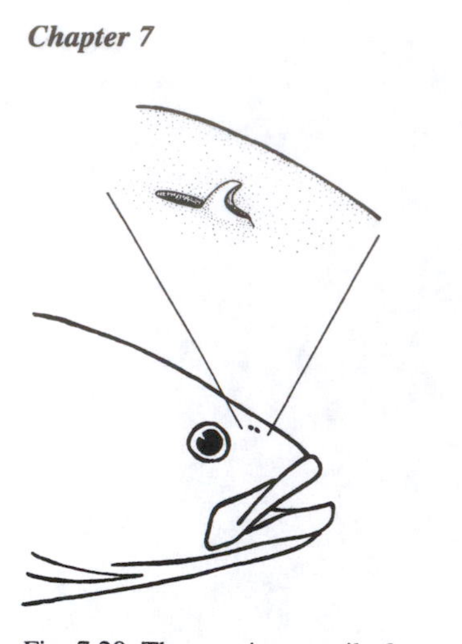

Fig. 7.20 The anterior nostril of many fishes is raised above the surrounding body surface, and a flap directs water into the nose.

7.9 Suspension Feeding

As we noted above, many organisms make a living by *suspension feeding*, that is, by separating food particles from the surrounding fluid. Rubenstein and Koehl (1977), in what has become a seminal work, noted that many suspension feeders manage to trap particles whose diameter is considerably smaller than the "mesh size" of the organs used for feeding. As an explanation for this phenomenon, they draw a comparison between biological filters and the filters used by engineers to separate aerosol particles (such as dust and smoke) from air. For example, the filter you might find in the cold-air intake of your furnace is formed of a loose reticulum of glass fibers with a spacing between fibers much larger than the diameter of a dust mote. Nonetheless, the filter is quite efficient at trapping dust. How do these sorts of filters work?

If small particles are to be captured, the filtration system must have three characteristics. First, the fluid in which particles are suspended must flow; this is the mechanism by which particles are delivered to the filter. Second, there must be solid elements that extend into flow. Finally, these "mesh elements" must be sticky. In other words, if a particle impacts upon a mesh element it is assumed that it adheres and is thereby trapped.

Given these prerequisites, Rubenstein and Koehl describe a variety of mechanisms whereby small particles can be brought into contact with mesh elements. For example, the trajectory of a particle as it passes through the filter can be affected by gravity so that it contacts a mesh element, or it can wander into the element by random Brownian motion or by its own motility. Or a particle can be attracted to the filter elements by electrostatic forces. However, these mechanisms are of minor biological importance compared to two others.

The first of these is straightforward. If the path of a particle as it moves through the filter is such that it passes within one particle radius of a mesh element, it contacts the element and is captured by *direct interception*. Rubenstein and Koehl provide a dimensionless index of the effectiveness with which this mechanism captures particles:

$$K_d = \frac{d_p}{d_c}, \tag{7.77}$$

where d_p is the diameter of the particle and d_c is the diameter of the cylindrical mesh element. The larger the particle or the thinner the cylinder, the more effective

[4] I thank Håkan Westerberg for bringing to my attention the fact that fish smell, along with the accompanying fluid-mechanical problems.

is the filtration. Note that this index is independent of the properties of the fluid or the specifics of the flow.

The second potentially important mechanism of particle capture is that of *inertial impaction*. Consider the situation shown in figure 7.21. A cylindrical mesh element (here seen end-on) extends into a steady current, causing streamlines[5] to bend around it. Imagine what happens to a particle as it moves along one of these streamlines. If the particle has the same density as the fluid, its path is bent as it flows around the cylinder and the particle behaves as if it were made of the fluid. Only if the streamline lies within one particle radius of the mesh element is the particle captured. If, however, the particle is more dense than the fluid, its inertia does not allow it to make as sharp a bend as that taken by the fluid. As a result, if the streamline is sufficiently close to the mesh element, the particle's excess density causes it to cross streamlines and impact on the cylinder. Thus, inertial impaction allows mesh elements to trap particles from a larger cross section of the flow. The process is a microscopic equivalent to the bugs that impact the windshield of your car as you cruise the highway.

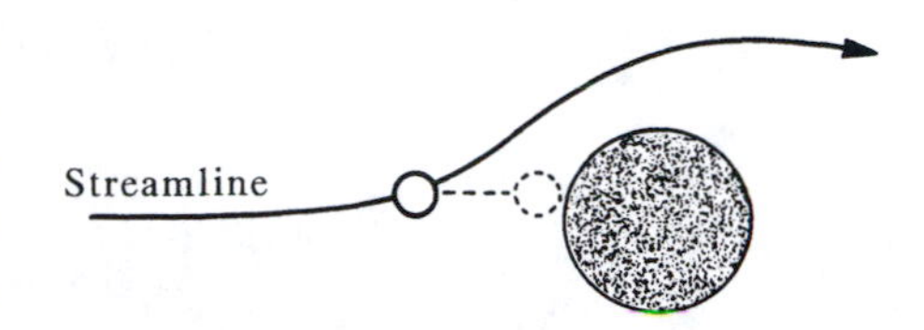

Fig. 7.21 If a particle is denser than the fluid in which it is suspended, it cannot follow an abrupt change in the direction of a streamline. As a result, the particle's inertia may cause it to impact on the catching surface of a filter.

Rubenstein and Koehl (based on theories presented by Fuchs 1964 and Pich 1966) proposed that the effectiveness of inertial impaction depends on the size of particles and mesh elements, the relative density of fluid and particles, the viscosity of the fluid, and the speed of the current. These factors are combined into a dimensionless index of filtering effectiveness:

$$K_i = \frac{d_p}{18d_c}\frac{\rho_e u d_p}{\mu}, \tag{7.78}$$

where ρ_e is the effective density of the particles.

In essence, this index is the product of a term proportional to the index of direct interception, K_d, and a term in the general form of a Reynolds number, which we may call the *particle Reynolds number*, Re_i:

$$Re_i = \frac{\rho_e u d_p}{\mu}. \tag{7.79}$$

Thus, the index of inertial impaction becomes

$$K_i = \frac{K_d Re_i}{18}. \tag{7.80}$$

For the purposes of this chapter, we are more interested in the summed effect of K_i (which depends on the properties of the fluid) and K_d (which does not), than we are in the magnitude of either index alone. Rearranging, we see that the overall index of filtration effectiveness is

$$K_o = K_d + K_i \tag{7.81}$$

$$= K_d\left(1 + \frac{Re_i}{18}\right). \tag{7.82}$$

In other words, when the particle Reynolds number exceeds 18, inertial impaction is a more effective means of filtration than is direct interception, and the overall effectiveness of filtration is more than twice that due to direct interception alone.

[5]A fluid-dynamic definition of a streamline is considered in chapter 13. For present purposes, the intuitive meaning of the word, the path that a fluid takes as it flows, should suffice.

Under what conditions will Re_i be large enough for inertial impaction to be important? As with terminal velocity, it is the effective density of particles that really governs the physics. Given particles of equal sizes moving in currents of equal velocity, Re_i for biological particles in air (where ρ_e = 1079 kg m^{-3}) is about 1100 times that in water (where ρ_e = 55 kg m^{-3}). For example, a particle 40 μm in diameter in air moving at 1 m s^{-1} has a particle Reynolds number of about 2400, much in excess of the critical value of 18. In constrast, the same size particle in water moving at the same velocity has a particle Reynolds number of only 2.2. And this example probably underestimates the typical difference in Re_i: 1 m s^{-1}, a mere breeze in air, is much faster than the currents typically found in the feeding apparatuses of marine organisms. At lower velocities, Re_i in water is lower still. Thus, we expect inertial impaction to be much more effective (and therefore much more important) in air than in water. In fact, the effectiveness of inertial impaction in air is so high that one might expect suspension feeding to be more widespread among terrestrial organisms than among those in the aquatic environment. How does this prediction compare with reality?

Despite its general reliance on direct interception alone, particle filtration is a wide spread phenomenon in aquatic environments. For example, direct interception has been implicated as the dominant mechanism in brittle stars, crinoids, zooanthids, black-fly larvae, spionid polycheates, cladocerans, and molluscan veliger larvae (LaBarbera 1984). The success of suspension feeding in water, despite its intrinsically lower efficiency, is likely due to the richness of the aquatic "soup"; even an inefficient filter may catch sufficient food.

In contrast, because there is often little in the way of suspended particulate matter available to be captured (for reasons we examined earlier), suspension feeding in air is rare despite its intrinsic effectiveness. There are, however, a few examples in which organisms apparently use inertial impaction as a means to remove aerosol particles from the air.

For example, the female cones of pine trees apparently capture pollen by inertial impaction (Niklas 1982a,b, 1987). Furthermore, because different species have slightly different particle Reynolds numbers, the design of cones allows for a selective filtration of pollen: the pollen captured by one species is slightly enriched in the fraction of pollen from that species relative to that of others.

In the remaining cases of terrestrial filtration, the particles captured are made of water in the form of fog.

Consider the plight of redwood and oak trees growing along the coast of California. Like all plants, these require a regular influx of water, but at this location it often does not rain between May and October. It is, however, often foggy. As the prevailing winds blow fog inland, the thin needles of the redwoods and the "Spanish moss"[6] epiphitic on oaks act as filters and fog droplets collect on them. This trapped water eventually drips to the ground under the crown, and has been shown to augment the trees' water intake substantially (Kerfoot 1968). It has also been noticed that those trees at the upwind edge of a stand, where the wind velocity is highest, capture the most fog. Broad-leafed trees in the same location are much less effective at filtering fog from the air.

The correlation of fog capture rates with the slenderness of mesh elements and with wind speed renders it likely that these coastal trees are using inertial impaction as a means for capturing fog. I know of no study, however, in which this hypothesis has been tested directly.

[6]On California oaks, what looks like Spanish moss is actually a lichen.

Other curious examples of fog capture can be found in desert environments. For example, in the Namib desert, rainfall is exceedingly scarce (a few mm per year), but the area is sporadically covered by fogs blown inland from the nearby coast. Several species of beetles have taken advantage of the fog as a source of water. For example, beetles in the tenebrionid genus *Lepidochora* construct shallow sand trenches in a line perpendicular to the direction of the prevailing wind. As the trench is dug, the sand removed is piled on either side, and it is these ridges that actually trap the fog. It can be hypothesized that the ridges act like mesh elements, diverting streamlines and causing fog droplets to impact on the sand inertially. After sufficient water has collected, the beetles suck it from the sand (Seely and Hamilton 1976).

Another example involves the beetle *Onymacris unguicularis*, which exhibits a behavior known as "fog basking." When fog is present, this insect moves from its usual position buried in a dune to a position on the dune surface at the crest of a ridge where the wind velocity is highest. It then assumes a head-down posture, lifting its abdomen up into the wind (fig. 7.22). Fog droplets collect on the legs and abdomen (by both direct interception and inertial impaction) and eventually trickle down to the mouth where they are imbibed. By this mechanism beetles can collect up to 34% of their body weight during a single basking session (Hamilton and Seely 1976).

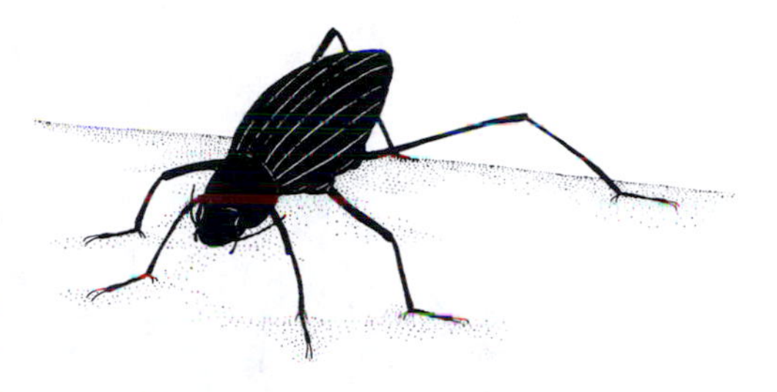

Fig. 7.22 A Namibian beetle (*Onymacris unguicularis*) in its "fog basking" posture.

Neither of these methods of filtration involves the sort of highly tuned feeding structures used by marine organisms or even those of pine cones, but the intrinsic effectiveness of inertial impaction in air allows these terrestrial "suspension feeders" to make do with less sophisticated machinery.

7.10 Summary . . .

The high effective density of terrestrial organisms, coupled with the low viscosity of air, means that terrestrial plants and animals have a high terminal velocity. As a result, the aerial plankton is much sparser than that found in lakes and oceans. The relatively low terminal velocity of animals in water may set an upper limit to the speed with which these creatures can walk.

Boundary layers at a given mainstream velocity are thinner in water than in air. As a result, it is harder for aquatic organisms to "hide" in the boundary layer than it is for their terrestrial cousins, but conversely, it is more difficult for small terrestrial organisms to be dispersed. Mites and scale insects may be "trapped" by the boundary layer surrounding leaves.

At high Reynolds and Schmidt numbers, the diffusive delivery of gases and nutrients is strongly affected by mainstream velocity; the faster the velocity, the more effective the delivery. As a result, it is often advantageous for an organism to sink or swim, thereby to create a relative velocity between itself and the surrounding fluid.

Suspension feeding in air is instrinsically more efficient than that in water due to the efficacy of inertial impaction, and a few terrestrial organisms effectively filter fog. But the concentration of aquatic plankton is sufficiently high that filter feeding is much more prevalent in water than in air.

7.11 . . . and a Warning

This chapter is only the briefest of introductions to the rich interaction between density and viscosity in nature. You are urged to consult both the original literature cited here as well as the texts by Vogel (1981), Grace (1977), and Denny (1988) for a more thorough discussion of this important subject. As always, the examples explored here, although correct within the limitations of their context, are not meant to be applied blindly. Before attempting to apply the information presented in this chapter to any specific biological example, you should carefully consult the original literature.

Chapter 8

Thermal Properties: Body Temperatures in Air and Water

In the preceding seven chapters, we have spent considerable time exploring the role of temperature as it affects the physics of fluids. Now it is time to examine the physics of heat itself as it applies to air and water. In the process, we will see how bumblebees and voodoo lilies keep warm, and why only large, terrestrial organisms have a problem keeping cool. We will calculate the heat lost when animals respire. The results help us to interpret why dolphins, whales, and seals still breathe air, and to appreciate the evolved ingenuity of warm-bodied tunas and sharks.

8.1 The Physics

8.1.1 *Specific Heat Capacity*

We begin by recalling (from chapter 3) that heat and temperature, although related, are not the same. Heat is the kinetic energy associated with the disordered motion of molecules. Temperature, on the other hand, is a measure of the average kinetic energy of the molecules in a material.

To give some tangibility to this difference, take a liter of water and, as an analogy to adding heat, add ten drops of blue dye. As a result, the water turns blue, the color being an analogue to temperature. Note that the quantity of dye and the solution's color have different properties. For example, if you take a 100 ml sample of the blue water, it contains only one-tenth as much dye as the entire liter, but still has the same color. Thus the size of the sample affects the *quantity* of dye present, but not the *quality* (color) it lends to the fluid.

Now let us consider heat itself. Again, heat is a measure of total thermal energy. The amount of heat contained in a given sample of material depends, therefore, on two factors. First, it depends on the mass and average speed of each molecule in the material because these define the kinetic energy. For a given average velocity, the more massive the molecules of a material, the greater the heat. Similarly, for a given average mass, the faster the molecules move, the greater the heat. Second, heat depends on the number of molecules in the sample. Obviously, if the kinetic energy of individual molecules is kept the same, the larger the sample, the more heat it contains.

In contrast, temperature, as a measure of *average* rather than *total* kinetic energy, is independent of the number of molecules present. The rarefied gas in the upper reaches of the earth's atmosphere provides an extreme example. There may be only a few hundred molecules in each cubic meter of this near-vacuum, but each of these few molecules typically possesses a substantial kinetic energy. For instance, the temperature of the atmosphere at an altitude of 1000 km is a torrid 1000 K. Although the temperature is quite high, a given volume of the material cannot contain much heat; there are simply too few molecules.

For a given material, the relationship between heat and temperature is measured by the number of joules that must be added to one kg of the material to raise its temperature one kelvin. This relationship is the material's *specific heat capacity*, or, for short, its *specific heat*, Q_s. For a gas, it matters whether specific heat is measured under conditions of constant pressure or of constant volume, the former being larger than the latter. For our purposes in this chapter, we are concerned with a gas (air) that under most biological conditions is free to expand, and therefore we use values for specific heat at constant pressure.

The specific heats of materials are quite variable. Hydrogen, for instance, has a very high specific heat, about 14,000 J kg^{-1} K^{-1}. Gold is on the other end of the spectrum with a specific heat of only 130 J kg^{-1} K^{-1}. In general, gases have higher specific heats than liquids or solids.

The specific heat of air is 1006 J kg^{-1} K^{-1}, a typical value for a gas. It varies negligibly across the biological range of temperature (table 8.1). Water, however, is an exception to the general rule. It has a very high specific heat for a liquid (table 8.2). In fact, with the exception of liquid ammonia, water has the highest specific heat of any room–temperature liquid, about 4200 J kg^{-1} K^{-1}, varying from 4218 at 0°C to 4179 at 40°C. It thus takes about four times as much heat to raise the temperature of a kilogram of water one degree as it does to raise the temperature of an equal mass of air.

Table 8.1 Properties relating to the transport of heat in air.

Property of Air	Units	Temperature (°C)				
		0	10	20	30	40
Q_s, specific heat capacity	J kg^{-1} K^{-1}	1006	1006	1006	1006	1006
$\mathcal{D}_H$, thermal diffusivity	$m^2\,s^{-1} \times 10^{-6}$	18.9	20.2	21.5	22.8	24.2
$\mathcal{D}_{O_2}$, diffusivity of O_2	$m^2\,s^{-1} \times 10^{-6}$	17.8	18.9	20.1	21.4	22.7
$\mathcal{K}$, thermal conductivity	$W\,m^{-1}\,K^{-1}$	0.0247	0.0254	0.0261	0.0268	0.0276
β, thermal expansivity	$K^{-1} \times 10^{-3}$	3.60	3.53	3.41	3.30	3.19
ν, kinematic viscosity	$m^2\,s^{-1} \times 10^{-6}$	13.2	14.8	15.3	16.2	17.2
μ, dynamic viscosity	$N\,s\,m^{-2} \times 10^{-6}$	17.2	17.7	18.2	18.7	19.1
Pr, Prandtl number		0.698	0.733	0.712	0.711	0.711
Le, Lewis number (O_2 in air)		0.94	0.94	0.93	0.94	0.94
ρ_f, density at 1 atmos.	$kg\,m^{-3}$	1.292	1.246	1.204	1.164	1.128

Sources: Armstrong 1979; Campbell 1977; Schlichting 1979.
Note: Actual values are obtained by multiplying the value listed for a given temperature by the units listed in the second column.

The relative similarity of specific heat between air and water can be misleading, however, because specific heat is measured on a per-mass basis. One cubic meter of air weighs only about 1.2 kg, while a cubic meter of water weighs 1000 kg. It thus takes about 3500 times as much heat to raise the temperature of a given *volume* of water one degree as it does to raise the temperature of the same volume of air.

8.1.2 *Thermal Conductivity*

Much of the biology of heat has to do with how energy is transported to and from organisms. Because heat energy is related to the random motion of molecules, we might guess that it is transported diffusively (chapter 6). Indeed, this turns out to be the case. Heat moves down a gradient of temperature in exactly the same fashion that molecules move down a gradient in concentration, and the rate at which heat moves depends on the *thermal conductivity*, $\mathcal{K}$ of the material through which heat is transported. This quantity is defined in a manner analogous to that of the molecular diffusion constant:

$$\mathcal{J}_Q = \frac{1}{A}\frac{dQ}{dt} = \mathcal{K}\frac{dT}{dx}, \tag{8.1}$$

where $\mathcal{J}_Q$ is the flux density of heat (in W m^{-2}), Q is heat (in joules), A is the area over which heat is transported, and dT/dx is the gradient of temperature in the direction of its transport. $\mathcal{K}$ has the units of W m^{-1} K^{-1}.

Table 8.2 Properties relating to the transport of heat in fresh water water and seawater (S = 35).

Property of Water	Units	Water	Temperature (°C) 0	10	20	30
Q_s, specific heat capacity	J kg^{-1} K^{-1}		4218	4192	4182	4179
$\mathcal{D}_H$, thermal diffusivity	m^2 s^{-1} × 10^{-6}		0.134	0.140	0.143	0.148
$\mathcal{D}_{O_2}$, diffusivity of O_2	m^2 s^{-1} × 10^{-6}		0.00099	0.00154	0.00210	0.00267
$\mathcal{K}$, thermal conductivity	W m^{-1} K^{-1}		0.5651	0.5867	0.6011	0.6157
β, thermal expansivity	K^{-1} × 10^{-3}	Fresh	−0.0680	0.0881	0.2067	0.3031
		Sea	0.0526	0.1668	0.2572	0.3350
ν, kinematic viscosity	m^2 s^{-1} × 10^{-6}	Fresh	1.79	1.31	1.01	0.80
		Sea	1.84	1.35	1.06	0.85
μ, dynamic viscosity	N s m^{-2} × 10^{-6}	Fresh	1790	1310	1010	800
		Sea	1890	1390	1090	870
Pr, Prandtl number		Fresh	13.4	9.36	7.06	5.41
		Sea	14.10	9.93	7.62	5.88
Le, Lewis number (O_2)			0.007	0.011	0.015	0.018
ρ_f, density at 1 atmos.	kg m^{-3}	Fresh	999.87	999.73	998.23	995.68
		Sea	1028.11	1026.95	1024.76	1021.73

Sources: Weast 1977; UNESCO 1987.

Note: When separate values are not given for fresh water and seawater, a property may be assumed to be similar in both media.

As materials go, the thermal conductivity of air is quite low, ranging from 0.0247 W m^{-1} K^{-1} at 0°C to 0.0276 W m^{-1} K^{-1} at 40°C. The thermal conductivity of water is about twenty-three times as great, ranging from 0.57 W m^{-1} K^{-1} at 0°C to 0.63 W m^{-1} K^{-1} at 40°C.

Although the thermal conductivity of water is large relative to that of air, it is small compared to that of many other materials (table 8.3). For instance, silver and copper conduct heat nearly seven hundred times better than water.

The thermal conductivity of body tissue is about 75% of that of free water, and is therefore a relatively good conductor of heat. In contrast, the thermal conductivity of fur is quite low, only about twice that of air.

Table 8.3 Thermal conductivity of some common materials.

Material	Thermal Conductivity $\mathcal{K}$, W m^{-1} K^{-1}
Silver	405.8
Copper	384.9
Aluminum	209.2
Steel	46.0
Glass	1.046
Water	0.586
Human tissue	0.460
Dry soil	0.335
Rubber	0.167
Wood	0.126
Animal fur	0.038
Air	0.025

Source: Schmidt-Nielsen (1979).

8.1.3 *Diffusivity of Heat*

The thermal conductivity of a fluid depends on three factors. First there is the diffusivity of heat in the material, $\mathcal{D}_h$, which, like the molecular diffusion coefficient, describes the tendency for *molecules* to wander from their starting point. Conductivity is also proportional to the material's density and specific heat, which together account for the amount of *energy* that is transported when molecules move. The product of these factors thus describe the ability of heat to move diffusively:

$$\mathcal{K} = \mathcal{D}_h \rho_f Q_s. \tag{8.2}$$

In air, the diffusivity of heat is very similar to the molecular diffusivity of oxygen or nitrogen (table 8.1). This makes intuitive sense, since it is the motion of oxygen and nitrogen molecules that are transporting the heat. Thus, the process of thermal diffusion in air is quantitatively, as well as qualitatively, similar to molecular diffusion.

The situation in water is quite different. The diffusivity of heat in water is much lower than that in air, as one might guess, but it is not nearly as low as the diffusivity of molecules. For example, at 10°C the diffusivity of heat in water is about one hundred times that of oxygen (table 8.1). This large diffusivity of heat relative to that of molecules is due to the manner in which the two are transported in liquids. Oxygen can be moved from point A to point B only if individual oxygen molecules carrom their way through the sea of water molecules between these two points. In contrast, heat can be passed from one molecule to the next. An energetic water molecule at point A can, like a billiard ball, strike another water molecule nearby, and in the process, transfer some or all of its kinetic energy. This second molecule can strike a third, and so forth. In this manner, the kinetic energy originally present at point A can be transported to B without any single molecule making more than a tiny fraction of the trip. The result is that heat diffuses in water two orders of magnitude faster than molecules. In summary, the transport of heat and molecules in water, although qualitatively similar, are quantitatively quite different.

The diffusivity of heat increases with increasing temperature. In air, the diffusivity is 28% higher at 40°C than at 0°C (table 8.1), while in water the increase is about 13%.

8.1.4 *The Prandtl and Lewis Numbers*

We now turn our attention to three dimensionless numbers that prove useful in predicting how heat is transported to and from living organisms.

Recall from chapter 5 that kinematic viscosity, ν, is a measure of the rate at which momentum diffuses in a fluid. For instance, the presence of a solid surface next to a moving fluid results in a decrease in the momentum of the fluid near the substratum. As a result, there is a velocity gradient adjacent to the substratum, the thickness of which is proportional to the square root of kinematic viscosity. The larger ν is, the farther this boundary layer extends into the fluid.

A similar situation applies to the distribution of heat in the vicinity of a solid object when the object is either hotter or colder than the moving fluid around it. Air or water next to the solid's surface is at a temperature very close to that of the solid, and the temperature gradually approaches that of the bulk fluid as the distance from the solid increases. In this manner, a gradient in temperature—the *thermal boundary layer*—is established, analogous to the velocity boundary layer described above. The more diffusive heat is in the fluid, the thicker is the thermal boundary layer.

The ratio of the thickness of these two boundary layers (velocity and heat) is a function of the kinematic viscosity and thermal diffusivity of the fluid (Bird et al. 1960):

$$\frac{\text{velocity boundary-layer thickness}}{\text{thermal boundary-layer thickness}} = \sqrt[3]{\frac{\nu}{\mathcal{D}_h}}. \tag{8.3}$$

The ratio $\nu/\mathcal{D}_h$, which is dimensionless, is known as the *Prandtl number*, *Pr*, named for Ludwig Prandtl, the originator of the notion of a boundary layer:

$$Pr = \frac{\nu}{\mathcal{D}_h}. \tag{8.4}$$

As we will see, the Prandtl number is used in the prediction of the rate at which heat is transported to or from an organism in a moving fluid.

The Prandtl number of air is about 0.7, varying from 0.698 at 0°C to 0.711 at 40°C. Thus, in air the velocity boundary layer is about 89% as thick as the thermal boundary layer.

The Prandtl number of water is considerably larger than that of air and varies substantially with temperature, decreasing from 13.4 at 0°C to 5.4 at 30°C. In other words, the velocity boundary layer in water is 1.8 to 2.4 times thicker than the thermal boundary layer.

Just as we have compared the diffusivity of momentum to that of heat using the Prandtl number, we can compare the diffusivities of molecules and heat (Weast 1977). In this case, the ratio of the two (yet another dimensionless value) is called the *Lewis number*, *Le*:

$$Le = \frac{\mathcal{D}_m}{\mathcal{D}_h} \tag{8.5}$$

The Lewis number for oxygen in air is very nearly 1 (table 8.1). This implies that the gradient in molecular concentration around an object in air is about the same as the thermal gradient around the object, should one exist. In fact, in air all three boundary layers that we have discussed (thermal, velocity, and molecular concentration) are roughly the same thickness, and conclusions regarding one are generally applicable to the others.

Because the diffusivity of molecules in water is small relative to that of heat, the Lewis number in water is similarly small. For oxygen, *Le* varies from 0.007 at 0°C to 0.018 at 30°C. In other words, the boundary layer of molecular concentration is one-fifth to one-quarter the thickness of the thermal boundary layer, which is itself only about a fifth the thickness of the velocity boundary layer. It should be apparent that in water, unlike in air, boundary-layer processes differ substantially depending on what is diffusing.

8.1.5 *The Grashof Number*

As we noted in chapter 4, the densities of air and water are both affected by temperature. Thus, when a solid object is placed in a fluid of a different temperature, the flow of heat into or out of the solid changes the density of the fluid in its immediate vicinity. As a result, the fluid is subjected to a buoyant force (either positive or negative) and begins to move. This temperature-induced motion (termed *free convection*) in turn affects the rate at which heat is exchanged between solid and fluid.

The tendency for free convection in a fluid depends on two factors. The greater the buoyant force, the faster the fluid convects. Conversely, the greater the viscous forces acting on the fluid, the greater its tendency to stay put. It is the ratio of these two factors, a dimensionless value known as the *Grashof number*, *Gr*, that governs the pattern of free convection.

We do not derive the Grashof number here, but it is a simple matter to provide an intuitive understanding of the variables from which it is formed. First we examine the factors that affect the buoyant force.

The greater the difference in temperature between a solid object and its surrounding fluid, the greater the effect is on the density of the fluid, and the greater the tendency to convect. Similarly, the more the fluid expands or contracts for a given change in temperature, the greater the effect on convection. The rate at which a material changes its volume as a function of temperature is known as its *thermal expansivity* or *coefficient of thermal expansion*, β, defined by the equation,

$$\beta = \frac{\Delta V}{V_0 \Delta T}, \tag{8.6}$$

where V_0 is the original volume of the material at a given temperature and ΔV and ΔT are the change in volume and temperature, respectively. β has the units K^{-1}.

Note that for an ideal gas,

$$\frac{\Delta V}{V_0} = \frac{\Delta T}{T_0}, \tag{8.7}$$

where T_0 is the absolute temperature at which V_0 is measured (chapter 3). As a result, for an ideal gas (of which air is a close approximation),

$$\beta = \frac{1}{T_0}. \tag{8.8}$$

The coefficient of thermal expansion cannot be so easily calculated for water; instead it must be empirically measured (see table 8.1).

Now, the coefficient of thermal expansion is a measure of the change in volume per original volume. To arrive at a value proportional to the overall volume of fluid affected by an object (which is closer to what we really want to know), β must be multiplied by a factor related to the size of the solid object. As before when dealing with dimensionless numbers, we chose a "characteristic length" of the object, ℓ_c, and use its cube as an index of the volume of fluid affected.

The product of temperature difference, thermal expansivity, and characteristic volume thus gives a value proportional to the overall change in volume in the fluid. To convert this to a measure of the buoyant force acting on the fluid, we need to multiply first by a factor related to the fluid's density (thereby giving a measure of the change in mass) and then by the acceleration of gravity (to give a force). Thus, the buoyant force acting on the fluid around an object is

$$\text{buoyant force} \propto g\rho_f \ell_c^3 \beta \Delta T, \tag{8.9}$$

where ΔT is the difference in temperature between the object and the surrounding fluid. Note that the sign of the buoyant force tells one whether the flow is upward or downward, but does not affect the transfer of heat.

For our present purpose, the viscous force acting to keep fluid in place is most conveniently expressed as the ratio of the square of dynamic viscosity to density:

$$\text{viscous force} \propto \mu^2/\rho_f. \tag{8.10}$$

As strange as this ratio might appear, it does indeed have the dimensions of a force.

Taking the ratio of the buoyant to the viscous forces, we arrive at last at the Grashof number,

$$Gr = \frac{g\rho_f^2 \ell_c^3 \beta \Delta T}{\mu^2}. \tag{8.11}$$

Noting that $\rho_f^2/\mu^2 = 1/\nu^2$, we see that

$$Gr = \frac{g\ell_c^3 \beta \Delta T}{\nu^2}. \qquad (8.12)$$

What is the relative magnitude of Grashof numbers in air and in water? The answer is somewhat complex because a number of factors need to be taken into account.

First, the thermal expansivity of air is much greater than that of pure water (fig. 8.1). In fact, at 3.98°C, the expansivity of fresh water is 0, and at this temperature air is infinitely more expansive. Unlike fresh water, the expansivity of seawater is always positive within the biological range of temperature. Nonetheless, β is much less than that of air. Thus, if thermal expansivity were the only factor to be taken into account, one would expect the Grashof number of air to be higher than that of water.

The kinematic viscosity of air, however, is an order of magnitude higher than that of water (table 8.1). Because this term is squared and appears in the denominator of the Grashof number, it tends to offset air's greater expansivity. At 20°C, for instance, the Grashof number of water is about fourteen times that of air (fig. 8.2). At lower temperatures, the Grashof numbers for the two fluids are more similar. Between 2° and 4°C, the Grashof number of air is higher than that of fresh water, but still less than that of seawater.

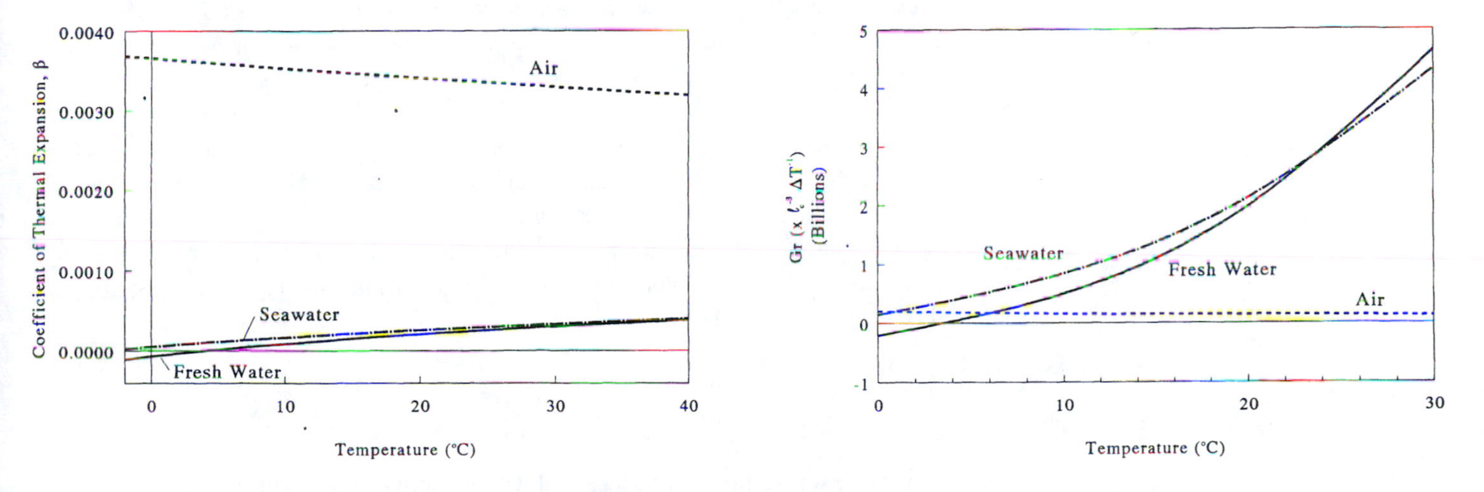

In summary, then, the tendency for free convection is generally stronger in water than in air, a contrast that increases with increasing temperature. The exception to this rule concerns fresh water at temperatures between 0° and 5°C.

Fig. 8.1 A small increase in temperature results in a much larger increase in the volume of air than of water. Note that the thermal expansivity of air decreases with increasing temperature, but the expansivity of water increases.

Fig. 8.2 Except at low temperatures, the Grashof number of an object in water is greater than that in air. The values on the ordinate are the Grashof number normalized to the size of an object (measured as its characteristic length cubed) and to the difference in temperature between the object and its surroundings. To obtain the actual Grashof number in a given case, multiply by $\ell^3\Delta T$.

8.2 Newton's Law of Cooling

Having now introduced all the pertinent thermal properties of air and water, our next task is to use these properties to predict the temperatures of organisms. We start by examining the process of heat exchange.

If a plant or animal is at a different temperature than the surrounding fluid, heat is exchanged at a rate that depends on three factors. First, the greater the difference in temperature between organism and fluid, the faster heat flows. This fact, inherent in eq. 8.1, should be obvious from experience; you get chilled faster if the air temperature is 0°C than if it is 30°C.

Second, the rate of heat transfer depends on the area of contact between the organism and its surroundings; the more area in contact with fluid at a different temperature, the faster heat can be exchanged. Again, this should be evident from experience. When you stand barefoot on a cold bathroom floor while brushing your teeth, you tend to minimize the area of contact with the floor by lifting your toes or balancing on your heels.

And finally, the rate of heat exchange is governed by the thermal and hydrodynamic properties of the fluid, all the factors we have so far explored in this chapter. These thermal and hydrodynamic properties are traditionally lumped together to form a *coefficient of heat transfer*, h_c.

Expressing these relationships as an equation, we see that

$$\text{rate of heat transfer (in watts)} = h_c A \Delta T, \qquad (8.13)$$

where ΔT is the difference in temperature between the organism and its surroundings, A is the area of contact between organism and fluid, and the coefficient of heat transfer has the units W m^{-2} K^{-1}. Eq. 8.13 is known as *Newton's law of cooling*. Using this relationship, we can specify the rate at which an organism gains or loses heat.

Or we could if we knew what h_c was. Therein lies the major problem in all studies of heat transfer. Newton's law of cooling is not so much a law as it is an equation by which h_c is defined. The precise determination of h_c is usually a matter of empirical experiment, and its value changes with the size and shape of the organism as well as with the thermal and hydrodynamic properties of fluids. As a result, it is seldom possible to specify h_c accurately for real organisms based on theory alone. In contrast, h_c values for simple shapes (such as spheres and cylinders) are predictable and can provide rough estimates of biologically relevant values for the coefficient of heat transfer. In this capacity, simple shapes can serve as a tool for exploring the thermal relationships among organisms, air, and water.

8.3 Estimating h_c

Because we are primarily concerned with a comparison between air and water rather than with a comparison between different organisms, we need not worry overly about the particular shape we use. In fact, it will best suit our purposes to choose the simplest shape possible, a sphere. In this case, we use the diameter of the sphere, d, as a characteristic measure of size.

Actually, the idea of using a sphere as a model organism isn't all that far-fetched—there are plenty of plants and animals in the world that are more nearly spherical than cylindrical or cubic. We will deal with cases where the use of a spherical model leads to misleading results as the need arises.

What then are the heat transfer coefficients for a sphere? Three cases are considered: conduction alone, free convection, and forced convection. In the latter two, we assume that the flow of fluid is sufficiently slow so that the boundary layer around the organism remains laminar.

8.3.1 *Conduction Alone*

If the fluid around the sphere is stagnant, heat can be transferred only by conduction. Such a situation is rare in the real world. Even if there is no wind or current, local heating or cooling of the fluid is likely to result in free convection. Nonetheless, we examine this case as a basis for comparison to other, more effective modes of heat transfer. For a sphere transferring heat by conduction alone (Bird et al. 1960),

$$h_{c,cond} = \frac{2\mathcal{K}}{d}. \tag{8.14}$$

This relationship is shown in figure 8.3.

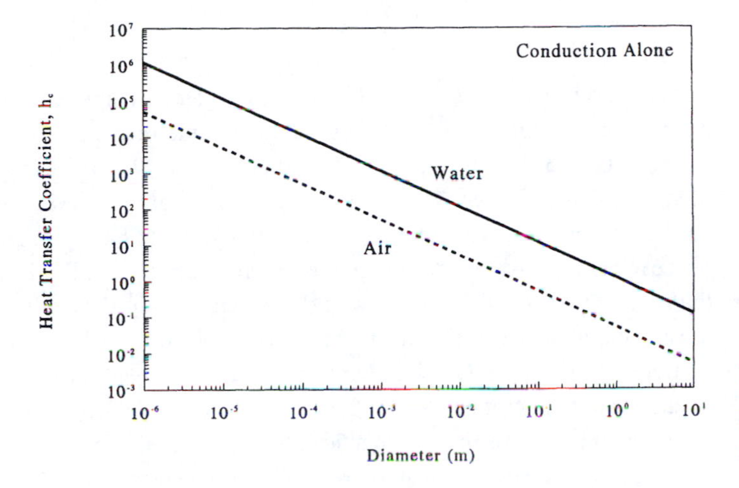

Fig. 8.3 Because the thermal conductivity of water is larger than that of air, the heat transfer coefficient for conduction from a sphere is larger in water than air.

Because we have specified that the fluid does not move, the heat transfer coefficient for an organism of a given size depends only on the conductivity of the fluid. And because the conductivity of water is about twenty-three times that of air, the conductive heat transfer coefficient for a sphere in water is twenty-three times that of the same sphere in air, independent of size.

Note, however, that the heat transfer coefficient itself is inversely proportional to the characteristic size of the organism. This is a general property of heat transfer, and it implies that larger organisms have a more difficult time gaining heat from, or losing heat to, their environment.

8.3.2 *Free Convection*

Next, we consider a situation in which the fluid around the sphere is moving due to free convection. In this case, the rate of heat transfer depends on the pattern of flow, for which the Grashof and Prandtl numbers are appropriate indices (Bird et al. 1960):

$$h_{c,free} = \frac{2\mathcal{K}}{d} + \frac{0.6\sqrt[4]{Gr}\,\sqrt[3]{Pr}\mathcal{K}}{d}. \tag{8.15}$$

This relationship is shown in figure 8.4. When the fluid velocity (indexed by Gr) goes to zero, this expression simplifies to that for conduction alone, $2\mathcal{K}/d$. Thus, for small organisms or small temperature differences (or both), heat exchange in the presence of free convection is little different from that by conduction alone.

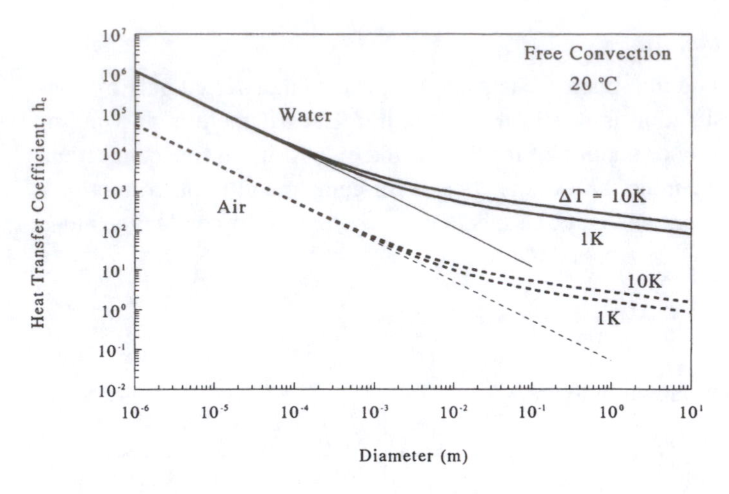

Fig. 8.4 The heat transfer coefficient for free convection (at 20°C) is larger in water than in air, and in each case it is larger than that for conduction alone (the thin, straight solid and broken lines).

Conversely, for large *Gr* ($> 10^6$) we can neglect the conductive portion of the expression without unduly affecting the magnitude of h_c. For example, at 20°C, the product $\ell^3 \Delta T$ must be greater than about 0.01 m^3 K for *Gr* in air to be greater than 10^6. Thus, a sphere 10 cm in diameter that maintains a ΔT of 10°C would have a Grashof number high enough for conduction to have a negligible effect on heat exchange. In water, the situation is less restrictive; $\ell^3 \Delta T$ need only be greater than about 0.001 m^3 K. In other words, in water a sphere 10 cm in diameter would have to maintain a ΔT of only 1°C to have *Gr* exceed 10^6.

Because the ratio of Grashof number in air to that in water varies substantially as a function of temperature (fig. 8.2), one must be careful when drawing conclusions regarding the relative rates of heat loss by free convection. For low temperatures (0° to 5°C), heat loss in air may be similar to, or even exceed, that in water. An extreme example occurs at 3.98°C, when the Grashof number of fresh water is zero and, as a result, there is no free convection. At this temperature, if the Grashof number in air is greater than 4.5×10^7, the heat transfer coefficient is larger in air than in water. At higher temperatures, however, free convective heat loss in water substantially exceeds that in air. At 20°C, for instance, the free convective heat loss in water is roughly one hundred times that in air.

8.3.3 *Forced Convection*

Finally, we consider the heat transfer coefficient for a sphere in a moving fluid, so-called *forced convection*. In this case, h_c depends on the characteristics of both the thermal and velocity boundary layers, and therefore on the Prandtl and Reynolds numbers (Bird et al. 1960):

$$h_{c,forced} = \frac{2\mathcal{K}}{d} + \frac{0.6\sqrt{Re}\,\sqrt[3]{Pr}\mathcal{K}}{d}. \tag{8.16}$$

This relationship is shown in figure 8.5.

Note that as the relative velocity between organism and fluid decreases, the Reynolds number goes to zero, and the heat transfer coefficient simplifies to that for conduction alone. Thus, for small organisms in very slow flow, heat exchange in the presence of forced convection is little different from that by conduction alone. Conversely, at relatively high *Re* (> 100) the purely convective portion of this expression is much larger than the conductive portion. As a result, at high *Re* we can safely neglect the conductive contribution to the heat transfer coefficient.

Reynolds numbers of this magnitude are quite common in biology. For example, a sphere 1 cm in diameter in a water current of 1 cm s^{-1} has an *Re* of 100.

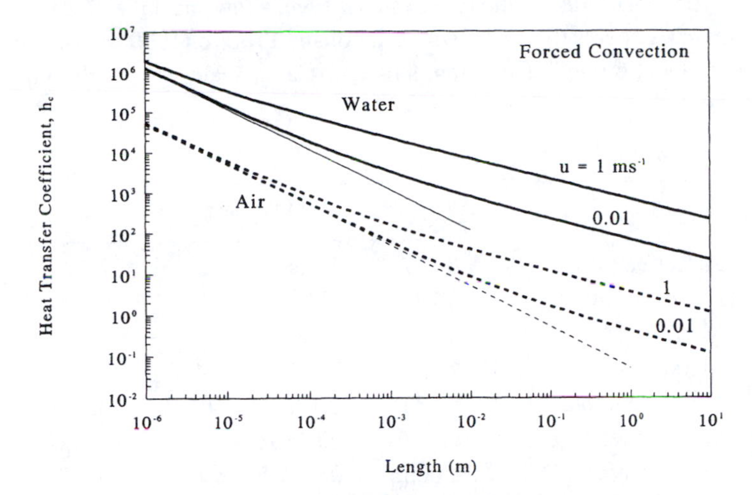

Fig. 8.5 The heat transfer coefficient for forced convection (at 20°C) is larger in water than in air, but in each case it is larger than that for conduction alone (the thin, straight solid and broken lines).

The Prandtl number of water is an order of magnitude higher than that of air, and for a given velocity and size of organism, Reynolds numbers in water are roughly fifteen times those in air (chapter 5) and the thermal conductivity of water is about twenty-three times that of air. As a result, at velocities such that the Reynolds number is greater than about 100, the forced-convective heat transfer coefficient of a sphere of a given size in water is larger than that in air by a factor of about 200. At equal Reynolds numbers (as opposed to equal velocity), h_c in water is roughly fifty times larger than that in air ($Re > 100$).

8.4 Body Temperature

What consequences do these differences in thermal properties have for living things? We begin with an exploration of the relationship between body temperature and metabolic rate.

As we noted briefly in chapter 4, the resting metabolic rate of animals varies allometrically with body mass, m (Schmidt-Nielsen 1979):

$$\text{metabolic rate (watts)} = Mm^{\alpha}, \qquad (8.17)$$

where the exponent, α, is always less than or equal to 1 and is usually about 3/4. M, the *coefficient of resting metabolic rate*, adjusts the equation for the intrinsic rate at which a particular organism burns its fuel. M has the units of (W $kg^{-\alpha}$).

M varies considerably among different animal groups (table 8.4; fig. 8.6). For example, passerine birds (songbirds, robins for instance) consume energy at rest at a rate about one thousand times that of representative marine invertebrates.

The scaling of metabolic rate has not been extensively studied in plants. Hemmingsen (1950), however, reports that the metabolic rates of seeds, pea seedlings, and leafless beech trees are similar to those of reptiles of equal mass, and α equals $3/4$. On the basis of this scant data, we treat plants as being metabolically similar to lizards (table 8.4).

Note that the values in table 8.4 are for *resting* metabolic rates. When an animal is active, its metabolic rate (and, therefore, its heat production) can rise substantially. For example, the metabolic rate of a flying bird is ten to twenty times that of the bird at rest, and the increase in metabolic rate for flying insects can be 100-fold or more. In general, the metabolic rates for plants are not subject to this type of variation, although, as we will see, a few specialized plants do indeed have what amounts to an "active" metabolism. Prothro (1979) reports that the maximal metabolic rates of both animals and plants scale with body mass with α equal to $3/4$.

Table 8.4 Allometric coefficients for resting metabolic rates (from Schmidt-Nielsen 1984 and Hemmingsen 1950).

Organism	M	M'	α
	$\mathrm{W\,kg^{-\alpha}}$	$\mathrm{l\,O_2\,s^{-1}\,kg^{-\alpha}}$	
Passerine birds	6.247	3.108×10^{-4}	0.724
Nonpasserine birds	3.792	1.887×10^{-4}	0.723
Mammals	3.390	1.687×10^{-4}	0.750
Marsupials	2.354	1.171×10^{-4}	0.737
Lizards and plants	0.378	0.1881×10^{-4}	0.830
Salamanders	0.038	0.01896×10^{-4}	0.660
Invertebrates	0.0056	0.002786×10^{-4}	0.750

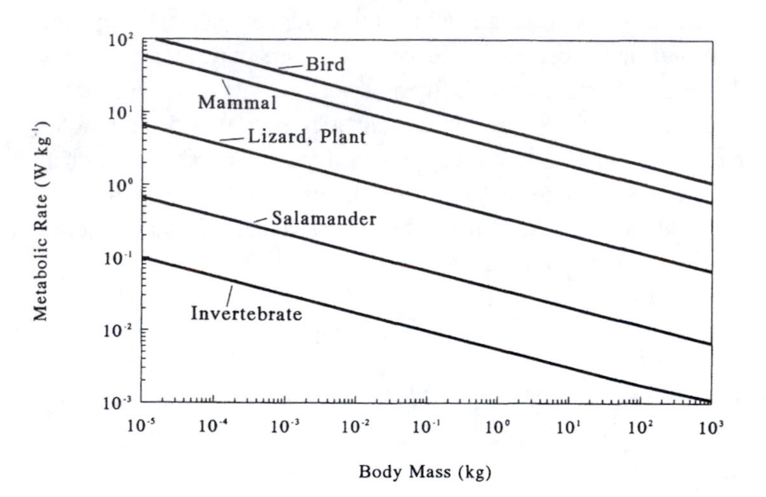

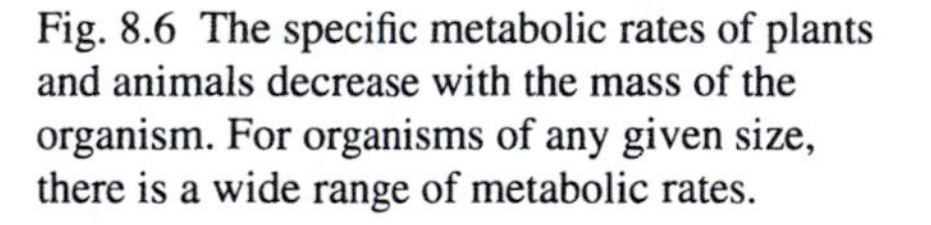

Fig. 8.6 The specific metabolic rates of plants and animals decrease with the mass of the organism. For organisms of any given size, there is a wide range of metabolic rates.

In summary, the resting metabolic rates of plants and animals vary over a broad range. This variation can have important consequences for the interaction among thermal properties of fluids, size of organisms, and body temperatures.

Before proceeding, we need to define carefully what we mean by body temperature. In a real organism, this is a tricky procedure because temperature can vary substantially from one part of the body to another. For example, the skin temperature on your arm is likely both to be quite variable and to be several degrees cooler than that inside your mouth. It is for these reasons that doctors take your "temperature" by placing the thermometer under your tongue. In general, temperature is highest in the central trunk of an animal and decreases toward the periphery. Here we avoid these complexities by using a simplified (and somewhat unrealistic) model for body temperature: we assume that temperature is the same throughout our spherical organism. This is equivalent to requiring that heat be extremely well "mixed" within the sphere, in other words, that the transport of heat within the organism is sufficient to erase any internal gradient in temperature.

We will also assume that the organism has no insulation; in other words that heat is freely exchanged between the surface of the sphere and the surrounding fluid.

Given these assumptions, the body temperature of a spherical organism can be calculated by assuming that the animal is at thermal equilibrium with its environment. To keep things simple, we assume that the only heat source available to the organism is its own metabolism. In making this assumption we are therefore ruling out many common external sources of heat such as radiation from the sun or conduction from a hot rock.

In this simplified scenario, the equilibrium temperature of the organism is that at which the rate of metabolic heat production is just equaled by the rate of heat loss to the surrounding fluid. The mathematics is quite simple. The volume of a sphere with diameter d is $\pi d^3/6$, and its mass is therefore

$$m = \frac{\rho_b \pi d^3}{6}, \tag{8.18}$$

where ρ_b is the average density of the body. By inserting this expression for body mass into eq. 8.17, we arrive at an expression for how total metabolic heat production varies as a function of the sphere's diameter:

$$\text{heat produced} = M\left(\frac{\rho_b \pi d^3}{6}\right)^{\alpha}. \tag{8.19}$$

For the sake of simplicity, we assume that $\alpha = 3/4$, a value that seems best to describe the overall allometry of animal and plant metabolism (Schmidt-Nielsen 1984; Hemmingsen 1950).

We know from Newton's law of cooling that the rate at which heat is transferred from sphere to fluid depends on the product of the area over which the two are in contact (in the case of a sphere, πd^2), the heat transfer coefficient, and the difference in temperature between body and fluid, ΔT:

$$\text{rate of heat transfer} = \pi d^2 h_c \Delta T. \tag{8.20}$$

For each of our three cases (conduction alone, forced convection, and free convection) we can substitute into eq. 8.20 the appropriate expression for the heat transfer coefficient, and, by assuming that metabolic heat is the only heat input available to the organism, equate the heat produced (eq. 8.19) with that transferred to the fluid (eq. 8.20):

$$M\left(\frac{\rho_b \pi d^3}{6}\right)^{\alpha} = \pi d^2 h_c \Delta T. \tag{8.21}$$

Solving for ΔT provides the answer we seek. To calculate body temperature in each case, one simply adds ΔT to the ambient fluid temperature.

8.4.1 *Temperature Set by Conduction Alone*

In the case of conduction alone,

$$\Delta T = 0.098 \frac{\rho_b^{3/4} d^{5/4} M}{\mathcal{K}}. \tag{8.22}$$

This result is shown graphically in figure 8.7A. Because exchange by conduction alone is the minimum rate at which heat is transported away from an object, these results represent the *maximum* body temperature possible in a given medium in the absence of insulation, and therefore deserve a detailed examination.

Chapter 8

Eq. 8.22 consists of two parts, one related to a thermal property of the medium (conductivity), and a second concerning factors that are biologically controlled (body density, size, and metabolic rate). We examine these parts separately.

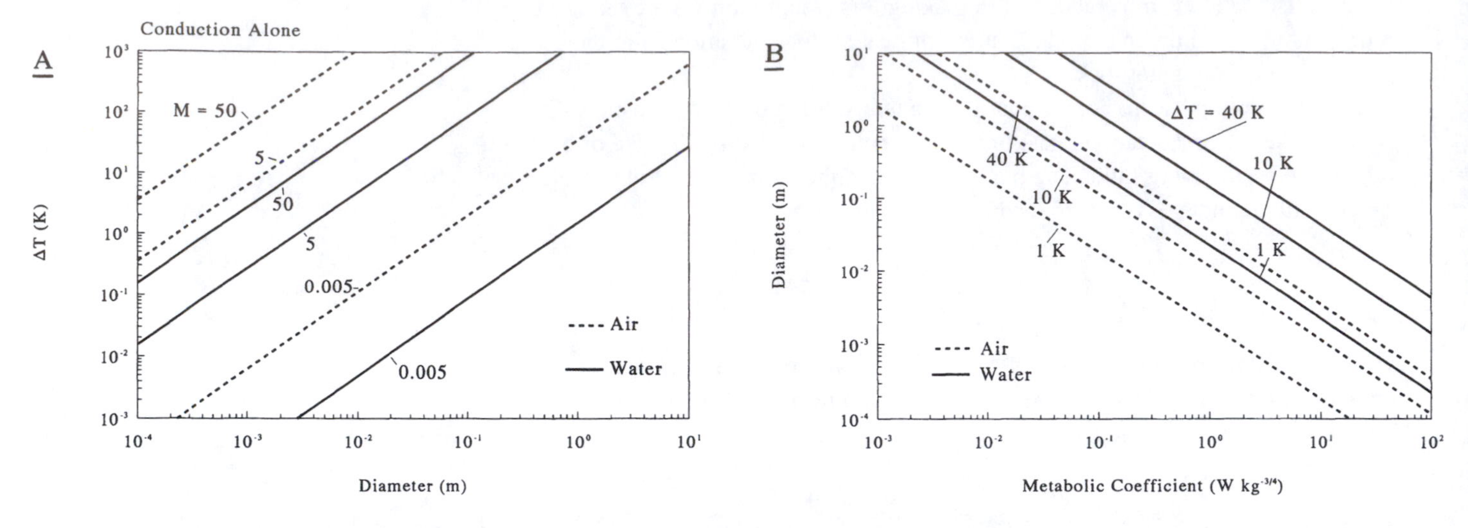

Fig. 8.7 The difference in temperature between an organism and its environment increases with the organism's size when heat is exchanged by conduction alone (A). The higher the metabolic rate, the higher the body temperature. The relationship of (A) can be reworked to predict the diameter of the sphere needed to maintain a certain body temperature given a certain metabolic rate, or the metabolic rate needed to maintain a certain body temperature given a certain diameter (B). Here a fluid temperature of 20°C is assumed and the density of the organism is 1080 kg m^{-3}.

When heat is transferred by conduction alone, the difference in temperature between the organism and its surroundings is inversely proportional to the conductivity of the fluid. As a result, organisms in water maintain a temperature about twenty-three times closer to that of their environment than do organisms in air.

The difference between the temperature of the body and that of the environment increases as the size of the organism increases, rising as $d^{5/4}$. In other words, it is easier for a large animal to stay warm than for a small one. Note that if metabolic rate scaled linearly with mass (i.e., $\alpha = 1$), then

$$\Delta T = 0.083 \frac{\rho_b d^2 M}{\mathcal{K}}, \tag{8.23}$$

and body temperature increases even faster with an increase in size than for a typical plant or animal. This has implications for colonial animals and other aggregations in which a large number of small, rapidly respiring organisms come together to form a large overall mass. In these organisms and aggregations, $\alpha \approx 1$ (Hughes and Hughes 1986), and ΔT should be predicted using eq. 8.23 rather than eq. 8.22.

ΔT also varies directly as a function of intrinsic metabolic rate, M. Results are shown in figure 8.8 for 1 kg organisms (ρ_b = 1000 kg m^{-3}, r = 0.13 m) with metabolic rates varying between that of a typical resting invertebrate and that of an actively flying bird. Note that ΔT is virtually zero for the low metabolic rate of invertebrates, salamanders, plants, and lizards, implying that in the absence of solar heat input the temperature of these organisms is essentially the same as that of their environment. This is why animals with these low metabolic rates are referred to as being "cold blooded" and why cucumbers keep their cool. At the other end of the metabolic spectrum, ΔT for a flying bird is quite high, implying that these and similar terrestrial organisms might actually overheat.

As an alternative way of looking at body temperature, we can use eq. 8.21 to calculate the size of organism necessary if a certain ΔT is to be maintained by a certain metabolic rate. Equating heat produced and heat lost (eqs. 8.19 and 8.20) and solving for d, we see that

$$d = 6.4\left(\frac{\Delta T}{M}\right)^{4/5}\frac{\mathcal{K}^{4/5}}{\rho_b^{3/5}}. \tag{8.24}$$

This relationship is shown in figure 8.7B. The more effectively the fluid conducts heat, the larger the size required. Because $\mathcal{K}$ is twenty-three times greater in water than in air, an animal would have to be $23^{4/5} = 12.3$ times larger in water to maintain the same body temperature, all other factors being equal.

And finally, equating eqs. 8.19 and 8.20 and solving for M tells us what metabolic rate is required to maintain a certain ΔT for a sphere of a given size:

$$M = 10.2\frac{\Delta T}{d^{5/4}}\frac{\mathcal{K}}{\rho_b^{3/4}}. \tag{8.25}$$

A spherical organism in water must respire about twenty-three times as fast as the same size of sphere in air to maintain its body at a given temperature above that of its surroundings.

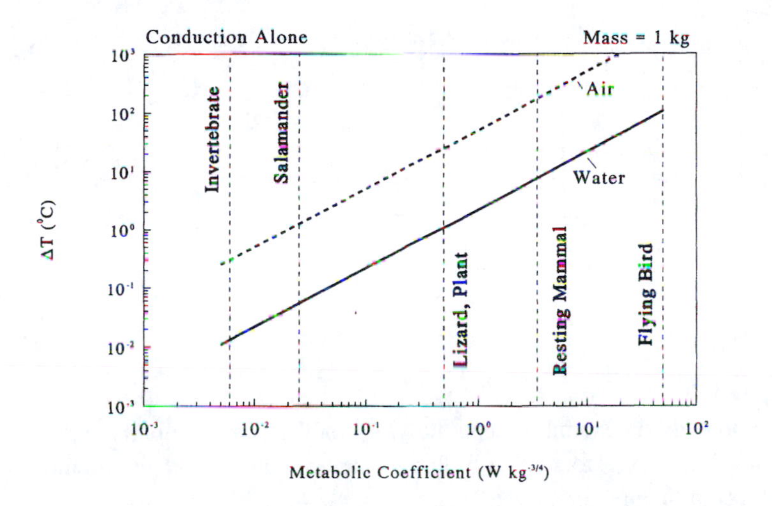

Fig. 8.8 The temperature difference between a 1 kg spherical organism (body density = 1080 kg m^{-3}) and its surroundings depends on the resting metabolic rate. Values shown here are for heat transfer by conduction alone. Temperatures are much higher for organisms in air than in water.

We now digress briefly, to return to the topic of hot-air balloons first discussed in chapter 4. We left this subject having concluded that for a hollow spherical organism to achieve neutral bouyancy in air, the body wall must be very thin (about 10^{-5} times the radius) and the internal air temperature must be about 8.5°C above ambient. We are now in a position to calculate the minimum metabolic cost of maintaining neutral buoyancy under these conditions.

Consider a "balloon" organism 10 cm in diameter with a body wall thickness of 0.5 μm. Inserting eq. 8.14 into eq. 8.13 and using the conductivity of air, we can calculate that the rate at which heat is transferred from the balloon to the air when the temperature inside the balloon is 8.5°C above that outside is 0.14 W. This may not seem like a lot of power, but all of it must be provided by the thin body wall, which has a mass of only 17 mg. In other words, the tissue in the body wall must produce heat at a rate of 8225 W kg^{-1} to keep the balloon aloft. This is a metabolic rate nearly two orders of magnitude larger than any found among existing organisms, providing yet another reason as to why biological hot-air balloons have never evolved.

8.4.2 *Temperature Set by Free Convection*

The body temperatures we have discussed so far are calculated assuming that heat is lost to conduction alone. If the fluid moves relative to the organism, heat is lost more effectively and ΔT is smaller.

Consider the case of a sphere in free convection where the Grashof number is large enough so that the heat lost by conduction is a small fraction of the overall. In this case, after expanding the expressions for *Gr* and *Pr*, we see that body temperature is

$$\Delta T = 0.41 d^{2/5} M^{4/5} \rho_b^{3/5} \frac{\mu^{2/15}}{\mathcal{K}^{8/15} \rho_f^{2/5} (g\beta)^{1/5} Q_s^{4/15}}, \tag{8.26}$$

a result shown in figure 8.9A. Note that when ΔT or d is small, *Gr* is likely to be small, and this equation is not an accurate description of reality.

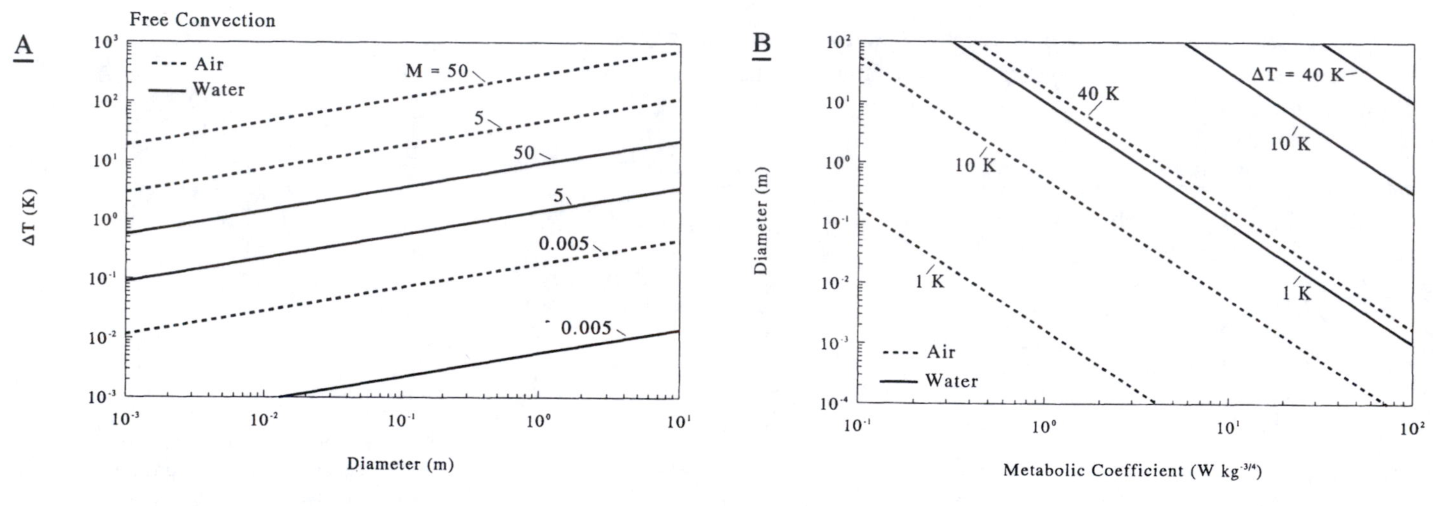

Fig. 8.9 The difference in temperature between an organism and its environment increases with the organism's size when heat is exchanged by free convection (A). The higher the metabolic rate, the higher the body temperature. The relationship of (A) can be reworked to predict the diameter of the sphere needed to maintain a certain body temperature given a certain metabolic rate, or the metabolic rate needed to maintain a certain body temperature given a certain diameter (B). Here the fluid temperature is assumed to be 20°C.

Eq. 8.26 may look somewhat forbidding, but its meaning is not difficult to grasp if its components are examined separately. Again there are two parts to the equation, one dealing with factors under biological control, and the other with the thermal properties of the medium.

As in the case of conduction alone, any increase in the size of the organism or its metabolic rate increases ΔT. Note, however, that ΔT scales differently with d than for pure conduction, increasing as $d^{2/5}$ rather than $d^{5/4}$. In other words, the size of an organism has a smaller effect on body temperature in the presence of free convection than when heat is exchanged by conduction alone. ΔT scales as $M^{4/5}$ rather than M, so any increase in metabolic rate results in a smaller increase in body temperature than would occur in purely conductive heat loss.

ΔT also depends on several properties of the medium. An increase in viscosity increases ΔT, but an increase in conductivity, density, thermal expansivity, or specific heat results in a decrease in ΔT. With each of these variables raised to a different power, it is not easy to see intuitively how their effects differ between air and water. It is a simple matter, however, to take the ratio of ΔT in air to that in water (again assuming $Gr > 10^6$):

$$\frac{\Delta T_a}{\Delta T_w} = \left(\frac{\mu_a}{\mu_w}\right)^{2/15} \left(\frac{\rho_{f,w}}{\rho_{f,a}}\right)^{2/5} \left(\frac{\mathcal{K}_w}{\mathcal{K}_a}\right)^{8/15} \left(\frac{\beta_w}{\beta_a}\right)^{1/5} \left(\frac{Q_{s,w}}{Q_{s,a}}\right)^{4/15}, \tag{8.27}$$

where the subscripts a and w refer to air and water, respectively. This result is presented in Figure 8.10. With the exception of temperatures between 0°C and 5°C where Gr is small and this relationship is invalid, ΔT is larger in air than in water. For example, at 20°C, ΔT is 38 times larger in air than in fresh water.

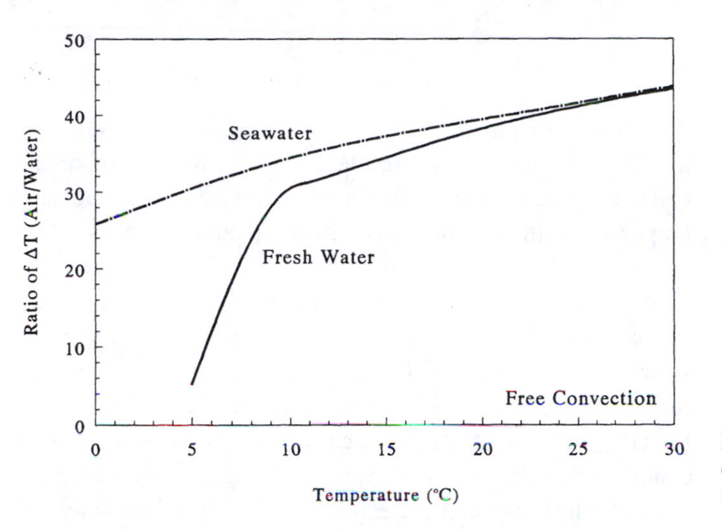

Fig. 8.10 The rate at which heat is lost by free convection is higher in air than in either fresh water or seawater.

This ratio can be somewhat misleading, however. We know already that ΔT is small due to conductive heat loss alone for all organisms except large, terrestrial birds and mammals and a few quirky aquatic organisms. Therefore, reducing ΔT by convection has little practical effect on most organisms, and the effects of free convective heat transfer are biologically notable only in the case of relatively large terrestrial organisms that have high metabolic rates. For instance, a spherical terrestrial invertebrate with a diameter of 10 cm, has a body temperature only 0.2°C above ambient if it loses heat just by conduction alone (Eq. 8.22). Even if free convection were a hundred times as effective at transporting heat, ΔT could not change by more than 0.2°C; a negligible difference. In contrast, a spherical passerine bird 10 cm in diameter has a ΔT of 11°C in free convection (Eq. 8.26), substantially lower than the 120°C maintained when heat is lost by conduction alone.

As for conduction, we can also calculate the diameter of sphere required to maintain a given ΔT in the face of free convection. For a given M,

$$d = 9.4\frac{\Delta T^{5/2}}{M^2 \rho_b^{3/2}} \frac{\mathcal{K}^{4/3} \rho_f (g\beta)^{1/2} Q_s^{2/3}}{\mu^{1/3}}. \tag{8.28}$$

Similarly, we can calculate the metabolic rate coefficient needed for a sphere of a given size to maintain a given ΔT:

$$M = 3.1\frac{\Delta T^{5/4}}{d^{1/2} \rho_b^{3/4}} \frac{\mathcal{K}^{2/3} \rho_f^{1/2} (g\beta)^{1/4} Q_s^{1/3}}{\mu^{1/6}}. \tag{8.29}$$

These results are shown in Figure 8.9B, and their exploration is left as an exercise for the reader.

8.4.3 *Temperature Set by Forced Convection*

Consider now the case of a sphere in forced convection at a Reynolds number high enough so that we can neglect heat transfer from conduction. Here,

$$\Delta T = 0.33 d^{3/4} M \rho_b^{3/4} \frac{\mu^{1/6}}{\mathcal{K}^{2/3} \rho_f^{1/2} Q_s^{1/3}} \frac{1}{u^{1/2}}. \tag{8.30}$$

As before, the apparent complexity of this equation can be dispelled by considering its parts separately. In this case there are three major components: (1) those factors under biological control (size, metabolism, and body density); (2) the thermal properties of the medium; and (3) the relative velocity between organism and fluid.

In forced convection, as with other forms of heat transfer, ΔT increases with increasing size. In this case, the difference in temperature between body and fluid scales as $d^{3/4}$ rather than $d^{5/4}$ as is the case for conductive heat exchange or $d^{2/5}$ for free convection. In other words, ΔT increases faster with organism size for forced convection than for free convection, but not as fast as for heat transfer by conduction alone. Note, however, that this refers to the rate of increase of ΔT, not to its absolute magnitude. Despite the difference in scaling, ΔT is usually much smaller for forced convection than for either free convection or conduction. This is because the winds that blow and currents that flow in the environment usually do so at rates greater than winds and currents induced by free convection.

Body temperature in forced convection increases directly with metabolic rate. However, heat transfer is so effective that in the face of moderate winds or currents even organisms with the high resting metabolic rates of birds and mammals have difficulty maintaining a body temperature much above that of the surrounding fluid (fig. 8.11A).

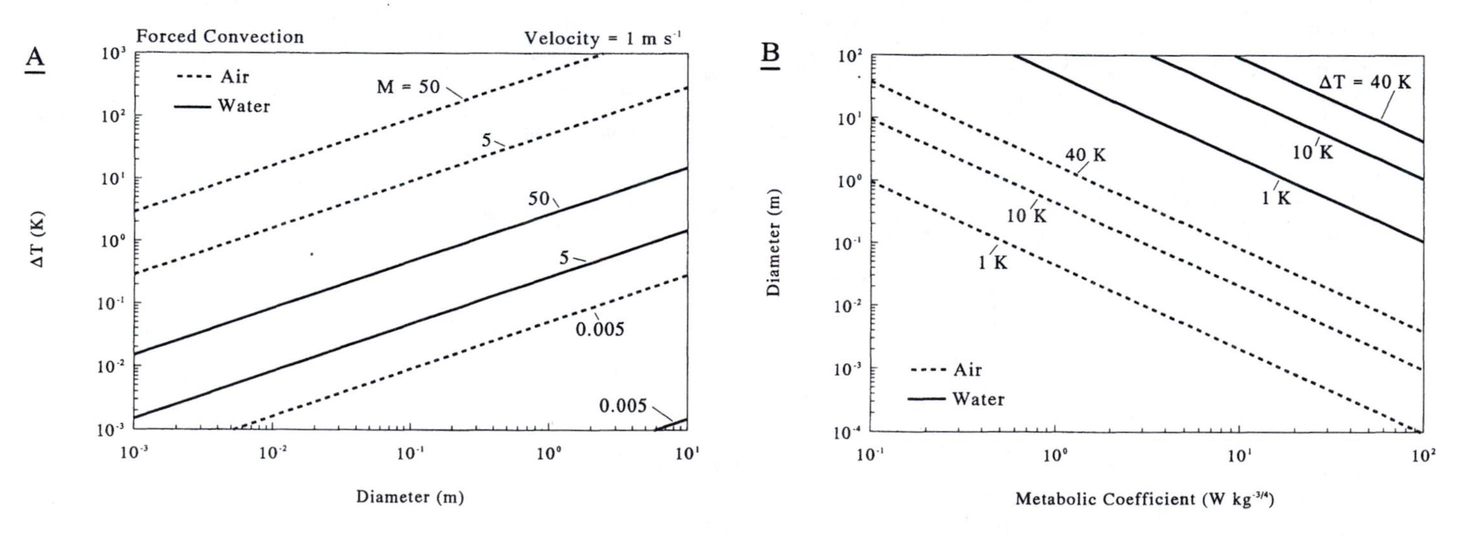

Fig. 8.11 The difference in temperature between an organism and its environment increases with the organism's size when heat is exchanged by forced convection (A). The higher the metabolic rate, the higher the body temperature. Here a velocity of 1 m s^{-1} is assumed. The relationship of (A) can be reworked to predict the diameter of the sphere needed to maintain a certain body temperature given a certain metabolic rate, or the metabolic rate needed to maintain a certain body temperature given a certain diameter (B). Here the fluid temperature is assumed to be 20°C.

Turning to the thermal properties of the medium, we are again presented with a series of parameters whose combined effect is difficult to grasp intuitively. We can, however, calculate the ratio of ΔT in air to that in water at a given fluid temperature and velocity (assuming $Re > 100$):

$$\frac{\Delta T_a}{\Delta T_w} = \left(\frac{\mu_a}{\mu_w}\right)^{1/6} \left(\frac{\rho_w}{\rho_a}\right)^{1/2} \left(\frac{\mathcal{K}_w}{\mathcal{K}_a}\right)^{2/3} \left(\frac{Q_{s,w}}{Q_{s,a}}\right)^{1/3}. \tag{8.31}$$

This ratio is always quite large (fig. 8.12). For example, at 20°C, ΔT is 190 times larger in air than in water.

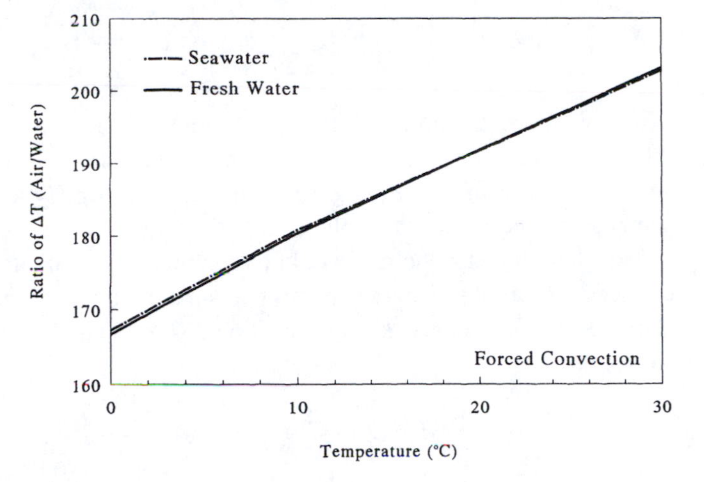

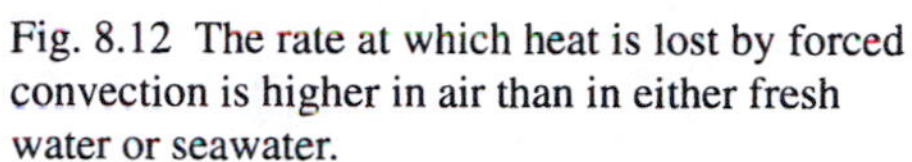

Fig. 8.12 The rate at which heat is lost by forced convection is higher in air than in either fresh water or seawater.

Finally, we note that ΔT is inversely proportional to the square root of velocity. The importance of this fact is best conveyed through an example. Consider a rough approximation to a human being: a terrestrial sphere with a diameter of 0.5 m, a mass of 70 kg, and a resting metabolic coefficient of 3 W $kg^{-3/4}$. The temperature of this "person" is shown as a function of wind speed in figure 8.13. A low fluid velocity results in a substantial decrease in ΔT, but the change in ΔT becomes smaller and smaller as velocity increases. In other words, if you are trying to stay cool on a hot day, a slight breeze will help a lot; a stiff wind will help more, but not as much as you might expect. This is why it is so rewarding to fan yourself in church on a hot day, and why ceiling fans, which seem barely to stir the air, nonetheless keep one surprisingly cool.[1]

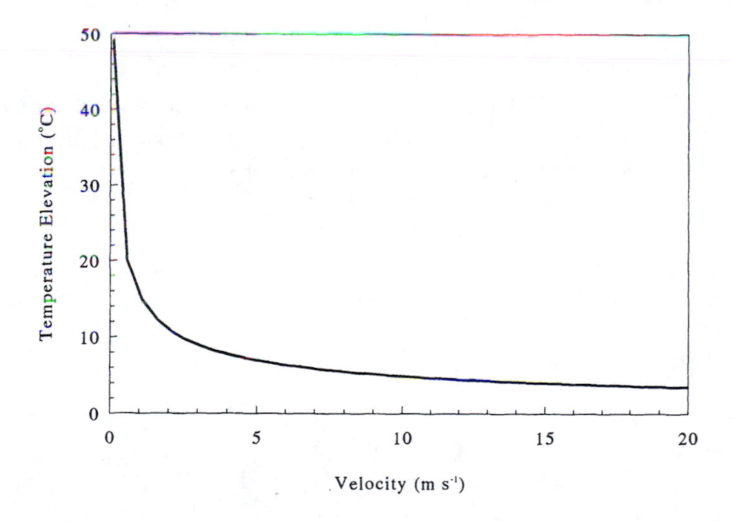

Fig. 8.13 The temperature of a 70 kg spherical human decreases with increasing wind speed. The rate of decrease is largest at low velocities.

We now consolidate these disparate factors and examine the effect of forced convection on body temperature. Consider first an extreme situation in air. An arctic mammal or bird is often called upon to maintain a ΔT of 40°C, a temperature differential that must be supported by an intrinsic metabolic rate of $M \approx 5$ W $kg^{-3/4}$. How big a sphere would be needed to maintain this ΔT in the face of

[1]This effect also involves evaporative cooling, a phenomenon discussed in chapter 14.

a 1 m s^{-1} breeze? Equating eqs. 8.19 and 8.20 and solving for d, we see that (fig. 8.11B)

$$d = 4.4 \frac{\Delta T^{4/3}}{M^{4/3}\rho_b} \frac{\rho_f^{2/3} Q_s^{4/9} \mathcal{K}^{8/9}}{\mu^{2/9}} u^{2/3}. \tag{8.32}$$

Inserting values appropriate for air, we calculate that a sphere nearly 0.4 m in diameter (a mass of nearly 40 kg) would be required. Thus an uninsulated animal smaller than a wolf would not be able to maintain its body temperature without somehow increasing its metabolic rate.

How high a metabolic rate would be required, for example, to keep a 10 cm diameter terrestrial sphere (analogous to a small rabbit, for instance) at a temperature 40°C above ambient? Again equating eqs. 8.19 and 8.20 and solving for M we see that

$$M = 3.0 \frac{\Delta T}{d^{3/4}\rho_b^{3/4}} \frac{\rho_f^{1/2} Q_s^{1/3} \mathcal{K}^{2/3}}{\mu^{1/6}} u^{1/2}. \tag{8.33}$$

After working through the calculation, we see that an arctic animal of this size would need a metabolic rate coefficient of 22 W kg$^{-3/4}$, high indeed, but well within the realm of possibility for an active bird or mammal.

If our rabbit wandered into a flowing stream, however, it would quickly die of hypothermia. Even an active bird or mammal (M = 20 W kg$^{-3/4}$) with a diameter of 1 m would maintain a body temperature only 0.3°C above that of the surrounding water in the presence of a 1 m s^{-1} current. An actively exercising spherical behemoth 10 m in diameter, with a mass greater than 500 tons, would have a body temperature only 6.5°C above that of the surrounding water. In forced convection, then, the ability to get rid of metabolic heat is seldom a problem, and it would appear that, in the absence of insulation or some internal physiological trickery, only large animals in air can maintain a body temperature substantially different from the temperature of their surroundings.

We are now in a position to compare body temperatures among cases where heat is lost by conduction alone, by free convection, and by forced convection (fig. 8.14). For simplicity, we arbitrarily chose one metabolic rate coefficient (M = 5 W kg$^{-3/4}$).

Fig. 8.14 For organisms with intrinsic metabolic rate coefficients of 5 W kg$^{-3/4}$, heat is lost more effectively in air by forced convection than by free convection (except at very large size), and more effectively by free convection than by conduction alone (A). The same pattern holds in water, but for any given size, the temperature difference between organism and surroundings is less (B).

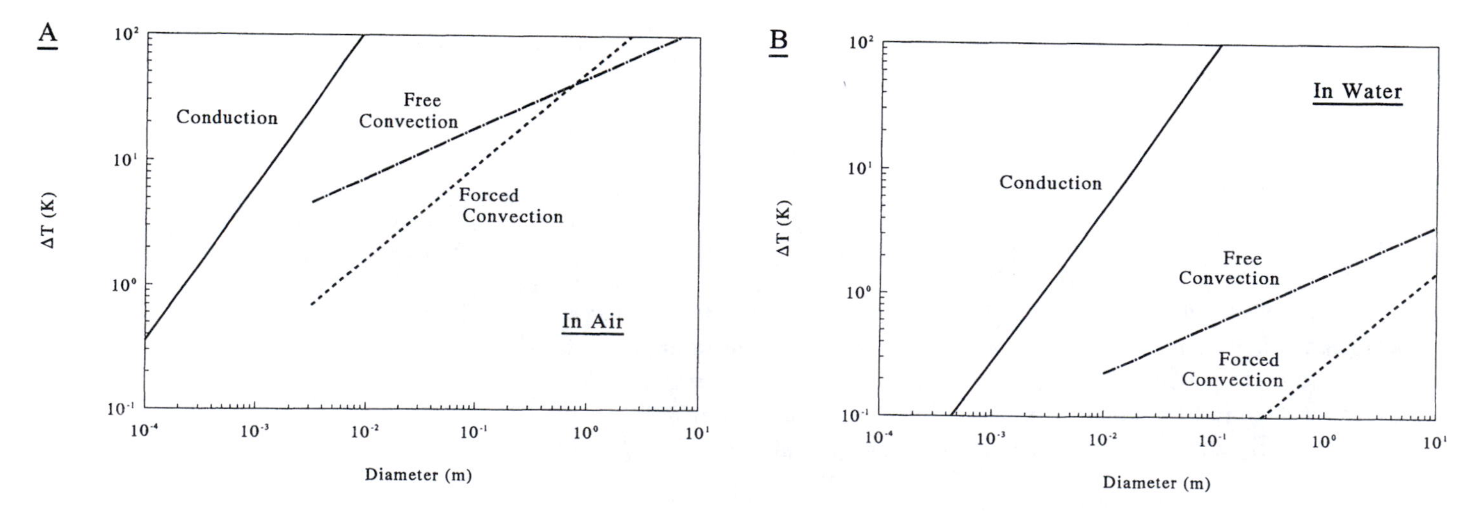

In air, body temperature is higher if heat is exchanged by conduction alone than for exchange by free convection, and higher for free convection than for forced convection at a velocity of 1 m s^{-1}. In water, the same rank order of body temperatures is found, but in the presence of convection (either free or forced), a substantial ΔT cannot be maintained at this metabolic rate.

In summary, we find that the difference in temperature between an uninsulated spherical organism and its surrounding fluid increases with any increase in the size of the organism or its metabolic rate. In air, this can lead to a problem for large organisms; it may be difficult for them to shed enough heat to maintain a livable body temperature unless they stand in a breeze. For small organisms in air, and virtually all organisms in water, resting metabolic rates present are insufficient to raise the body temperature substantially above that of the environment.

Note that these conclusions are drawn on the basis of a simple model: a spherical organism with no insulation and a well-stirred interior. It is now time to compare these results to the real world. At times the conclusions drawn from our model are at odds with reality, but far from being a problem, these cases serve to highlight important physiological tricks by which plants and animals cope with the thermal properties of air and water.

8.5 Of Shrews, Skunk Cabbages, and Dinosaurs

8.5.1 *Keeping Warm*

First, let us consider the situation in air.

As we have seen, at small sizes, an elevated body temperature can be maintained only by a very high metabolic rate. This fact has interesting biological consequences. Consider, for instance, bumblebees. The flight muscles of these insects need to be at 30°C or above in order to provide enough power for flight (Heinrich 1979). On a cold morning, where the air temperature is, say, 10°C, this means that ΔT must be 20°C before the bee can take wing. Nonetheless, bees fly on cold days. What metabolic rate must they sustain?

The flight muscles of a bumblebee are contained in a roughly spherical thorax about 5 mm in diameter, and are heated by their own metabolism, as we have assumed in our exploration here. It is therefore reasonable to use eq. 8.22 to calculate that in air a sphere 5 mm in diameter needs a metabolic rate coefficient of at least 30 W $kg^{-3/4}$ to maintain ΔT at 20°C. This translates to a specific metabolic rate of 200 W kg^{-1} for a 0.5 g bee,[2] and experiments have shown that bees indeed have such high metabolic rates. This extraordinarily high metabolic rate coefficient is equivalent to that of a typical flying bird, and the absolute rate is considerably higher! It is a tribute to the metabolic and respiratory machinery of bees and other flying insects that this magnitude of metabolism is possible. Note, however, that this high metabolic rate is closely tied to activity on the part of the bee. It must raise its metabolic rate well above that at rest (by shivering, for instance) to heat itself up. As soon as it becomes inactive, its body temperature returns to that of the ambient air.

There is a flip side to this puzzle. If a flying bee can heat its muscles to 30°C on a cold day, it seems likely that it could overheat on a hot day. Indeed this

[2]From eq. 8.17 we know that the bee produces heat at a rate of

$$\text{rate of heat production} = Mm^{\alpha} = 30 \times 0.0005^{3/4} = 0.1\text{W}. \tag{8.34}$$

One-tenth of a watt per 0.5 grams is the equivalent of 200 W kg^{-1}.

is the case, and bumblebees have evolved mechanisms that allow them to dump heat effectively from the thorax to the abdomen, from where it is shed to the air (Heinrich 1979). In this situation, the small size of the animal is an advantage, allowing for effective heat loss on hot days.

Two other examples of small, hot animals bear consideration. Shrews and hummingbirds have masses of only 1 to 10 g, yet can maintain body temperatures that are 20° to 30°C above ambient. They do so through the use of three stratagems. First, like bumblebees, they have unusually high metabolic rates. The metabolic rate coefficient for hovering hummingbirds may be as high as 65 W $kg^{-3/4}$ (Lasiewski 1963), which translates to 280 W kg^{-1} for a 3 g bird.[3].

Second, they live in air. It would be virtually impossible for animals of this size to maintain an elevated body temperature in water.

And finally, they have evolved effective forms of insulation. The fur on a shrew and the fluffy feathers on the body of a hummingbird serve to trap a layer of stagnant air around the animal's body. This stagnant air transfers heat primarily by conduction, a process we have seen to be much less effective than transfer by convection.

The presence of insulation means that these and many other organisms cannot be accurately modeled by the procedures we have used here. We have assumed that the entire body of the organism is at the same temperature. But the presence of a layer of insulation means that the surface of an animal in contact with the surrounding fluid (the outer edge of the insulation) can be at a substantially different temperature than that of the body inside. The thicker the insulation, the more gradual the gradient of temperature at the periphery of the organism, and thereby the lower the rate of heat transfer. Bees have also taken advantage of this effect. The thorax is covered with a dense pile that serves as an effective insulator.

Despite their morphological adaptations, shrews and hummingbirds live continually on the edge of hypothermia. To maintain their body temperature, they must eat constantly (several times their body mass in food per day). To fast, even for a day, is to die, and rest is possible only when accompanied by a reduction in body temperature. Both hummingbirds and shrews go into torpor at night, lowering their body temperature to within a few degrees of that of the surrounding air. In this respect, small terrestrial birds and mammals behave much like flying insects.

Even more surprising than bees, shrews, and hummingbirds are plants that maintain an elevated temperature. Several members of the family Araceae, for instance the voodoo lily (fig. 8.15A) and the common philodendron, have inflorescences that heat themselves to a temperature of 35° to 45°C (Meeuse 1966) for several hours during the day. This temperature can be maintained even if the air temperature is as low as 4°C (Nagy et al. 1972). Now, the hot part of the philodendron inflorescence is a cylinder only 3 to 4 cm in diameter formed from sterile male flowers, and the calculations made above would therefore suggest that the metabolic rate of these flowers must be extraordinarily high. Indeed they are. At a temperature of 37°C, the flowers produce 170 W kg^{-1}, a metabolic rate approaching that of a hovering hummingbird (250 W kg^{-1}) of similar mass.[4]

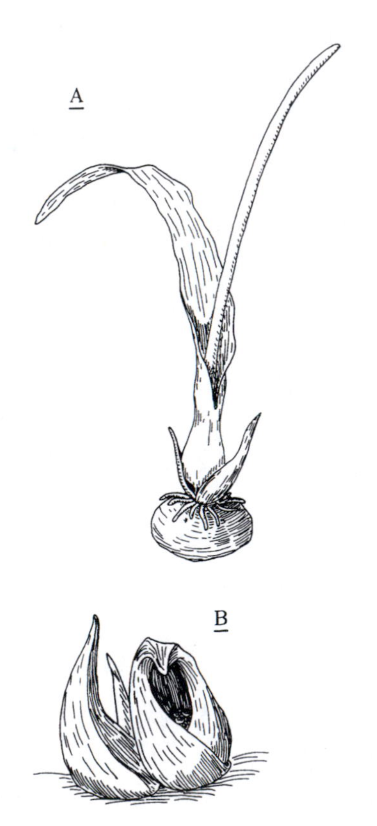

Fig. 8.15 The voodoo lily (A) and the skunk cabbage (B) are both capable of maintaining an elevated body temperature.

[3]Note that a small fraction (about a fifth) of this metabolic energy goes to do mechanical work in keeping the bird aloft, rather than to heat. Therefore, this estimate of the *thermal* metabolic rate is a slight overestimate.

[4]Note that this is a metabolic rate, not the metabolic rate coefficient. To achieve a metabolic rate of 280 W kg^{-1}, M must be about 65 W $kg^{-3/4}$ for a 3 g hummingbird.

Why would a plant heat its flowers? For many of them, it seems that the function of the hot inflorescence is to volatilize the odoriferous compounds produced by the inflorescence. These substances (which are described as having a smell reminiscent of dung, urine, or decaying meat) attract a variety of insects that assist in pollination. The inflorescence is heated at the appropriate time relative to the maturation of male and female flowers so that self-pollination is avoided.

In one plant, the skunk cabbage of the eastern United States (fig. 8.15B), an elevated temperature is maintained for as much as two weeks as the plant "burns"the starch stored in its roots (Knutson 1974). These plants flower early in the spring, and the maintenance of a warm inflorescence can help to avoid freezing. In this respect these bizarre plants behave much like mammals and birds.

The biochemical basis for heat production in these plants is an intriguing story, but one that will not be explored here. You may wish to consult Meeuse (1975) or Laties (1982).

What about aquatic organisms? In accordance with our predictions, I know of no aquatic plant that maintains an elevated body temperature, but the general conclusions presented above with respect to animals seem at odds with reality. If, as we predict, a 500-ton sphere with the metabolic rate of a mammal can maintain a body temperature only 6.5°C above ambient, how can we account for whales, which weigh maybe 100 tons? Even worse, how can we acount for dolphins, seals, sea lions, and otters that weigh only a fraction of a ton? These aquatic mammals often inhabit arctic waters where they maintain a ΔT of about 30°C. Still more difficult to explain are aquatic birds that may weigh only a kg, yet nonetheless maintain a ΔT of nearly 40°C. And finally, there are a few fish (tunas and large sharks) that maintain a core body temperature several degrees above that of the ambient water.

The answer to these discrepancies between theory and fact rests on three points. First and foremost, aquatic mammals and birds have solved the problem of heat loss in a fashion analogous to that shown by small, terrestrial animals. The blubber of porpoises and whales acts as insulation in much the same fashion as the fur or pile of a shrew or bee. The feathers of most aquatic birds are well oiled, allowing them to trap and maintain a layer of stagnant air around the animal when under water. The low thermal conductivity of air allows the feathers to act as an effective insulator. Thus most warm aquatic animals maintain an elevated temperature by having effective insulation.

Second, porpoises and whales maintain their high body temperature by having a resting metabolic rate that is about twice that of other mammals (Irving 1969), and they spend a large fraction of their time actively swimming, which serves further to increase their heat production. Although aquatic birds have a metabolic rate similar to that for terrestrial birds, this rate is nonetheless high (fig. 8.6). Thus, some of the ability of aquatic animals to maintain a high body temperature is, like that of a bee or a skunk cabbage, due to an unusually high metabolic rate.

The third mechanism for maintaining an elevated body temperature in water has evolved in a few tunas and sharks. This intriguing example of physiological trickery—a countercurrent heat exchanger—is described later in this chapter in the context of the thermal cost of respiration.

In spite of these exceptions, the general conclusion we reached on the basis of a simple model holds true. Aquatic organisms are generally at the same temperature as their surroundings.

8.5.2 *Keeping Cool*

So far we have only considered the problem of staying warm. A different set of problems is faced by large terrestrial organisms. Once they attain a mass in excess of a few hundred kilograms, the ability of even forced convection to draw off their metabolic heat is stretched near the limit. According to our predictions (eq. 8.30), a spherical "cow" weighing 600 kg would have a body temperature nearly 60°C above ambient, even when standing in a 1 m s^{-1} wind, and would quickly heat up sufficiently to cook itself if the breeze died. How, then, to explain the presence of cows and other large terrestrial mammals with high metabolic rates? In fact, the problem extends to large animals even if they have only moderate metabolic rates. For example, Alexander (1989) estimates that large dinosaurs such as *Brachiosaurus* would have a body temperature nearly 60°C above ambient even with the relatively low metabolic rate typical of extant lizards. How did these extinct giants keep cool?

First, most large terrestrial animals aren't spherical. Changing body shape to include protuberances such as legs, arms, and ears increases the area of contact between the hot animal and the cooler environment, and thereby increases the rate at which heat can be shed. The large ears of African elephants are an excellent example of this strategy. When the animal is hot, the blood vessels in its ears dilate, thereby increasing the rate at which body heat is transported to the ears. The ears are then flapped to increase convective cooling. In this respect, ears act like the cooling vanes on a motorcycle engine, effectively dumping heat into the air. Dinosaurs may have used a similar strategy. Farlow et al. (1976) suggest that the bony plates on the back of *Stegasaurus* (fig. 8.16) acted to dissipate body heat by convection.

Fig. 8.16 *Stegasaurus* may have used its dorsal dermal plates as a means of transferring excess heat to its environment.

This factor of shape can also help to explain why leaves stay at ambient temperature unless strongly heated by the sun. The cooling of leaves by free convection has been elegantly studied by Vogel (1970).

Second, we note that the largest terrestrial mammals (elephants, hippopotamuses, and rhinoceroses) do not have fur, and therefore have less insulation than smaller mammals. In fact, there is a general trend among ungulates for larger animals to have less fur (Louw and Seely 1982). Similarly, large dinosaurs are thought to have lacked any obvious form of insulation (Alexander 1989).[5]

The third, and more important, cooling stratagem used by terrestrial animals involves a mechanism for getting rid of metabolic heat that we have not yet touched upon, *evaporation*. By losing water to the surrounding air, either by sweating or panting, terrestrial animals can dump immense amounts of heat. This phenomenon is so important that it warrants its own separate discussion, and we return to it in chapter 14.

8.6 The Thermal Cost of Respiration

Let us now return to animals for whom getting rid of heat is not a problem, but for whom staying warm is. This includes most small birds and mammals that require a body temperature of 35° to 40° C, and a variety of other organisms, such as dolphins and whales, fish, reptiles, and flying insects, for whom a high body

[5]One does wonder, however, about woolly mammoths. They were larger than present-day elephants but quite hirsute. Alexander (1989) supposes that the presence of fur on such large mammals was an adpatation to the cold environment in which they lived.

temperature, although not required, is certainly advantageous. We have briefly explored a few tactics through which these organisms can decrease the rate at which they lose heat, but have ignored one important mechanism by which they seemingly *must* lose heat—respiration. Every time a hot animal takes a breath or passes water over its gills to acquire oxygen, it seems likely that it also loses heat to its environment. How important is this heat loss, and how does its importance compare between terrestrial and aquatic environments?

We begin by examining the supposition that heat must be lost to acquire oxygen. At the microscopic level of an individual alveolus in a lung, tracheole in an insect abdomen, or filament in a gill, the transfer of oxygen to the blood from the surrounding fluid is accomplished by molecular diffusion. The rate of this transfer is therefore governed in large part by the molecular diffusion coefficient. Similarly, the transport of body heat from the blood to the environment is accomplished by thermal diffusion, and the rate is governed by the thermal diffusivity.

Consider an animal breathing air. As we noted earlier in this chapter, the diffusivity of heat in air is virtually the same as the diffusivity of oxygen ($Le = 0.94$). Therefore, if the flux of air through the lung or tracheole is such that the concentration of oxygen in the blood or hemolymph can come to equilibrium with the concentration of oxygen in the air, the temperature of the blood (or hemolymph) and air must also come to equilibrium. In other words, the blood (or hemolymph) and air come to the same temperature. In practice, because of the greater specific heat of water, this means that the air in the lung or tracheae is heated to nearly the temperature of the body, requiring an expenditure of energy on behalf of the organism. Similarly, substantial amounts of heat are exchanged as water flows past gill filaments. I suppose that it would be possible in theory to construct a membrane that was permeable to oxygen but still acted as an effective insulator, but in reality, such membranes do not seem to exist in animals.

We can assume, then, that in the absence of any physiological trickery, aquisition of oxygen is generally accompanied by the loss of heat. What, exactly, is the thermal cost of respiration? And what tricks are possible to avoid it? We first examine respiration in air.

8.6.1 *Breathing Air*

To begin, we return to a convention used in chapter 4 and note that rates of metabolic energy expenditure can be expressed as liters of oxygen (at STP) per second per kilogram of body mass. Here we assume that the consumption of one liter of oxygen results in the expenditure of 20.1 kJ of energy. This conversion is shown in table 8.4 for the allometric equations describing the metabolic rates for various animals from invertebrates to birds, allowing us to calculate the rate at which resting animals consume oxygen.

We next recall that oxygen forms only 20.95% of air, so that for each liter of oxygen consumed, 4.8 liters (4.8×10^{-3} m^3) of air must be heated to body temperature. The thermal cost of heating this air is equal to the product of the volume of air heated, the density of air (at STP), the specific heat of air, and the temperature increase. Expressed as an equation,

$$\text{thermal cost (joules per liter O}_2) = 0.0048\rho_{a,STP}Q_s\Delta T. \qquad (8.35)$$

The density of air at STP is 1.292 kg m^{-3} and Q_s is 1006 J kg^{-1} K^{-1}, so the thermal cost is 6.2 ΔT J per liter.

Multiplying the thermal cost per liter of oxygen by the rate at which oxygen is consumed, we arrive at the overall thermal cost of respiration:

$$\text{thermal cost (watts)} = 6.2\Delta T M' m^{\alpha}, \tag{8.36}$$

where M' is the coefficient of resting metabolic rate expressed in liter s^{-1} $kg^{-\alpha}$ from table 8.4.

The thermal cost of respiration can be expressed as a fraction of total resting metabolism by dividing eq. 8.36 by Mm^{α} as given by table 8.4. Because $M'/M = 4.98 \times 10^{-5}$ liters of oxygen per joule, this leads to the simple relationship,

$$\text{fractional thermal cost of respiration } = 0.00031\Delta T. \tag{8.37}$$

For example, it costs a 100 kg arctic mammal 1.2 W to heat the air it breathes when the air temperature is 0°C. This is only about 1% of the 107 W of total resting metabolism. At a higher ambient temperature, the cost is even less.

In arriving at this conclusion, we have tacitly assumed that the organism can absorb all the oxygen that enters the lungs. However, mammals can utilize only about 25% of the oxygen they breathe in and birds perhaps 33% (Schmidt-Nielsen 1979). As a result, the actual thermal cost of respiration is three to four times that calculated above, that is, about 3% to 4% of the total resting metabolism at 0°C. Even this higher figure is quite small, allowing us to conclude that the thermal cost of respiration in air is likely to be negligible.

This conclusion pertains only to the cost of heating the air an animal breathes. In reality there is also a thermal cost due to the water that is evaporated in the process, and this evaporative heat loss can be substantial. We will return to this subject in chapter 14, where we will see that heat exchange in the noses of many animals can reduce the thermal cost of respiration.

8.6.2 *Breathing Water*

What is the thermal cost of respiring in water? Here we are presented with several important differences. First, unlike concentration in air, the concentration of oxygen in water varies both with temperature and salinity (fig. 8.17). For example, there are 10.22 liters of O_2 in a cubic meter of fresh water at 0°C, but only 6.35 liters m^{-3} at 20°C. At 20°C there are 5.17 liters of oxygen per cubic meter of

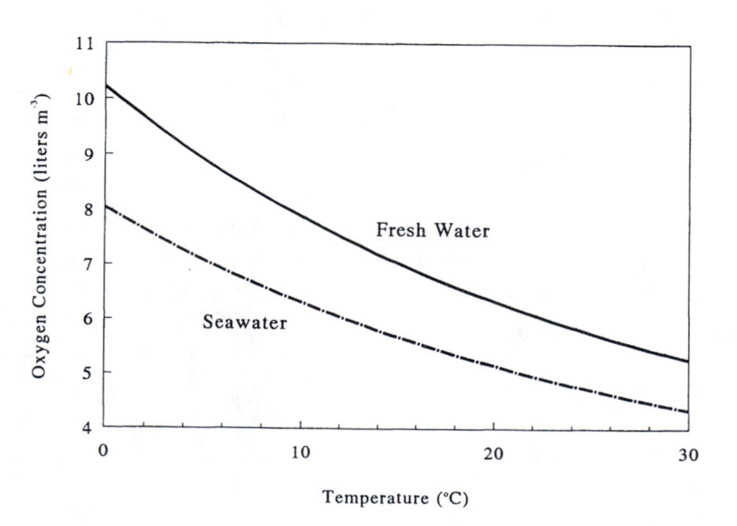

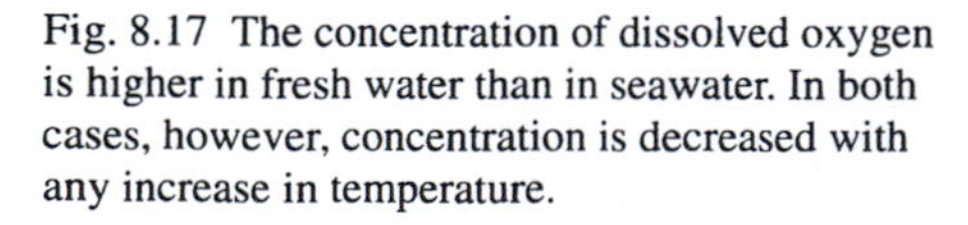
Fig. 8.17 The concentration of dissolved oxygen is higher in fresh water than in seawater. In both cases, however, concentration is decreased with any increase in temperature.

seawater, only 81% of that in fresh water at the same temperature. In general, the higher the temperature or the salinity, the lower is the concentration of oxygen.

The oxygen concentrations shown in figure 8.17 are in terms of the volume of gas per m^3 of liquid measured at one atmosphere for each temperature. These volumes can be converted to volume at STP:

$$\text{volume of } O_2 \text{ (STP) per m}^3,\ V = [O_2]\frac{273}{T}, \tag{8.38}$$

where oxygen concentration is expressed as liters of oxygen per cubic meter of water at temperature T. The volume of water (in m^3) that blood must come into contact with to acquire a liter (at STP) of oxygen is simply $1/V$, a value that varies from about 0.1 to 0.2 over the biological temperature range.

As a result, blood must contact 100 to 200 liters of water (with its high Q_s) to acquire 1 liter of oxygen, and it seems likely that, in the process, blood is cooled to the temperature of the surrounding water. Thus the thermal cost of acquiring oxygen from water is approximately that entailed in reheating blood that has been cooled to ambient temperature. The volume of blood cooled depends on the oxygen-carrying capacity of the blood. The blood of mammals, for instance, can carry about 200 STP liters of oxygen for every cubic meter of blood, the exact figure depending on factors such as pH (Schmidt-Nielsen 1979). In other words, 0.005 m^3 of blood must be reheated for every liter of oxygen consumed. The thermal cost of this reheating is

$$\text{thermal cost (joules per liter of } O_2) = 0.005\rho_{bl}Q_{s,bl}\Delta T, \tag{8.39}$$

where ρ_{bl} is the density of blood and $Q_{s,bl}$ is its specific heat. Assuming that the specific heat and density of blood are similar to those for seawater,

$$\text{thermal cost (joules per liter of } O_2) = 2.1 \times 10^4 \Delta T. \tag{8.40}$$

Multiplying this value by $M'm^\alpha$ gives the approximate thermal cost of respiration in water:

$$\text{thermal cost (watts)} = 2.1 \times 10^4 M'm^\alpha \Delta T. \tag{8.41}$$

Consider a hypothetical example. It could conceivably be advantageous for a warm-blooded animal such as a dolphin to breathe water instead of air. Such an adaptation would remove the necessity for the animal to return periodically to the water's surface, thereby increasing the time available in which to hunt food. However, if a 100 kg dolphin swimming in 7°C water were to breathe water and still maintain a body temperature of 37°C, it would expend energy at a rate of 3361 W just to heat its respiratory water. This is more than thirty times greater than its resting metabolic rate of 107 W! It suddenly becomes clear why marine mammals and birds continue to breathe air, and why water-breathing organisms (such as fish) are seldom much warmer than their watery surroundings.

We can express the fraction of resting metabolic rate used to heat respiratory water in a general fashion by dividing eq. 8.41 by the resting metabolic rate from table 8.4. Again, $M'/M = 4.98 \times 10^{-5}$ liters per joule, so that

$$\text{fractional thermal cost of respiration } = 1.05\Delta T. \tag{8.42}$$

The cost may actually be a bit higher since fish utilize only 80% to 90% of the oxygen in the water passing through the gills (Schmidt-Nielsen 1979). According to this calculation, the only way to keep the respiratory cost to a small fraction of the resting metabolic rate is to keep ΔT to a small fraction of a degree.

8.6.3 *Countecurrent Heat Exchange*

One might be tempted to conclude on this basis that it would be extravagently inefficient for water-breathing animals to be substantially warmer than their surroundings, leading in turn to the prediction that warm-blooded fish should not have evolved. There is, however, a loophole in this reasoning—a loophole through which several species of fish have blithely swum.

The calculations made above assume that the water passing through a gill is heated to the same temperature as blood in the gill. This is very likely true, but it does not mean that the blood in the gill need be at the same temperature as the blood in the rest of the body.

Consider, for instance, the situation shown in figure 8.18. The artery delivering blood to the gill filament runs parallel to a vein returning from the gill, and the two are in close physical contact. We can assume that because it has just passed through the gill, the blood in the vein at point A is at the temperature of the ambient water. Now what happens if the blood entering the artery at point B is warmer than ambient? As this warm blood flows toward the gill, it loses heat to the cold blood in the vein. In fact, if sufficient time is allowed before the blood in the artery gets to the gill filament, it arrives there at a temperature very nearly equal to that of the water, having given up all its excess heat to blood in the vein. As a result of this *countercurrent heat exchanger*, a small ΔT is presented to the gills and the thermal cost of respiration is largely avoided.

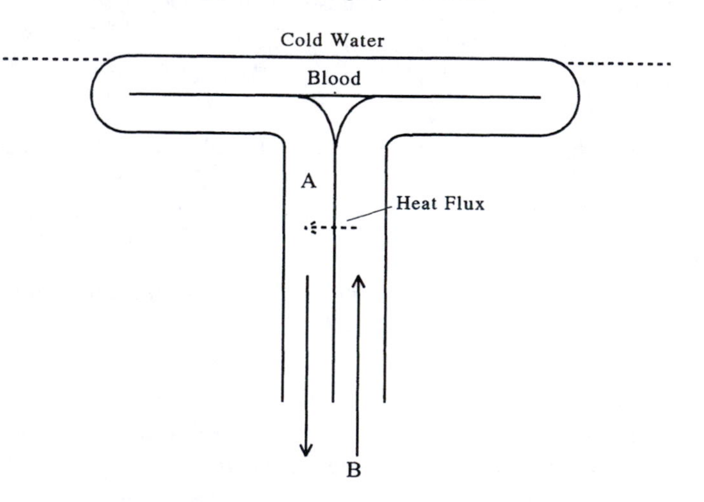

Fig. 8.18 Warm blood entering a countercurrent heat exchanger loses heat to blood returning from the gill. As a result, body heat is conserved.

But wait, you might ask, if heat can flow from the artery to the vein, why can't oxygen from the newly oxygenated blood in the vein flow into the oxygen-depleted blood in the artery? If this happened to any great degree, little oxygen would be delivered from the gills to the body, negating the respiratory function of the gill. The situation is saved, however, when we recall that in water heat is more than one hundred times as diffusive as oxygen. In other words, the Lewis number is small (table 8.2). It is thus possible to design a countercurrent system that exchanges virtually all the excess heat in the blood while exchanging very little of the oxygen.

This type of countercurrent heat exchanger is used by fish such as tunas and lamnid sharks to maintain their swimming muscles, heart, and digestive tract at temperatures several degrees above that of their surroundings (Carey and Teal 1966, 1969; Carey et al. 1971; Block 1991). This elevated body temperature allows the fish to swim at a higher velocity than they otherwise could, and thereby (presumably) to catch more food. Of course, catching more food becomes a necessity due to the higher metabolic rate associated with maintaining a warm body, but we can only assume that the trade-off has been to the fish's advantage.

8.7 Summary . . .

The thermal conductivity of water is about twenty-three times that of air. As a result, it is difficult for aquatic organisms to maintain a body temperature different from their surroundings. Those that do must be relatively large and well insulated, and even then must use tricks such as countercurrent heat exchangers or lungs to avoid the thermal cost of breathing water. The use of a countercurrent heat exchanger by some fish is feasible only because the diffusivity of heat in water is much greater than the diffusivity of oxygen.

In the terrestrial environment, small organisms can maintain an elevated body temperature, but only by having an extraordinarily high metabolic rate or an outside source of heat such as the sun. In contrast, large terrestrial mammals and birds are unique in that they may actually have trouble losing heat fast enough to maintain a livable temperature. In general, this problem is solved through evaporative cooling, a process that will be explored in chapter 14.

8.8 . . . and a Warning

This chapter has utilized a simple model to explore the biological consequences of the thermal properties of air and water. You are reminded that the body temperatures predicted by this model are intended for comparative purposes only, and are not meant to represent accurately those of real organisms whose shapes are likely not spherical and who may sport coats of insulation. In the real world the sun also shines, animals may radiate heat to their surroundings, and the organism may well be in contact with hot rocks or ice. All of these effects do violence to the assumptions of our model and can substantially affect body temperature. The field of thermal biology is an active and exciting one, and if you are interested in learning more about the thermal consequences of environmental media you should consult texts such as Campbell (1977), Monteith (1973), Gates (1980), or Nobel (1983).

Chapter 9

Electrical Resistivity and the Sixth Sense

Anyone who has spilled coffee into a computer or had rain leak through the roof onto the television has had personal experience with the fact that water is more electrically conductive than air; all those circuits that worked just fine when dry are suddenly shorting out in worrisome fashion. If this contrast in conductivity can make such a difference for manufactured electrical devices, what consequences has it had for the electrical machinery of living things? In particular, we consider how the resistivity of the medium affects the ability of animals to detect the electrical currents flowing from other organisms.

This ability, a sixth sense, might be of considerable use. For example, the contraction of muscles entails substantial movement of ions, and therefore must be accompanied by an electrical signal. If this signal can be detected at a distance by other animals, it could serve as a warning to prey that a predator is present or a clue to the predator that prey is nearby. If conditions are such that visual clues are not available (at night or in the dark depths of the ocean, for instance), these electrical cues might be the only clues as to the presence of other animals.

In fact, the ability to detect electrical signals is fairly common among aquatic organisms. Sharks and rays, for instance, are able to detect the electrical activity produced by the muscular contractions of breathing fish, and can use this information to locate flounders and plaice even when these flatfish are buried in the sediment (Kalmijn 1974). A variety of electric fishes use pulses of current as a means of communication (Bullock and Heiligenberg 1986), and the platypus can sense the electrical signal of its prey (Scheich et al. 1986).

In contrast, the ability to detect electrical signals at a distance appears to be entirely lacking in terrestrial organisms, suggesting that there is something about life in a gaseous medium that is not amenable to this sensory capability. The vast difference in electrical resistivity between the two media is an obvious candidate as the culprit, a possibility we now explore.

9.1 The Physics

9.1.1 *Ohm's Law*

Before examining the biological effects of electrical conductivity, we need to review the basic physics of electrical circuits.

Consider the simple circuit shown in figure 9.1A. A battery provides a source of electrical potential energy, $\mathcal{V}$, measured in volts, V, and the effect of this potential is to drive a current of charges around the circuit. The magnitude of the current, I, is determined by how many charges pass a given point in the circuit per unit time. In SI, the number of charges is measured in coulombs, and currents (in coulombs per second) have the units of amperes. Note that electrical potential energy depends on the number of charges present. Thus, volts have the units of joules per coulomb.

For simple circuits such as this, the ratio of voltage to steady current is set by the circuit's resistance, R. Thus,

$$R = \frac{\mathcal{V}}{I}. \tag{9.1}$$

This simple relationship is known as *Ohm's law* and forms the basis for understanding much about how electricity functions. In honor of Georg Simon Ohm and his law, electrical resistance is measured in *ohms*, Ω.

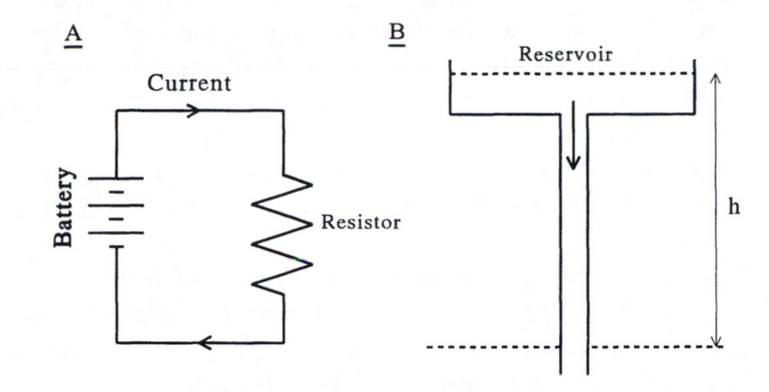

Fig. 9.1 A simple electrical circuit (A) is in many respects analogous to a reservoir of water flowing downhill through a pipe (B). The rate at which electrical current flows is set by the ratio of electrical potential energy to the circuit's resistance. The rate at which water flows is set by the ratio of gravitational potential energy to the pipe's resistance.

A hydraulic analogy may help to make sense of Ohm's law. An elevated reservoir of water (fig. 9.1B) has the potential for causing water to flow downhill. In this sense, the gravitational potential energy of water in a reservoir is analogous to the voltage potential of a battery. Note that both gravitational potential and voltage potential are relative measures. The gravitational potential energy of water in the reservoir is measured relative to some reference water level as shown. Similarly, the potential energy of a battery is measured relative to a reference level, traditionally taken to be the voltage at the battery's negative terminal. Thus, a 9V battery has a voltage potential at its positive terminal of 9 volts relative to the negative terminal.

If we connect a pipe between our elevated reservoir and water at the reference level, water flows down through it. The volume of water passing through a given cross section of the pipe in unit time is a measure of the current driven by gravity, analogous to the electrical current driven by our battery. Now, the rate at which water flows from a reservoir of given elevation depends on the resistance of the pipe. The smaller the pipe's diameter, the greater its resistance (chapter 5), and the less the volume per time that flows. Similarly, the greater the electrical resistance of a circuit, the less the electrical current that flows in response to a given voltage.

Voltage, current, and resistance are measures of the macroscopic electrical properties of a circuit. For example, we can pick a piece of wire from the shelf and measure its total resistance, but this gives us little information about how electrons actually flow through the material from which the wire is constructed. To understand the movement of electrons at a more basic level, we need to relate the macroscopic properties of the circuit to the microscopic physics of the situation. For example, in this chapter we are concerned with how the properties of air and water control the flow of electrical current in biologically relevant circuits, and we therefore must deal with the microscopic electrical properties of these media.

Fortunately, the transposition from macroscopic to microscopic properties is relatively straightforward. At a microscopic level, Ohm's law can be stated as

$$\psi = \frac{\mathrm{E}}{j}, \tag{9.2}$$

where ψ is the *resistivity* of the material in which the current is conducted,[1] E is

[1]Resistivity is traditionally given the symbol ρ. However, to avoid confusion with mass density, we use ψ instead.

the *electrical field strength* at a point in the material, and j is the local *current flux density*. Let us briefly examine each of these quantities, beginning with E.

We know from Newton's first law of motion that an object, even an exceedingly small one like an electron, moves only in response to an applied force. Thus, to start electrons moving through a material, there must be a force applied to them. This force is a result of the presence of an electrical field. Just as mass feels a force directed toward the center of the earth in response to the earth's gravitational field, an electron, due to its charge, experiences a force in response to an electrical field. The strength of a gravitational field, g, is a measure of the force per *mass* exerted by the field, whereas the strength of an electrical field, E, is measured as the force exerted per *charge*.

Continuing the analogy between electricity and hydraulics, we recall from Chapter 3 that the gravitational potential energy per mass of water is gh, where h is the vertical distance from the reference level. In using this relationship, we have tacitly assumed that g is constant. In a situation where g varies substantially over the distance h, the gravitational potential energy must be described as

$$\text{gravitational potential energy} = \int_0^h g\,dy. \tag{9.3}$$

Similarly, the voltage potential energy of a single charge is

$$\mathcal{V} = \int_0^\ell \mathrm{E}\,dx, \tag{9.4}$$

where ℓ is the distance from the charge to the zero reference point of the electrical field. Note that for both gravitational and voltage potential energy we have tacitly assumed that the object in question is moved in the direction along which the field acts. For instance, the earth's gravitational field is directed vertically. As a result, only vertical displacement of a mass affects its potential energy; horizontal motion has no effect. Similarly, the voltage potential energy of a charge is affected only when the charge has a component of displacement in the direction of the electrical field.

In the simple situation of a battery connected to a cylindrical wire, the strength of the electrical field is constant and its direction is parallel to the wire. Thus, the voltage difference between two points on the wire separated by a distance ℓ is

$$\mathcal{V} = \mathrm{E}\ell. \tag{9.5}$$

The electrical field has the units of force per charge; in SI units, newtons per coulomb or, equivalently, volts per meter.

Current flux density, j, is simply the amount of current flowing through a given small area, A, of the material, where the area is taken to lie in a plane perpendicular to the direction of electron movement. In other words,

$$j = \frac{I}{A}. \tag{9.6}$$

Current flux density has the units of amperes per square meter.

And finally, electrical resistivity, ψ, is a measure of a material's intrinsic resistance to the flow of electrical current. Combining eqs. 9.2, 9.5, and 9.6, we see that

$$\psi = \frac{E}{j} = \frac{\mathcal{V}/\ell}{I/A}. \tag{9.7}$$

But $\mathcal{V}/I = R$ (eq. 9.1), so that

$$R = \psi\frac{\ell}{A}. \tag{9.8}$$

In other words, the higher the resistivity (a property of the material from which an object is constructed), the higher the resistance. But resistance also increases with the length of the object, and decreases with an increase in cross-sectional area. For example, if two pieces of copper wire have the same diameter, the longer piece has the greater resistance. If the two pieces have the same length, the piece with the greater diameter has the lower resistance.

Rearranging eq. 9.8, we see that

$$\psi = R\frac{A}{\ell}, \tag{9.9}$$

indicating that resistivity has the units of Ω m.

It is common in some fields (physical oceanography, for instance) to deal with the *conductivity* of a material, χ, rather than its resistance:

$$\chi = \frac{1}{\psi}. \tag{9.10}$$

For example, we noted in chapter 4 that the practical salinity of seawater is determined by measuring conductivity.

Next, we note that because a force is required to push electrons through a length of resistor, energy is expended as current flows through a circuit. The greater the current, the greater the rate at which energy is expended, and the greater the power of the circuit. The appropriate expression for electrical power can be derived by a simple, dimensional consideration. Current has the units of coulombs per second, so to get from current to power (joules per second) we need to multiply by something that has the units of joules per coulomb. This is simply voltage. Thus,

$$P = \mathcal{V}I. \tag{9.11}$$

Using Ohm's law (eq. 9.1), we can also say that

$$P = I^2R \tag{9.12}$$

$$P = \frac{\mathcal{V}^2}{R}. \tag{9.13}$$

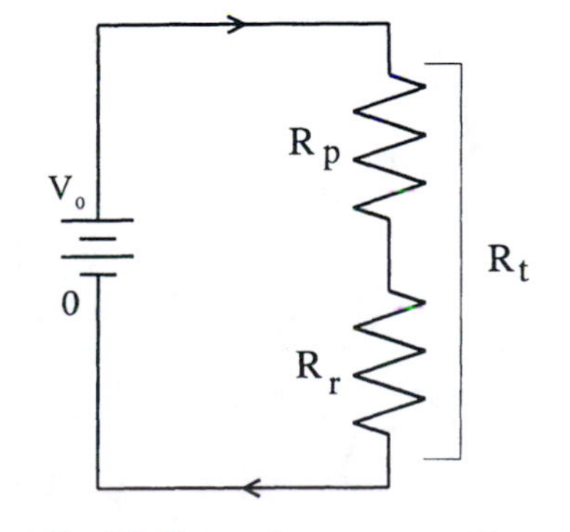

Fig. 9.2 Two resistors connected in series act as a voltage divider.

And finally we explore the manner in which voltage varies as electrons move through a resistor. Consider the situation shown in figure 9.2, a resistor connected between a source of electrical potential energy ($\mathcal{V}_0$ = 10V) and a point at 0V. Before an electron enters the resistor, we know that it has a 10V potential for movement, and we know that by the time it reaches the other end of the resistor, this potential has been entirely dissipated. What, then, is the voltage potential of the electron somewhere in the middle of the resistor? The voltage potential at any point is proportional to the fraction of the overall resistance through which the electron has yet to pass. Thus if R_p is the resistance through which the electron has moved and R_r is the remaining resistance through which it must move before arriving at 0V,

$$\mathcal{V} = \mathcal{V}_0 \frac{R_r}{R_p + R_r}. \tag{9.14}$$

In other words, a series of resistors hooked end to end divide up the overall voltage according to the fraction that each resistor contributes to the total resistance.

9.2 Electrical Resistivity of Air and Water

The electrical resistivity of air and water are about as different as any two properties can be. The electrical resistivity of dry air at sea level is approximately $4 \times 10^{13} \Omega$ m (Weast 1977). In contrast, the resistivity of seawater ($S = 35$) is about $2 \times 10^{3} \Omega$ m. Thus, air is 20 billion times more resistive than seawater.

In seawater, electrical current is carried by ions (primarily Na^+ and Cl^-) rather than by bare electrons. The concentration of these ions is a function of both salinity and temperature, and empirical measurements provide a method for relating these factors (see table 9.1). The resistivity of seawater is graphed in figure 9.3A and that of fresh water in figure 9.3B, both as a function of temperature. The higher temperature or salinity, the lower the resistivity.

Table 9.1 Calculation of the conductivity of water as a function of temperature and salinity (Poisson 1980).

First, one calculates the ratio, K, of conductivity at a given temperature T (in °C) to the conductivity at 15°C for a salinity $S = 35$:

$$K = 0.6765836 + 2.005294 \times 10^{-2}T + 1.110990 \times 10^{-4}T^2 - 7.26684 \times 10^{-7}T^3 + 1.3587 \times 10^{-9}T^4. \tag{a}$$

This value is then used to calculate conductivity at temperature T and salinity S:

$$\chi(T, S) = \frac{S}{35}(0.042933K) + S(S - 35)\big(B_0 + B_1 S^{1/2} + B_2 T + B_3 S + B_4 S^{1/2} T + B_5 T^2 + B_6 S^{3/2} + B_7 TS + B_8 T^2 S^{1/2}\big), \tag{b}$$

where

$$B_0 = -8.647 \times 10^{-6} \quad B_5 = -1.08 \times 10^{-9}$$
$$B_1 = 2.752 \times 10^{-6} \quad B_6 = 2.61 \times 10^{-8}$$
$$B_2 = -2.70 \times 10^{-7} \quad B_7 = -3.9 \times 10^{-9}$$
$$B_3 = -4.37 \times 10^{-7} \quad B_8 = 1.2 \times 10^{-10}.$$
$$B_4 = 5.29 \times 10^{-8}$$

The resistivity at this temperature and salinity is the inverse of $\chi(T, S)$:

$$\psi(T, S) = \frac{1}{\chi(T, S)}. \tag{c}$$

The resistivity of pure water is actually quite high. For example, laboratory-grade distilled water has a resistivity of about $2 \times 10^9 \Omega$ m, about one ten-thousandth that of dry air. However, pure water is seldom found in nature. Most rivers and lakes have a salinity of 0.01 to 0.1 (Hutchinson 1957) and therefore these fresh waters have a resistivity of 1 to $9 \times 10^6 \Omega$ m.

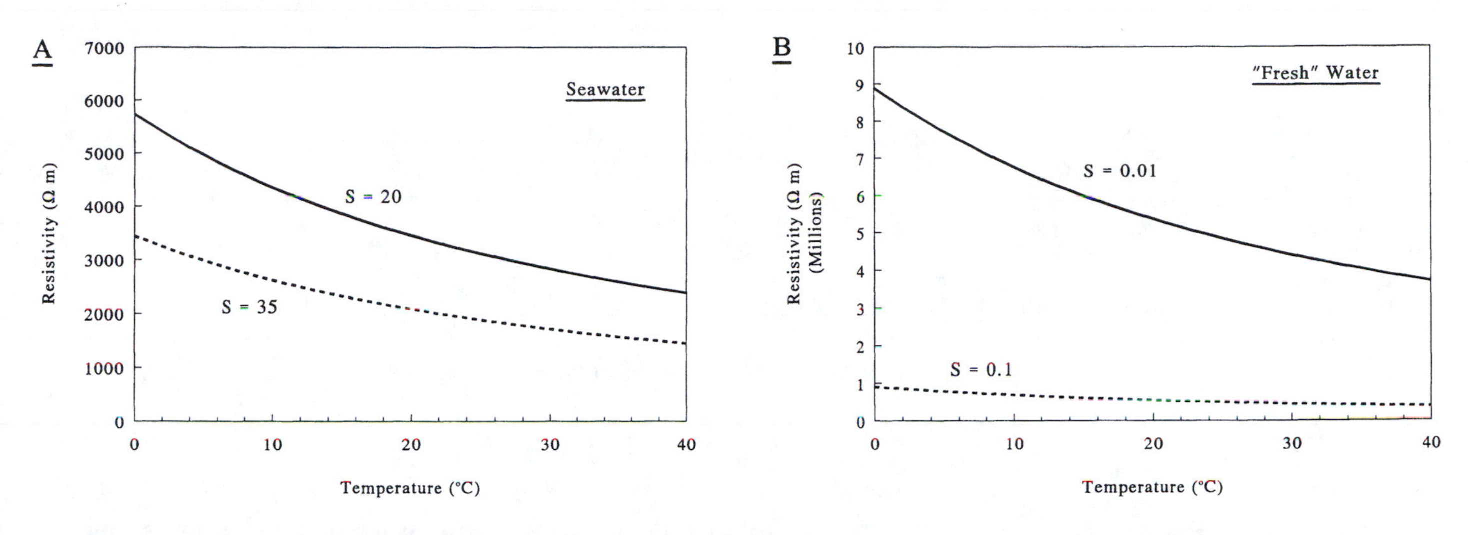

Fig. 9.3 The resistivity of seawater decreases with increasing temperature and salinity (A). Similarly, the resistivity of fresh water (which always has some dissolved solids) decreases with increasing temperature (B).

9.3 Sensing Electrical Activity at a Distance

9.3.1 *The Field*

We are now in a position to explore the effects of differing environmental resistivity on the ability of animals to sense electrical activity in their environment. The mathematics describing this sixth sense is not difficult, but it is distractingly tortuous. For this reason the basic derivations have been relegated to an appendex at the end of the chapter and only the pertinent results are presented in the text.

There are two steps in this exploration. First we examine the voltage an animal must produce to broadcast a signal with a given power. We then discuss how a biological detector can use this power to sense the presence of an electrical field.

Consider a simple situation in which we have two spheres, one at a positive voltage with respect to the other so that current flows from the positively charged sphere to the negatively charged sphere through the intervening medium (fig. 9.4). This simple circuit is an appropriate model for the sort of biological circuits that might broadcast an electrical signal to the environment. For example, in the process of breathing, an animal contracts muscles. As a result, electrical current produced at one point may leak into the surrounding medium and travel to a point somewhere else on the body where the potential is momentarily more negative. To give our calculations some tangibility, let us assume that our two spheres are a model for a flatfish (a plaice, for instance) as it lies hidden beneath a thin layer of sand, or for a field mouse hidden among weeds.

In light of this circuit's role as a producer of an electrical field, we refer to this system of two spheres as *the source* of an electrical signal. Note that this use of the term "source" refers to the signal, not the electrical current. One sphere is the source of current, and the other the sink, but the two taken together are the source of the signal that is broadcast into the medium.

Let us suppose that this signal is to be detected by a predator traveling over the source. For simplicity, we confine our investigation to locations directly above the center of the source (fig. 9.4). In other words, we ask what strength of signal the plaice or field mouse presents to a shark or an owl a distance y directly overhead.

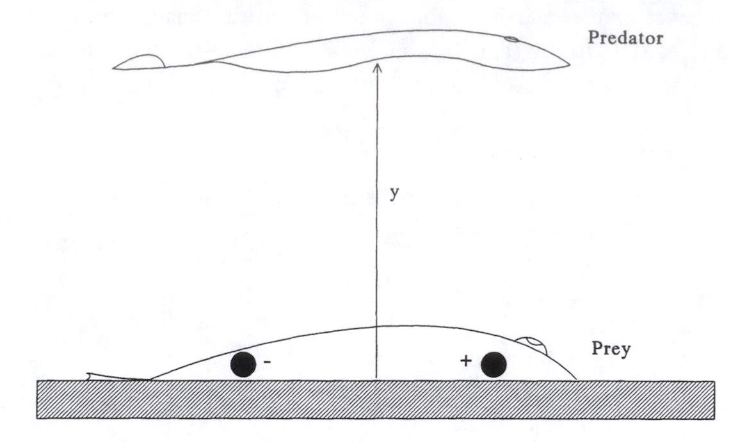

Fig. 9.4 The muscular activity of a flatfish creates a voltage difference between two areas on the body (here modeled as spheres). The resulting electrical signal can potentially be sensed by a predatory skate swimming overhead.

It can be shown (eq. 9.47) that the field at the predator is

$$E = \frac{r_s \mathcal{V}_s \ell_s}{2} [y^2 + (\ell_s/2)^2]^{-3/2}, \tag{9.15}$$

where r_s is the radius of the source spheres, $\mathcal{V}_s$ is the voltage produced by the source, and ℓ_s is the separation between source spheres. Note than when $y \gg \ell_s/2$,

$$[y^2 + (\ell_s/2)^2]^{-3/2} \approx y^{-3}. \tag{9.16}$$

Thus, when the height of the predator above the prey is large relative to the length of the prey, eq. 9.15 simplifies to

$$E = \frac{r_s \mathcal{V}_s \ell_s}{2y^3}. \tag{9.17}$$

Eqs. 9.15 and 9.17 tell us that the field presented to a predator depends on the voltage produced by the prey and is independent of the medium's resistivity. If we assume that muscular contractions in air are accompanied by approximately the same voltages as contractions in water, this suggests that the fields produced by aquatic and terrestrial animals are similar. Why then can aquatic animals detect electrical signals at a distance, but terrestrial animals cannot?

9.3.2 *Power*

We find the answer when we explore what it is that is being detected. Consider the situation when you try to tune a radio to a particular station. If the station is powerful, you receive a strong signal. For once, common vernacular and the precise language of physics agree—it is the *power output* of the radio station (rather than its field strength) that determines your ability to hear the news. This is why radio stations advertise the wattage of their signal.

Why is power rather than field strength the parameter of importance? Because, for a signal to be detected, work must be done. Before the news can come out of the speaker, electrons in a radio's antenna must be made to move, and this movement requires the expenditure of energy. In turn, the energy to move the antenna's electrons must be provided by the distant radio station. The more power broadcast by the station, the more power that can be delivered to your radio, and the stronger the signal is.

The same holds true for biological signals. The ability of an organism to detect electrical activity at a distance depends on the electrical power that can be delivered to its detectors. As a result, we now try to make a connection between the field produced by a source, and the power that is broadcast.

We begin by recalling that electrical power is equal to $\mathcal{V}^2/R$ (eq. 9.13). Thus, if we can calculate the resistance of the source (R_s), we can calculate the power broadcast. Now, it can be shown (eq. 9.44) that when the radius of the source spheres (r_s) is small relative to the distance between spheres (a situation we assume to hold for our source), the source resistance is

$$R_s \approx \frac{\psi}{2\pi r_s}. \tag{9.18}$$

Thus,

$$P_s \approx \frac{2\mathcal{V}_s^2 \pi r_s}{\psi}. \tag{9.19}$$

In other words, if aquatic and terrestrial animals produce the same voltage, the power broadcast by the terrestrial animal is smaller by a factor of 20 billion. Therein lies the first reason why biological electrical signals are difficult to detect in air.

For example, Kalmijn (1974) reports that a wide variety of aquatic animals produce voltages of 10 μV and a few, such as bony fishes and gastropods, produce 100 to 500 μV. When animals are wounded, the voltages they produce may increase substantially. For instance, intact crustacea produce about 50 μV, but injured animals produce up to 1250 μV. Let us take 100 μV as a representative value of the voltage signals produced by aquatic animals. We assume that the resistivity of seawater is 2×10^3 Ω and (arbitrarily) that r_s is 2 mm. Thus, a typical marine animal broadcasts a signal with a power of about 6×10^{-14} W. This is a very low power, but it is positively herculean compared to the 3×10^{-24} W broadcast by an equivalent terrestrial organism.

9.3.3 *The Detector*

We can explore the process of signal detection further by examining what happens at the detector itself. For the sake of simplicity we use a detector with a simple geometry similar to that of the source (fig. 9.5A): two spheres in contact with the medium. These spheres each have a diameter of r_d, are separated by a distance ℓ_d, and are oriented such that the line connecting spheres is parallel to the electrical field (in this case, horizontally). The center of the detector is located at a distance y from the center of the source. As before, we assume that $r_d \ll \ell_d$.

Because the geometry of the detector is similar to that of the source, we know immediately that the resistance to current flow between spheres due to the medium alone is $R_m = \psi/(2\pi r_d)$ (eq. 9.44). However, in addition to the resistance from the medium, we may also assume that there is some connection between the detecting spheres within the animal's body, and that this internal connection itself has a resistance, R_d.

Given this geometry, the physics of detection can be represented by a simple circuit diagram (fig. 9.5B). In essence, one replaces the source and the medium with

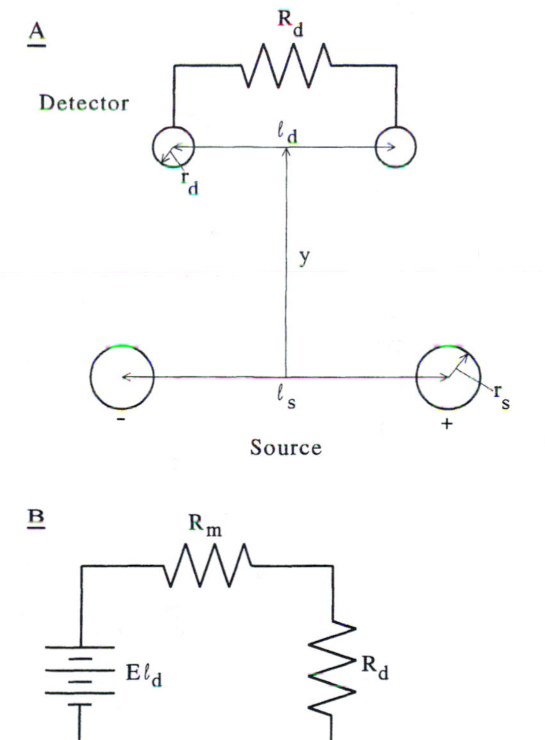

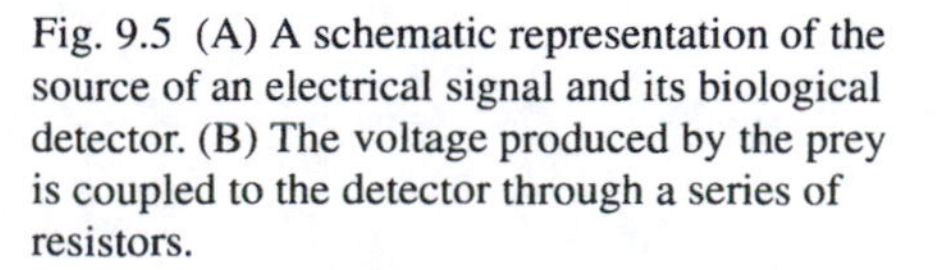

Fig. 9.5 (A) A schematic representation of the source of an electrical signal and its biological detector. (B) The voltage produced by the prey is coupled to the detector through a series of resistors.

a battery and a single resistor, and this is then connected in series to the detector.[2] The source produces an electrical field, the strength of which is E (eq. 9.15) at the detector. Integrating E along the line connecting the detector spheres, we calculate that a voltage, $\mathcal{V} = \mathrm{E}\ell_d$, is present in the medium between the detector's spheres. This voltage is shown as the battery in figure 9.5B, and it has the potential to do work on the detector.

The current that flows through the detector in response to this voltage is set by the total resistance, $R_m + R_d$. Thus,

$$I_d = \frac{\mathcal{V}}{R_m + R_d}. \tag{9.20}$$

Although there is a voltage $\mathcal{V}$ in its vicinity due to the local electrical field, not all of this voltage can be sensed by the detector because of the presence of R_d. In effect, the organism is part of a voltage divider, such that the voltage across the detector, $\mathcal{V}_d$, is

$$\mathcal{V}_d = \mathcal{V}\frac{R_d}{R_m + R_d}. \tag{9.21}$$

We are now in a position to calculate the power that the source transfers to the detector. Recalling (eq. 9.11) that $P = \mathcal{V}I$, we see that

$$P_d = \mathcal{V}_d I_d = \frac{(\mathrm{E}\ell_d)^2 R_d}{(R_m + R_d)^2}. \tag{9.22}$$

Inserting our expression for E from eq. 9.15 and replacing R_m with $\psi/(2\pi r_d)$, we arrive at our final answer:

$$P_d = \frac{(\mathcal{V}_s \ell_s \ell_d r_s)^2}{4} \frac{R_d}{\left(\frac{\psi}{2\pi r_d} + R_d\right)^2} [y^2 + (\ell_s/2)^2]^{-3}. \tag{9.23}$$

For our present purpose, the important part of this equation is the expression involving the resistance of the medium and that of the detector:

$$\text{resistance term} = \frac{R_d}{\left(\frac{\psi}{2\pi r_d} + R_d\right)^2}. \tag{9.24}$$

Note that this term is small when R_d is either very large or very small, and because the rest of the equation is multiplied by this term, the power received by the detector is small if R_d is either too large or too small. This implies that there is an optimal value for R_d. By taking the derivative of the resistance term and setting it equal to zero, it can be shown that this term is maximal when $R_d = R_m$. In this case,

$$\text{resistance term} = \frac{\pi r_d}{2\psi}, \tag{9.25}$$

and the maximum power the detector can receive is

$$P_{d,opt} = \frac{\pi(\mathcal{V}_s \ell_s \ell_d r_s)^2 r_d}{8\psi} [y^2 + (\ell_s/2)^2]^{-3}. \tag{9.26}$$

[2] You might wonder why the source couples to the detector in series rather than in parallel. The answer lies with *Thévenin's theorem* regarding the simplification of resistive networks. For a complete explanation, see any standard text in electrical engineering (e.g., Yorke 1981).

If $y \gg \ell_s/2$, eq. 9.26 simplifies to

$$P_{d,opt} = \frac{\pi(\mathcal{V}_s \ell_s \ell_d r_s)^2 r_d}{8\psi y^6}. \quad (9.27)$$

Therefore, even when the detector is optimally "tuned" to the resistivity of the medium, the power that can be conveyed to the detector varies inversely with the medium's resistivity. In other words, for a given source voltage, the power delivered to a detector in seawater is 20 billion times that delivered in air. Therein lies another reason why detection of distant electrical signals is highly unlikely in the terrestrial environment.

Actually, the situation is even worse than is suggested by eq. 9.26. To be optimally tuned, a detector's internal resistance must equal that due to the surrounding fluid. In seawater, with its low resistivity, this does not pose a problem. In air, however, the internal resistance of an optimally-tuned detector would need to be on the order of $10^{13}\Omega$, far higher than is likely for biological detectors. As a result, detectors in air would effectively short-circuit the very signals they were trying to detect.

Viewed from a different perspective, eq. 9.26 should give us a healthy respect for those animals that detect electrical signals in water. The power available to a detector scales approximately with the inverse *sixth* power of distance from the source. In other words, increasing one's distance from the source by a factor of two decreases the signal available by a factor of sixty-four. It is a tribute to the neural machinery of aquatic organisms that they can effectively detect signals that are attenuated this rapidly.

Eq. 9.26 gives one final clue as to how animals such as fish detect electrical signals. Note that the power available to the detector increases with the square of the distance between detector spheres. Perhaps this is one reason why fish have arrayed their electrical detectors as they have. For instance, the grotesquely widened heads of hammerhead and bonnethead sharks place the shark's electrical detectors (the *ampullae of Lorenzini*) as far apart as is practical. Bony fishes array their electrical sensors (the *lateral line organs*) in a long line along the side of the body. In either case, if the detectors were confined to a typically shaped head (as is the case for many senses), the short baseline available would decrease the power delivered to the electrical sensors.

9.4 Directional Dependence

The calculations we have performed here apply only to the electrical signal directly above the prey. What is the nature of the signal at other points in space? This question can be answered by calculating for a given x, y coordinate the field strength due to each source sphere and adding the strengths together. We take as a biologically appropriate example a source that produces 100 μV, with r_s = 2 mm, and ℓ_s = 20 cm. The results are shown in figure 9.6, where the line traces the location of a field of 10 μV m^{-1}.

A field of a given strength extends farthest from the source along the axis connecting spheres and is minimal perpendicular to this axis. As a result, the calculations made above predict the minimum field strength of the signal.

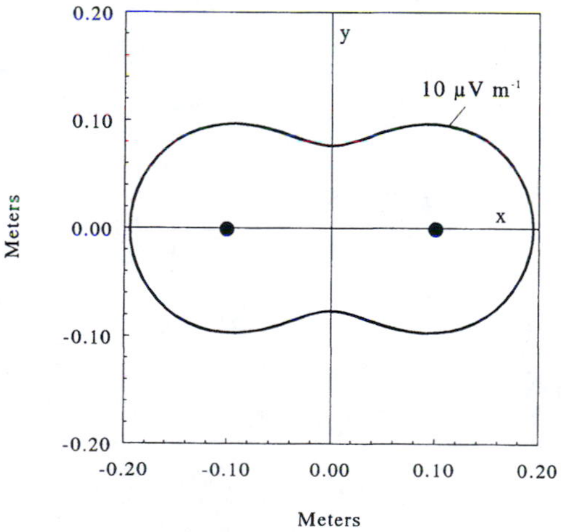

Fig. 9.6 The strength of a voltage signal produced by a dipole source is strongest along the line connecting spheres. Shown here are those points where the field strength is 10 μV m^{-1}.

9.5 Electrical Detection in Water

Having explored the physics of electrical detection, we are now in a position to examine critically the performance of aquatic animals. What abilities do marine animals have, and over what distances are these abilities effective at locating prey?

The most sensitive animals are fishes (sharks, rays, and the common eel), and they can first detect fields when they exceed 6.7 to 10 μV m^{-1} (Kalmijn 1974).

Let us assume as before that prey produce a voltage of 100 μV and that r_s is 2 mm. If the prey is 20 cm long, the field strength directly above the source is as shown in figure 9.7. A field of 10 μV m^{-1} extends about 8 cm from the center of the animal. Thus, our "typical" marine animal could be detected when the predator was 8 to 10 cm above the prey. Given the spatial variation in field strength shown in figure 9.6, we can predict that the limit of detection is somewhat larger if the predator approaches the prey obliquely. Experiments reported by Kalmijn (1974) show that the effective detection distance is similar to this prediction. For example, one species each of shark and ray were able to detect and orient toward a buried plaice from a distance of 10 to 15 cm.

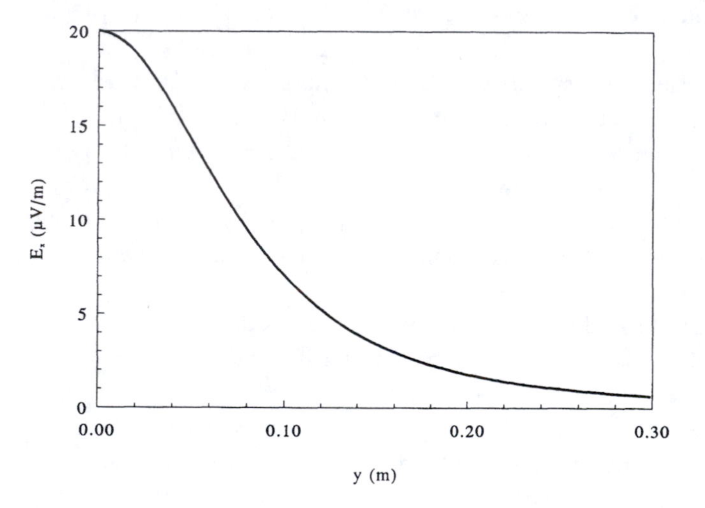

Fig. 9.7 The electric field strength produced by a dipole source decreases with distance from the source. Here the field is measured along a line perpendicular to the axis connecting the spheres and passing through a point half way between them.

If instead of a typical marine organism we choose a wounded one, the voltage produced might increase by a factor of ten. The distance to which a given field strength extends, however, increases by only a factor of $\sqrt[3]{10} = 2.15$. Thus, even the most sensitive shark must approach to within about a third of a meter of a typical injured animal before the prey can be detected.

In summary, the sensitivity of some aquatic animals is sufficient to allow them to sense electrical activity at distances of a few centimeters. This ability may be highly advantageous when searching in the dark or when hunting buried or hidden prey. However, the rapid attenuation of the electric field insures that electroreception can never replace vision or sound as a means for sensing the environment at a great distance.

9.6 Other Uses of an Electrical Sense

In this chapter we have explored the ability of one animal to sense the electrical activity of another. Electroreception can be put to other uses well. For example, the African electric fish *Gymnarchus niloticus* (fig. 9.8A) uses specially modified muscles to produce a weak electrical field around its body, and Lissman and Machin

(1958) have shown that the fish can use this field to sense inanimate objects in its vicinity. The presence of an object distorts the fish's electrical field, a distortion that can be sensed by the fish and used as an aid to navigation in the muddy waters inhabited by this species. This sense is useful, however, only for detecting objects nearby. The rapid attenuation of the electrical field insures that objects farther than a few centimeters from the fish cannot be detected.

The need to detect the distortion of the electrical field places constraints on the locomotion of electric fish. If the animal were to undulate its body while swimming, the resulting undulation of the electric field could interfere with signals from the environment. In response to this problem, *Gymnarchus* has evolved an unusual mode of locomotion: it swims by undulating its elongated dorsal fin rather than its body. A similar strategy has evolved in the South American electric fishes (e.g., *Electrophorus electricus*, fig. 9.8B). In this case, however, it is an elongated anal fin on the ventral side of the body that undulates.

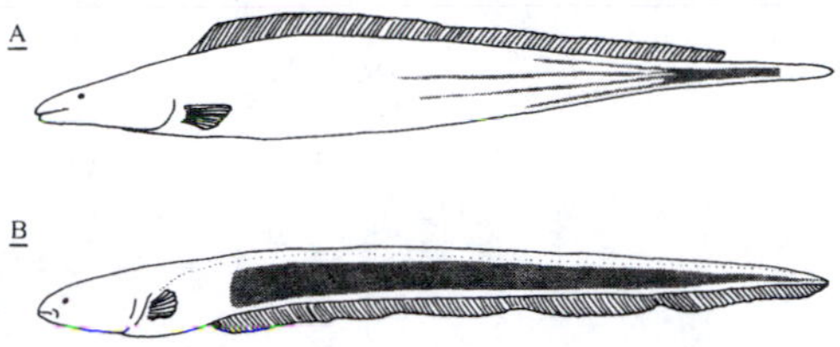

Fig. 9.8 To avoid confusion from sensing changes in their own electrical fields, electric fish such as *Gymnarchus niloticus* (A) and (B) *Electrophorus electricus* swim using undulations of the centerline fins. Reprinted from Bond (1979) by permission of Saunders College Publishing.

Electrical fields can also be used to stun prey (Bond 1979). For example, the electric ray *Torpedo californicus*, the electric catfish *Malapterurus electricus*, and the electric eel *E. electricus* have evolved highly specialized electric organs that can produce 200 to 650V! Once again, however, the attenuation of the electric field insures that these voltages are effective as a weapon only over short distances. For example, the ray typically waits for prey to swim overhead, whereupon it raises its fins to encircle the prey. Only then is the shock released.

And finally, it is possible, in theory, for aquatic organisms to detect the electrical field created when seawater (a conductor) flows in the earth's magnetic field. This ability would allow the organism to orient and navigate in the absence of visual cues. A thorough discussion of this possibility can be found in Kalmijn (1974, 1984).

9.7 Summary . . .

The high resistivity of air makes it very difficult for a biological source of voltage to broadcast power to a distant detector. Even when optimally tuned to its environment, a terrestrial detector can receive only one twenty-billionth of the power available to a detector in seawater, and optimal tuning appears unlikely. These large handicaps have apparently been sufficient to inhibit the evolution of electrical detectors in terrestrial animals. Even in seawater, the signals produced by animals are quite small and can be detected only over a short distance.

9.8 . . . and a Warning

As always, the results presented in this chapter should not be extrapolated beyond their intended scope. Our model using paired spheres is only a very crude approximation to the actual form of biological sources and detectors. As a consequence, the results derived here can be, at best, only a rough estimate of the actual process of signal production and detection. For a report on the empirical measurement of fields around fish, you should consult Knudsen (1975). We also have only considered sources of steady current, whereas many biological sources would perhaps be better thought of in terms of pulsed or alternating current. If you want to pursue these avenues of thought, you should consult Fessard (1974) or Bullock and Heiligenberg (1986).

9.9 Appendix

9.9.1 *Calculating the Field*

We begin our analysis by examining the electrical field that is produced by a small sphere as it pumps current radially into a medium of uniform resistivity ψ (fig. 9.9A). As current flows out of the sphere at rate I, the current flux density at a distance r from the sphere's center is

$$j = \frac{I}{4\pi r^2}, \tag{9.28}$$

where $4\pi r^2$ is the area of the spherical surface at radius r through which the current flows.

Given this expression for j, we can utilize eq. 9.2 to quantify the field strength at r:

$$\mathrm{E} = \psi j, \tag{9.29}$$

$$\mathrm{E}(r) = \frac{\psi I}{4\pi r^2}. \tag{9.30}$$

If instead of pumping current into the medium, the sphere sucks current out of the medium, the current is $-I$ and the sign of the field is reversed.

Although a single sphere produces a simple field, it is biologically more relevant to consider a circuit consisting of two spheres (fig. 9.9B) where each sphere produces its own field and the two fields interact additively. As we have seen, this circuit can serve as a model for a biological source of an electrical signal.

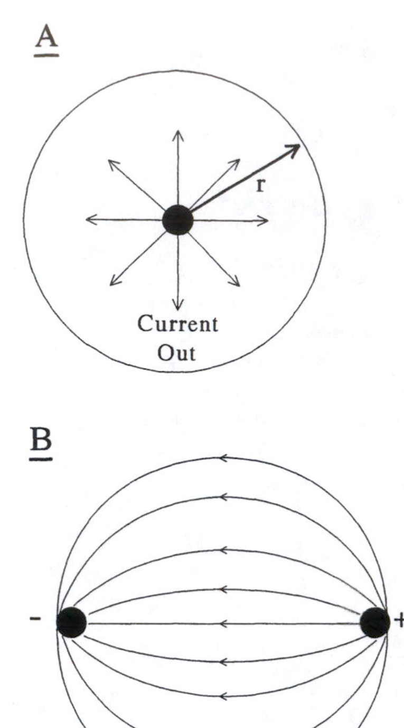

Fig. 9.9 When only one charged sphere is present, current flows radially (A). The presence of a second sphere of opposite polarity causes lines of current flow to bend (B).

To add the fields from two spheres, we must specify the dimensions of the system. For simplicity, we locate our spheres on the x axis, straddling the origin, and separate them by a distance ℓ_s (fig. 9.10A). Thus the positive sphere (+) lies at $x = \ell_s/2$, and the negative sphere $(-)$ lies at $x = -\ell_s/2$. Given this geometry, the distance from (+) to an arbitrary point x,y in the environment is

$$r_{(+)} = \sqrt{y^2 + (x - \ell_s/2)^2}, \tag{9.31}$$

and similarly, the distance from $(-)$ is

$$r_{(-)} = \sqrt{y^2 + (x + \ell_s/2)^2}. \tag{9.32}$$

Given these expressions for the radial distances from each sphere, we may rewrite eq. 9.30 for the positive sphere as

$$\mathrm{E}_{(+)} = \frac{\psi I_s}{4\pi r_{(+)}^2}. \tag{9.33}$$

For the negative sphere,

$$\mathrm{E}_{(-)} = -\frac{\psi I_s}{4\pi r_{(-)}^2}. \tag{9.34}$$

Our task now is to sum these two fields at a point of interest. For simplicity, we choose a point directly above the origin, i.e., on the line $x = 0$. We initially approach this question through a graphic examination of the fields produced by (+) and $(-)$ (fig. 9.10A).

The field of each sphere is directed radially: outward for (+) and inward for (−). Because a point directly above the origin is equidistant from both spheres, the strengths of the two fields (represented by the length of the line in fig. 9.10A) are the same.

Each field can now be decomposed into its x-directed and y-directed components (fig. 9.10B). It is apparent that the y-directed component of (+) is equal in strength but opposite in direction from that of (−). As a result, the y-directed components of the two fields cancel each other.

The x-directed components are also of equal magnitude, but in contrast to the y-directed components, they point in the same direction, and thereby reinforce each other. Thus, the overall field acts in the x-direction and has a magnitude equal to the summed components of the two x-directed fields.

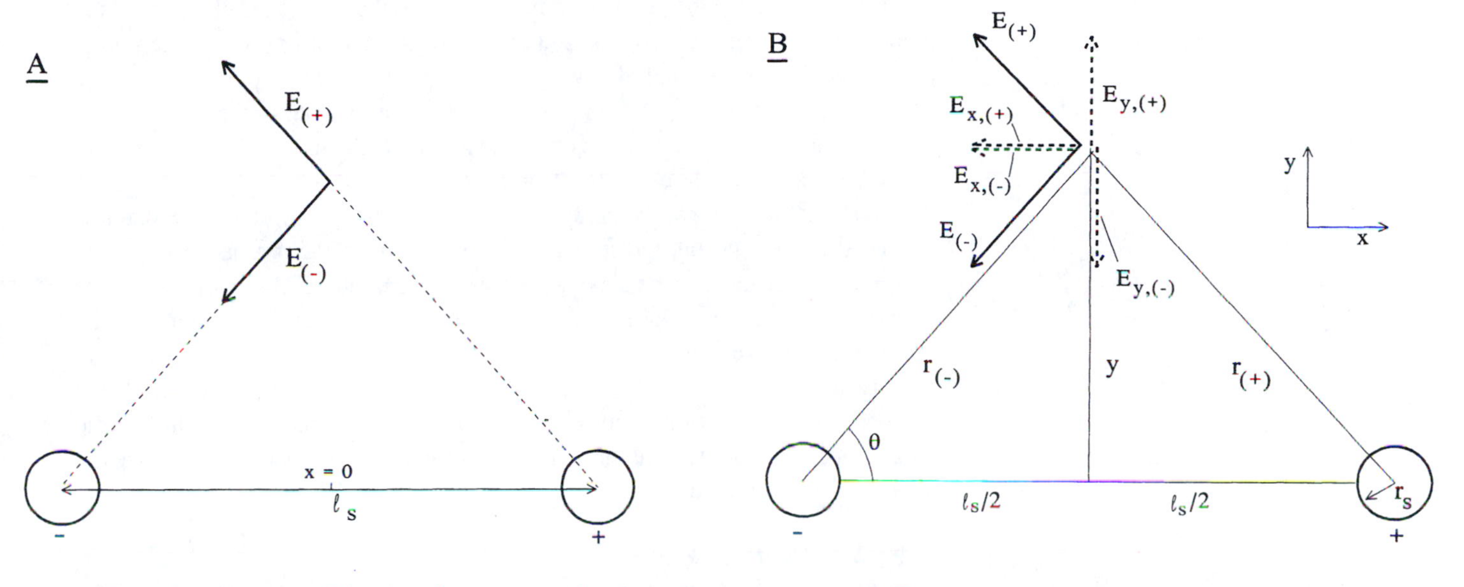

Fig. 9.10 The net field strength at any point is the vector sum of the individual strengths from each sphere (A). The directional components of field-strength vectors can be added to calculate the net field strength (B). Note that the y-directed fields cancel.

From the geometry of the situation, we note that the x-directed component, E_x, is equal to $\mathrm{E}\cos\theta$. Thus the x-directed component due to the positive sphere is

$$\mathrm{E}_{x,(+)} = \mathrm{E}_{(+)}\,\frac{-\ell_s/2}{r_{(+)}}. \tag{9.35}$$

Similarly, the x-directed component due to the negative sphere is

$$\mathrm{E}_{x,(-)} = \mathrm{E}_{(-)}\,\frac{\ell_s/2}{r_{(-)}}. \tag{9.36}$$

Expressing E as in eq. 9.30 and noting that we are examining the field at a point where $x = 0$, we see that

$$\mathrm{E}_{x,(+)} = -\frac{\psi I_s \ell_s}{8\pi r_{(+)}^3}. \tag{9.37}$$

Similarly,

$$\mathrm{E}_{x,(-)} = -\frac{\psi I_s \ell_s}{8\pi r_{(-)}^3}. \tag{9.38}$$

Similarly, by setting $x = 0$ in eqs. 9.31 and 9.32, we see that $r_{(+)} = r_{(-)} = \sqrt{y^2 + (\ell_s/2)^2}$, where y is the distance above the source at which the predator is located (fig. 9.10B). Substituting this value into eqs. 9.37 and 9.38, and adding $E_{x,(+)}$ and $E_{x,(-)}$, we arrive at our final answer:

$$E_x(y) = -\frac{\psi I_s \ell_s}{4\pi} [y^2 + (\ell_s/2)^2]^{-3/2}. \tag{9.39}$$

Now, when $y \gg \ell_s/2$, $[y^2 + (\ell_s/2)^2]^{-3/2} \approx y^{-3}$, and eq. 9.39 becomes

$$E_x(y) \approx -\frac{\psi I_s \ell_s}{4\pi y^3}. \tag{9.40}$$

In other words, the x-directed component of the electrical field (which in this case is the only component) acts in the negative x direction, as we deduced from figure 9.10A. Reversing the polarity of the spheres would reverse the direction of the field. The field strength increases with increasing current produced by the source (as one might expect), and also with increasing resistivity of the medium. The field strength decreases with approximately the cube of distance from the origin.

Where has all this calculation gotten us? Eq. 9.39 tells us that the field strength above the source increases in direct proportion to the product of environmental resistivity, ψ, and the source current, I_s. This means that if a biological source produces a fixed current, the strength of the resulting field at a distance y is 20 billion times greater in air than in seawater! Alternatively, a source in seawater would have to produce 20 billion times the current of a source in air to produce the same field at a given distance. Why, then, isn't it much easier to detect electrical activity in air?

To answer this question we must relate the current produced by the source to the voltage necessary to produce this current. This will tell us what voltage is needed to produce a given field strength.

9.9.2 *Calculating the Resistance*

Before we can relate current to voltage, we must calculate the resistance between the two source spheres. From Ohm's law (eq. 9.1) we know that $R = \mathcal{V}/I$, and from eq. 9.4 we know that by integrating E between the two spheres we can calculate $\mathcal{V}$. There might seem to be some question as to what path we integrate along, but this turns out not to matter, and we can simplify the situation considerably by integrating along the x axis, in which case the y-directed component of the field strength is 0.

We implement this strategy by returning to our initial equations for the x-directed component of the field (eqs. 9.33 and 9.34). We insert into these equations the expressions for $r_{(+)}$ and $r_{(-)}$ from eqs. 9.31 and 9.32 and set $y = 0$. Adding the resulting expressions for $E_{x,(+)}$ and $E_{x,(-)}$, we see that the net field strength per source current is

$$\frac{E_x}{I_s} = \frac{\psi}{4\pi}\left[\frac{1}{(x - \ell_s/2)^2} + \frac{1}{(x + \ell_s/2)^2}\right]. \tag{9.41}$$

Integrating along the x axis, we arrive at an expression for resistance:

$$R = \frac{\mathcal{V}}{\mathcal{I}} = \frac{\psi}{4\pi}\int_{-\frac{\ell_s}{2}+r_s}^{\frac{\ell_s}{2}-r_s}\left[\frac{1}{(x - \ell_s/2)^2} + \frac{1}{(x + \ell_s/2)^2}\right] dx, \tag{9.42}$$

where r_s is the radius of the source spheres.

Carrying out this integration, we find that

$$R = \frac{\psi}{2\pi r_s}\left(\frac{2r_s - \ell_s}{\ell_s - r_s}\right). \tag{9.43}$$

If $r_s \ll \ell_s$, then

$$R \approx \frac{\psi}{2\pi r_s}. \tag{9.44}$$

For the sake of simplicity, we assume that the spheres in our source have radii small compared to their separation and use eq. 9.44 when a resistance is called for.

9.9.3 *The Field Revisited*

This result immediately allows us to calculate the voltage that the source must produce to result in a given current. Recalling that $\mathcal{V} = IR$ (eq. 9.1),

$$\mathcal{V}_s = \text{source voltage} = \frac{I_s \psi}{2\pi r_s}. \tag{9.45}$$

This means that to produce a given source current, 20 billion times more voltage is needed in air than in seawater. This factor tends to offset the greater field produced in air for a given source current. Expressing source current in terms of source voltage, we see that

$$I_s = \frac{2\pi r_s \mathcal{V}_s}{\psi}. \tag{9.46}$$

Inserting this expression into eq. 9.39 and dispensing with the negative sign (which only tells us the direction of the field), we arrive at the conclusion that the magnitude of the field strength is

$$\text{E} = \frac{r_s \mathcal{V}_s \ell_s}{2}[y^2 + (\ell_s/2)^2]^{-3/2}. \tag{9.47}$$

Chapter 10

Sound in Air and Water: Listening to the Environment

We are now at the second of three chapters in which we examine the roles that air and water play in how animals perceive the world around them. This chapter deals with sound: how it is produced, propagated, and received.

Aside from light (which we explore in the next chapter), sound is the major source of information available to animals about the location and nature of objects in their environment. We will see how animals (both terrestrial and aquatic) use sound to estimate the size, range, and velocity of objects. Although terrestrial and aquatic animals use sound to similar ends, the means by which sound is detected and analyzed must take into account the differences between sound in air and in water. For example, the speed of sound is much higher in water than in air, a fact that makes it more difficult for dolphins to locate fish from echoes than for bats to locate moths. Conversely, it is intrinsically more difficult for terrestrial animals to hear than it is for aquatic animals, a difference that has led to the evolution of ingenious mechanisms for sound detection among insects and vertebrates.

10.1 The Physics

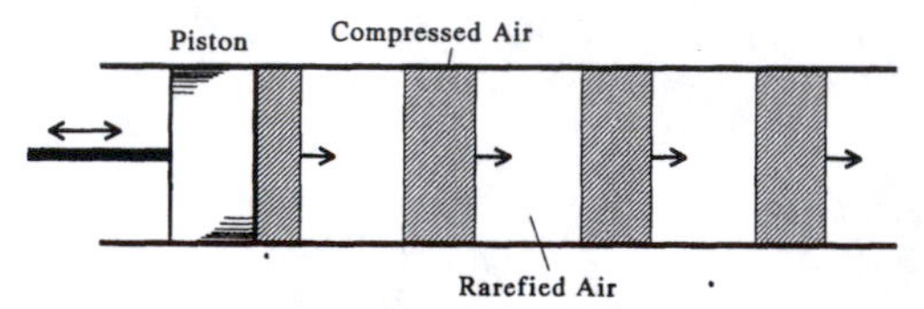

Fig. 10.1 An oscillating piston alternately compresses and rarefies the air in a pipe, thereby producing sound.

Consider the apparatus shown in figure 10.1, a piston filling the end of a long, open-ended cylinder. The piston can be driven rapidly back and forth by a motor. How does the fluid in the cylinder respond to the piston's motion?

We begin our analysis with both piston and fluid at rest and then start the motor. As the piston moves into the cylinder, it displaces the fluid in front of it. The fluid has mass, however, and its inertia opposes this displacement. As a result, the pressure rises in front of the cylinder, and this increase in pressure is accompanied by a compression of the fluid.[1] The compression is not confined to the immediate vicinity of the piston, however. As the pressure rises in front of the piston, the fluid adjacent to the piston exerts a force on fluid farther along the cylinder, and a wave of compression propagates to the right. At the end of its stroke, the piston reverses direction. The fluid's inertia again opposes the change, creating an area of low pressure and rarefaction to the right of the piston which, like the preceding area of compression, propagates along the cylinder. These alternating areas of compression and rarefaction traveling through the fluid are an example of the type of longitudinal mechanical wave that is called *sound*.

Before we can explore the biological roles of sound, we need to examine several aspects of its physics. To begin, we return to the apparatus of figure 10.1 and consider a simple example. Again we start with piston and fluid at rest. The motor is then turned on, but only briefly, producing a single wave of compression that propagates to the right along the cylinder. How fast does this wave travel? In other words, what is the speed of sound?

Our task is made simpler if we change our perspective. Until now, we have stood stationary in the laboratory, watching the wave of compression move off to our right. We could, however, run to the right along the cylinder, and if we

[1] This compression is due largely to the fact that the piston moves in a confined space. In the absence of the surrounding cylinder, air would tend to go around the piston rather than compress in front of it unless the piston were moving at a substantial fraction of the speed of sound or the piston were quite large (a substantial fraction of a sonic wavelength).

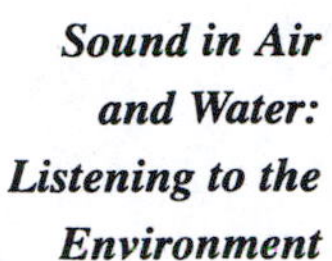

paced ourselves correctly, we could just keep up with the wave. In this moving (or *Lagrangian*) frame of reference, the wave of fluid compression is stationary and the cylinder moves to our left at the speed of sound. The situation is analogous to a train passing a station. If you sit stationary on the platform, the train (analogous to the sound wave) passes you moving north at, say, 60 kilometers per hour (kph). If, however, you move with the train (analogous to running alongside the cylinder at the wave speed), the station appears to move to the south at 60 kph.

The wave of compression, as viewed from our moving frame of reference, is shown in figure 10.2. The cylinder and the air in it (with the exception of the single compressed wave) move to our left at speed c. Because in our frame of reference the compressed wave is stationary, c must be equal to the speed of the wave through still air, though in the opposite direction.[2] Our problem, then, is to calculate the magnitude of c. We do this by considering what happens to fluid as it moves from right to left through the area of compression.

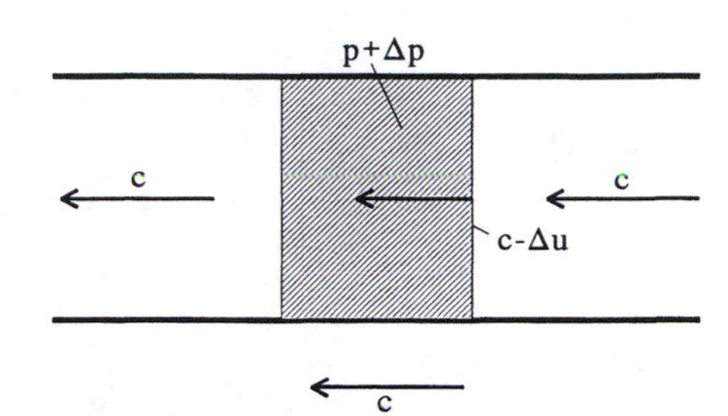

Fig. 10.2 As air enters a wave of compression, it slows down and its pressure increases. This change in speed and pressure can be used to calculate the speed of sound.

Before it encounters the wave, fluid is at ambient pressure, p. Within the wave, the pressure is higher by an amount Δp and returns to p after fluid has passed through the wave. Now, as a small volume of fluid first encounters the wave of compression, it slows down. It is this reduction in velocity that causes the compression and thereby the increase in pressure. The situation is analogous to cars traveling along a highway. If a line of equally spaced cars, all traveling at 60 kph, encounters a zone where the speed limit is 40 kph, the spacing between cars decreases as their speed is reduced.

We denote the magnitude of reduction in fluid velocity by Δu, and note that the change is to a lower velocity. Thus, the fluid velocity within the area of compression is $c - \Delta u$.

We know from Newton's first law of motion (chapter 3) that if the velocity of the fluid changes, a force must be acting upon it. In this case, the force causing the reduction in velocity is the increased pressure in the sound wave acting over the cross-sectional area of the cylinder, A. Therefore, the force acting on a volume of fluid as it enters the wave is ΔpA.

On what mass of fluid does this force act? Let us consider the fluid that enters the wave in a short period of time, Δt. Before entering the wave, this fluid fills a length of the cylinder equal to $c\Delta t$ and therefore has volume $Ac\Delta t$. If the fluid outside the compressed area has density ρ_0, this volume of fluid has mass $\rho_0 Ac\Delta t$.

We know that this mass is slowed as it enters the wave of compression by an amount $-\Delta u$, and, given our definition for the volume of fluid affected, we see that the change in velocity occurs over time Δt. Thus, the deceleration of the fluid upon entering the wave is $-\Delta u/\Delta t$.

We are now in a position to calculate the magnitude of c. From Newton's second law (chapter 3) we know that force is equal to mass times acceleration. We can therefore equate the force acting on our volume of fluid, with the product of its mass and acceleration as calculated above:

$$\Delta pA = \rho_0 Ac\Delta t \frac{-\Delta u}{\Delta t}. \tag{10.1}$$

[2]We use the symbol c (rather than u) to denote the wave *celerity*, that is, the speed at which the sound wave moves. This allows us to differentiate between the speed of the wave and the speed of the fluid in the wave.

Canceling terms and rearranging, we see that

$$\rho_0 c^2 = \frac{-\Delta p}{\Delta u/c}. \tag{10.2}$$

Now, upon entering the area of compression, the original volume of our fluid particle ($V_0 = Ac\Delta t$) is reduced by the amount,

$$\Delta V = A\Delta u\Delta t. \tag{10.3}$$

Thus, the change in volume, as a fraction of the original volume, is

$$\frac{\Delta V}{V_0} = \frac{A\Delta u\Delta t}{Ac\Delta t} = \frac{\Delta u}{c}. \tag{10.4}$$

We can then substitute $\Delta V/V_0$ for $\Delta u/c$ in eq. 10.2, with the result that

$$\rho_0 c^2 = \frac{-\Delta p}{\Delta V/V_0}. \tag{10.5}$$

The second half of eq. 10.5 ($-\Delta p/(\Delta V/V_0)$) should look familiar. This is the definition of the bulk modulus of the fluid, B_s, as explained in chapter 4. Thus we see that

$$c^2 = \frac{B_s}{\rho_0} \tag{10.6}$$

$$c = \sqrt{B_s/\rho_0}. \tag{10.7}$$

In other words, the speed of sound for any fluid is equal to the square root of the ratio of bulk modulus to density.

This immediately allows us to estimate the speed of sound in water. The bulk modulus of water is about 2×10^9 Pa (table 10.1) and its density is about 1000. Therefore,

$$c \approx \sqrt{\frac{2 \times 10^9}{1000}} = 1414\,\mathrm{m\,s^{-1}}. \tag{10.8}$$

Precise values for the speed of sound in water are given in table 10.2.

The situation is slightly more complex for gases such as air. One might be tempted to recall from chapter 4 that for an ideal gas at constant temperature,

$$\frac{\Delta p}{p_0} = \frac{-\Delta V}{V_0}, \tag{10.9}$$

in which case $B_s = p_0$ and $c = \sqrt{p_0/\rho_0}$. In fact, this is the conclusion Newton reached in his analysis of the speed of sound. Unfortunately, it is wrong. For most sound waves, where the medium is compressed and rarefied many times per second, the gas in an area of compression does not have time to come to thermal equilibrium with it surroundings. In other words, when air is compressed by a sound wave, it heats up.[3] Conversely, where rarefied, it cools. Under these conditions, in which no heat enters or leaves the gas as it compresses or expands

[3] The same thing happens in water, but the large specific heat capacity of water renders the change in temperature negligibly small.

Table 10.1 The isothermal bulk modulus of fresh water and seawater at various temperatures (calculated from UNESCO 1987).

T (°C)	Bulk Modulus (Pa $\times 10^9$)	
	Fresh Water	Seawater (S = 35)
0	1.9652	2.1582
10	2.0917	2.2695
20	2.1790	2.3459
30	2.2336	2.3924
40	2.2604	2.4128

Table 10.2 The speed of sound in air (from Weast 1977), fresh water, and seawater (S = 35) (from UNESCO 1983).

T(°C)	Speed of Sound (m s^{-1})		
	Fresh Water	Seawater	Dry Air
0	1402.4	1449.1	331.5
10	1447.3	1489.8	337.5
20	1482.3	1521.5	343.4
30	1509.1	1545.6	349.2
40	1528.9	1563.2	354.9

(a situation known as *adiabatic* compression or expansion), eq. 10.9 is not valid. Instead, one must use the equation for adiabatic compression, in which case

$$\frac{\Delta p}{p_0} = -\gamma_s \frac{\Delta V}{V_0}, \tag{10.10}$$

where γ_s is the ratio of the specific heat capacity of the gas at constant pressure to that at constant volume (Feynman et al. 1963). Thus,

$$c = \sqrt{\gamma_s p_0 / \rho_0}. \tag{10.11}$$

For dry air, $\gamma_s = 1.402$.

It is useful to take this calculation one step farther. From the behavior of an ideal gas (chapter 4), we know that

$$p_0 V_0 = N \Re T, \tag{10.12}$$

where N is the number of moles of gas present in volume V_0, $\Re$ is the universal gas constant (= 8.134 Joules mol^{-1} K^{-1}), and T is absolute temperature. Now the density of this gas is

$$\rho_0 = \frac{N\mathcal{M}}{V_0}, \tag{10.13}$$

where $\mathcal{M}$ is the molecular weight of the gas, in kg mole^{-1}. Note that the average molecular weight of air is 0.0288 kg (chapter 4).

Substituting eq. 10.13 into eq. 10.12 and rearranging, we see that

$$\frac{p_0}{\rho_0} = \frac{\Re T}{\mathcal{M}}. \tag{10.14}$$

This allows us to restate the equation for the speed of sound in air:

$$c = \sqrt{\frac{\gamma_s \Re T}{\mathcal{M}}}. \tag{10.15}$$

In other words, the speed of sound in air is independent of density and pressure and depends only on temperature. At 20°C, for instance,

$$c = \sqrt{\frac{1.402 \times 8.314 \times 293.15}{0.0288}} = 344 \text{ m s}^{-1}, \tag{10.16}$$

only about one-quarter the speed of sound in water. Values for the speed of sound in air at different temperatures are given in table 10.2.

You may recall from chapter 8 that temperature is proportional to the kinetic energy of molecules, $m\langle u^2 \rangle/2$, where m is the mass of each molecule and $\langle u^2 \rangle$ is the mean square velocity. As a result, the speed of sound in air, which is proportional to the square root of temperature, is proportional to $\sqrt{\langle u^2 \rangle}$, the rms velocity of gas molecules. This makes intuitive sense. A wave of compression or rarefaction cannot move through a gas any faster than the molecules themselves move. In fact, the speed of sound in air is about 70% of the root mean square speed of individual molecules (Feynman et al. 1963).

We return to the comparison of the speed of sound in air and water in the next section, but before we do, we must finish our exploration of the physics of sound in general.

As we have seen, sound waves consist of alternating compressions and rarefactions that are related to the displacement of molecules. Where molecules come together, the pressure is high, and where they move apart, the pressure is low. It will be valuable to describe in some greater detail the interrelationship between the movement of molecules and the resulting pressure.

During the passage of a sound wave, molecules are displaced from the position they hold in still air. This displacement, Δx, is along the axis of sound propagation and in most cases is sinusoidal. That is ,

$$\text{displacement, } \Delta x = \mathcal{A} \sin(2\pi f t), \tag{10.17}$$

where $\mathcal{A}$ is the amplitude of the displacement, t is time, and f is the *frequency* of the sound. Frequency has the dimensions of cycles per second, for which the SI unit is the Hertz, Hz.

This relationship is depicted in figure 10.3. At the beginning of each cycle (where ft is an integer), fluid particles return to the same point. In other words, there is no net movement of the fluid as sound moves through it.[4]

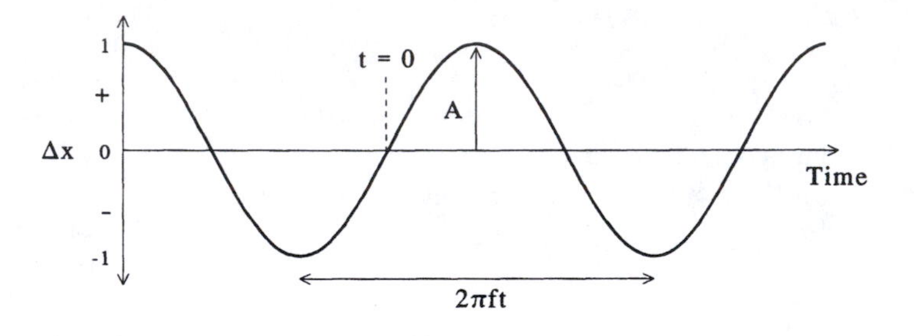

Fig. 10.3 At a single position in space, the displacement of a fluid particle varies sinusoidally through time as sound waves pass by.

The behavior shown in figure 10.3 is that of fluid particles at one position as a function of time. Alternatively, we can look at particle displacement at one time as a function of distance along the direction of propagation. In this case, the appropriate equation is

$$\Delta x = \mathcal{A} \sin\left(\frac{2\pi x}{\lambda}\right), \tag{10.18}$$

where x is a measure of distance and λ is the *wavelength* of the sound. This perspective is depicted in figure 10.4. Wherever x is a multiple of λ the particle displacement is the same.

These two perspectives on displacement—one at a single time, the other at a single position—are complementary and can be combined into an all-purpose equation:

$$\Delta x = \mathcal{A} \sin\left(\frac{2\pi x}{\lambda} - 2\pi f t\right). \tag{10.19}$$

The appearance of the negative sign in this equation can cause confusion. Think of it this way. If you are at the origin in figure 10.4 and sound waves are moving to the right, the waves in the positive x direction are those that passed you some time ago. In other words, for a particular location *on the waveform* (such as the

[4]Actually, there is a very slight net movement of fluid in the direction of wave propagation (see Morse and Ingard 1968), but for most biological purposes this transport is negligible.

point labeled 1) you would have to go *back* in time to experience that part of the waveform when it was at the origin. Alternatively, you could get to the same spot on the wave by going *forward* along the x axis. Time and distance act in opposite directions with respect to position on the waveform, hence the negative sign.

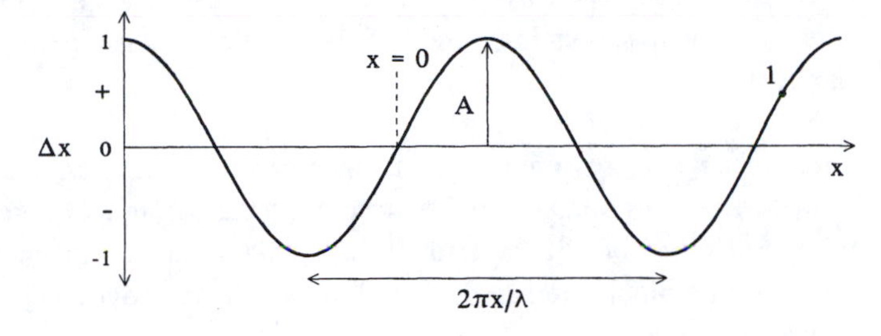

Fig. 10.4 At a single instant in time, the sonic displacement of fluid particles varies sinusoidally as a function of distance along the direction of wave propagation.

Let us digress for a moment to explore the relationship between f and λ. We know that the wave travels one wavelength every cycle, and that there are f cycles per second. As a result, the speed at which the wave travels is the product of frequency and wavelength:

$$c = \lambda f. \tag{10.20}$$

Rearranging, we see that

$$\lambda = \frac{c}{f}. \tag{10.21}$$

In other words, if you know the speed of sound (as calculated by eq. 10.7 or 10.15) and its frequency, you can calculate the wavelength. For example, the wavelength of sound with $f = 1000$ Hz is about 1.4 m in water and about 0.3 m in air. We return to the subject of wavelength in the next section.

From eq. 10.19 we can calculate the velocity of particles. Because velocity, u, is the derivative of Δx with respect to time,

$$u = -2\pi f \mathcal{A} \cos\left(\frac{2\pi x}{\lambda} - 2\pi f t\right). \tag{10.22}$$

The relationship between displacement and velocity is shown in figure 10.5. The two are 90° (= $\pi/2$ radians) out of phase. When (or where) the displacement is maximal, the velocity is 0, and vice versa.

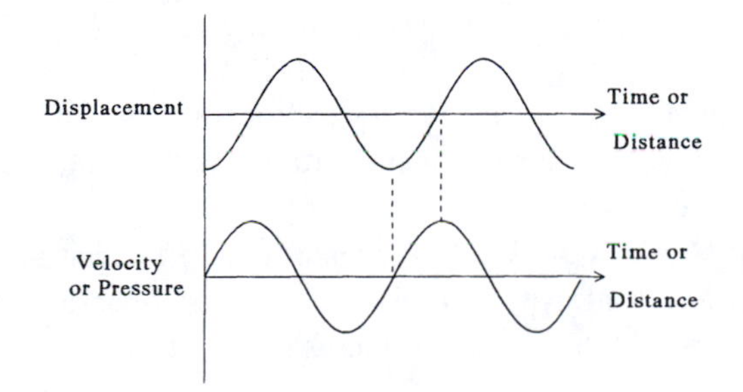

Fig. 10.5 In a sound wave, both velocity and pressure are 90° out of phase with particle displacement. Where displacement is maximal, velocity and pressure are 0 and vice versa.

Although we do not derive it here, there is also a relationship between the change in pressure associated with a sound wave, Δp, and the amplitude of particle displacement (see Morse and Ingard 1968 or Clay and Medwin 1977):

$$\Delta p = -2\pi f \rho_0 c \mathcal{A} \cos\left(\frac{2\pi x}{\lambda} - 2\pi f t\right), \tag{10.23}$$

where ρ_0 is the fluid density in the absence of acoustic compression or rarefaction. This relationship shown in figure 10.5. Where velocity is maximal the pressure is maximal.

Now, the velocity is maximal where the displacement is zero, so this means that the change in pressure is maximal where the displacement is zero. This makes intuitive sense. A point at zero displacement separates areas where fluid is moving in opposite directions. If the fluid on either side of the point is moving toward the point, the pressure is increased. If fluid on both sides moves away from the point, the pressure is decreased.

Note from eqs. 10.22 and 10.23 that

$$\Delta p = \rho_0 c u. \tag{10.24}$$

This relationship is useful because it is analogous to Ohm's law for the flow of current in electrical circuits (chapter 9). Here we use the deviation in pressure as an analog to voltage, and particle velocity as an analog to electrical current. Thus, the quantity $\rho_0 c$ acts as the acoustical equivalent of electrical resistance. As such, $\rho_0 c$ represents the tendency for the fluid to impede the passage of sound, and for this reason $\rho_0 c$ is referred to as the *characteristic impedance* or *specific acoustic resistance* of the fluid.[5]

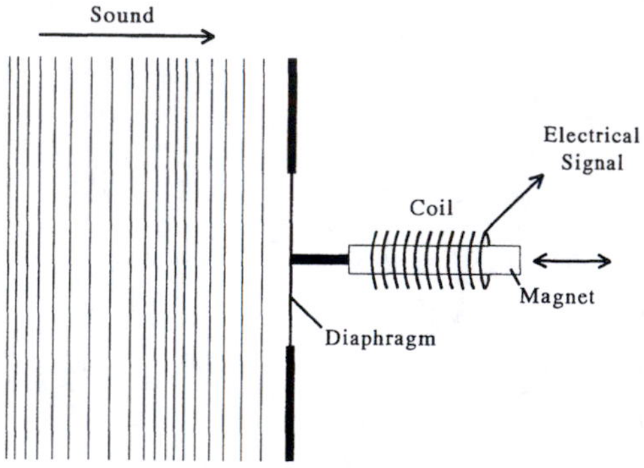

A final aspect of the physics of sound requires our attention. As sound moves through a fluid, it transports energy. For example, if a small disk is placed with its flat face perpendicular to the path of a sound wave, the oscillating pressure acting on the disk results in an oscillating force that can do work (fig. 10.6). For instance, a magnet attached to the disk moves in response to the force on the disk and, by moving relative to a coil of wire, can induce an electrical current. Thus, sound produced at a distant source delivers power to the disk that can be used to move electrons. This is the physical mechanism that allows a microphone (a device similar to the disk-magnet-coil apparatus described here) to transduce sound.

Fig. 10.6 Sound waves impinging on the diaphragm of a microphone cause a magnet to oscillate within a coil. This in turn produces an electrical translation of the sound.

The rate at which energy is transported by sound waves per area upon which the sound impinges is known as the *intensity*, $\mathcal{I}$, of sound. It can be shown that

$$\mathcal{I} = 2\rho_0 c \pi^2 f^2 \mathcal{A}^2 \tag{10.25}$$

$$= \frac{(\Delta p_{max})^2}{2\rho_0 c}, \tag{10.26}$$

where Δp_{max} is the maximal pressure deviation. $\mathcal{I}$ has the SI units of W m^{-2}. Note that intensity is proportional to the *square* of displacement amplitude or of pressure deviation.

When dealing with sound, relative intensity is often more important than absolute intensity. For instance, in cases concerning human hearing one often wants to know how loud a sound is relative to some well-known standard; the absolute

[5] Acoustic resistance and impedance are not always the same. There are situations in which resistance forms only one part of a more complicated expression for the overall impedance (see Morse and Ingard 1968). However, for the simple planar waves we deal with here, the two are identical.

intensity has less practical importance. For example, the fact that a sound is particularly loud is typically conveyed by comparing it to the din of a boiler factory or the roar of a jet engine. To accommodate this requirement for a relative measure, the "loudness" of sound is measured in decibels, dB:

$$\text{dB} = 10 \log_{10} \frac{\mathcal{I}}{\mathcal{I}_{ref}}, \tag{10.27}$$

where $\mathcal{I}$ is the sound intensity of interest and $\mathcal{I}_{ref}$ is a reference intensity chosen to fit the particular situation. In the study of acoustics, it is common practice to use as a reference the intensity of sound that is just audible to the average human, a value usually taken to be 10^{-12} W m^{-2}. Given this reference, sound at the threshold of hearing has a loudness of 0 dB.

To recover actual intensity values from decibels, one works backward through eq. 10.27:

$$\mathcal{I} = \mathcal{I}_{ref} 10^{\text{dB}/10}. \tag{10.28}$$

In some cases it is convenient to calculate decibels based on sound pressure. For example, the voltage signal derived from many microphones is proportional to acoustic pressure rather than intensity. Noting that intensity is proportional to the square of pressure (eq. 10.26), we see that

$$\text{dB} = 10 \log_{10} \left(\frac{p}{p_{ref}}\right)^2 = 20 \log_{10} \frac{p}{p_{ref}}, \tag{10.29}$$

and pressure can be recovered from decibels by again working backward:

$$p = p_{ref} 10^{\text{dB}/20}. \tag{10.30}$$

The reference value of pressure must be specified for each case.

Before leaving the physics of sound, it should be noted that the relationships presented here apply to sounds far away from the disturbance that caused them, in what is called the *far field*. Very near a vibrating object, in the *near field* the relationships among displacement, velocity, and pressure are different from those presented here. However, the strange properties of the near field dominate only within about 0.2 wavelengths of the source (Alexander 1968). In exploring the biology of sound, we will not encounter a situation where the near field is of importance, and thus are free to use equations that apply only to the far field.

10.2 Sound in Air and Water

The speed of sound in fresh water is about 1450 m s^{-1}, about 4.3 times that in air (table 10.2). Speed increases with increasing temperature from 1402 m s^{-1} at 0°C to 1529 m s^{-1} at 40°C, a variation of about 9% relative to the smaller value (fig. 10.7). The increase in speed is due both to the increase in bulk modulus with temperature and to the concomitant decrease in density. The speed of sound in seawater is about 3% higher than that in fresh water, due to the higher bulk modulus of seawater (table 10.1).

The values cited here are for water at one atmosphere. The speed of sound in water increases slightly as the pressure is increased, amounting to about 0.016 m s^{-1} per meter of depth (Clay and Medwin 1977).

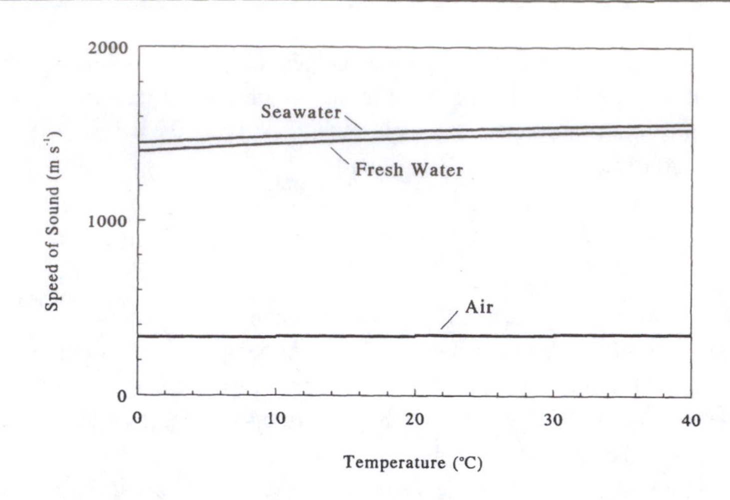

Fig. 10.7 The speed of sound is much higher in water (either salty or fresh) than in air, and increases slightly with temperature in both media.

The speed of sound in air is only about 23% that in water, varying from 331 m s^{-1} at 0°C to 355 m s^{-1} at 40°C (table 10.2; fig. 10.7). In this case, the increase in the speed of sound is due primarily to the increase in molecular velocity and amounts to about 7% over the biological range of temperatures. Because sound in air is independent of density (eq. 10.15), the speed of sound (at a given temperature) is largely independent of altitude.

The speed of sound in both water and air is slow compared to that in many solids. For example, the speed of sound in metals typically varies from 3500 to 5200 m s^{-1}, and the speed of sound in granite is 6000 m s^{-1} (Resnick and Halliday 1966).

Because the speed of sound is about 4.3 times greater in water than in air, the wavelength at any frequency is larger in water than in air by a factor of 4.3 (see eq. 10.12). Examples of wavelengths at different frequencies are given in table 10.3.

Table 10.3 The wavelengths of sound at 20°C for various frequencies.

Frequency (Hz)	Wavelength (m)		
	Fresh Water	Seawater	Dry Air
1	1482	1522	343
10	148.2	152.2	34.3
100	14.82	15.22	3.43
1,000	1.482	1.522	0.343
10,000	0.1482	0.1522	0.0343
100,000	0.0148	0.0152	0.0034
1,000,000	0.00148	0.00152	0.00034

Due to both the greater wave speed and greater density of water, the specific acoustic resistance of water is about 3500 times that of air (table 10.4). While the specific acoustic resistance of water increases with increasing temperature by about 7% over the biological temperature range, that of air decreases by about 7%.

Table 10.4 The specific acoustic resistance of air and water.

T (°C)	Specific Acoustic Resistance (kg m^{-2} s^{-1})		
	Fresh Water	Seawater	Dry Air
0	1.402×10^6	1.490×10^6	428.3
10	1.447	1.530	420.5
20	1.480	1.558	413.5
30	1.503	1.579	406.5
40	1.517	1.594	400.3

Now, the specific acoustic resistance appears in the expressions for acoustic intensity. In one case (eq. 10.25), the square of the displacement amplitude is multiplied by acoustic resistance (among other things) to give intensity. Because the characteristic impedance of water is about 3500 times that of air, to have the same intensity the displacement amplitude of molecules in water must be only about one-sixtieth that in air. But intensity is also determined by the *ratio* of the square of acoustic pressure to acoustic resistance (eq. 10.26), so that at a given intensity of sound the pressure amplitude in water must be sixty times that in air. We will return to these relationships when we consider how fish hear.

As sound moves through a fluid, some acoustic energy is lost to viscous processes and eventually dissipated as heat. As a result, the intensity of sound is attenuated. The rate of attenuation (expressed here in dB per km) depends both on the frequency of the sound and on various properties of the medium. The attenuation characteristics of air and water are shown in figures 10.8 and 10.9, respectively.

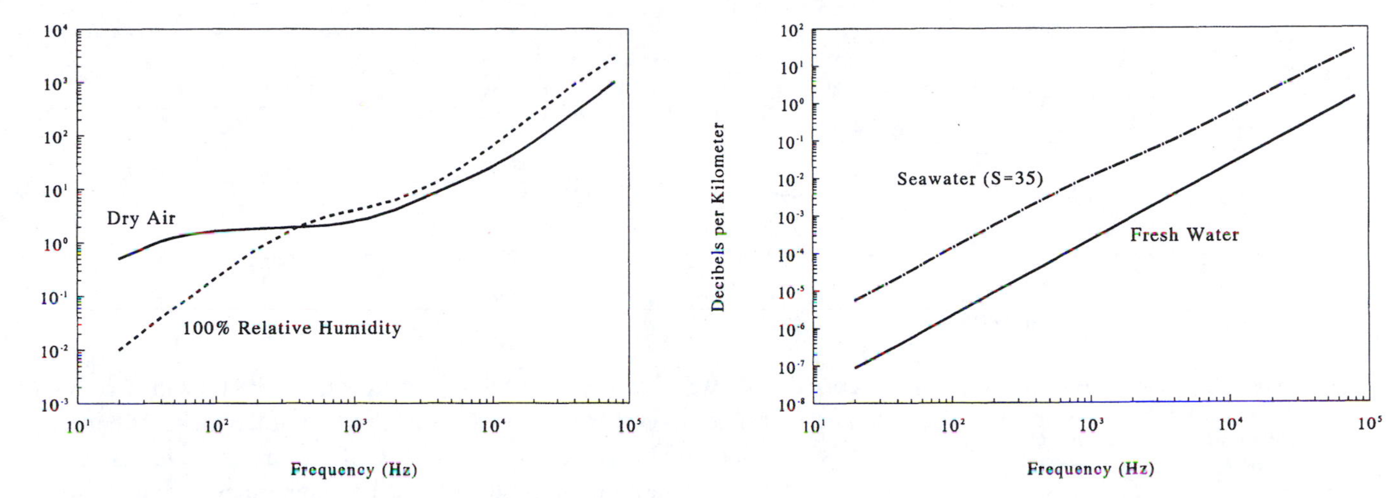

Fig. 10.8 Low-frequency sounds are attenuated less effectively by air than are high frequencies. At frequencies above about 400 Hz, humid air attenuates sound more effectively than does dry air. Below 400 Hz, however, the attenuation of sound in humid air is less than that in dry air.

Fig. 10.9 Low-frequency sounds are attenuated less effectively by water than are high frequency sounds. The borate ions in seawater cause sound to be attenuated more effectively than in fresh water.

High-frequency sounds are attenuated much more rapidly in air than are low frequencies. This explains why distant thunder is heard as a low rumble while thunder nearby is a higher-frequency crash. Note, however, that the ability of air to transmit low-frequency sound is strongly dependent on the relative humidity. Dry air attenuates sound of f = 20 Hz about fifty times more effectively than does very humid air (fig. 10.10).

Water transmits sound much more effectively than does air. For instance, at f = 20 Hz, sound is attenuated nearly 6 million times faster in dry air than in fresh water (fig. 10.11). Note, however, that seawater attenuates sound much more rapidly than does fresh water. The attenuation of sound by seawater at frequencies most relevant to biology (10 to 10,000 Hz) is due largely to the interaction of sound energy with borate anions in the water (Clay and Medwin 1977). These molecules have a mode of vibration with a resonant frequency in the range of these sounds. As a result, an appreciable fraction of the acoustic energy is diverted into internal vibrations of borate molecules rather than into the displacement of water and is eventually lost as heat. At higher frequencies (10 to 1000 kHz), sound energy is attenuated through interaction with sulphate anions.

The values cited here for both air and water represent minimum values of attenuation. As we will see, the presence of particulate matter (which, for water, includes bubbles) can drastically increase the rate of attenuation.

10.3 The Information in Sound

We are now in a position to explore how the acoustic characteristics of air and water have affected life. We begin with a brief discussion of the implications of sonic wavelength.

Chapter 10

In this chapter we explore how animals use sound to provide information about their environment. But the wavelengths of sounds that animals can hear place severe constraints on the type of information that can be perceived and the rate at which information can be brought up to date.

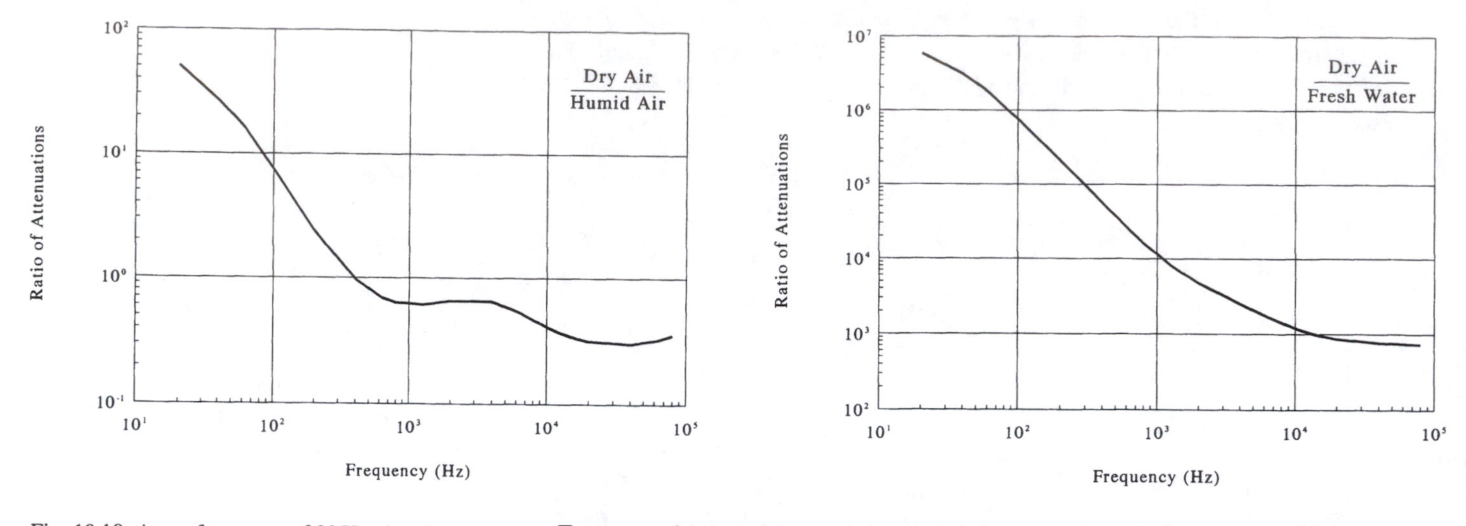

Fig. 10.10 At a a frequency of 20 Hz, dry air attenuates sound fifty times more effectively than does humid air, a fact of potential importance to elephant communication. At 10, 000 Hz, however, humid air attenuates sound about seven times as effectively as does dry air.

Fig. 10.11 Air attenuates sound much more effectively than does water.

For example, we will see later in this chapter that to reflect a sound effectively an object must be larger than the sound's wavelength. For the highest frequency sounds typically used by animals (about 10^5 Hz), this means that objects cannot be easily detected if they are smaller than about 1.5 cm in water or 3 mm in air.

The same constraint of wavelength applies to light, except that a typical beam of light has a wavelength on the order of 10^{-7} m. Thus it is possible in theory for an animal to detect with its eyes objects that are 100,000 times smaller than it can detect with its ears.

Constraints of size apply to the organs that detect sound and light as well. Eyes resolve an image of the environment by sampling the light incident on the retina with a closely packed array of sensory cells. The smaller the cells, the greater the number of samples the eye has of the image, and the more detail that can be perceived. The size of retinal cells is ultimately limited by the wavelength of light (chapter 11), but this size is quite small, about three times the wavelength (1.5 μm). This allows an eye of reasonable size to sample an image at thousands of different locations.

The same type of constraints apply to an ear attempting to form an acoustic image of the environment. To sample such an image, the ear would need an array of sensors, each several wavelengths in maximum dimension. But because the wavelength of sound is so large compared to that of light, an image-forming ear would be impractically large. For example, if a sensor were three times the wavelength of sound, at a frequency of 10^5 Hz in air each sensor would be 1 cm in diameter. Thus an ear 10 m in diameter would be needed to sample a transect through an acoustic image at one thousand different spots.

This constraint on size has proven daunting, and no animal has evolved an ear capable of directly producing an acoustic image. Instead, ears are point detectors of sound. Much of this chapter is devoted to an exploration of how animals have worked within this constraint to use sound as a source of information. We begin with the subject of echoes.

10.4 Echoes

When an acoustic wave encounters an object with a characteristic impedance different from that of the medium, some of the sound energy is deflected. If the object is large compared to the wavelength of sound, much of this deflected energy "bounces off" the object, and the redirected sound is referred to as a *reflection*. When the object is small compared to the wavelength, much of the wave "wraps around" the object in a process called *diffraction*. As a result of diffraction, however, sound energy is transmitted in many directions away from the object, and the sound is said to be *scattered*. Reflection is one special case of scattering. Scattered sound that returns to its source is known as an *echo*.

Echoes can provide animals with information regarding their surroundings. The classic example is the bat, which emits short chirps of high frequency sound and listens for the scattered acoustic signal. A similar system is used by some cave-dwelling birds (oil birds and swiftlets) to navigate in the dark, and by dolphins, porpoises, and whales to locate objects under water. The information provided by sonic echoes also provides the basis for sonar systems used to find and track gamefish and submarines, and to map the ocean floor. The basic ideas behind the biological implementation of echolocation are well known (Griffin 1986; Nachtigall and Moore 1986), and we will not explore them here in any great depth. Instead we concentrate on how biological echolocation systems must operate differently in air and water.

Our first example involves the size of an object that can be detected. As mentioned above, the manner in which sound is scattered depends on the size of an object relative to the acoustic wavelength. Because wavelengths in water are 4.3 times those in air at a given frequency, we can guess that an object of a given size will scatter sound differently in water than in air.

The pattern in which an object scatters sound depends on the shape of the object and, for objects with a complex shape, is difficult to predict. We can, however, take a simple shape and see how its scattering characteristics differ between air and water. As has been common in previous chapters, we chose a sphere as our model. In this case, our choice has some biological validity. Many of the prey items commonly pursued by porpoises and whales, for instance, are acoustically spherical. For example, we will see that it is the swim bladder of fish that is responsible for most sound scattered from these animals, and as pressure vessels these bladders are often roughly spherical.

So, how is sound scattered from spheres? We confine ourselves initially to considering rigid spheres for which the circumference is less than the wavelength of sound; in other words, cases in which $2\pi r/\lambda < 1$, where r is the radius of the sphere. This restriction should not be too confining. For example, at a frequency of 60 kHz (a common echolocating signal for both bats and dolphins), the wavelength in air is about 5.5 mm, restricting us to spheres with a diameter less than about 1.8 mm. Some of the flying insects hunted by bats have effective body sizes smaller than this. In water, with its longer wavelength, we may consider objects up to 4.3 times this size, 7.2 mm. At lower frequencies, larger spheres fit the bill.

Given this restriction on size, the intensity of the echo emanating from an object is (Morse and Ingard 1968)

$$\mathcal{I}_s = \mathcal{I}_0 \frac{256\pi^4 r^6}{9\lambda^4 \ell^2}, \tag{10.31}$$

where $\mathcal{I}_0$ is the intensity of sound impinging on the sphere, ℓ is the distance from the center of the sphere at which the echo is received, and we have ignored the directional variation of the scattered sound and only considered sound scattered directly back toward the source.[6]

There are two important points to be gained from this equation. First, the intensity of the echo is inversely proportional to the *fourth* power of wavelength. For scattering from a given size sphere, the echo in air is therefore $4.3^4 \approx 342$ times as strong as the echo in water. As a consequence, animals using echolocation in water must either hunt larger prey to receive an audible echo, or must emit higher frequencies than animals in air. Bats use frequencies in the range of 22 to 154 kHz in their echolocation cries (Simmons and Grinnell 1986), while dolphins and whales use clicks which have peak energies at frequencies in the range of 30 to 150 kHz (Popper 1980). Thus, it is doubtful that aquatic animals in general have adjusted their frequencies sufficiently to allow the location of objects as small as those that can be hunted by terrestrial echolocators.

The second point raised by eq. 10.31 is the strong dependence of echo strength on the size of the sphere; $\mathcal{I}_s$ increases with the *sixth* power of radius. In practical terms, this means that if a variety of sizes are present, only the largest can be detected. For example, $2^6 = 64$, so sixty-four 1-mm-diameter spheres would be needed to give the same echo intensity as one 2-mm-diameter sphere. Seven hundred and twenty-nine 1-mm-diameter spheres would be required to give the same echo as a single 3-mm-diameter sphere. In practice, this means that small fish or insects can hide from an echolocating predator simply by being near a somewhat larger prey item.

It also means that bats can effectively locate even small insects on a foggy night, and that dolphins can locate small fish in muddy water. Consider, for example, fine mud with a particle radius of 10 μm suspended in water. Working through the calculation, we see that at a frequency of 60 kHz,

$$\frac{\mathcal{I}_s}{\mathcal{I}_0} = \frac{7.4 \times 10^{-21}}{\ell^2}. \tag{10.33}$$

In other words, due to their small size, sediment particles are virtually transparent to sound. Because of this, sound in murky waters may have the same functional importance as light in clear waters or in air. This fact has been of apparent importance in evolution. For example, the freshwater dolphins of the muddy Amazon River have a highly-effective echolocating capability (Popper 1980).

Note that eq. 10.31 applies only when $2\pi r/\lambda < 1$. When $2\pi r/\lambda$ is greater than about 10, the scattered intensity is

$$\mathcal{I}_s = \mathcal{I}_0 \frac{r^2}{4\ell^2}. \tag{10.34}$$

In other words, for spheres large relative to the wavelength of sound, the intensity of the echo scales directly with the projected area of the object and is independent

[6]The complete equation is (Morse and Ingard 1968):

$$\mathcal{I}_s = \mathcal{I}_0 \frac{16\pi^4 r^6}{9\lambda^4 \ell^2}(1 - 3\cos\theta)^2, \tag{10.32}$$

where θ is the angle between the direction of propagation of the incident and that of the scattered sound. For sound scattered back toward the source, $\theta = 180°$.

of wavelength. In this case, it would be much more difficult for small prey to hide among their larger cousins.

For objects such that $1 < 2\pi r/\lambda < 10$, the scattered intensity fluctuates by a factor of about three depending on the precise ratio of circumference to wavelength (fig. 10.12), but it never differs drastically from the dependence on area alone as observed at larger sizes.

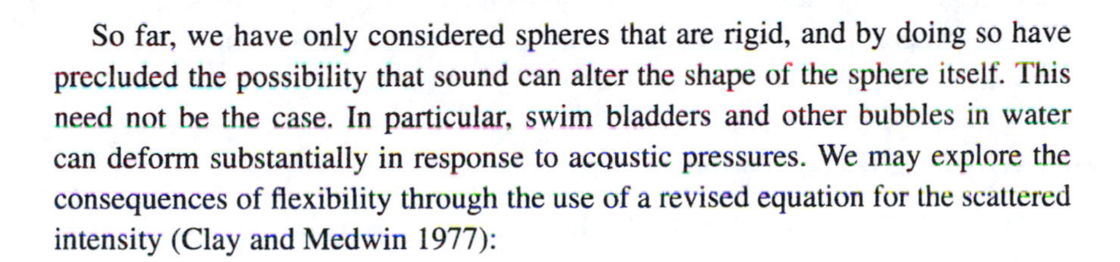

Fig. 10.12 The intensity of scattered sound decreases drastically as the circumference of a scattering particle is decreased below the wavelength of the incident sound. If the circumference is greater than ten times the wavelength, however, the intensity of scattered sound (per projected area) is independent of particle size. (After Clay and Medwin 1977)

So far, we have only considered spheres that are rigid, and by doing so have precluded the possibility that sound can alter the shape of the sphere itself. This need not be the case. In particular, swim bladders and other bubbles in water can deform substantially in response to acoustic pressures. We may explore the consequences of flexibility through the use of a revised equation for the scattered intensity (Clay and Medwin 1977):

$$\mathcal{I}_s = \mathcal{I}_0 \frac{16\pi^4 r^6}{\lambda^4 \ell^2}\left[\left(\frac{k_1 - 1}{3k_1}\right)\left(\frac{k_2 - 1}{2k_2 + 1}\right)\right]^2. \tag{10.35}$$

Here k_1 is defined as

$$k_1 = \frac{c_s^2 \rho_s}{c_f^2 \rho_f}, \tag{10.36}$$

where the subscript s refers to values for the sphere and f to values of the fluid.

The parameter k_2 is defined as the ratio of sphere density to fluid density:

$$k_2 = \frac{\rho_s}{\rho_f}. \tag{10.37}$$

In eq. 10.35 we have again ignored the directional variation in $\mathcal{I}_s$, and only consider the sound returned directly toward the source. As before, it is assumed that the circumference of the sphere is small compared to the wavelength.

In many respects eq. 10.35 (for compliant spheres) is similar to eq. 10.31 (for rigid spheres). $\mathcal{I}_s$ varies directly with r^6 and inversely with λ^4. The compliance of the sphere can, however, have a major effect. Let us consider two cases likely to be encountered in the real world: an aqueous sphere in air (an analogue to a flying insect) and an air-filled bubble in water (an analogue to a swim bladder). The two spheres are the same size and are ensonified by sound of the same frequency and intensity. How do the two echoes compare?

Inserting the appropriate values for the speed of sound and the density of the spheres and media, we find that the square of the term in brackets in eq. 10.35 (the term relating to the compliance of the sphere) is thirty-five times larger in water than in air. In other words, because an air bubble is compressible while water is not, a bubble is a much more effective scatterer of sound. This is why the swim bladders of fish make such good acoustic targets for both dolphins and commercial echolocating "fish finders."

This effect is not sufficient, however, to offset the fact that, at a given frequency, the wavelength in water is 4.3 times that in air. Due to effects of wavelength, the first fraction in eq. 10.35 is about $4.3^4 \approx 340$ times larger in air than in water at a given frequency. As a result, the scattered intensity in air is still about ten times that in water for objects of the same size at the same frequency.

We will return to the surprising properties of bubbles when we consider sound reception later in this chapter.

Before leaving the subject of echoes, let us consider one final twist in their biological use: to locate an object through the use of echoes, it may not be necessary to receive an echo from the object itself. For example, several species of tropical bats catch fish by flying just above the surface of lakes and dipping their feet into the water at the proper moment. Apparently the bats locate fish by the echoes they receive from the small surface waves that fish create (Suthers 1965). As we will see later in this chapter, it is virtually impossible for bats to receive echoes from deeply submerged fish themselves. In chapter 13 we will consider the mechanism by which surface waves are produced.

10.5 Attenuation of Sound

We noted above that air attenuates high-frequency sound much more rapidly than low frequencies. Thus, if a terrestrial animal wishes to communicate via sound over long distances, it is best to use the lowest frequency possible. This strategy is indeed used by elephants. Langbauer et al. (1991) have shown that groups of female elephants and calves can coordinate their movements when separated by up to 4 km (about 2.5 miles) by "talking" to each other using sounds composed of frequencies ranging from 15 to 35 Hz. These low-frequency conspecific calls can also be used by males to locate distant females in estrus. Langbauer et al. do not report the humidity during their experiments, but note that they were carried out toward the end of the dry season. If the humidity was indeed low during these measurements, they may substantially underestimate the distance over which elephants could communicate when there is more moisture in the air.

We return to the subject of elephant's hearing in the next section when we consider the localization of sound.

The relatively low attenuation of sound by water allows aquatic animals to communicate over much larger distances than those available to their terrestrial cousins. For example, whales and dolphins are thought to be able to communicate over many hundreds of kilometers using sound (Payne and Webb 1971). The utility of long-distance sonic communication has not escaped human notice. The U.S. Navy, for instance, uses hydrophones as a means of tracking distant surface ships and submarines.

10.6 The Localization of Sound

It is of obvious advantage to an organism to be able to tell the direction and distance to a source of sound. This applies to male elephants attempting to locate a mate, to predators who use sound to find their prey, and to prey who might very much want to know the location of their pursuers.

10.6.1 *Direction*

There are four basic strategies by which an animal can tell the direction from which a sound arrives. In the first, the animal senses direction by sensing the axis along which fluid particles are displaced in the sound wave. For example, when sound comes from a source to the right of an animal, the air or water molecules in the animal's vicinity move back and forth along a left-right axis. In contrast, sound arriving from the front causes motion along an anterior-posterior axis. The ability to distinguish between these two sound sources requires an ear that responds to the displacements associated with sound waves rather than to pressure. As we will see later in this chapter, many fish have this ability, and they can therefore sense the direction to a source of sound by directly measuring sonic displacements (Hawkins and Myrberg 1983). This ability is unusual among animals, however, and we now turn our attention to strategies that can be used by animals whose ears sense pressure rather than displacement.

In the first of these strategies, the animal senses direction by comparing the intensity of sound reaching its two ears. This ability is based on two mechanisms. First, if the sound arrives from the right, the right ear is closer to the source than is the left, and the sound intensity at the right ear is therefore larger. Second, when the sound arrives from the right, the presence of the body attenuates the sound reaching the left ear, and the reverse is true if the sound arrives from the left. For either mechanism, the more closely the animal faces directly toward or away from the sound source, the more similar the intensity reaching the two ears.

There are three problems inherent in this strategy of localization. First, the attenuation of sound with distance from the source is small over the distance between an animal's ears. For example, even in air at a frequency of 10,000 Hz, sound attenuates at a rate of only about 70 dB per kilometer. For ears separated by a distance of 1 to 100 cm, this amounts to an intensity difference of only 0.0007 to 0.07 dB. An even smaller difference is found at lower frequencies in air and at all frequencies in water.

The second problem concerns an ambiguity regarding direction. An object 45° forward of the right ear may be confused with an object 45° to the rear. The ambiguity is often worst when the source of sound lies in the organism's median plane. In other words, a sound originating directly in front of the organism results in the same intensities reaching both ears, but so will a sound originating directly behind the organism or directly above. These ambiguities can be minimized if the external part of the ear (the *pinna* or ear lobe) is directional in its ability to receive sound. For instance, the ear lobes of human beings are directional "antennae" for sound (Shaw 1974) and are angled out from the head by about 30°, facing forward. As a result, the ears are more sensitive to sounds arriving from in front than from above or behind. The ambiguities inherent in localizing sounds by comparing intensity can also be minimized if the animal moves its head slightly while listening.

The third problem is less easily circumvented. When an object such as a head is small compared to the wavelength of sound, it does not cast a distinct sound "shadow." This is why small objects are so inefficient in scattering sound. It also means that when wavelengths are long (i.e., when frequencies are low or the speed of sound is high), the head may not act as an effective blocker of sound energy, and the intensity of sound reaching the two ears is similar regardless of the direction from which it arrives. For instance, a rule of thumb (from Lord Rayleigh) purports that a substantial sound shadow is cast by the head only when the circumference of the head exceeds $\lambda/4$. For example, at a frequency of 100 Hz, sound in air has a wavelength of greater than 3 m, and the head of a human being (r = 8.75 cm, circumference = 0.55 m) does little to cast a sound shadow on the "downstream" ear (fig. 10.13; Gulick et al. 1989). As a result, you and I cannot use relative intensity as an effective clue as to the direction of low-frequency sources of sound. The head of an elephant, which has a circumference of perhaps 4.5 m, casts a substantial sound shadow for frequencies as low as approximately 25 Hz, helping to explain how these animals can use low-frequency sounds to locate each other.

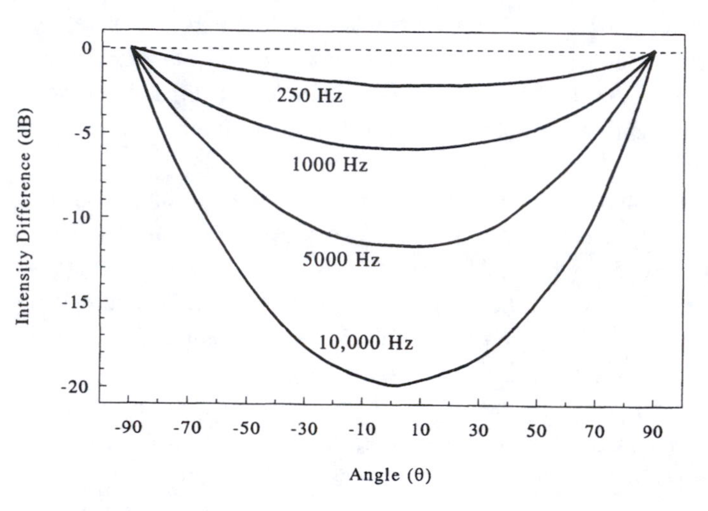

Fig. 10.13 The "sound shadow" cast by a human head is greatest for sounds arriving from the right (or left), an angle, θ, of 0° (or 180°). The difference is zero for sounds arriving from in front (90°) or behind (−90°). The difference in intensity between ears is greater the higher the frequency of the sound. (Redrawn using data reported by Gulick et al. 1989)

The problem of casting a sound shadow extends to higher frequencies in water because of the higher wavelengths there. For instance, at 100 Hz, the wavelength in water is 14.3 m, and even the heads of whales will barely cast an acoustic shadow.

Despite the problems with this strategy, it can be used as part of an effective system of sound localization. In owls, for instance, sound localization is of paramount importance because these animals hunt at night and therefore cannot rely on visual cues. In evolutionary response, the feathers of the face of a barn owl are asymmetrically arranged to serve as directional acoustic antennae: the right ear preferentially hears sounds arriving from the right and above; the left preferentially hears sounds arriving from the left and below (fig. 10.14; Knudsen 1981). The directional properties accruing from the feathers are demonstrably important; birds who have had their facial feathers shaved off are much less adept at localizing sound. Even in this well-tuned system, the ability to localize sound by relative intensity has its limitations. The directionality of the sound reception system is best at high frequencies where the wavelength is smaller than the dimensions of the face. At low frequencies, the system does not work nearly so well.

Fig. 10.14 Tha facial ruff of a barn owl acts as a directional receptor for sound. The right ear preferentially hears sound from the right and above, the left ear from the left and below. (Drawn from a photograph in Konishi 1975)

The second strategy for localizing sound from measured pressures involves the difference in arrival time of a sound at the two ears. Suppose that the ears are separated by a distance x. If the source of sound is to the right, the right ear receives the sound x/c seconds before the left ear. The difference in arrival time between ears is directly related to the angle from which the sound arrives (fig. 10.15). If we define a sound arriving from the right as having an angle, θ of 0, and the distance to the sound source is much greater than the distance between ears, then

$$\Delta t = \frac{\Delta x}{c} \approx \frac{x \cos \theta}{c}. \tag{10.38}$$

Note that θ is measured relative to a line pointing to the right and can lie in any plane. Therefore θ defines a conic surface on which the sound source can lie.

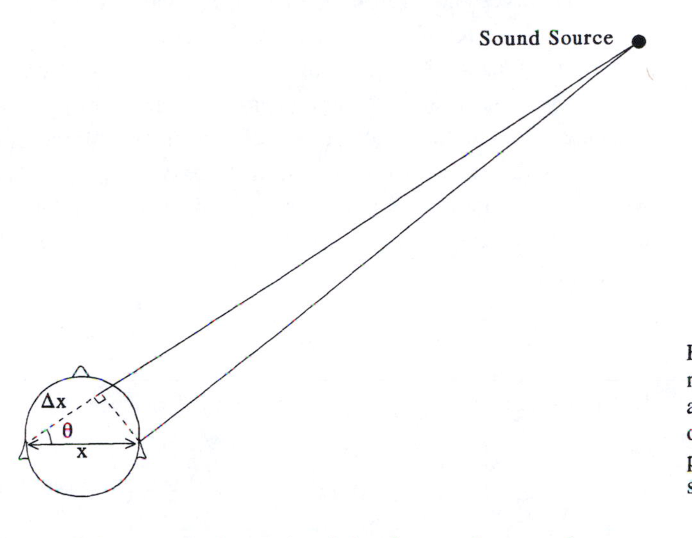

Fig. 10.15 Unless a sound source is in the median plane of the head, sound must travel a greater distance (here approximated as Δx) to one ear than the other. This difference in distance provides a clue as to the location of the sound source.

If the organism "knows" the speed of sound and the distance between its ears, and can measure Δt, it can calculate the angle to the sound source:

$$\theta \approx \arccos \frac{c\Delta t}{x}. \tag{10.39}$$

This mechanism has the advantage of being independent of frequency. Again, however, there is a possible ambiguity—sounds from different locations on the cone defined by θ have the same lag time. This ambiguity must be resolved using other information, such as relative intensity.

The critical problem with this method, however, involves the very short intervals that must be measured. The maximum time delay between ears is x/c. For a bat, with $x \approx 5$ cm, this delay is only about 0.15 ms given the velocity of sound in air. Short as this time is, it is long enough to be detected by the nervous system, and arrival time appears to provide a major clue when bats localize sound (Suga 1990).

The problem is worse for smaller animals. For instance, the ears of a hummingbird or locust may be separated by less than a centimeter, and the time of arrival of sound may differ between ears by only 0.03 ms. This is an exceedingly short time, one that would be difficult for an animal to measure directly. Some insects, birds, and amphibians have solved the problem, however, through an ingenious strategy that we will examine later in this section.

For a human being x is about 20 cm, and Δt is about 0.6 ms, an interval that our nervous system can detect and utilize.

Timing the arrival of sounds is much more exacting in water, however. Because the speed of sound in water is 4.3 times that in air, the interval between the arrival of sound at the two ears is reduced by a factor of 4.3. Thus, in water, the maximum arrival time interval for a human being is about 0.14 ms, an interval which apparently strains the ability of our nervous system to measure. SCUBA divers, for instance, report that it is difficult to localize sound underwater. Fish are not bothered by the small difference in arrival time inherent in the aquatic environment. As we will see, they can directly measure the direction of a sound source.

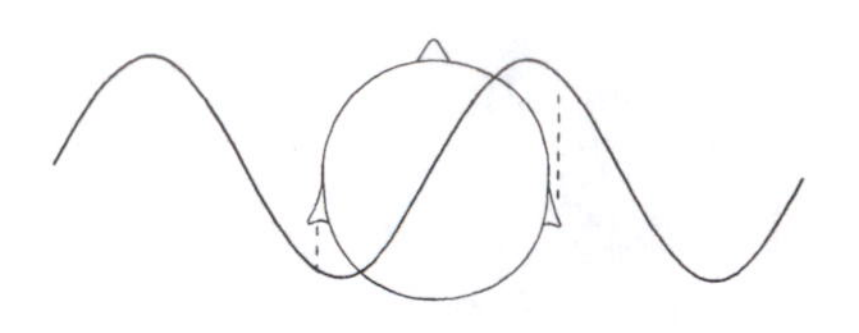

Fig. 10.16 Ears sample sound at different points in a wave. In other words, the phase of the sound at the two ears is different.

The third strategy of directional location via pressure measurements involves detecting the phase shift between sounds heard by the two ears. Consider the series of sound waves shown in figure 10.16. As waves pass an animal, sound arrives first at the "upstream" ear, and slightly later at the opposite ear. As a result, the right and left ears are, at any time, listening to different parts of the sound wave. In other words, the sounds the two ears hear are out of phase, and the difference in phase (like the difference in time of arrival) is a direct function of the position of the sound source relative to that of the ears. The phase of a point on the wave form is the product of 2π and the fraction of a wavelength that has passed since the beginning of the previous wave (fig. 10.16). Thus, the phase difference between ears, ϕ, is

$$\phi = \frac{2\pi\Delta x}{\lambda}. \tag{10.40}$$

Recalling that $\lambda = c/f$ and that $\Delta x = x\cos\theta$, we see that

$$\phi = \frac{2\pi f x\cos\theta}{c}. \tag{10.41}$$

If the organism "knows" the frequency of sound, the speed of sound, and the distance between its ears, and can measure the phase shift, it can calculate θ:

$$\theta = \arccos\frac{\phi c}{2\pi f x}. \tag{10.42}$$

There are three problems with this strategy of sound localization. First, there is the ever-present ambiguity—sounds from different locations can have the same phase shift—and the ambiguity must be removed by using other cues.

Then there are two problems related to frequency. First, the phase shift between ears depends on the frequency of sound (eq. 10.41). To understand why this is so, compare the waves shown in figures 10.17A and 10.17B. The distance between ears is the same in both cases, but because the frequency is higher in figure 10.17B, the phase shift is larger. As a result, direction can be inferred only if there is a mechanism compensating the phase shift for the frequency of sound.

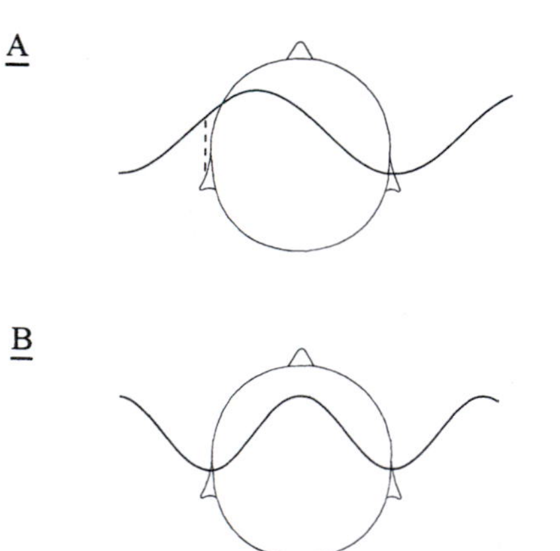

Fig. 10.17 The phase shift between ears is dependent on the frequency of sound. The right ear in (A) samples sound at the same point in a wave as that in (B). But because the sound in (B) has a higher frequency, and therefore a shorter wavelength, the phase shift between ears is larger in (B) than in (A).

Second, if the wavelength of the sound is an exact multiple of the distance between ears (as in fig. 10.17B), phase shift can be an ambiguous indicator of direction. For instance, if the distance between ears is exactly equal to the wavelength, sound arriving from the right reaches the right ear exactly one cycle before reaching the left. As a result, the two ears hear sound that is in phase, just as they would for a sound arriving from the median plane. This potential ambiguity is most likely for high-frequency sounds for which $x \ll \lambda$. At the other end of the spectrum, for low frequency sounds, the difference in phase between the two ears may be too small to detect. This latter problem is more severe in water where wavelengths are longer, and where phase shifts are therefore smaller.

We now return to the problem of sound localization by small animals. We noted above that due to the small distance between their ears, these animals may have a problem in detecting the difference in the time of arrival of sound. This problem has been solved in at least some small animals through the evolution of a physical connection between ears (fig. 10.18B). Any pressure acting on one ear is transmitted through an air–filled "pipe" to the other ear,[7] and the *tympanic membranes* (the "ear drums") of the two ears act in concert (fig. 10.18B). As a result, the ear drum is displaced only when the pressure is different between ears, and this type of "pressure-difference" detector responds directly to the phase difference between ears. By thus relieving the neural circuitry of the burden of measuring small differences in the time of sound arrival, a pressure-difference phase detector allows small animals such as crickets, locusts, amphibians, and birds to localize sound.

Recall that the speed of sound, and thereby the wavelength, in both air and water increases by 7% to 9% over the biological range of temperatures. As a result, any of the mechanisms for sound localization that rely on knowing c or λ must be compensated for temperature. For example, the phase shift between ears for sound at a given angle θ and frequency f is 7% smaller at 40°C than at 0°C in air. Without knowledge of the temperature of the medium, this difference in phase shift would be interpreted as a difference in angle (see Eq. 10.42). For example, when $\theta = \pi/4$, a 7% change in the speed of sound would give an apparent change in direction of about 0.073 radians (roughly 4°).

Despite the inherent problems of determining direction, animals combine the methods described above and thereby perform admirably. For example, bats and owls have been shown to localize sounds within 1° to 2°, and dolphins have similar directional acuity. Humans, cats, and opossums can localize sounds within 1° to 6° (Lewis 1983). These abilities are a tribute to the ability of the nervous system to assimilate complex data.

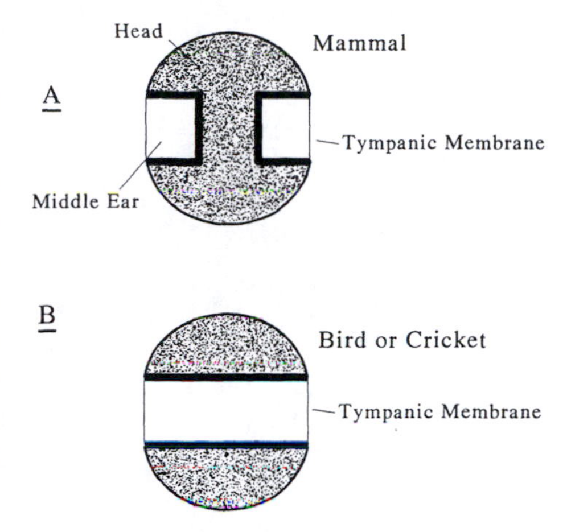

Fig. 10.18 The ears of mammals (A) act separately; there is no effective sound path between them. In contrast, the ears of some insects, amphibians, and birds (B) have an air-filled channel between ears that causes the two tympanic membranes to act as a unit. This acoustic coupling of the two ears allows them to detect the pressure *difference* between ears, and thereby to detect the phase shift.

10.6.2 *Range*

In addition to being able to discern the direction of a sound, it is often handy to know the distance to its source. The simplest mechanism for range determination is used by both bats and dolphins and involves measuring the interval between when a burst of sound is released by the animal and when an echo arrives. The product of this interval and the speed of sound in the medium is equal to twice the distance from the source to the scatterer (it takes half the time for sound to travel to the scatterer, the other half to travel back). Thus,

$$\text{distance, } \ell = \frac{tc}{2}, \tag{10.43}$$

where t is the time between chirp and echo.

The return time of an echo depends on the speed of sound, and therefore can be affected by temperature. Because the speed of sound increases with increasing temperature, animals will underpredict the distance to objects as the temperature rises unless their neural mechanisms are temperature compensated. For instance, a bat whose uncompensated ranging mechanism is "calibrated" for 0°C could underestimate the distance to an object by about 7% at 40°C. The situation is similar in water.

The only other major problem with this mechanism of range determination is the necessity to discriminate between very short periods. Rearranging eq. 10.43

[7]This is one case where sound quite literally could go in one ear and out the other. This trick would require sound with a wavelength less than the distance between ears, however.

we see that a small change in distance, $\Delta\ell$, results in a very small change in time, Δt:

$$\Delta t = \frac{2\Delta\ell}{c}. \tag{10.44}$$

For example, a change in distance of 1 m in air involves a change in time of echo arrival of only 5.8 ms. In water, the problem is 4.3 times worse. A change in distance of 1 m changes the time of echo arrival by only about 1.4 ms.

Despite these constraints, bats can perceive differences in distance to a scatterer of about 1 cm (Simmons and Grinnell 1986), a phenomenal performance that implies a temporal acuity of about 60 μs! In this respect, dolphins may be less adept. Experiments on captive animals suggest that they can time the arrival of an echo to an accuracy of only about 24 ms, implying a range acuity of only 17 m (Popper 1980).

In this section we have explored only the simplest mechanism for acoustically determining the range of a sound source. Other mechanisms exist. Many bats, for instance, include a frequency modulated (FM) segment in each of their echolocation signals. During this segment, the frequency of the emitted sound drops smoothly through a range of about 10 kHz. Relying on theoretical treatments of sonar signals, several authors have suggested that these FM segments can provide information about the distance between bat and prey. We will not explore this mechanism here, but you are invited to consult Nachtigall and Moore (1986).

10.7 The Doppler Shift

When we considered echolocation earlier in this chapter, we tacitly assumed that the animal producing the sound and the object producing the echo were both stationary. Of course this need not be the case, and when either the sound source or the scattering object moves, information regarding their relative motion is contained in the resulting acoustic signal.

To see how this works, consider the situation shown in figure 10.19A. A stationary source at point A produces sound with wavelength λ, emanating in all directions. These waves are received by an animal at point B. If the receiving animal is stationary, how many sound waves pass by in a given time? Because waves travel with velocity c and each wave is λ meters long, c/λ waves arrive at the receptor per second.

Fig. 10.19 The frequency of sound perceived by an animal is shifted by the relative motion between the animal and the source of sound. This Doppler shift depends on whether the animal is moving relative to a stationary source (A), or the source moves relative to a stationary animal (B).

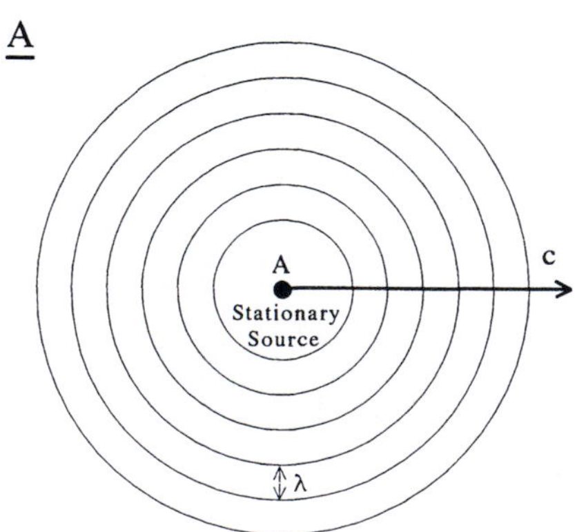

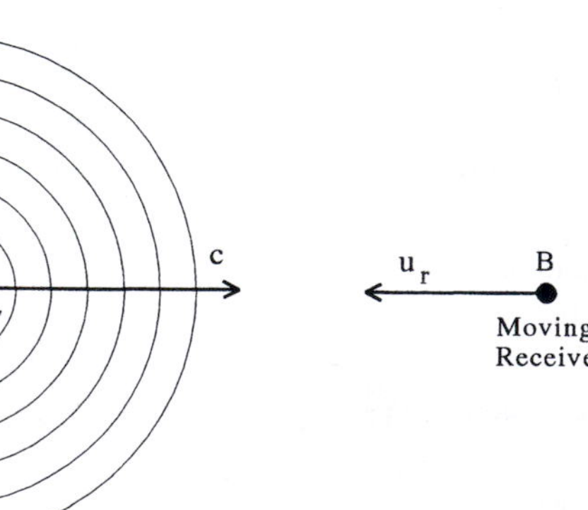

Now imagine what would happen if the sound waves emanating from the source could somehow be "frozen" in place. In this hypothetical case, because source, receiver, and waves are all stationary, no waves pass the receiver. If, however, the receiver were to move toward the source, it would pass through the stationary waves. A receiver moving at velocity u_r would encounter u_r/λ waves per second.

We now "thaw" the sound waves, allowing them to move as before with velocity c. In this case, how many waves pass the receiver in a given time as it moves toward the source? It encounters c/λ waves per second due to the speed of the waves themselves, and an additional u_r/λ waves per second due to its own motion. As a result, the frequency of sound heard by the moving animal is

$$f' = \frac{c + u_r}{\lambda}. \tag{10.45}$$

But we know from eq. 10.21 that $\lambda = c/f$, where f is the frequency emitted by the source. Substituting this expression for λ in eq. 10.45 we see that

$$f' = f\frac{c + u_r}{c}. \tag{10.46}$$

In other words, the sound heard by a receiver as it moves toward the source is higher than that emitted. Just the opposite would be true if the receiver were moving away from the stationary source. In this case,

$$f' = f\frac{c - u_r}{c}. \tag{10.47}$$

If, in each case, we subtract the emitted frequency from that received, we see that the change in frequency, Δf, is

$$\Delta f = f\frac{\pm u_r}{c}, \tag{10.48}$$

where the shift is positive when the receiver is moving toward the source and negative when it is moving away. As a result, the sound emanating from the receiver contains information about the receiver's motion. This shift in the frequency of sound is one example of the phenomenon known as a *Doppler shift*, named in honor of Christian Johann Doppler (1803–1853).

Before exploring the biological uses of Doppler shifts and the effects on them of air and water, we need to finish our examination of relative motion. So far, we have only considered the case of a receiver moving relative to a stationary source. What happens if the source moves relative to a stationary receiver (fig. 10.19B)? The effect is similar, but subtly different.

As the source moves, it tends to catch up with the sound waves it has produced. For example, if the sound emitted has frequency f, one wave moves a distance c/f before the next wave is produced. During this time, the source, moving with velocity u_s, travels a distance u_s/f in the direction of sound propagation. As a result, the effective wavelength of sound in front of the source is $\lambda' = (c/f) - (u_s/f)$. Noting from eq. 10.21 that $f = c/\lambda$, we see that the frequency of sound that reaches a stationary receiver, f', is

$$f' = c/\lambda' = f\frac{c}{c - u_s}, \tag{10.49}$$

and the sound reaching the receiver is higher than that emitted by the amount,

$$\Delta f = f\frac{u_s}{c - u_s}. \tag{10.50}$$

When the source is moving away from a stationary receiver, the sound reaching the receiver is lower than that emitted by the amount,

$$\Delta f = f\frac{-u_s}{c + u_s}. \tag{10.51}$$

Thus, the frequency of sound heard by the receiver increases either if the receiver is moving toward a stationary source or if the source is moving toward a stationary receiver. Frequency is shifted down if the source and receiver are moving apart. However, even if the relative velocities are the same in each case (in other words, $u_r = -u_s$), the shift in frequency is not. Movement by the source shifts the sound more than movement by the receiver if the two objects are converging, less if they are moving apart (fig. 10.20).

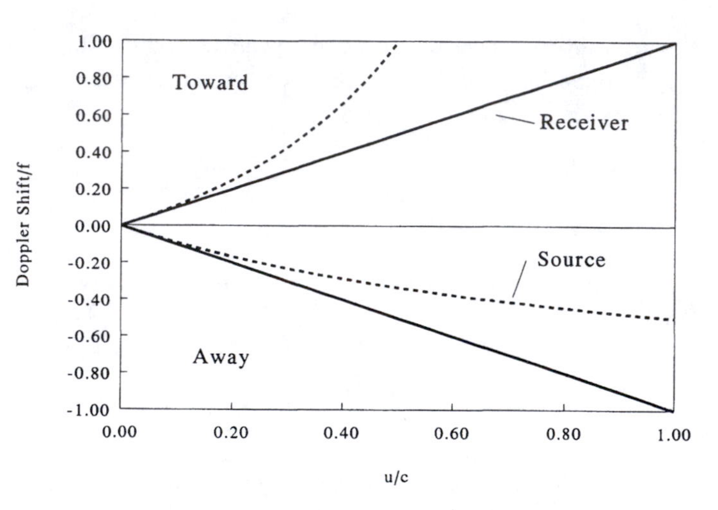

Fig. 10.20 The Doppler shift differs depending on whether the source or the receiver moves, but the difference is slight for speeds less than 10% of the speed of sound.

This strange lack of symmetry is apparent, however, only when the relative speed of source and receiver is an appreciable fraction of the speed of sound. Because the fastest organisms (flying birds) move at about 0.1 c at most, the Doppler shift is negligibly different whether it is the source or receiver that is moving.

When both receiver and source move, the frequency heard by the receiver is

$$f' = f\frac{c + u_r}{c - u_s} \tag{10.52}$$

if the objects are moving toward each other. If they are moving away,

$$f' = f\frac{c - u_r}{c + u_s}. \tag{10.53}$$

In all cases, velocities are measured relative to the stationary medium.

Doppler shifts are used by a variety of animals to glean information from their environment. For example, bats can detect the Doppler shift of the echoes returning to them from prey or other objects, and they use this information to tell them how fast they are approaching their next meal or the nearest tree. The frequency of the echo can also be modulated by the beating wings of the insect prey. As the wings move toward the bat, the frequency of the echo increases; as they move away, the frequency decreases. As a result, the motion of the wings imposes a "warble" to the echo (Schnitzler et al. 1983; Kober 1986). The frequency of this warble (in other words, the frequency with which the echo changes frequency) is proportional to the insect's wing-beat frequency, and can therefore provide the bat with information as to the type of prey it is chasing. A mosquito, for instance, has a different wing-beat frequency than a moth.

Note, however, that the shift in frequency of an echo depends on the speed of sound. For instance, an insect moving toward a bat at 1 m s^{-1} shifts the frequency of a 60 kHz signal by about 180 Hz. A fish moving toward a porpoise at 1 m s^{-1} shifts a 60 kHz signal by only about 40 Hz. As a result, it is more difficult to use Doppler shifts to detect motion in water than in air.

The information contained in the Doppler shift is also affected by temperature. The higher the temperature, the higher the speed of sound and the lower is the Doppler shift for a given velocity of source or receiver. Just as bats and dolphins must compensate their ranging apparatus for changes in the temperature of the medium, they must also compensate the neural mechanism by which they interpret velocity in terms of Doppler shifts.

10.8 Hearing

Finally, let us briefly examine the mechanism by which sound is sensed by animals.

The situation is somewhat analogous to the problem we dealt with in chapter 9 regarding the detection of an electrical field. As sound waves move through a medium, they transport energy that can be used to do work on an appropriate detector. Biological sound detectors come in two basic varieties.

In fish, the modified cilia of special sensory cells (*hair cells*) are in contact with a small calcareous ball (the *otolith*) (fig. 10.21). The otolith is denser than its surrounding fluid and is free to move. As sound waves pass through the sensory apparatus, the hair cells and their attached tissues (which have a density very near that of water) are displaced just as the fluid molecules in the medium are displaced, but the inertia of the denser otolith causes it to lag behind. As a result, the sensory cilia are bent along the direction of sound propagation, and this displacement is transduced into nervous impulses that convey acoustic information (including, directional information) to the brain. Thus, fish hear because their sensory apparatus is displaced as sounds pass by. With the exception of those fish that have a swim bladder, the application of a change in hydrostatic pressure alone (with no displacement) does not affect the hearing apparatus. We will consider the special case of fish with swim bladders later in this chapter.

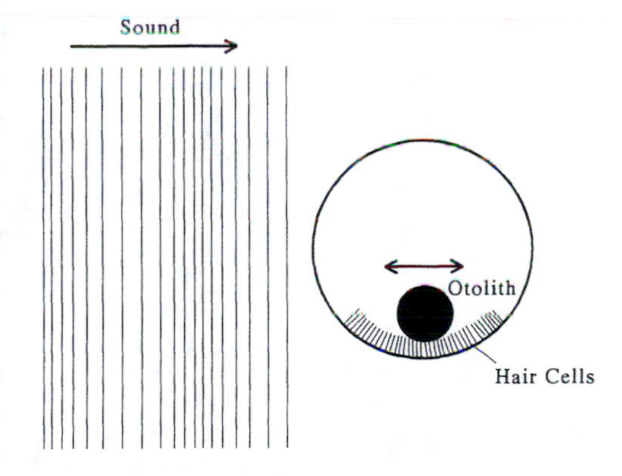

Fig. 10.21 As sound impinges on a fish, the dense otolith remains stationary relative to the hair cells beneath it. The resulting deflection of the hair cells produces a nervous signal containing information about sound. Note that sounds arriving from different directions cause the hair cells to bend in different directions. Thus, this type of displacement-sensitive (rather than pressure-sensitive) sound receptor can directly sense the direction of a source of sound.

A few other animals have organs that respond to acoustic displacement. For example, the cercal hairs on the abdomen of a cricket respond to the motion of the air around them, as we saw in chapter 7.

In most animals, however (including amphibians, reptiles, birds, mammals, and some insects), it is a change in pressure, rather than displacement, that is heard. The arrival of sound waves at the "ears" of these organisms applies an oscillating pressure to the membrane generally known as the *tympanic membrane*. This pressure, acting over the area of the membrane, results in a force which, by a

variety of means, is used to move the sensory cilia of hair cells. As these cilia are bent, appropriate nerve impulses are relayed to the brain. Although it is ultimately a displacement of cilia that is sensed, this motion results from the application of a pressure.

These brief descriptions cannot convey the beauty and complexity of the auditory apparatus, and you are urged to consult the reviews by van Bergeijk (1967), Wever (1974), Griffin (1986), Ewing (1989), and Lewis (1983) for expositions on the evolution of hearing. However, our concern here is not with the method by which sound is transduced to the language of neurons, but rather with the physics by which power is transported to the sensory machinery. As was the case for electrical signals, the presence of sound waves can be detected only if sufficient power is delivered. Therein lies a problem for terrestrial animals.

To see how the phenomenon works, we explore what happens when sound waves move from one medium to another. Consider, for instance, planar waves of sound in air propagating at a right angle toward the flat surface of a lake (fig. 10.22). Directly at the interface between air and water, two criteria must be met. First, the pressure in the air at any time must equal the pressure in the water. This is a result of Newton's third law of motion—the force with which the air pushes on an area of the interface is resisted by an equal and opposite force within the water. Now, we know from experience that when sound in air meets a water surface, an echo results. Therefore, the pressure at the air-water interface has three components: there is the pressure of the sound wave incident upon the interface, p_i; there is the pressure in the reflected echo, p_r; and there is the pressure of any sound that is transmitted into the water, p_t. Due to the equality of pressure cited above,

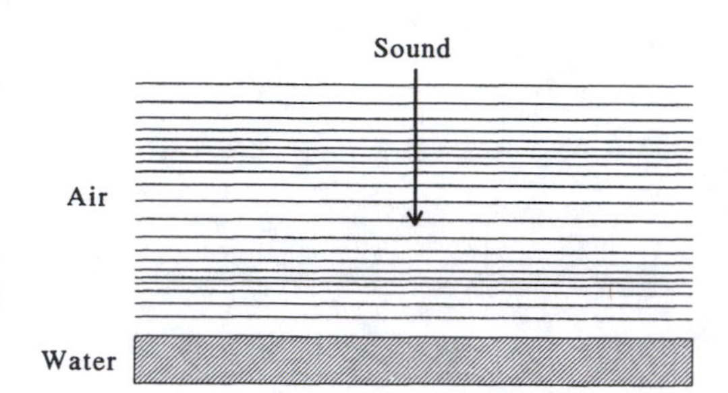

Fig. 10.22 Sound impinging on the water's surface from above can either be reflected or transmitted.

$$p_i + p_r = p_t. \tag{10.54}$$

We also can deduce that the displacement of air molecules at the interface must equal the displacement of the adjacent water molecules. If this were not the case, air would move away from the water, leaving behind a region of vacuum, or air would be forced into the water. Either case is absurd, so we know that the displacement of air and water are the same. Again, there are three components to the situation: the displacement amplitude of air due to the incident wave, $\mathcal{A}_i$; the displacement amplitude due to the reflected wave, $\mathcal{A}_r$; and the displacement transmitted to the water, $\mathcal{A}_t$.

To figure out how to tote up these displacement components, we consider an extreme example. If sound were to impinge on a perfectly rigid surface, no displacement could be conveyed to the solid, $\mathcal{A}_t$ would be 0, and $\mathcal{A}_r$ would equal $\mathcal{A}_i$. This can be the case only if

$$\mathcal{A}_i - \mathcal{A}_r = \mathcal{A}_t. \tag{10.55}$$

We thus have two equations (eqs. 10.54 and 10.55) that much be satisfied at the interface. Recalling from eq. 10.23 that $p = 2\pi f \rho_0 c \mathcal{A}$, we can solve eq. 10.54 in terms of displacement amplitudes, with the result that

$$\mathcal{A}_t = \mathcal{A}_i \frac{2\rho_a c_a}{\rho_a c_a + \rho_w c_w}, \tag{10.56}$$

where the subscripts a and w refer to air and water, respectively. In other words, the displacement amplitude of sound transmitted into water depends on the incident amplitude in air and on the relative characteristic impedances of air and water.

Now, at 20°C, the characteristic impedance of air is about 412 kg m^{-2} s^{-1}, and that of water is about 1.5×10^6 kg m^{-2} s^{-1} (table 10.4). As a result, the amplitude of sound waves transmitted to water is only about 1/1800th that in air.

Recalling from eq. 25 that $\mathcal{I} = 2\pi^2 f^2 \rho_0 c \mathcal{A}^2$, we can calculate that

$$\mathcal{I}_t = \mathcal{I}_i \frac{4\rho_a c_a \rho_w c_w}{(\rho_a c_a + \rho_w c_w)^2}. \tag{10.57}$$

Inserting the appropriate values for air and water, we see that the intensity of sound transmitted into water is about 0.1% of that in air. In other words, when a planar sound wave in air impinges perpendicularly onto a water surface, 99.9% of the acoustic power is reflected back into the air. This is one reason why bats, flying in air, cannot detect objects below the surface of a lake or pond. Very little of their signal can get into the water, and any resulting echo is similarly attenuated when it travels back through the interface into air.

The drastic attenuation of signals moving from air into water poses a problem for terrestrial organisms as they attempt to hear sounds in their gaseous environment. Terrestrial animals, like all organisms, are constructed primarily from water, and the cells that transduce sound energy to nervous impulses are bathed in a watery fluid. Because of the resulting high acoustic impedance of the body, sound energy in the surrounding air tends to reflect rather than to be transmitted to the acoustic sensing apparatus. This mismatch of impedances limits the sensitivity of sound detectors in air in the same way in which a resistance mismatch limits the sensitivity of terrestrial electrical detectors.

Unlike electrical detectors, which seem not to have evolved in terrestrial organisms, a variety of mechanisms have evolved to improve the efficiency with which sound energy is delivered to the hair cells of terrestrial animals. In general, these involve structural mechanisms that tend to minimize the difference in impedance between air and ear. The ears of human beings (and other mammals) are an illustrative example.

The cells in a mammal that actually sense sound are housed in the watery medium of the cochlea, and therefore have a high characteristic impedance. However, the fluid of the cochlea is connected to the outside environment through a peculiar mechanical apparatus. There are two openings in the bony case of the cochlea. One, the round window, is covered by a flexible membrane that separates the fluid interior of the cochlea from the air in the eustachian tube. This window provides "volume relief" so that the fluid in the cochlea can move when pushed upon. The other opening, the oval window, is also covered by a membrane, but in this case the membrane is attached to a bony plate, the *stapes* (fig. 10.23). The stapes in turn attaches to the *incus*, which attaches to the *malleus*, which attaches to the tympanic membrane. The fact of note here is that the tympanic membrane has a much greater area (about 70 mm^2) than does the footplate of the stapes (3.2 mm^2). As a result, sound pressure acting over the large area of the tympanic membrane is concentrated on the small area of the oval window, in effect decreasing the apparent impedance of the ear by a factor of $70/3.2 \approx 22$. Further, the sound impinging on the ear drum's 70 mm^2 is collected over an area of about 1800 mm^2 by the ear lobe. Thus, if we assume that the ear lobe effectively transmits to the

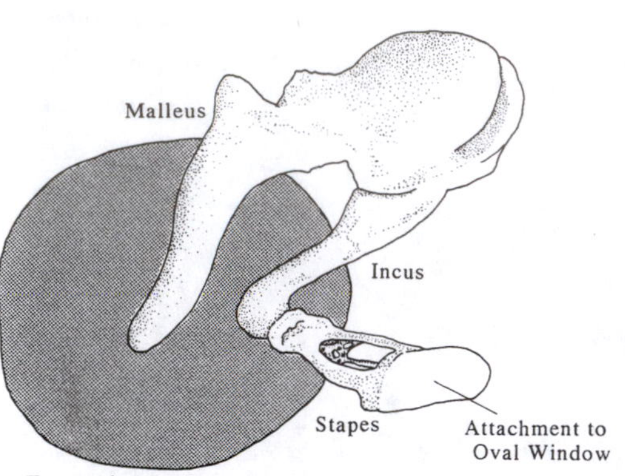

Fig. 10.23 The middle ear of mammals consists of a series of levers that transmits the motion of the tympanic membrane (the "ear drum") to the oval window of the cochlea. In the process, the displacement associated with a given sound is decreased but the force is concomitantly increased, helping to reduce the impedance difference between air and the inner ear.

ear drum all sound falling upon it, the apparent impedance of the ear is decreased by a factor of $1800/3.2 \approx 563$.

Yet a third factor needs to be taken into account. The malleus, incus, and stapes form a system of levers such that the movement of the tympanic membrane is reduced in amplitude by about 23% when it reaches the oval window (Yost and Nielsen 1977). As with any lever system, a reduction in displacement is accompanied by an amplification of force (Alexander 1968), and force on the ear drum is amplified by about 30% at the oval window. In effect, this reduces the apparent acoustic impedance of the ear by another 30%, for an overall reduction of about 730. This does not make up for the entire disparity in impedance between air and water (a factor of nearly 3600), but it does substantially increase the transmission of sound power to the sensory cells in the cochlea.

One final factor lowers the apparent impedance of the sensory apparatus. The presence of the round window allows the fluid in the cochlea to move much more freely than if it were encased entirely in bone. This freedom of movement reduces the impedance yet another 4-fold. Taken together, the various mechanisms described here insure that the effective impedance of the ear differs from that of air by only about 25%.

Similar mechanical strategies have evolved in many animals, all in apparent response to the necessity of matching the ear's acoustical impedance to that of the surrounding air (van Bergeijk 1967).

Table 10.5 Specific acoustic resistances for various tissues.

Tissue	Specific Acoustic Resistance ($kg\,m^{-2}\,s^{-1}$)
Seawater	1.560×10^6
Krill	1.572×10^6
Copepod	1.500×10^6
Fish flesh	1.572×10^6
Fish bone	11.680×10^6

Source: Calculated from data given by Clay and Medwin (1977).

Animals in water do not have this problem. Because the materials of the body have characteristic impedances close to that of water (table 10.5), sound energy has little difficulty in being transmitted to internal sensory cells. The relatively high impedance of bone leads to some reflection, but at a planar interface between water and bone about 44% of the incident acoustic impedance is transmitted (eq. 10.57). Because of the inherent impedance match between flesh and water, many fish need no special apparatus for conducting sound from their environment to the sensory apparatus buried deep in the head; sound simply passes through the body.

This raises an interesting question. Dolphins and whales evolved from terrestrial organisms, and as a result, have ear structures "designed" to match the impedance of their ears to that of air. How do these structures function in water? It is likely that the reception of sound via the tympanic membrane and the middle-ear bones is largely bypassed in cetaceans (for instance, the malleus is only loosely attached to the ear drum), and dolphins rely instead on sound transmitted directly to the bony case of the inner ear through a channel of fat in the lower jaw (Popper 1980). Thus, upon their return to an aquatic environment, mammals reevolved acoustic sensory strategies appropriate to the medium.

Despite the ease with which sound energy is transmitted into bodies in water, at least one mechanism has evolved for amplifying the acoustic signal available to sensory cells. Consider a bubble of air in water. If the circumference of the bubble is small relative to the wavelength of sound, we have already seen that very little of the sound in the water is scattered by the bubble (eq. 10.31). In other words, very little of the pressure signal in the water is reflected by the bubble, implying that the pressure in the bubble must be very nearly that in the surrounding water. Thus,

$$p_a \approx p_w. \tag{10.58}$$

Recalling from eq. 10.23 that $p = 2\pi f \rho_0 c \mathcal{A}$, we conclude that in the case of a small bubble ,

$$2\pi f \rho_a c_a \mathcal{A}_a \approx 2\pi f \rho_w c_w \mathcal{A}_w, \tag{10.59}$$

where $\mathcal{A}_w$ is measured well away from the bubble.

Rearranging and canceling terms, we see that

$$\frac{\mathcal{A}_a}{\mathcal{A}_w} \approx \frac{\rho_w c_w}{\rho_a c_a} \approx 3600. \tag{10.60}$$

In other words, the displacement of fluid particles in the vicinity of a bubble is very much greater than the displacement in the bulk of the surrounding fluid.

This expression is only a rough approximation because the dynamic response of the bubble has not been taken into account. A more precise treatment of the subject shows that the displacement amplification by a bubble depends on the ability of a bubble to deform, and this in turn depends on the frequency of sound and on the size of the bubble. Deformation is greatest at the bubble's resonant frequency, f_r:

$$f_r = \frac{1}{2\pi r}\left(\frac{3\gamma_s p_0}{\rho_w}\right)^{1/2}, \tag{10.61}$$

where p_0 is the ambient pressure (1×10^5 Pa at the water's surface) and r is the bubble's radius (fig. 10.24). Recall that γ_s is the ratio of specific heat capacity of air measured at constant pressure to that at constant volume.

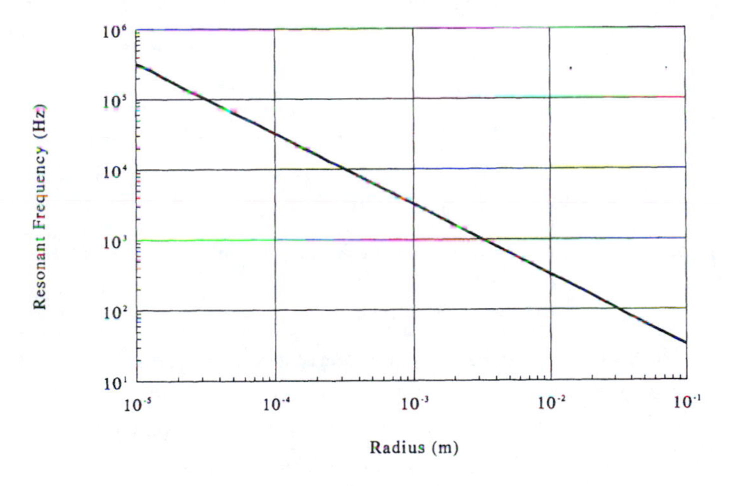

Fig. 10.24 The resonant frequency of an air bubble in water decreases with an increase in frequency.

The displacement amplification depends on how close the sound's frequency is to the resonant frequency (Sand and Hawkins 1973):

$$\frac{\mathcal{A}}{\mathcal{A}_0} = \frac{2\pi f a c_w \rho_w}{3\gamma_s p_a \sqrt{\left[1-\left(\frac{f}{f_r}\right)^2\right]^2 + \left(\frac{f}{f_r q}\right)^2}}, \tag{10.62}$$

where $\mathcal{A}$ is the actual displacement amplitude in the vicinity of the bubble, $\mathcal{A}_0$ is the amplitude in the absence of the bubble, and q is a parameter related to how much acoustic energy is lost to viscous processes. The larger the q, the less damped is the bubble. This expression is graphed in figure 10.25 for a variety of q's and bubbles of two different radii.

Now, the swim bladder of a fish is essentially an air bubble with a q of 2 to 4 (Sand and Hawkins 1973). As a result, the swim bladder can act as a displacement amplifier for sounds of a range of frequencies, an effect used by a variety of fish to enhance their ability to hear. In some, such as the Clupeiformes (herrings and cods, for instance), a small anterior extension of the swim bladder lies adjacent to the acoustic sensory apparatus. A variety of fish have small pockets of gas which, during development, detach from the swim bladder and lie next to the sensory apparatus. In the Ostariophysi (e.g., minnows and catfish), a series of small bones (the *Weberian ossicles*) connect the swim bladder to the structure containing the otolith. In this fashion, the amplified displacement of the swim bladder is transmitted to the sensory cells. In all these cases, acoustic pressure acting on the swim bladder is transduced into a displacement acting on the sensory apparatus. In this respect, the swim bladder in fish acts much like the tympanic membrane in the ears of terrestrial animals. Experiments have shown that the frequency response of hearing in at least some fish follows roughly that predicted for a bubble the size of the swim bladder (Alexander 1966; Sand and Hawkins 1973), and elegant experiments by Fay and Popper (1974) demonstrate that over a large range of frequencies hearing in these fish is sensitive to pressure rather than displacement.

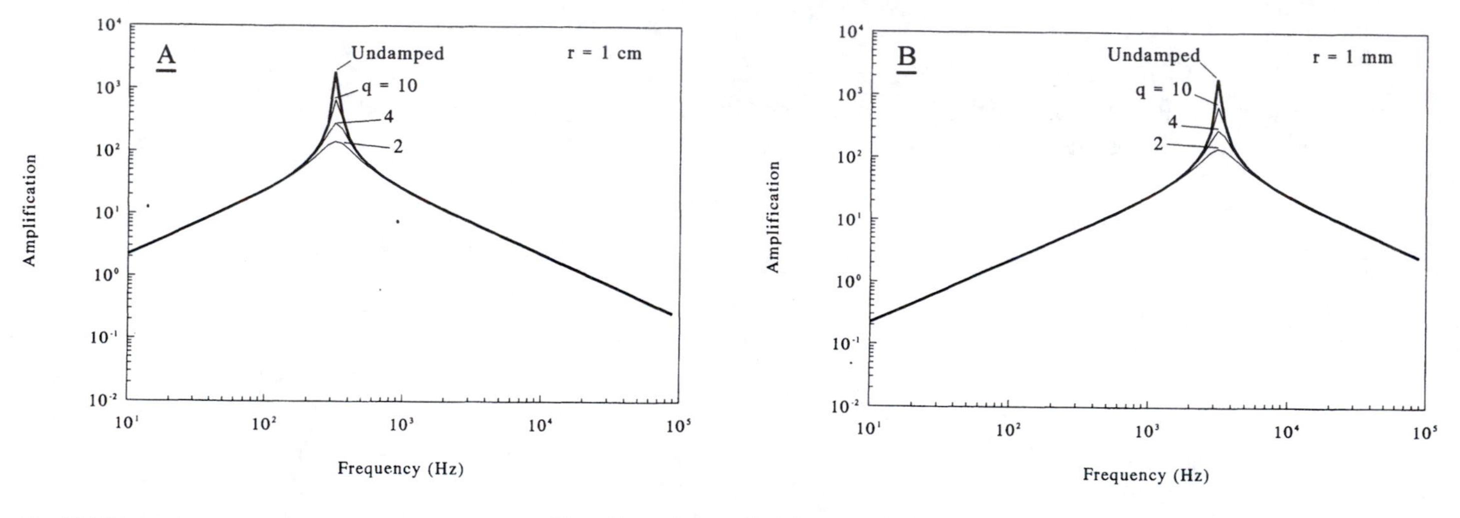

Fig. 10.25 Near its resonant frequency, an air bubble amplifies particle displacements. The amplification depends on how "damped" the bubble is, the lower q is the more damped the bubble and the lower the amplification. Note that at frequencies much higher than the resonant frequency, the bubble actually reduces the particle displacement. (A) A bubble with a radius of 1 cm. (B) A bubble with a radius of 1 mm.

Note from figure 10.25 that at frequencies far removed from the resonant frequency, the presence of a bubble may actually deamplify the displacement amplitude. Thus, a fish with a swim bladder 1 cm in radius would have more difficulty hearing sounds above 10 kHz than if it did not have a swim bladder. This is probably not a problem, since most fish do not hear sounds above about 1000 Hz, for reasons that are not entirely clear but are, in any case, unrelated to the swim bladder (Tavolga 1971).

The situation is reversed for smaller swim bladders. At a swim bladder radius of 1 mm, high frequencies are amplified quite well, but frequencies below 100 Hz are deamplified (fig. 10.25). This is the situation in the cod, for which the upper limit to hearing is around 400 Hz (Sand and Hawkins 1973). Why have a swim bladder whose peak amplification occurs at frequencies too high to hear? The answer becomes apparent when we plot separately the displacement amplitude of the swim bladder and that of the water in its absence (fig. 10.26). At a constant intensity, the displacement amplitude in the absence of the swim bladder decreases with increasing frequencies as required by eq. 10.25. In the presence of the swim bladder, however, the displacement amplitude is virtually independent of frequency across the range of frequencies that the fish hears. Only well above the fish's auditory range does amplification by the swim bladder offset the decrease in amplitude due to an increase in frequency. The "flat" response of the swim bladder at low frequencies may make it easier for the fish's nervous system to interpret complex sounds.

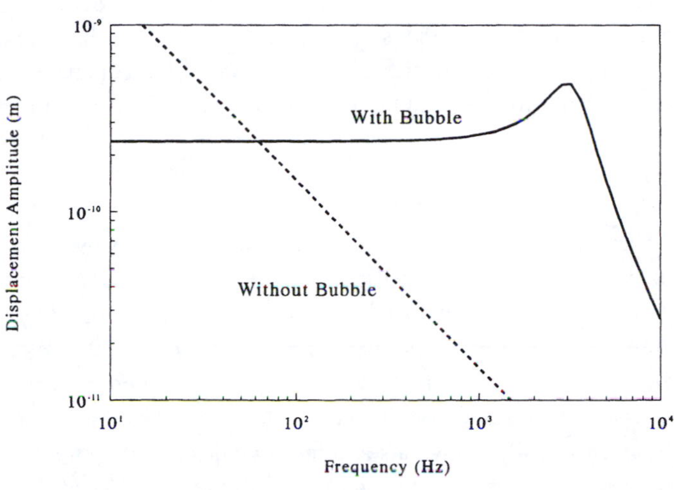

Fig. 10.26 At frequencies below the resonant frequency, a bubble serves to reduce the dependence of particle amplitude on the frequency of the sound. Thus, the presence of a swim bladder can help to give a fish's ear a "flat" frequency respose.

Before leaving the subject of sound at the interface between air and water, we should note one final phenomenon. When we calculated the intensity of sound penetrating into water from a source in air, we assumed that sound approached the water at a right angle. This direction is traditionally referred to as a zero *angle of incidence*, where the angle of incidence is measured relative to a line perpendicular to the plane of the interface (fig. 10.27). It can be shown that the greater the angle of incidence, the less the power transmitted from one medium to the other. At a critical angle,

$$\text{critical angle} = \arcsin \frac{c_a}{c_w}, \tag{10.63}$$

all of the acoustic energy in air is reflected parallel to the air-water interface, and none enters the water. Thus, for angles of incidence greater than about 13°, no sound power can be transmitted from air to water. This is yet another reason why it is virtually impossible for a flying bat to detect a stationary object underwater. Unless the bat is nearly above the object, literally none of its acoustic signal enters the water, insuring that no echo can be heard.

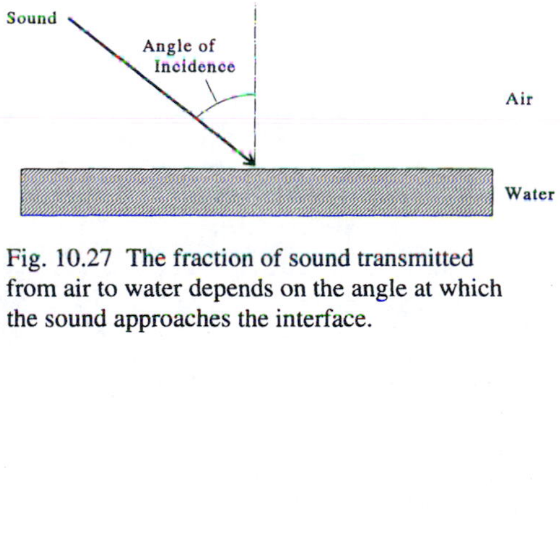

Sound
Angle of Incidence
Air
Water

Fig. 10.27 The fraction of sound transmitted from air to water depends on the angle at which the sound approaches the interface.

10.9 Summary . . .

Both air and water attenuate high-frequency sounds faster than low-frequency sounds, and air attenuates sound much more rapidly than does water. As a result, terrestrial animals attempting to communicate over long distances must use low-frequency signals.

The speed and wavelength of sounds in water are approximately 4.3 times those in air at a given frequency. It is consequently more difficult for organisms in water to detect echoes from small objects and to detect from the Doppler shift in frequencies the relative speed of these objects. The relatively long wavelengths of sound in water make it more difficult to localize sound sources from pressure measurements, but fish can localize sounds by directly measuring the particle displacements accompanying sound waves.

Sound may have the potential to be of greater biological importance in water than in air. Low-frequency sounds travel more than a million times farther in water than in air, and for that bulk of the ocean which lies below the photic zone as well as in turbid lakes and rivers, sound may take the place of light as the best source of information about the environment. Furthermore, because the characteristic impedance of animal tissues is close to that of water, it is inherently easier for aquatic organisms to detect acoustic signals than it is for terrestrial organisms, and strategies such as swim bladders can actually amplify the sound reaching the ears.

10.10 . . . and a Warning

The discussion of acoustics in this chapter has focused on a few biological consequences of the physical differences between air and water. In the process, we have glossed over many of the details of sound reception by animals and have virtually ignored the intriguing neural methods by which acoustic information is processed to provide a functional picture of the environment. For example, in discussing the information available from echoes, we have only looked at single frequencies, while many bats and most whales and dolphins include a variety of frequencies in their calls and can apparently make sense of the very complicated echoes that result. The many, possible effects of fluid physics on these interpretations have not been touched. Therefore, the discussion here should be viewed as a peculiarly biased introduction of the fascinating subject of bioacoustics, and you are urged to pursue the topic further by consulting texts on the biology of sound such as Griffin (1986), Nachtigall and Moore (1986), Lewis (1983), Gulick et al. (1989), and Ewing (1989), and texts on theoretical acoustics such as Morse and Ingard (1968) and Kinsler and Frey (1962).

Chapter 11

Light in Air and Water

In chapter 9 we explored the properties of a time-invariant electrical field and noted that its effects are confined to an area quite close to the field's source. The results can be profoundly different, however, if instead of remaining constant, the strength and polarity of the field vary rapidly through time. A changing electrical field induces around itself a changing magnetic field, which in turn creates a changing electrical field. These two interacting and self-perpetuating fields (together described as an electromagnetic wave) proceed away from their source, propagating outward in all directions. The speed of this propagation is exceedingly high, about 3×10^8 m s^{-1}. In fact, as James Clerk Maxwell (1831–1879) elegantly proposed, the speed at which an electromagnetic wave moves is exactly the speed of light, leading to the conclusion that light is simply one form of electromagnetic radiation. In this chapter, we explore the interactions of light with air and water, and the consequences for plants and animals.

In that we deal with electrical fields, our exploration of the optical properties of air and water is an extension of chapter 9. The approach we take is quite different, however. Whereas only a few organisms can sense a distant static electrical field, and then only in water, the ability to sense and utilize electromagnetic radiation is nearly universal among plants and animals. Plants use light as a source of energy for photosynthesis, and animals use light as a means of obtaining information about their surroundings. Both plants and animals use light as a source of heat. Because these abilities are so widespread and important, they have received considerable attention from scientists, and the the mechanisms by which organisms interact with electromagnetic radiation have been well described (see, for example, Autrum 1979, 1981; Davson 1972; Gates 1980; Horridge 1975; Nobel 1983; Waterman 1981). In this chapter we focus instead on the ways in which air and water affect electromagnetic radiation before it arrives at the cells (or parts of cells) where it is absorbed and utilized.

11.1 The Physics

11.1.1 *The Electromagnetic Spectrum*

Electromagnetic radiation is classified according to the frequency, f, with which its electric and magnetic fields oscillate. For example, radio waves, a type of electromagnetic radiation, have frequencies that typically range from about 10^6 Hz (for AM broadcasts) to 10^8 (for FM and TV broadcasts). The frequencies of visible light are very much higher. Green light, for instance, has a frequency of about 6×10^{14} Hz. For these exceedingly high frequencies, it is often more convenient to deal with the wavelength of the radiation rather than with its frequency. The two are related in the same manner as for sound waves (chapter 10).

$$\lambda_{vac} = \frac{c}{f}, \tag{11.1}$$

where c is the speed of propagation in a vacuum, approximately 299,800 km s^{-1}. For instance, green light ($f = 5.77 \times 10^{14}$ Hz) has a wavelength of about 5.2×10^{-7} m, traditionally expressed as 520 nanometers (nm). The speed of propagation is slower in air and water than in a vacuum, so, for a given frequency, the wavelength is shorter in these media.

Much of the terminology applied to electromagnetic radiation is based on human perception. The human eye is sensitive to electromagnetic radiation with λ_{vac} varying between about 400 nm (perceived as violet) and 740 nm (perceived as red). The other colors of the visible spectrum lie between these limits (table 11.1). Radiation with wavelengths shorter than that of violet light is termed *ultraviolet* (UV), and that with wavelengths longer than that of red light, *infrared* (IR). The relative sensitivity of the human eye to these wavelengths is shown in figure 11.1. In bright light, visual information is obtained primarily from *cone cells* in the retina, and these are most sensitive to yellow-green light. In dim light, the eye relies on *rod cells* rather than cones, and these are most sensitive to blue-green light.

Table 11.1 The visible spectrum of light. The frequencies and energies cited refer to the representative wavelengths given in column 3.

Color	Approximate Wavelength Range (nm)	Representative Wavelength (nm)	Frequency (Hz)	Energy J
Ultraviolet	below 400	254	11.80×10^{14}	7.82×10^{-22}
Violet	400 to 425	410	7.31×10^{14}	4.85×10^{-22}
Blue	425 to 490	460	6.52×10^{14}	4.32×10^{-22}
Green	490 to 560	520	5.77×10^{14}	3.82×10^{-22}
Yellow	560 to 585	570	5.26×10^{14}	3.49×10^{-22}
Orange	585 to 640	620	4.84×10^{14}	3.20×10^{-22}
Red	640 to 740	680	4.41×10^{14}	2.92×10^{-22}
Infrared	above 740	1400	2.14×10^{14}	1.41×10^{-22}

Note: Wavelengths are those measured in a vacuum.

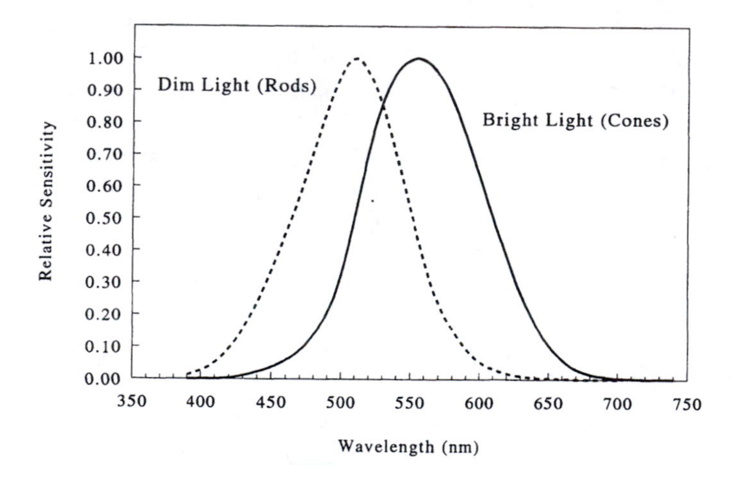

Fig. 11.1 In bright light human eyes rely primarily on cones (which are most sensitive to yellow-green light), whereas in dim light they use rods (which are most sensitive to blue-green light).

Note that the terms "visible," "infrared," and "ultraviolet" apply only to human perception. Other animals can perceive radiation with wavelengths both shorter and longer than those perceived by human beings. For example, insects and birds can readily see ultraviolet light (Burkhardt 1989; Burkhardt and Maier 1989) and some snakes can detect infrared light (Bullock and Cowles 1952; Gamow and Harris 1973). For these organisms ultraviolet and infrared are just as visible as violet and red are to us. Although plants do not "see" in the same sense that animals do, they do respond to a broad spectrum of electromagnetic radiation in their pattern of growth and their ability to photosythesize. In this chapter we use the term "light" in a generic sense to refer to any electromagnetic radiation that can be perceived and utilized by an organism.

Virtually all of the light important to plants and animals is produced by the sun. As a "black body"[1] radiator with an apparent surface temperature of about 5800 K, the sun produces radiation with a broad spectrum of wavelengths (fig. 11.2).

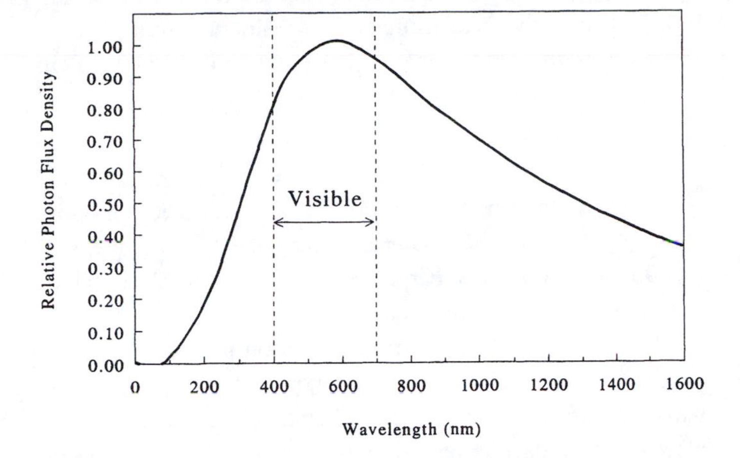

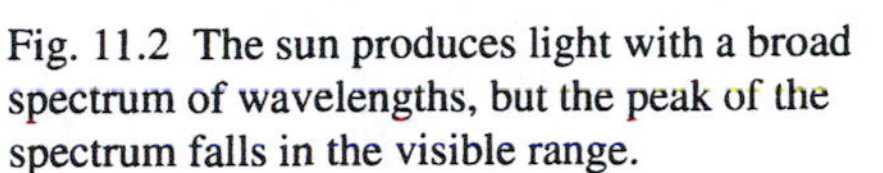
Fig. 11.2 The sun produces light with a broad spectrum of wavelengths, but the peak of the spectrum falls in the visible range.

By analogy to sound (chapter 10), we express the *intensity* of light as the rate at which radiant energy is transported to a given area on a flat surface held perpendicular to the direction of propagation. The average intensity of solar radiation at the top of the earth's atmosphere, the so-called *solar constant*, is about 1360 W m^{-2}. The instantaneous intensity varies by $\pm$ 3.5% through the course of a typical year due to variations in the sun's surface temperature and to the varying distance of the earth from the sun.

11.1.2 *Scattering and Absorption*

Light interacts with matter in two fundamental ways—it can be either absorbed or scattered. These two processes are related, and it will be worthwhile to examine them in some detail.

As we have seen, an electromagnetic wave is formed, in part, of an oscillating electrical field. As a result, it can exert a force on the charged particles in atoms—the electrons and protons (see chapter 9). Thus, when a light wave impinges on an atom or a molecule, it causes electrons and protons to vibrate. Now, the resting mass of an electron is 1/1855 that of a proton, and, being less bulky, electrons are much more easily moved by an applied force. Consequently, it is primarily through the movement of electrons that an electric field affects an atom.

[1]A "black body" is an ideal object that absorbs all electromagnetic radiation incident upon it. When such a body is heated, it emits radiation with a distribution of wavelengths that depends on its temperature. This distribution is accurately described by *Planck's spectral distribution equation* (Resnick and Halliday 1966),

$$\frac{d\mathcal{J}_{P,\lambda}}{d\lambda} = \frac{k_1 \lambda^{-5}}{e^{k_2/\lambda T} - 1}, \tag{11.2}$$

where $\mathcal{J}_{P,\lambda}$ is the radiant energy flux density associated with a given wavelength (W m^{-2}). The constants k_1 and k_2 have the values,

$$k_1 = 2\pi c^2 h \tag{11.3}$$

$$k_2 = \frac{ch}{k}, \tag{11.4}$$

where c is the speed of light, h is Planck's constant (6.626×10^{-34} J s), and k is Boltzmann's constant (1.38×10^{-23} J K^{-1} per molecule). The derivation of this equation was one of the important initial steps in the formulation of quantum mechanics. For a complete explanation, see any standard physics text.

The oscillation of electrons in response to an incident light wave in turn leads to the formation of secondary electrical and magnetic fields, and thereby to the propagation of another electromagnetic wave. In other words, the energy of incident light is *scattered* by its interaction with matter. The wave length of the scattered light is usually the same as that of the incident light.[2]

The scattered wave propagates in all directions, but its intensity varies such that

$$\mathcal{I} \propto \sin^2 \theta, \tag{11.5}$$

where θ is the angle between the axis along which electrons vibrate and the direction of propagation for the scattered wave (fig. 11.3A). Intensity is maximal when $\theta = 90°$ (i.e., in a plane perpendicular to the axis along which electrons vibrate) and is zero along the vibrational axis.

Now, the axis of an electron's vibration is parallel to the electric field of the incident wave, and this field is perpendicular to the direction in which the incident wave propagates. Thus, if an incident light wave travels horizontally with its electrical field oscillating in a vertical plane, the light scattered from a molecule has a maximal intensity in a horizontal plane.

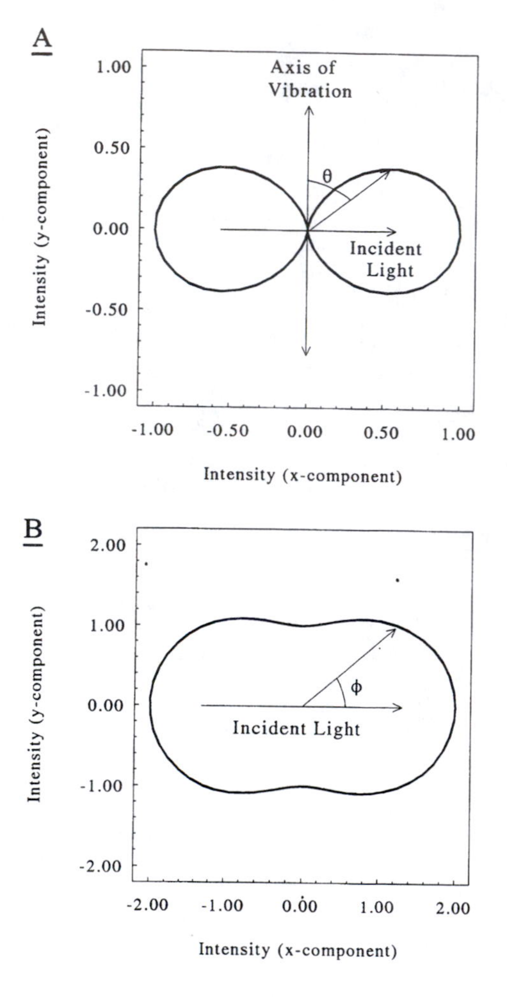

Fig. 11.3 A single wave of light scattered from an atom is most intense in the direction of the incident wave and has no intensity at all along the direction in which electrons vibrate, perpendicular to the incident wave (A). When many waves are scattered, the average intensity is still greatest in the direction of the incident beam, but does not go to zero perpendicular to the incident beam (B).

Note that the pattern of figure 11.3A applies only to the light scattered from a single incident wave. A typical beam of light contains many waves, and their electrical fields are randomly oriented. When one averages the light scattered from these many waves, a different pattern is obtained (fig. 11.3B). In this case, the intensity of scattered light is symmetrical about the direction of the incident light's propagation rather than about the axis of electron vibration, and it never goes to zero:

$$\mathcal{I} \propto 1 + \cos^2 \phi, \tag{11.6}$$

where ϕ is the angle between the incident light and the direction of observation (fig. 11.3B).

The process by which atoms scatter light is in many ways analogous to the way in which sound is scattered by small particles (chapter 10). True to this analogy, the ability of an atom to scatter light depends on the wavelength of the radiation; the shorter the wavelength, the more effective the scattering. In fact, the relative intensity of scattered light is proportional to λ^{-4} just as we saw for sound (Tanford 1961):

$$\mathcal{I}_s(\phi) = \mathcal{I}_0 \frac{2\pi^2(1 + \cos^2 \phi)(dn/d\rho)^2 \mathcal{M}\rho}{N\lambda^4 \ell^2}, \tag{11.7}$$

where $\mathcal{I}_0$ is the intensity of the incident light and $\mathcal{I}_s(\phi)$ is the intensity of the scattered light viewed from an angle ϕ relative to the direction of incident light. $\mathcal{M}$ is the molecular weight of the scattering molecule, n is the refractive index (a concept discussed later in this chapter), ρ is the density of the fluid, N is Avagadro's number, and ℓ is the distance from the scattering molecule at which the observation is made. The relative intensity of scattered light is shown in figure 11.4 as a function of λ_{vac}. Because short wavelengths are scattered more effectively, blue light is scattered more strongly than is red light.

[2]There are a few materials that emit a fraction of their scattered light at twice or three times the frequency of the incident light. But these materials are rare, and I know of no biological significance to this strange optical property.

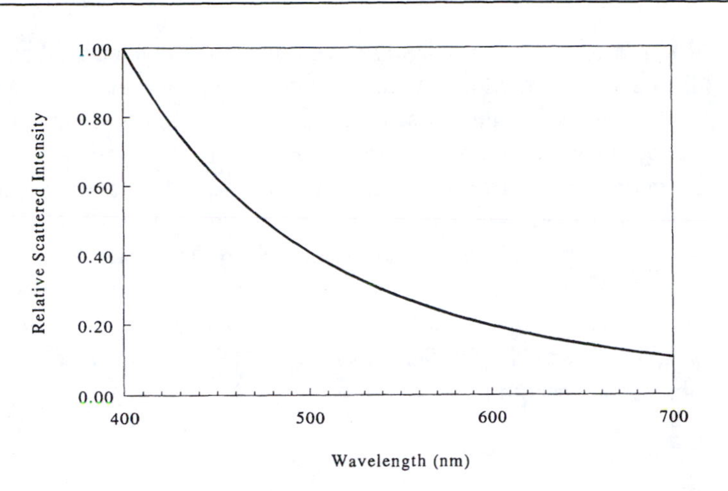

Fig. 11.4 The intensity of red light scattered from an atom or small molecule is only about 10% of the intensity of blue light scattered from the same particle.

The scattering described by eq. 11.7 applies only to particles that are small relative to the wavelength of light—atoms and small molecules. Larger particles (such as dust, fog, and bacteria) also scatter light, but in a much more complicated fashion. Scattering by these relatively large particles is called *Mie scattering* after Gustav Mie, who first derived its mathematical description in 1908. Mie scattering can be important both in air and water, but because it depends largely on factors that are not inherent to the fluid, we will not pursue the subject here. You are urged to consult texts on the subject such as Jerlov (1976).

Under special circumstances, the energy of a light wave incident upon an atom is entirely absorbed instead of being scattered. This occurs when the energy of the incident light exactly matches one of several critical energies for electrons in the atom or molecule. In this case, rather than vibrating within a given energy level, an electron jumps to a higher energy level, a process that does not involve the emission of a new electromagnetic wave. As a result, all the energy of the incident wave is retained (at least temporarily) by the absorbing atom or molecule.

What determines whether a given light wave matches a critical level for a given material? Critical energy levels are set by the nature of individual atoms and the molecule of which they are a part. The prediction and description of these levels is one of the major triumphs of quantum mechanics, a subject that we will not explore here. For our purposes it is sufficient to note that for any atom or molecule, the critical energy levels can, in theory, be predicted. Given these levels, we then need to ask what it is about a particular electromagnetic wave that determines whether its energy matches one of these levels. This in turn determines whether the wave will be absorbed or scattered.

In answering this question, we are forced to deal with one of the fundamental problems of physics. To this point we have described light as consisting of electromagnetic waves. In this description waves are continuous—as long as the electrical field of the light's source continues to oscillate, waves are produced and propagated. Under certain circumstances, however, light behaves as if it is formed of a series of particles called *photons*, and each photon acts as if it were formed of light with a particular wavelength. The energy, W_λ, associated with the photon depends on this wavelength:

$$W_\lambda = hf = hc/\lambda_{vac}, \tag{11.8}$$

where λ_{vac} is the wavelength in a vacuum and h is *Planck's constant*, 6.626×10^{-34}

J s. Thus, a photon of blue light, with its short wavelength, has a higher energy than a photon of red light with its relatively long wavelength (table 11.1).

It is an interesting consequence of this relationship that for a given critical energy in an atom, only a certain wavelength of light serves to boost electrons from one energy level to another. As a result, only those particular wavelengths that correspond to critical energies are absorbed, and all other wavelengths are scattered. These absorbed wavelengths form dark lines in the spectrum of light passing through a material, and are an important means by which the composition of materials can be determined. For instance, the composition of interstellar gas, which would be difficult to sample directly, is nonetheless known from its absorption spectrum.

11.1.3 *Attenuation*

The effects of absorption and scattering combine to determine the rate at which light energy is attenuated as a beam travels through a particular medium. In general, a beam traveling through a distance x loses a particular fraction of its energy, leading to the conclusion that light intensity decreases exponentially with distance traveled. This relationship can be expressed as *Lambert's law*:

$$\mathcal{I}(x) = \mathcal{I}_0 e^{-\alpha_\lambda x}, \tag{11.9}$$

where $\mathcal{I}_0$ is the original intensity of the incident light, $\mathcal{I}(x)$ is its intensity after traveling a distance x through the medium, and α_λ is the *attenuation coefficient*. Note that α_λ can vary with wavelength.

11.1.4 *Why Is Light So Important?*

Before leaving the subject of light energy, let us digress briefly to ask a basic question: Why is light so important to life? One possible answer concerns the amount of energy a material gains if it absorbs a photon.

Consider, for instance, a water vapor molecule in the atmosphere. As we have seen (chapter 3), the average kinetic energy of this molecule is $(3/2)kT$ or, at 20°C, about 6×10^{-21} J.

Now, water strongly absorbs red light with a wavelength of λ_{vac} = 680 nm. One photon of such light has an energy of hc/λ_{vac} or about 3×10^{-19} J. When this photon is absorbed, all its energy is transferred to the water molecule. In other words, by absorbing a photon of light, the energy of a water molecule is increased, on average, by a factor of fifty.

This is an extraordinary increase in energy, one that the molecule is unlikely to attain by any other means. For example, we can calculate from Boltzmann's equation the probability that a molecule will, just by chance, have an energy, W, equal to fifty times the average kinetic energy at a given temperature:

$$\text{Probability} = e^{-\frac{W}{kT}} = e^{-\frac{50\times 3kT/2}{kT}} = e^{-75} = 2.7 \times 10^{-33}. \tag{11.10}$$

In other words, only about three molecules in 10^{33} would be expected to attain such a high energy by chance. Now, 10^{33} molecules is 1.66×10^{9} moles, or for water, a volume of about 30,000 m^3. Thus, at 20°C, only one water molecule in 10,000 cubic meters of ocean has the thermal energy equal to that of a molecule that has just absorbed a photon of red light. This is about as close as physics can get to asserting that this energy is never obtained by chance alone.

Alternatively, one can compare the energy gained from light absorption to that gained by chemical means. Now the universal currency for chemical energy in organisms is adenosine triphosphate (ATP), and the energy gained from the hydrolysis of ATP to adenosine diphosphate (ADP) is about 7×10^{-20} J per molecule. Thus, the hydrolysis of more than four ATP molecules is required to yield the same energy gained from the absorption of just one photon of red light.

The very high energies made available by the absorption of light allow chemical reactions, such as photosynthesis, to take place that would otherwise be impossible. For this reason, if no other, the absorption of light is important to plants and animals.

11.1.5 *Refraction*

We noted above that the speed of light varies with the material through which the light propagates. It is fastest for light in a vacuum, and slower in all other materials. One consequence of this property is that a beam of light can change direction when passing from one material to another, a process known as *refraction*. To understand why this is so, we consider the situation shown in figure 11.5.

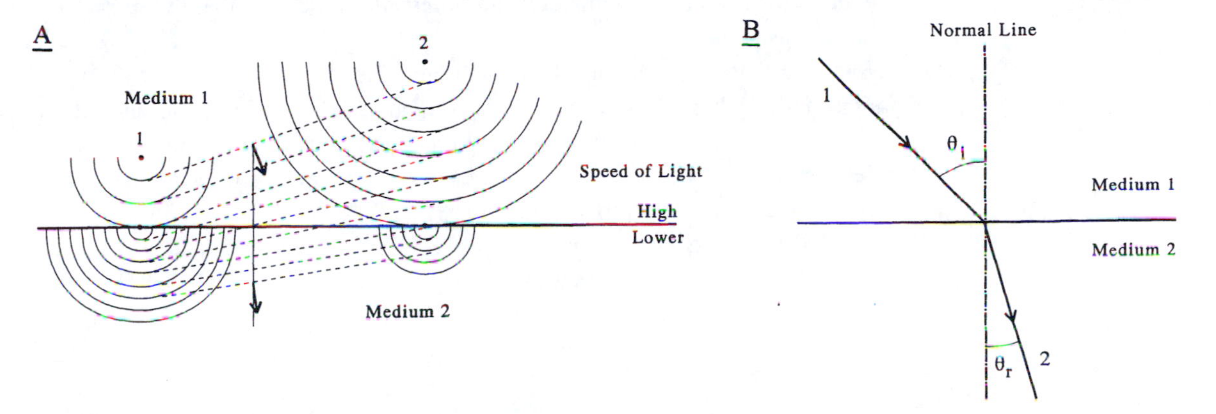

Fig. 11.5 When light moves into a medium with a higher refractive index (that is, a lower speed of light), it is refracted (A). The change in direction can be described in terms of "light rays" and their angle relative to the normal line (B).

Light is emitted simultaneously from two points located some distance above the interface between media. The speed of light in the upper medium (medium 1) is greater than that in the lower (medium 2). The light emitted by each source takes the form of an expanding hemispherical wave, here shown in two dimensions as a series of semicircles. Now, according to a principle first stated by Christian Huygens (1629–1695), if light behaves as a wave, its overall propagation can be traced through time by following a line that connects the wavefronts from the two sources (the dashed lines in fig. 11.5A).

As waves propagate through the first medium, they all move at the same speed, and the line connecting wavefronts maintains a constant angle relative to the interface. In other words, light moves in a constant direction. As light strikes the interface, it is scattered by molecules in the second medium, producing new waves, and as before, we can follow the path of the scattered light by following the line connecting wavefronts.

Because the line connecting our source points is oriented at an angle to the interface, the light from point 1 arrives at the interface before that from point 2. In the interval, the scattered light from the first source travels a shorter distance than that from the second because it moves through the "slower" medium. If we examine the situation just as the second wave front reaches the interface, we see that the line connecting wavefronts in the second medium lies at a different angle to the interface than did the analogous line in the first medium. Because light moves slower in the second medium, the direction of its propagation is more nearly perpendicular to the interface. In other words, light is refracted.

This refractive change in direction would be reversed if the light source were placed in medium 2. In this case, because light moves from a medium where its speed is slow to one where it is fast, the direction of propagation would be deflected to become less nearly perpendicular to the interface.

It will be useful to translate these results into more conventional terms. Instead of following the propagation of light by tracking wavefronts, we can instead draw a line perpendicular to the wavefront, a so-called *ray*, and use this to indicate the direction of propagation. In this translation, the situation of figure 11.5A becomes that of figure 11.5B. The ray in medium 1 strikes the interface at a point and moves off in a different direction in medium 2. The orientation of rays is described relative to a line that is perpendicular to the interface at the point where the ray strikes it. The angle between this *normal* line and the incident ray is the *angle of incidence*, θ_i, similar to that we defined for sound in chapter 10. The angle between the refracted ray and the normal line is the *angle of refraction*, θ_r. Note that the incident ray and the normal line define an *incident plane*, and that the refracted ray lies in this plane.

Note also that the direction of rays is reversible. In other words, if light in figure 11.5B originates in medium 2, traveling along ray 2 toward the interface, it follows ray 1 when it emerges into medium 1. As a result of this reversibility, the designation of which ray is considered incident and which refracted depends solely on the direction of propagation.

The change in direction as light changes speed is a strong argument that under such circumstances light acts as a wave. In the early 1600s, Willibrod Snell showed that there is a simple relationship between the speed of light in two media and the angles of incidence and refraction:

$$\frac{\sin\theta_i}{\sin\theta_r} = \frac{u_1}{u_2}, \tag{11.11}$$

where θ_i and θ_r are the angles of incidence and refraction, respectively, and u_1 and u_2 are the speeds of light in the first and second medium. This relationship is known as *Snell's law*.

The ratio u_1/u_2 is called the *refractive index* of the second medium with respect to the first, and is given the symbol $n_{2,1}$. Often, the refractive index is measured with respect to a vacuum, in which case the subscript is traditionally dropped. For example, light travels 40% faster in a vacuum than in a material with a refractive index, n, of 1.4.

Snell's law (eq. 11.11) can be restated in terms of the refractive index:

$$\frac{\sin\theta_i}{\sin\theta_r} = n_{2,1} = \frac{n_2}{n_1}. \tag{11.12}$$

The refractive indices of a variety of materials are shown in table 11.2. Note that these indices are valid only for the specified wavelength, in this case 589 nm. In general, the index of refraction of a material decreases with increasing wavelength. As a result, red light is refracted through a smaller angle than is blue light. This is the basis for the classic experiment whereby white light is split by a prism into its component colors.

As we have seen, when light moves from a medium with high refractive index to one with lower (e.g., from water to air), the refracted ray is diverted from the normal. At a critical angle of incidence, θ_c, the refracted ray lies perpendicular to the normal—that is, parallel to the interface—and it never emerges from the medium in which it is first propagated. At angles of incidence greater than θ_c, light is reflected from the interface back into the medium of higher refractive index, a situation known as *total internal reflection*. It is because of this kind of reflection that light can be efficiently carried down a fine glass fiber surrounded by air, and this is the basis for the "fiber optics" used by doctors and phone companies.

The critical angle of incidence can be found by substituting an angle of 90° for θ_r in eq. 11.12, with the result that

$$\theta_c = \arcsin n_{2,1}. \tag{11.13}$$

Note that there is no critical angle for light moving from a medium of low refractive index to one with higher refractive index.

Table 11.2 The refractive indices of various common materials measured at $\lambda_{vac} = 589$ nm.

Medium	n
Vacuum	1.0000
Air	1.0003
Water	1.33
Ethyl alcohol	1.36
Fused quartz	1.46
Glass, crown	1.52
Glass, dense flint	1.66

Note: This is the wavelength of a particular line (the D line) in the emission spectrum of sodium, and therefore is a handy wavelength for making measurements of a refractive index.

11.1.6 *Reflection*

When a ray of light strikes the interface between media of different refractive index, some of its energy is transmitted and refracted as we have seen above. However, some energy is not transmitted and is instead *reflected*. There are two aspects of reflection that require our attention.

We first note that when light is reflected, the angle of reflection (again measured relative to the normal line) is equal to the angle of incidence (fig. 11.6). This relationship tells us the direction of the reflected ray, but does not tell us how its intensity compares to that of the incident ray.

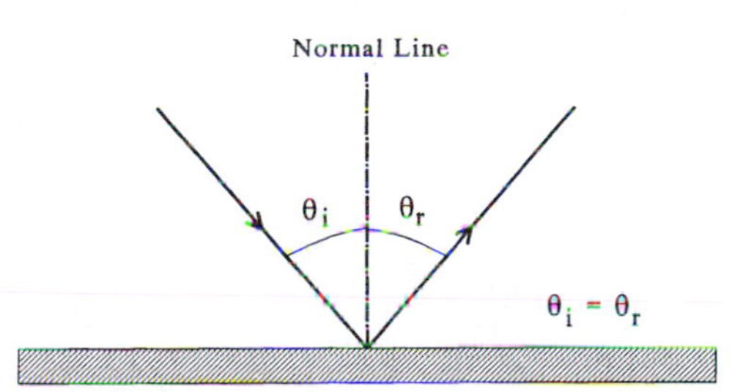

Fig. 11.6 When light is reflected from a surface, the angle of incidence (θ_i) is equal to the angle of reflection (θ_r).

When we explored the properties of sound in chapter 10, we found that sound energy incident on a planar interface between media could be both transmitted and reflected, and that the relative intensity of the transmitted and reflected sound depended on the difference in acoustic impedance between the media. An analogous situation applies to light. If a beam of light is propagated along a normal line toward the planar interface between two media, the ratio of the intensity of the reflected light to that of the incident light is

$$\frac{\mathcal{I}_r}{\mathcal{I}_0} = \left(\frac{n_2 - n_1}{n_2 + n_1}\right)^2, \tag{11.14}$$

where n_1 and n_2 are the refractive indices of the media that form the interface. Both indices are measured relative to a vacuum. Consider an example. At high noon light from the sun shines vertically onto the surface of a lake. In this case the interface is between air ($n = 1.0003$) and water ($n = 1.33$), and only about 2% of the light energy is reflected back into the air. The remaining 98% is transmitted into the lake.

Note that eq. 11.14 implies that the reflection of light from an interface is symmetrical. If 2% of incident light is reflected when light passes from air to water, 2% is also reflected when light passes from water to air.

Note also that eq. 11.14 applies only when the angle of incidence is zero. The fraction of light reflected from an interface increases as the angle of incidence increases (List 1958):

$$\frac{\mathcal{I}_r}{\mathcal{I}_0} = \frac{1}{2}\left(\frac{\sin^2[\theta_i - \arcsin(\theta_i/n)]}{\sin^2[\theta_i + \arcsin(\theta_i/n)]} + \frac{\tan^2[\theta_i - \arcsin(\theta_i/n)]}{\tan^2[\theta_i + \arcsin(\theta_i/n)]}\right), \quad (11.15)$$

where in this case it is assumed that air is one of the media and n is the refractive index of the second medium.

Up to an angle of 50° (0.87 radians) or so, the increase in reflection is small, but at larger angles most light is reflected regardless of the difference in refractive index between media (fig. 11.7).

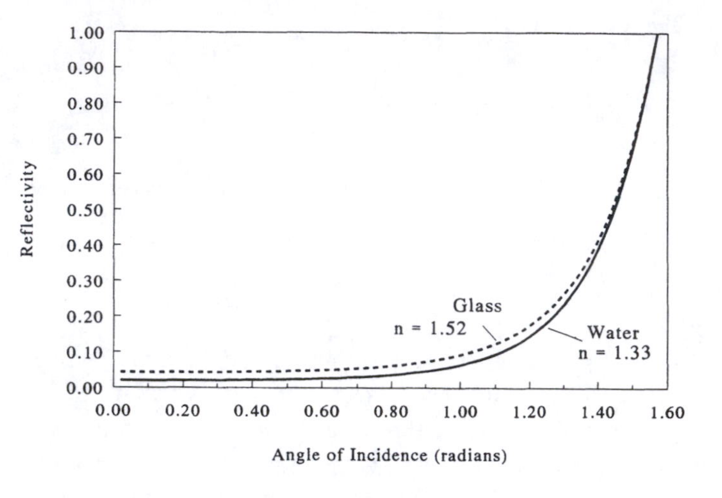

Fig. 11.7 The reflectivity of a surface (i.e., the fraction of incident light that is reflected) increases with an increase in the angle of incidence. This accounts for the glare one encounters when sunlight shines at a large angle of incidence on glass or water.

11.2 The Optical Properties of Air and Water

We now turn our attention to the optical properties of air and water. Two of these—refractive index and attenuation coefficient—have the most striking biological consequences.

11.2.1 *Refractive Index*

The refractive index of air is approximately 1.0003, only slightly different from that of a vacuum (table 11.2). As with all materials, the refractive index of air decreases with an increase in wavelength, but the effect is minimal (fig. 11.8). For instance, the refractive index for violet light (λ = 410 nm) is only 0.0006% higher than that of red light (λ = 680 nm).

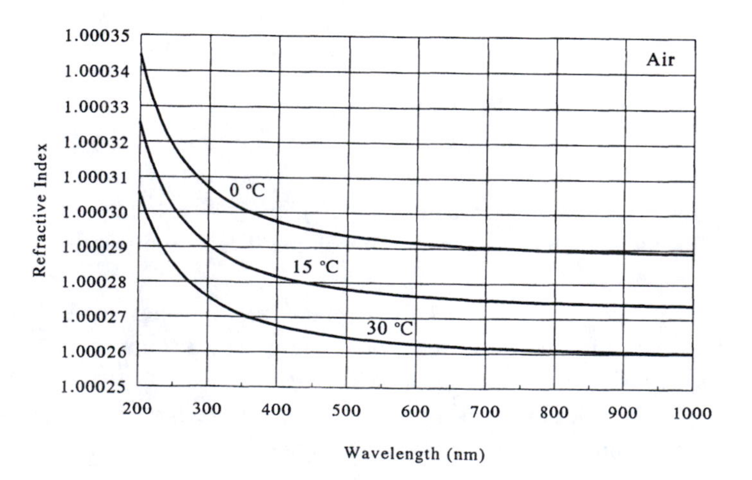

Fig. 11.8 The refractive index of air decreases with both increasing wavelength and temperature, but the variation is quite small.

Light in Air and Water

At constant pressure, the refractive index of air decreases slightly with an increase in temperature, but the effect is small over the biological range (fig. 11.9). The n for air at 40°C is only 0.004% lower than that at 0°C. The presence of water vapor in the air lowers its refractive index slightly (fig. 11.9), but the effect is generally negligible. The index of refraction increases slightly with an increase in pressure (fig. 11.10), but again, the effect is slight.

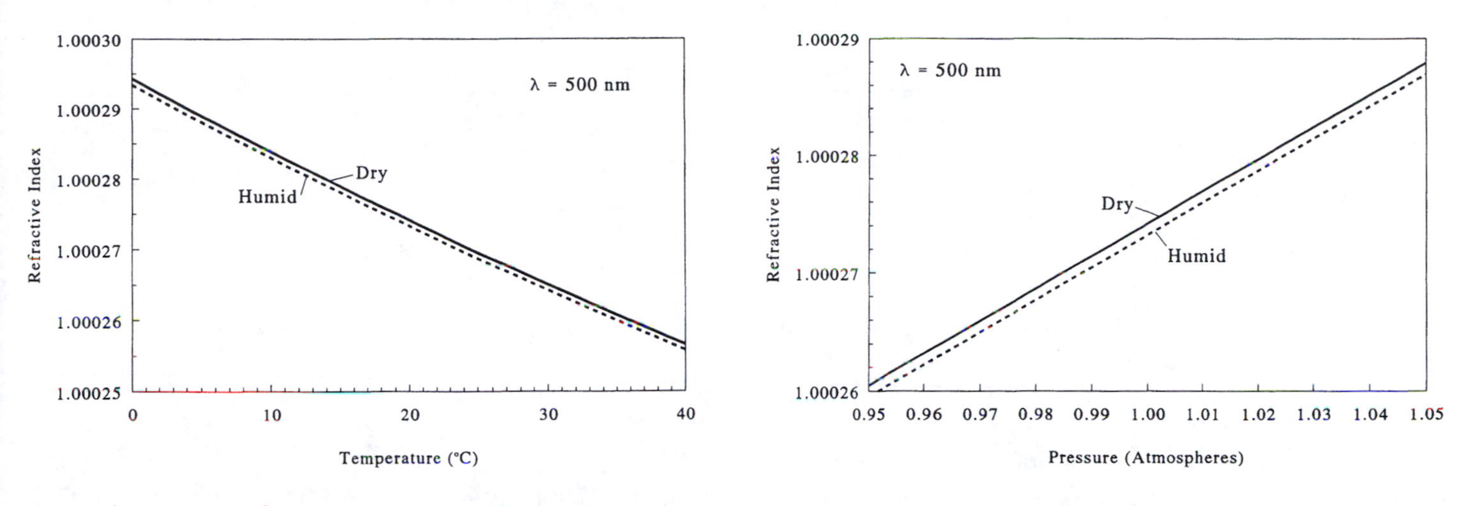

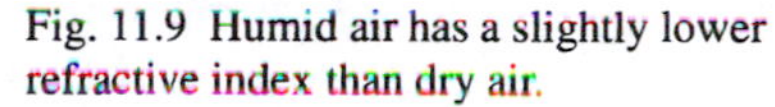

Fig. 11.9 Humid air has a slightly lower refractive index than dry air.

Fig. 11.10 The refractive index of air increases with increasing pressure, but the effect is small over the range of typical barometric pressure at sea level.

In summary, the refractive index of air is nearly the same as that of a vacuum, and, within the biologically relevant range of parameters, not does change substantially regardless of the factor considered.

In contrast, the refractive index of water is quite high, varying from about 1.34 for blue light to 1.33 for red light (fig. 11.11). In other words, light travels about one-third faster in air than it does in water. The refractive index of aqueous solutions varies both with temperature and with salinity: the higher the temperature, the lower is n; the higher the salinity, the higher is n (fig. 11.12). Neither effect is of great magnitude. For example, the variation in n with temperature is only about 0.15% across the biological range, and the variation with salinity is only about 0.6% for $S = 0$ to 40.

The biological consequence of these differences in the refractive index are discussed in section 11.4.4.

Fig. 11.11 The refractive index of water decreases with increasing wavelength.

Fig. 11.12 The refractive index of water increases with increasing salinity and decreases with increasing temperature.

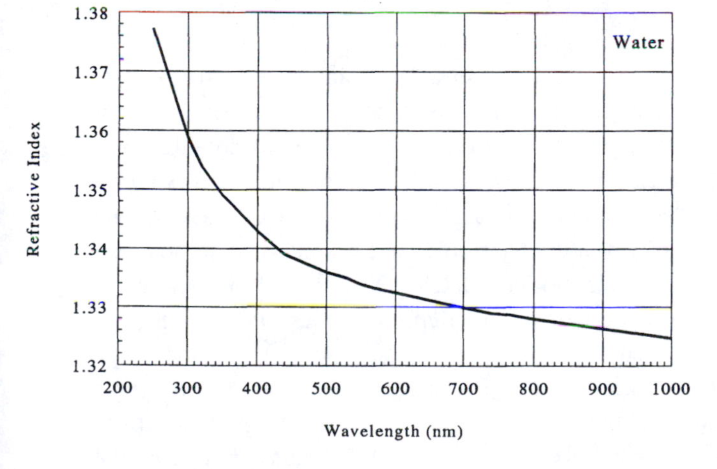

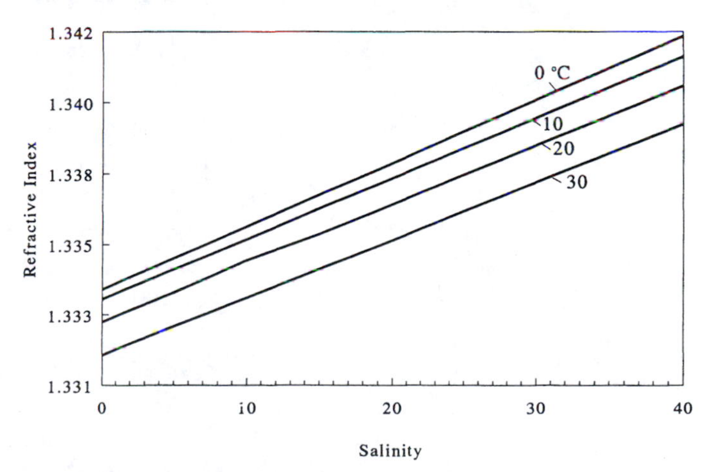

11.2.2 *Absorption and Scattering by Air and Water*

Air and water are strikingly different in their ability to attenuate light. The attenuation coefficients for dry air and pure water are shown in figure 11.13. Note that the attenuation coefficient is presented on a logarithmic scale. Depending on wavelength, the attenuation coefficient of water is 100 times to 10 million times larger than that of air. In other words, water is much less transparent to light than is air.

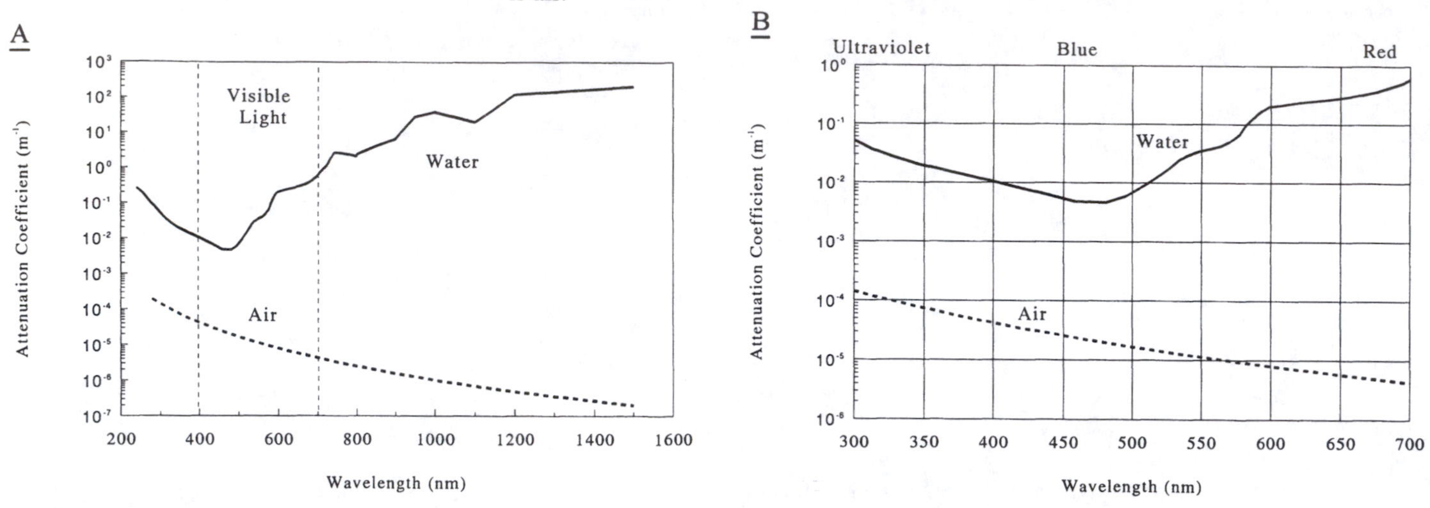

Fig. 11.13 Light is attenuated much more effectively by pure water than by air (A). The attenuation coefficient of water is lowest for blue light, and is higher for red light than for ultraviolet light. (The data cited here for air, and for water at wavelengths in excess of 800 nm, are taken from List 1958. The remaining water data are taken from Shifrin 1988.) Note that water in the natural environment is likely to have higher attenuation coefficients than those cited here. (B) An enlargement of (A), showing the visible range of light in greater detail.

It is surprising to find that the attenuation coefficient of pure water in the visible and ultraviolet is a matter of some debate (Shifrin 1988; Jerlov 1976). At short wavelengths the presence of even minute amounts of particulate matter in a sample can substantially increase attenuation, and, as a result, the measurements of different researchers often disagree. The values shown in figure 11.13 are those suggested by Shifrin (1988) on the basis that they are the lowest reported and were therefore likely to have been made on the cleanest samples. Water in the natural environment is likely to have higher attenuation coefficients than those shown here.

11.3 Consequences of Attenuation

11.3.1 *Air*

Before solar radiation can interact with plants and animals, it must first pass through the atmosphere. In the process, it is both scattered and absorbed, and, as a result, the light reaching an organism is different from that emitted by the sun.

As we have noted, different molecules absorb light at different wavelengths. Of particular importance in the atmosphere are ozone (O_3), carbon dioxide, and water vapor. Ozone absorbs light in the ultraviolet, and its presence in the upper atmosphere drastically reduces the amount of UV reaching the earth's surface. Both carbon dioxide and water vapor absorb light at various wavelengths, but primarily in the red and infrared.

As a result of atmospheric absorption by these molecules, light at the earth's surface has a different spectral quality than that arriving at the top of the atmosphere (fig. 11.14). For example, only about 28% of the photons emitted by the sun lie in the visible range, but 45% of those reaching the earth's surface are visible. Note that the spectral characteristics of light shown in figure 11.14 are not exactly those that would be predicted from the attenuation coefficients presented in figure 11.13. These coefficients were measured in dry air, and therefore lack the peaks in attenuation due to water vapor.

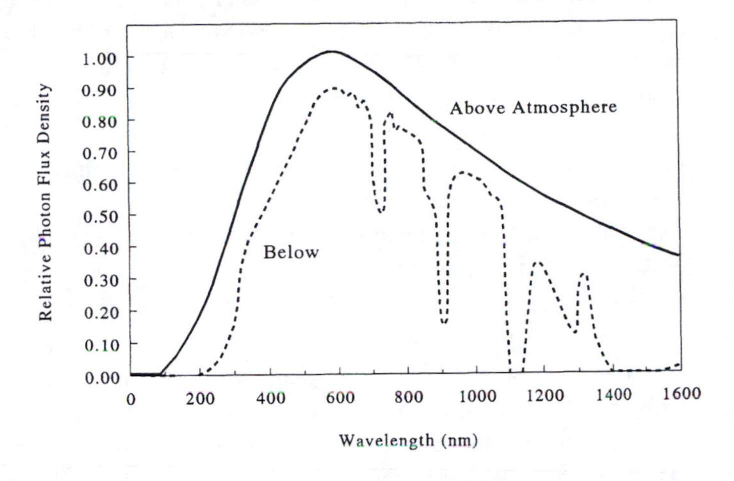

Fig. 11.14 Light reaching the earth's surface has a different spectral quality than that at the top of the atmosphere due to the absorption of various wavelengths by ozone, water, and carbon dioxide. (From *Biophysical Plant Physiology and Ecology* by Park. S. Nobel. Copyright ©1983 by W. H. Freeman and Company. Reprinted by permission)

Note also that air attenuates blue light more effectively than red light (fig. 11.13B). This fact can help to explain why sunsets are red. As the sun nears the horizon the optical path through the atmosphere becomes longer. At sunset, the path that sunlight must travel through air is sufficiently long to attenuate most of the blue light, leaving the remaining light with a reddish hue.

11.3.2 *Water*

Because water absorbs light much more strongly than does air, light can penetrate only a relatively small distance into an aqueous medium. This effect is shown graphically in figure 11.15. Light traverses 50 m of air with virtually no loss of intensity. But in traversing the same thickness of water, intensity is severely attenuated. Only 78% of the incident blue light is still present, for instance, and virtually none of the incident red light.

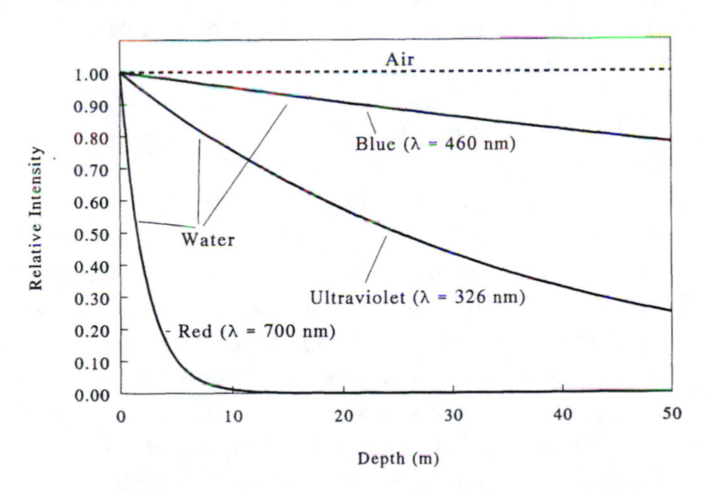

Fig. 11.15 Visible light passes through 50 m of air virtually unattenuated, whereas the intensity of light passing through 50 m of water is drastically reduced. This reduction is most noticeable for red light and least for blue light. Note that at a depth of 50 m about 25% of incident ultraviolet light is still present.

The attenuation of light by water has several biological consequences. First, the rapid attenuation of light means that aquatic plants can photosynthesize effectively only if they are near the surface. Even plants specifically adapted to low light levels cannot photosynthesize effectively at a depth much below about 200 m. Now, the oceans have an average depth of about 4000 m. Thus, the attenuation properties of light preclude about 95% of the earth's water from being used as habitat for photosynthesizing plants.

Second, rapid attenuation renders sight inefficient as a means of long-distance communication in an aqueous medium. This effect is well known to SCUBA divers, who often describe the water in which they dive in terms of the maximum distance at which they can reliably see an object. Even when the water is clear of plankton and sediment, the "visibility" in the ocean is seldom greater than 30 m. We have already noted (chapter 10) that under these circumstances sound, rather than light, is often the preferred mechanism by which animals gather information about their surroundings.

Note also that the attenuation coefficient of water varies with wavelength (fig. 11.13A). Attenuation is high in the UV and IR, and is minimal for light at visible wavelengths. Given that life initially evolved in an aqueous medium, it may not be a coincidence that "visible" light corresponds to those wavelengths for which water is most transparent. The same argument can be applied to the pigments used by plants to capture light for photosynthesis. All of the major photsynthetic pigments (chlorophylls, carotenoids, and phycobilins) absorb light in the range of 400 to 700 nm, the range at which water is least attenuating (Nobel 1983).

Even within the visible range, the attenuation coefficient of water varies substantially (fig. 11.13B); red light is attenuated much more strongly than blue light. Again this effect is well known to SCUBA divers, who note that the apparent color of objects changes rapidly with depth. For instance, a camera case that is bright red at the surface appears gray at a depth of only a few meters because all of the available red light has been absorbed by water above (see fig. 11.15). The rapid absorption of red light has had evolutionary consequences for plants. Because chlorophyll *a* (the most common photosynthetic pigment) absorbs strongly at a wavelength of 680 nm (red light), it is a relatively ineffective means for gathering light at depth. However, plants which live deep beneath the water's surface have accessory pigments (carotenoids and phycobilins) that absorb at shorter wavelengths.

The complete story of how plants adjust their light-gathering apparatus to the spectral quality of the incident light is fascinating, but lies well outside the purview of this book. You are invited to consult Nobel (1983) or other texts on plant physiology for a more complete discussion.

11.3.3 *Why Is the Sky So Blue?*

The attenuation effects we have discussed so far have been due to the combined action of absorption and scattering. The separate effects of scattering are worthy of mention, however, because they help to explain the color of both the sky and the sea. If sunlight were not scattered by air, we would see light only when looking directly at the sun. This is the case on the moon, for instance, which does not have an appreciable atmosphere. Air *does* scatter light, however, with the result that when we look at a point in the sky away from the sun we see light that has been scattered. Because short wavelengths are scattered much more effectively than long wavelengths (eq. 11.7), the color of this scattered light is blue.

The same conclusion applies to water. If one were to dive into an ocean of pure water and look in a direction away from the sun, the water would appear to have a bluish tinge because this is the wavelength that is most effectively scattered. However, for reasons we will not delve into here, liquids are much less effective at scattering light than are gases (Tanford 1961) so pure water does not appear nearly as blue as the sky. Note that the blue color of the sea when viewed from the air is largely a result of the reflected image of the blue sky. On an overcast day, for instance, the sea surface often appears gray. Similarly, the blue one sees at depth in the ocean is primarily due to the lack of red light and the presence of particles that reflect the remaining blue light back to one's eye.

11.4 Vision

We now consider the mechanisms by which animals see, and how these differ between air and water.

Vision consists of a series of events. First, light from the environment is selectively sampled to form an image. The means of sampling may be a simple pinhole placed in the path of the light or a complex system of lenses and mirrors. Once an image is formed, it must be detected. To this end, cells containing structures sensitive to light are placed such that the image falls upon them. There are many different types of sensory cells in animals, and only a few will be discussed here. Once light is detected by a sensory cell, the event is transduced to an electrical signal in the nervous system. The signals from various sensory cells are then analyzed by the brain to complete the process of vision. In this section we concern ourselves only with the initial steps in this process: the formation of the image and how the spatial properties of the image interact with those of the sensory cells. These are the stages of vision that are most profoundly affected by the optical properties of air and water.

11.4.1 *Pinhole Optics*

We begin with a brief analysis of how an image is formed. The trick is to insure that light coming from a given position in the environment can reach only one spot on the surface formed by the sensory cells. Consider the situation shown in figure 11.16A. A sensory surface, which we will generically refer to as a *retina*, faces two point sources of light, each emitting the same amount of radiant energy. The light from source 1 radiates in all directions, and unless somehow altered, illuminates the entire retina. The same is true for source 2. If one of the sources is extinguished, the total light energy falling on the sensory surface will be halved, but the retinal cells have no way of telling whether it was source 1 or source 2 that no longer shines. In other words, without some means for sorting out the direction from which light arrives, an organism can sense only the overall brightness of its environment.

The simplest means by which light can be manipulated to form a coherent image is to place a small hole some distance in front of the retina (fig. 11.16B). If the hole is small enough, light from source 1 can fall on only a limited portion of the retina, while light from source 2 falls on a different spot. An image is thus formed. Within certain limits,[3] the smaller the hole, the more precisely the direction is defined from which light arrives, and the clearer is the image.

Pinhole eyes are used by a wide variety of invertebrates. In general, however, the diameter of the pinhole approaches that of the eye (e.g., the pigmeted cup eye of a planarian), and directional resolution is low. These eyes can sense the difference between light and dark and whether light is coming from the left or right, but little

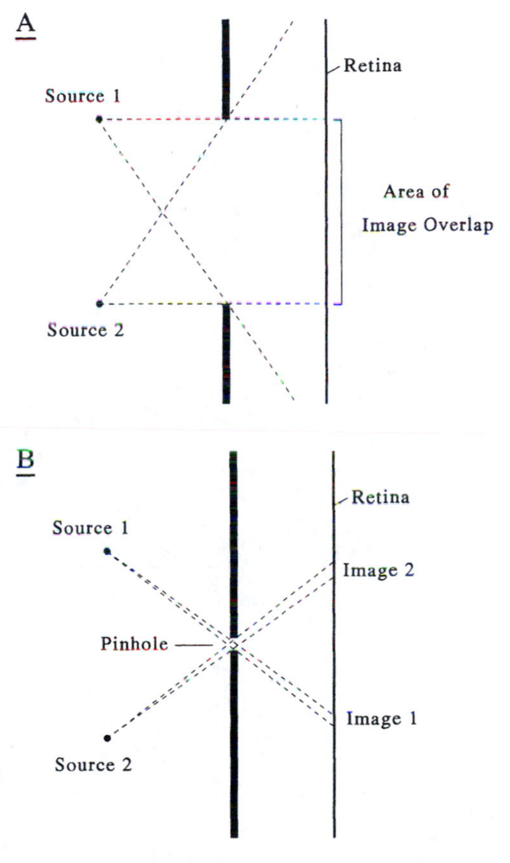

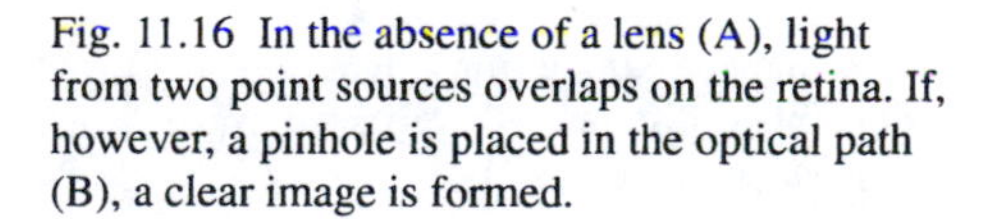

Fig. 11.16 In the absence of a lens (A), light from two point sources overlaps on the retina. If, however, a pinhole is placed in the optical path (B), a clear image is formed.

[3]If the hole is too small, light entering one side can interfere with light entering the other side. In such a case, the image of the hole can be a complex diffraction pattern of alternating light and dark rings. These problems occur, however, only for holes with diameters that approach the wavelength of light, far smaller than the pinholes used by animal eyes.

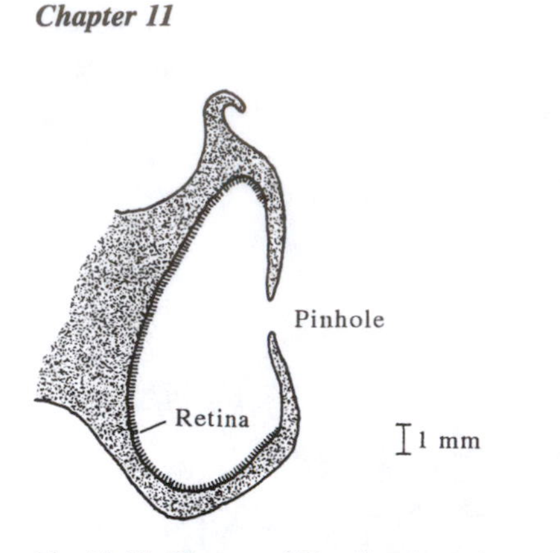

Fig. 11.17 The eye of *Nautilus*, here shown in cross section, uses a pinhole lens to project a clear image on the retina. (Redrawn from Young 1964 by permission of Oxford University Press)

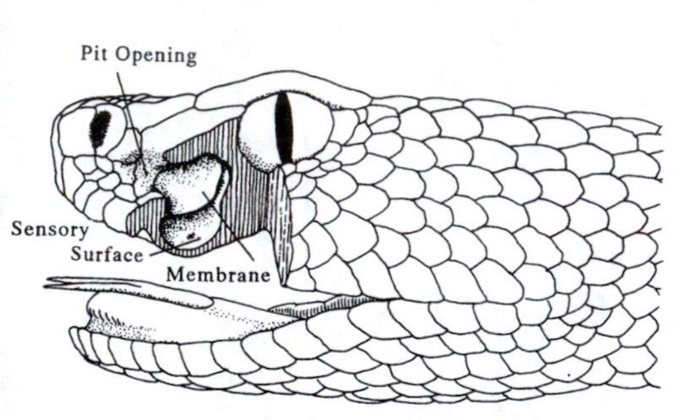

Fig. 11.18 The pit organs of some snakes use a pinhole lens (albeit one of large relative diameter) to form an infrared image of their surroundings. (Redrawn from Gamow and Harris 1973. Original figure copyright ©1973 by Scientific American, Inc. All rights reserved)

more. Apparently the only pinhole eye that forms a clear image is that used by *Nautilus* (fig. 11.17).

Before leaving the subject of pinhole eyes, we will examine one last, strange example. Two families of snakes, the boas and the pit vipers, can detect infrared light with a wavelength of 10 μm or longer. These are the wavelengths emitted by warm objects such as mice and birds, and the ability to sense infrared light allows these snakes to hunt their prey in the complete absence of visible light.

The infrared sensors of pit vipers appear to act like pinhole eyes. The sensory cells are contained within two "pits," one on either side of the head (fig. 11.18). The side walls of each pit act like a pinhole, limiting the angle from which infrared light may enter the pit. The fields of view for the two pits cross in front of the head. By adjusting the direction of the head until both pits "see" the same brightness, the snake can point itself toward the prey (Gamow and Harris 1973).

The sensory cells in snakes directly detect the increase in temperature as they absorb infrared radiation. This capability is possible in air (which is both reasonably transparent to IR radiation and has a low thermal conductivity) but would be impossible in water. Not only does water rapidly absorb IR, thereby limiting the distance over which detection would be feasible, but the high thermal conductivity of water would not allow surface sensory cells to be heated (see the discussion in chapter 8).

11.4.2 *The Lens*

Pinhole eyes have an intrinsic problem. In the process of forming a clear image, they eliminate most of the available light. As a result, the image is very dim. This problem can be circumvented, however, through the use of a lens.

Consider the situation shown in figure 11.19. A spherical interface (here shown in cross section) separates media of different refractive indices. The height of this "lens" is arbitrarily set equal to its radius. A point source of light lies in the left hand medium (medium 1), which has a low refractive index, n_1. The retina lies to the right of the interface in medium 2, which has a relatively high refractive index, n_2. The center of curvature of the interface lies at point C, and a line connecting the center of the retina and the center of curvature defines the optical axis of the system. Let us follow what happens to rays of light from the source as they impinge on the lens.

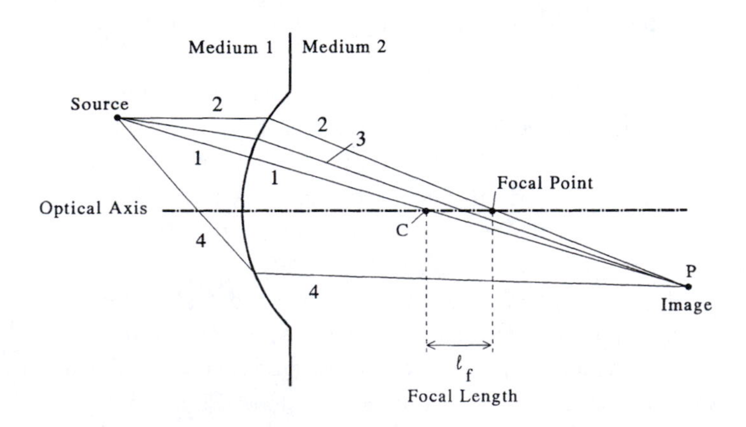

Fig. 11.19 Light striking a simple lens at different points is differentially refracted to form an image at point P.

Light is emitted in all directions from the source, but we may choose a few specific rays to get an idea as to what happens in the "eye." Light which propagates along a line joining the source and the center of curvature of the lens (ray 1) strikes the interface at a normal angle, and is therefore not refracted. In contrast, light propagating along a line parallel to the optical axis (ray 2) strikes the lens at an angle and *is* refracted. Because medium 2 has a higher refractive index than medium 1, ray 2 is bent to lie more closely parallel to its normal line. In other words, for the situation shown here, ray 2 is refracted downward. A consideration of the geometry of the situation shows that the refracted ray 2 crosses the optical axis at the *focal point* of the lens. Because ray 2 has been bent downward, it crosses ray 1 at a point, P.

If we consider a third ray originating at the source, lying between rays 1 and 2, we find that it strikes the lens at a smaller angle of incidence than does ray 2, and is therefore refracted less. In fact, the amount of refraction is just sufficient to send ray 3 exactly toward point P. A fourth ray lying below ray 1 strikes the lens at a very high angle of incidence and is strongly refracted, in this case upward. Once again the orientation of the interface is just sufficient to direct ray 4 through point P.

In summary, then, the action of a spherical interface of this sort is to direct all the light from a point source to a single spot, P, on the sensory surface. This spot is the *image* of the source. A second source would form a similar image, but at a different spot on the retina. In this fashion, spherical interfaces (and other more complex lenses) form images by redirecting light rather than by throwing it away. As a result, a much brighter image is formed than that which can be attained using a pinhole. Virtually all image-forming eyes use lenses of some sort, and it is to the properties of this type of image formation that we confine our attention. The discussion that follows is drawn largely from Land (1981), and you are invited to consult this source for additional details.

11.4.3 *The Eye*

A generalized cameralike eye, consisting of an aperture, a lens, and a retina, is shown in figure 11.20A. This is the kind of eye found in worms, gastropods, cephalopods, spiders, fish, amphibians, reptiles, birds, and mammals. The second major eye type, an *ommatidium*, is shown in figure 11.20B. An ommatidium forms one facet of a *compound eye*, the type of eye found primarily in the arthropods.

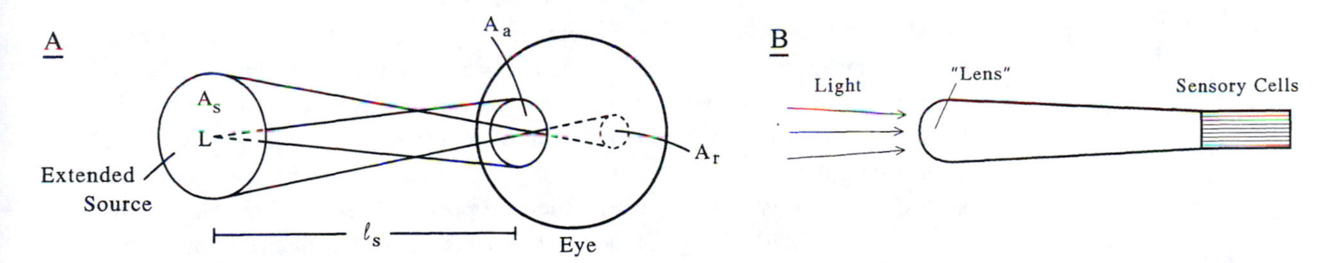

Fig. 11.20 An extended source of light forms an extended image on the retina of a cameralike eye (A). In contrast, an ommatidium (B) projects on its sensory cells light from only a very restricted solid angle. In either case, the brightness of the image depends on the f-number of the eye, as explained in the text. (redrawn from Land 1981 by permission of Springer-Verlag, New York, Inc.)

In terms of their optical system, the only major difference between the cameralike eye and the ommatidium[4] is the spatial extent of their retinas. In the cameralike eye, the retina extends over a large area so that many sensory cells fall within the image cast by the lens. In the ommatidium, only a few sensory cells are present,

[4]To be specific, the ommatidium shown here is part of an *apposition compound eye*, in which each eye facet focuses an image on its own set of receptors. Some insects have *superposition compound eyes*, in which many facets act together to project a single image on a broad expanse of sensory cells. Although different in structure from a cameralike eye, these superposition compound eyes are optically similar to them, and for this reason they are not treated separately here.

and these sample only a small portion of the image. Although this difference is important for how visual information is presented to the nervous system, it is immaterial to our exploration of the effects of air and water. Here we focus our attention on just the cameralike eye, and simply note that the same arguments apply to compound eyes as well.

Consider the cameralike eye of figure 11.20A. In this case, the object viewed by the eye is an extended rather than a point source. We may think of the object, however, as being formed from an aggregation of point sources. The overall brightness of the object, termed its *radiance*, depends therefore on two factors: (1) the rate at which radiant energy is emitted by each of its points, and (2) the number of points that make up the object, that is, its area.

First consider the rate at which energy is transmitted from individual points. Each point produces energy at a rate of P watts and this energy propagates in all directions. As a result, at distance r from the point, the radiated energy is spread over a spherical surface with an area of $4\pi r^2$. Thus, the farther light is from its point of emission, the more area over which its energy is spread.

To quantify the amount of light energy falling on a retina at distance r from a point source, we need to know the fraction of the overall surface at radius r that is covered by the receiving area, A. This fraction is known as the *solid angle* (measured in steradians) and is equal to A/r^2. Because the radiant energy falling on any receiving surface depends on its solid angle, we measure the overall radiance of an extended source as watts per square meter of the source (a measure of the total rate at which energy is emitted by all points in the source) per steradian. Thus, to calculate the rate at which radiant energy falls on a given area of retina, one multiplies the radiance of the source first by the source area and then by the solid angle of the retina with respect to the source.

The extended source in figure 11.20A lies at a distance ℓ_s from the aperture of the eye. If this circular aperture has area A_a, its solid angle relative to the source is A_a/ℓ_s^2. If the source has area A_s and radiance L, the rate at which radiant energy falls on the aperture is

$$P_a = \frac{LA_sA_a}{\ell_s^2}. \tag{11.16}$$

All of the light that falls on the aperture is focused on the retina. If the light source is far from the eye (which in nature is usually the case), the image of the source is in focus at a distance behind the lens very nearly equal to the lens' focal length, ℓ_f. Note that the focal length is measured from what would be the center of curvature for the simple lens we dealt with in figure 11.19. For more complex lenses, this point is more correctly called the *posterior nodal point* (see Jenkins and White 1957). The image of the source covers an area A_r on the retina, as can be determined by drawing lines from the edges of the source through the nodal point. As a result of these considerations, we can calculate that the brightness of the image, i.e., the *radiant flux density* (in W m^{-2}) falling on the retina, is

$$\mathcal{J}_{P,r} = \frac{P_a}{A_r} = \frac{LA_sA_a}{\ell_s^2 A_r}. \tag{11.17}$$

We next note from the geometry of figure 11.20 that as long as ℓ_s is large compared to the dimensions of the eye, the solid angle subtended at the eye by the source is nearly equal to the solid angle subtended by the retinal image of the source at the nodal point. In other words,

$$\frac{A_s}{\ell_s^2} \approx \frac{A_r}{\ell_f^2}. \tag{11.18}$$

Substituting this result into eq. 11.16 we see that the rate at which light energy falls on the aperture is

$$P_a = \frac{LA_aA_r}{\ell_f^2}, \tag{11.19}$$

and the radiant flux density on the retina is

$$\mathcal{J}_{P,r} = \frac{LA_a}{\ell_f^2}. \tag{11.20}$$

If the aperture of the eye (the pupil) is circular with a diameter d, its area is $\pi d^2/4$. Thus,

$$\mathcal{J}_{P,r} = \frac{\pi}{4}L\left(\frac{d}{\ell_f}\right)^2. \tag{11.21}$$

In other words, the brightness of the image at the retina depends on the radiance of the source and the square of the ratio of d to ℓ_f. This is an important fact, one worth dwelling upon for a moment: *the brightness of the image delivered to the retina depends not on the size of the eye's pupil, but on the ratio of the aperture to the focal length*. Thus, an eye can have a large pupil but still project a dim image on its retina if the focal length is long.

Now, the inverse of the ratio d/ℓ_f is known as the *f-number*. This is the same f-number by which the aperture opening of a camera is measured. For instance, when you set the aperture on your camera to $f/4.5$, you manipulate the iris such that the aperture diameter is $1/4.5$ times the lens's focal length. As every photographer knows, the higher the f-number (i.e., the smaller the aperture relative to the focal length) the dimmer the image projected on the film and the longer the required exposure. The f-numbers on a typical camera lens vary from about 1.8 to 22.

The f-number of the human eye with its pupil wide open is about 2, roughly the same as that of a bee ($f/2.4$). As a result, the radiant flux density of the image in a bee's eye is about the same as that in yours or mine, even though the aperture of a single facet in the bee's compound eye has as diameter only 1/280 that of the human eye.

The dependence of the brightness of an image on the f-number underlies one of the important biological effects of the difference in the refractive index between air and water. Consider, for instance, the simple "lens" of figure 11.19, a spherical interface. It can be shown that the focal length of this lens is (Jenkins and White 1957)

$$\ell_f = \frac{n_2 r}{n_2 - n_1}, \tag{11.22}$$

where r is the radius of curvature of the interface, n_1 is the refractive index of the medium outside the eye, and n_2 is the refractive index of the medium inside. For a fixed radius of curvature, the greater the difference between n_1 and n_2, the shorter the focal length and the greater the radiant flux density at the retina.

11.4.4 *Eyes in Air and Water*

The inside of all eyes is filled with either a watery fluid, a vitreous gel, or cytoplasm. In each case, the refractive index is about 1.34, very nearly the same as that of seawater. In air, then, the difference between n_1 and n_2 is about 0.34, and the focal length for this type of simple lens (shown in fig. 11.19) is about four times its radius of curvature. The aperture diameter for the eye in figure 11.19 was arbitrarily set equal to the radius of curvature, so the f-number for this eye is about 4. In other words, this simple lens is capable of projecting an image that is less bright than that projected by a typical camera lens.

If this same eye is submerged in fresh water, the refractive index of the outside medium, n_1, becomes 1.33. In this case, the difference between n_2 and n_1 is only 0.01 (as compared to 0.34), and the focal length of the lens is now 134 times its radius. Thus, in water the eye of figure 11.19 would have an f-number of 134!

This high f-number implies that the brightness of the projected image is much lower in water than in air. From eq. 11.21 we calculate that for the simple optical system of figure 11.19, an object of a given radiance appears about 1100 times brighter in air (where the f-number is 4) than in water (where the f-number is 134). It is clear that the difference in refractive index between air and water can have profound effects on a visual system.

In either air or water, the simple lens of figure 11.19 could stand some improvement. The f-number of the lens in air is higher than that of a cheap camera, for instance, and the f-number in water is clearly unacceptable. In response, both terrestrial and aquatic animals have evolved schemes for shortening the focal length of their eyes.

The simplest mechanism is to build a lens with a higher refractive index than that of cytoplasm. Most animal lenses are built from proteins such that their refractive index is about 1.5 (Land 1981), similar to that of glass. This seems to be the practical limit for biological materials, however; the lenses of aquatic animals do not have appreciably higher refrative indices than those of their terrestrial relatives.

Second, the lens can be built with two refracting surfaces rather than just one (fig. 11.21). In this case, light is refracted as it moves from the surrounding water into the lens and is then refracted a second time as it leaves the lens and moves into the fluid filling the eye. For this type of lens, the focal length is described by the equation,

$$\frac{1}{\ell_f} = \frac{n_2 - n_1}{n_3 r_1} + \frac{n_3 - n_2}{n_3 r_2} - \frac{\ell_d (n_2 - n_1)(n_3 - n_2)}{n_3 n_2 r_1 r_2}, \tag{11.23}$$

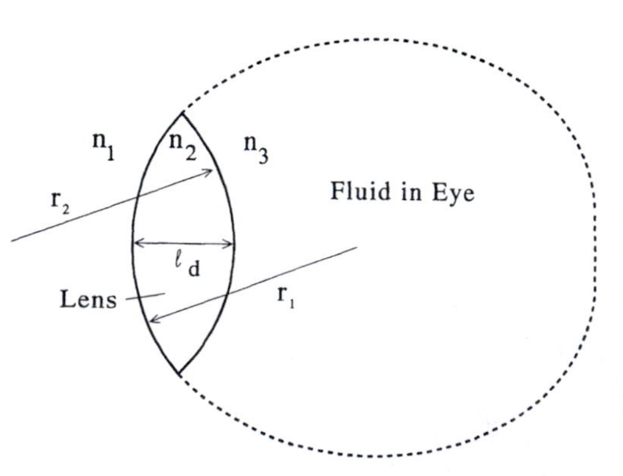

Fig. 11.21 Lenses can be made more effective by having two (instead of one) curved interfaces between materials of a different refractive index. The image-forming capabilities of the thin lens shown here are described in the text.

where n_1 is the refractive index of the fluid outside the eye, n_2 is that of the lens, and n_3 is that of the fluid in the eye (Jenkins and White 1957). The thickness of the lens, ℓ_d, is measured as the distance between the front and back surfaces along the optical axis.

Note that the front surface of the lens curves in the opposite direction from the back surface. By convention the radius of curvature of the front surface, r_1, is deemed to be positive, and that of the back surface, r_2, is negative. Note that when $n_2 = n_3$, eq. 11.23 simplifies to the expression we obtained previously for a simple spherical interface.

Through the course of evolution, terrestrial animals have adjusted the radii of curvature of their two-sided lenses to provide eyes with f-numbers of roughly 2, a design that projects an image with acceptable brightness. But aquatic organisms, constrained by the high refractive index of water, have had to stretch to the limit the design of their lenses. For lenses made from a homogeneous material, the shortest practical focal length (and therefore the highest f-number) is obtained when the lens is a sphere, and the lenses of virtually all aquatic organisms are spherical.

In a spherical lens $r_1 = -r_2$, $\ell_d = 2r_1$, and if the fluid inside the eye has the same refractive index as the surrounding water, $n_1 = n_3$. Under these typical circumstances, eq. 11.23 simplifies to

$$\ell_f = \frac{rn_1}{2(n_2 - n_1)}. \tag{11.24}$$

Even with a spherical lens, the focal length is quite large. For example, if n_2 is 1.52 and the lens is submerged in water ($n_1 = 1.33$), $\ell_f = 3.5r$. Even if the pupil of an eye containing such a lens is made as large as possible (that is, it has a diameter equal to that of the lens, $d = 2r$), the f-number for the eye is 1.75.

An f-number of 1.75 is considerably better than $f/134$, and is comparable to the f-numbers of camera lenses, but it would still be relatively ineffective at projecting a bright image. For instance, a spherical lens ($n_2 = 1.52$) in air would have a focal length equal to $0.96r$, and (for an aperture equal to the lens diameter) would have an f-number of 0.48. A lens of $f/0.48$ forms an image that is 13.3 times brighter than a lens of $f/1.75$ (see eq. 11.21). Thus, as a result of the low refractive index of air, terrestrial animals can, if needed, form lenses with considerably better light-capturing qualities than those of aquatic animals.

Although spherical lenses are good at capturing light (and especially so in air), most terrestrial animals do not use them. Why not? If they are good enough for aquatic organisms, why wouldn't they be all that much better for animals in air? The explanation seems to lie with the ability of these lenses to form a clear image.

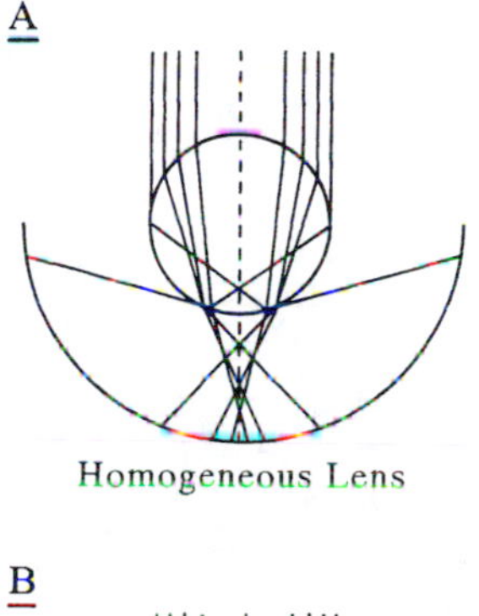

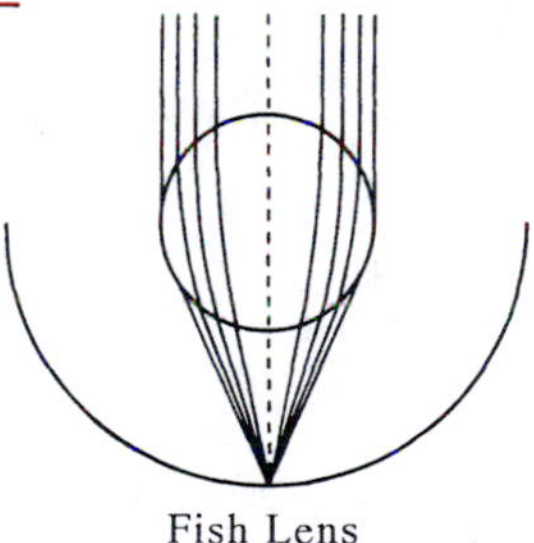

Fig. 11.22 A spherical lens made from a homogeneous material (A) produces a severe spherical aberration. This problem has been solved in fish through the use of a lens of variable refractive index (B). (Redrawn from Pumphrey 1961. by permission of Cambridge University Press)

Problems with Spherical Lenses

An ideal lens focuses parallel rays of incident light on a single focal point regardless of where on the lens these rays fall. A spherical lens (or any other spherical refracting surface) does a good job of focusing light only for parallel rays that lie close to the optical axis (so-called *paraxial rays*) (fig. 11.22A). These rays have a small angle of incidence. Rays falling farther from the optical axis have a larger angle of incidence and are focused closer to the lens (fig. 11.22). As a result of this *spherical aberration*, the image formed on a retina by a spherical lens can be quite fuzzy. When the central portion of an image is brought into focus, the periphery is unfocused, and vice versa.

The simplest way to deal with the problem of spherical aberration is to insure that only paraxial rays reach the retina. There are two ways in which this can be accomplished. If the radius of curvature of the lens is large, and only a small fraction of a sphere is used (a so-called *thin lens*), light from objects in front of the eye always has a small angle of incidence, and spherical aberration is minimal. The drawback, of course, is that because r is large, the focal length of the lens is increased (eq. 11.23). This is likely not to be a problem for terrestrial organisms. If the lens is made from a material with a high refractive index (e.g., 1.5), the difference in refractive index between the lens and the surrounding air is sufficient for a thin lens to produce a reasonably short focal length and an acceptable f-number, and many terrestrial animals use thin (or nearly thin) lenses.

Aquatic organisms cannot use this option, however. Due to the high refractive index of water, the radius of curvature of the lens must be kept as small as possible to yield a practical focal length and f-number. Given a fixed r, the simplest solution to the problem of spherical aberration is to allow light to pass only through the central portion of the lens. By using an iris to block the peripheral rays that would be out of focus, "stopping down" the lens in this fashion indeed reduces spherical aberration. Of course, by decreasing the aperture diameter, this procedure also increases the f-number. In other words, for a fish the simple solution to spherical aberration involves reducing the light available to the retina. Since the principal reason for using a spherical lens in the first place is to *increase* the brightness of the image, this strategy is unlikely to produce acceptable results.

In response to this problem, fish have evolved a trick for avoiding spherical aberration without having to resort to stopping down their eyes. The material from which a fish lens is constructed has a refractive index that varies from 1.52 along the optical axis to 1.34 at its periphery (Land 1981) (fig. 11.22B). There are two consequences of this variation. First, light entering the lenses near the periphery is refracted less than light entering near the optical axis. This counteracts the tendency of the lens to focus these peripheral rays at a shorter focal distance. Thus, the anisotropic nature of the lens material corrects for spherical aberration.

There is a second bonus to this lens design. Any light that does not enter the lens exactly at the optical axis is refracted continuously as it passes through the lens. For instance, light entering to the right of the optical axis is immediately refracted to the left, as it would be in an isotropic material. In a fish lens, however, as the light moves toward the optical axis it moves continuously into material with a higher refractive index, and is therefore refracted further, with the handy result that the lens has a shorter focal length. A wide variety of spherical fish lenses have a focal length approximately equal to 2.55 times their radius, substantially shorter than the 3.5 we calculated earlier for an isotropic lens. The ratio of $\ell_f/r = 2.55$ is known as *Matthiessen's ratio* after L. Matthiessen, who first described this property of fish lenses in 1880 (Land 1981).

If the aperture of a fish lens is equal to its diameter, a focal length corresponding to Matthiessen's ratio corresponds to an f-number of about 1.3, a value similar to that of the lenses found in many cameras.

Not all fish have spherical lenses, and the exceptions serve to highlight the optical differences between air and water. For instance, mudskippers (fish in the genus *Periophthalmus*) spend much of their time out of water, and even when the body is in water, the eyes, which are mounted high on the head, may extend into the air. The lens in a mudskipper's eye is considerably flattened compared to that of a more typical fish, as befits its comparatively terrestrial existence (Bond 1979).

The four-eyed fish of the West Indies (*Anableps*) is an even better argument for the ability of evolution to respond to the physics of air and water (fig. 11.23). This fish spends its life at the water's surface where the dorsal half of the eye extends into the air, and the eye has evolved a bizarre shape where the iris is pinched horizontally in the middle to form two apertures. The single lens is elongated and oriented such that light from below the water's surface passes through the lens in the direction of its shorter radius of curvature, and light from above the interface passes through the lens along a direction having a larger radius of curvature. In effect, the fish does have four eyes, two adapted for terrestrial vision and two for aquatic vision. The Indian mullet, *Rhinomugil corsula*, has a similar adaptation, but in this case the fish has its body axis vertical when the eyes are at the surface, so the pinch in the iris is rotated 90° to that in *Anableps* (Bond 1979).

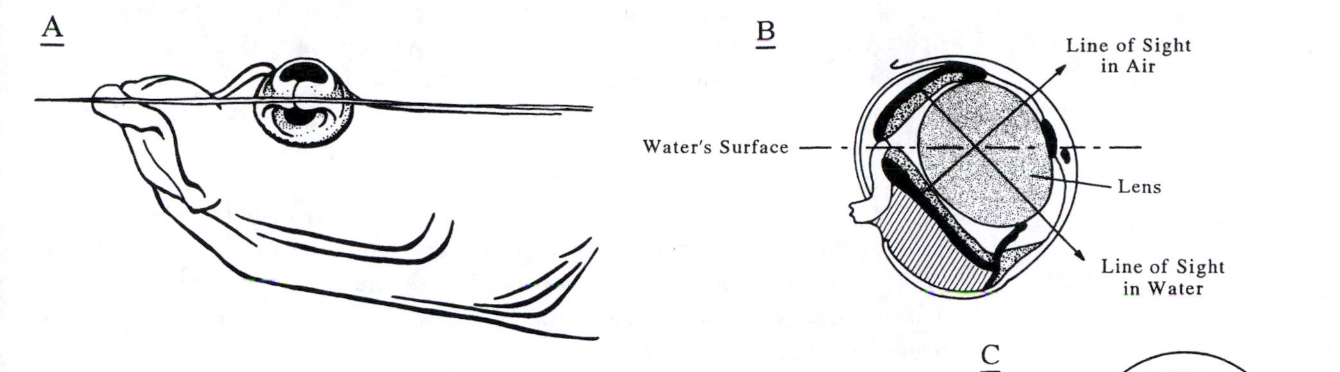

Fig. 11.23 The four-eyed fish (*Anableps*) spends much of its time at the water's surface with its eyes half in and half out of the water (A). The eye has two separate visual systems (B), optically separated by a pinched-in iris (C). [(A) redrawn from Bond 1979 by permission of Saunders College Publishing; (B) redrawn from Walls 1942 by permission of the Cranbrook Institute of Science]

Whirligig beetles (beetles of the genus *Gyrinus*) are an insect analog of *Anableps*. These beetles swim at the water's surface while hunting for prey and need to see both in air and in water. In apparent response, the compound eyes of these insects have split in half, one half lying on top of the head and the other on the bottom. The light receptors of the two eyes are different (Wachmann and Schroer 1975), though the functional differences between the aerial and aquatic eyes are not entirely clear.

Seals, sea lions, and other semiaquatic mammals face a problem similar to that of *Anableps* or whirligig beetles: they need to see well in both water and air. However, for these animals the evolved solution is quite different. Seals' eyes are highly sensitive, as befits their need to hunt fish in the dim light of the ocean depth, and they have strongly refractive lenses to provide a reasonable f-number in an aquatic medium. It might be thought that this lens would require substantial adjustment to yield a well-focused image in air. However, when the seal is out of water the iris contracts so that the pupil is a minuscule point. This avoids overloading the retina in the bright light of a terrestrial day. It also allows the pupil to act as a pinhole lens, providing the seal with a clear image of its surroundings (Walls 1942; Schusterman 1972). Thus seals circumvent the main problem associated with pinhole optics—a dim image—by having a sensitive retina.

Cormorants have solved the problem differently. These diving birds hunt fish visually, and therefore must be able to see clearly in water. They also fly, an activity that is unforgiving of the visually impaired. When the bird is in air, the lens is shaped like that of any other bird. However, when the bird enters the water, the muscular iris wraps around the lens and squeezes it into a nearly spherical shape, appropriate for the new medium (Walls 1942). Turtles use the same stratagem.

One marine animal has solved the problem of attaining a short focal length in a fashion different from all others. Scallops have approximately sixty small eyes that dot the edge of their mantle. These eyes are notable both for their iridescent blue-green color and for the fact that they use a mirror to focus light rather than a lens (fig. 11.24). Light entering the eye passes through the retina onto a spherical mirror at the back of the eye. The shape of this mirror then brings the image into focus at the retina. Presumably the out-of-focus light passing through the retina on its way to the mirror forms a more-or-less constant background on top of which the in-focus image is added. The result is a coherent (if low contrast) image that the scallop can use to survey its environs (Land 1965, 1978). Because a mirror does not rely on a change in refractive index to bring light into focus, the scallop's eye can achieve an admirably short focal length (Land 1981) and an f-number of about 0.6!

The optical design of the scallop eye extends even beyond its ability to gather light effectively. A strangely-shaped lens lies in front of the retina (Fig. 11.24). The apparent function of this lens is to slightly change the direction from which light approaches the mirror, and thereby correct for the mirror's spherical aberration (Land 1981).

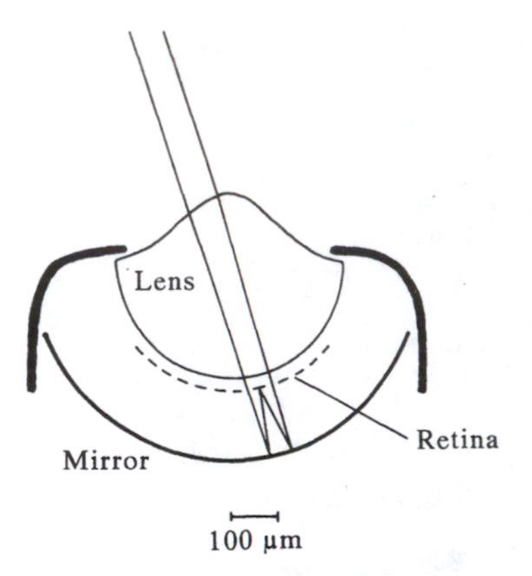

Fig. 11.24 The eye of a scallop uses a combination of a lens and mirror to achieve a very low f-number. (Redrawn from Land 1965 by permission of the Physiological Society)

The mirror used by a scallop is different from the one you have in your bathroom. It, like the mirror sheen of fish scales, relies on the ability of light to interfere with itself, a fact that can account for the brilliant color of the eye. However, the physics of interference is a topic we will not discuss here. You may wish to consult a standard physics text (such as Resnick and Halliday 1966) on the subject of interference mirrors.

11.4.5 *Resolving Power*

The ability to project a bright image onto the retina is not the only problem that must be solved by eyes. Once a sufficiently bright image is formed, it is necessary to sample the image with appropriate resolution.

The resolving power of an eye can be expressed in terms of the eye's ability to distinguish a pattern of alternating light and dark stripes. If the image of a dark stripe is projected onto one sensory cell and the image of a light stripe is projected onto the adjacent cell, the eye can detect the stripe's presence. If the spacing between stripes in the image is any smaller, each cell has the image of both light and dark stripes projected on it and the presence of stripes cannot be detected reliably. For instance, a grid of black and white stripes that is easily resolved when near one's eye appears as a uniform gray when held some distance away.

Now, the spacing of stripes in the image depends both on the actual size of stripes on the object being viewed and on the distance of the object from the eye. A wide stripe held far from the eye can cast the same size image as a thin stripe held close. In this respect, it is the angle subtended by a stripe (rather than its absolute size) that is important. For the situation shown in figure 11.25, the angle subtended by a pair of black and white stripes is d/ℓ, where d is the combined width of the pair (one black and one white stripe) and ℓ is the distance from the eye. This angle is equal to θ, the angle subtended by the stripes' image on the retina. When θ is equal to the angle subtended by just two cells, the stripes can just be perceived. Thus the minimal angle that the eye can resolve is

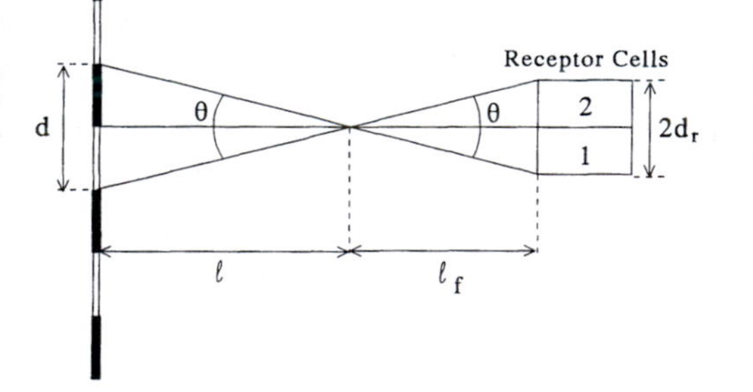

Fig. 11.25 The resolving power of an eye depends on the visual angle subtended by two retinal cells.

$$\text{minimal angle} \approx \frac{2d_r}{\ell_f}, \tag{11.25}$$

where d_r is the center-to-center spacing between sensory cells in the retina (approximately the same as the cell diameter). The *resolving power* of the eye (measured in radians^{-1}) is the inverse of this angle:

$$\text{resolving power} = \frac{\ell_f}{2d_r}. \tag{11.26}$$

The higher the resolving power of the eye, the smaller the object that the animal can resolve at a given distance.

Note that a need for resolving power places different constraints on the eye than does the need for a bright image. In this case, the *longer* the focal length, the better. Because the ratio of focal length to aperture size is often fixed by the need to provide a sufficiently bright image, the options are limited for how an increase in resolving power may be obtained.

The simplest means is to increase the overall size of the eye. By increasing both focal length and aperture, the brightness of the image can be maintained while allowing for an increase in resolution. This tactic has been used by animals that require an eye that is both highly effective at gathering light as well as one that has sufficient resolving power to allow the animal to hunt its moving prey—sharks and squids, for example. In fact, the largest eyes known are those of deep-sea squids, which may reach 37 cm in diameter (Bullock and Horridge 1965).

The alternative way to attain high resolving power is to decrease the size of the sensory cells. The smaller the cells, the smaller the angle they subtend in an image, and the greater their ability to resolve stripes. There is, however, a lower limit to the practical size of sensory cells, set by the way light is conducted within the cell.

After light enters a sensory cell, it is carried along the cell's length by total internal reflection. As it reflects its way down the cell, the light is gradually absorbed by visual pigments, thereby triggering the complex chemical mechanism by which light energy is transduced to a nervous signal. If, however, the diameter of the cell approaches the wavelength of light (0.4 to 0.7 μm), strange things can happen. At these very small sizes, the light wave may not be totally confined to the cell, even though on average it is totally internally reflected. Instead, certain modes of transmission can allow a substantial fraction of the light energy to move parallel to, but slightly outside of, the cell. If the light energy is actually transmitted *outside* the cell, it cannot be effectively absorbed by molecules *in* the cell, and, further, it is susceptible to leakage into adjacent cells. Thus, when these modes of transmission predominate, adjacent cells may be unable to distinguish between the light falling upon them, and resolution is reduced. In this fashion, the ability of the cell to contain light sets the lower limit to the size of sensory cells.

The predominance of deleterious modes of light transmission can be predicted through the use of an index (Snyder 1975a,b):

$$\Psi = \frac{\pi d_r}{\lambda}\sqrt{n_2^2 - n_1^2}. \tag{11.27}$$

Here d_r is the diameter of the sensory cell, n_2 is the refractive index of the cell's interior, and n_1 is the refractive index of the surrounding medium. When Ψ is less than 2 to 3, less than 50% of the light energy travels inside the cell, and the cell is inefficient in detecting light. Thus, d_r must be kept sufficiently large to keep Ψ greater than 3.

Land (1981) cites figures for the refractive indices inside sensory cells that range from 1.37 to 1.41. The extracellular space probably has a refractive index of 1.34, and we use as an example a wavelength of 500 nm. Given these values, Ψ is less than 3 when d_r is 1.1 to 1.6 μm. This then is the practical lower limit to the size of sensory cells. For a given required resolving power and f-number, this size sets the lower limit for the size of the eye.

Consider, for instance, a typical retina in which sensory cells are 1.5 μm in diameter. Using this retina, what size of eye is needed to resolve black and white stripes each 1 mm wide at a distance of 10 m? In this case, the ratio of d ($= 2\times10^{-3}$ m) to ℓ is 2×10^{-4}, and this ratio is equal to the ratio of $2d_r$ ($= 3\times10^{-6}$ m) to ℓ_f. As a result, the focal length of the eye must be 1.5 cm, and the overall diameter of the eye would be about 2 cm. This is approximately the size of a human eye, and, indeed, on a good day I can just barely resolve 1 mm stripes at a distance of 10 m.

Note that the lower limit for the size of sensory cells could be decreased if it were air rather than an aqueous fluid that surrounded the cells. In this case, n_1 would equal 1.0 rather than 1.34, and, if the internal refractive index is 1.4, the cell need only have a diameter of 0.5 μm to have Ψ greater than 3. In other words, if visual sensory cells could be surrounded with a layer of air, the resolving power of an eye of a given size could be increased by a factor of two to three. Alternatively, the eye could retain its existing resolving power but decrease its size by a factor of two to three. Although this strategy could work in theory, I know of no animal in which it has been utilized.

11.5 Refraction at the Water's Surface

Let us consider for a moment what the aerial environment looks like to an aquatic organism. From Snell's law (eq. 11.12) we know that

$$\frac{\sin \theta_a}{\sin \theta_w} = 1.33, \tag{11.28}$$

where θ_a is the angle of incidence in air and θ_w is the angle of incidence in water. This means that if a fish looks at the surface with angle θ_w, its line of sight in the air beyond the surface lies at θ_a:

$$\theta_a = \arcsin(1.33 \sin \theta_w). \tag{11.29}$$

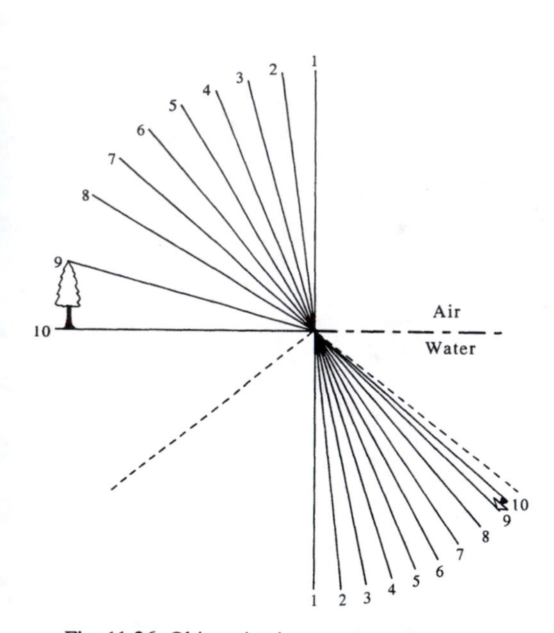

Fig. 11.26 Objects in air can appear displaced and distorted when viewed by a fish. For example, a fish looking a mere 47° from the vertical can see a tree on the horizon, but the tree appears much shortened. Looking farther from the vertical (dashed line), the fish sees a reflected image of objects under water.

This relationship is shown in figure 11.26, and it leads to some strange effects. For example, an object in the air directly above the fish is located in space just where it appears to be located to the fish's eye, and it is undistorted. In contrast, an object off the vertical is actually closer to the horizon than it appears to the fish, and it appears to be vertically compressed. That is to say, an image that subtends a given angle on the retina corresponds to a larger angle in air.

Another way of expressing this fact is to note that the slope of the line ($d\theta_a/d\theta_w$) becomes steeper the larger θ_w is. At an angle θ_w of about 48° (the critical angle for an air-water interface), the fish's line of site in air is parallel to the water's surface. In other words, the fish need only look at an angle of about 48° to the vertical to see objects on the horizon. At larger θ_w, light reaching the eye has been internally reflected from the underside of the water's surface, and the fish sees an upside down image of the aquatic environment around it.

This pattern of refraction has potential biological consequences. For instance, plants and corals in water can never receive direct sunlight at an angle any greater than 48° from the vertical, whereas terrestrial plants can get light at dawn or dusk that arrives from a nearly horizontal direction. One might expect, therefore, that the photosynthetic structures of aquatic organisms would have a different spatial arrangement than that of terrestrial organisms. This is indeed the case, but not in the way this argument would predict. The leaves of terrestrial plants typically have their photosynthetic cells only on the top surface. In contrast, many aquatic plants have photosynthetic cells on both the upper and lower surfaces of their leaves or fronds. Thus, aquatic plants seem adapted to receive light from all directions, whereas terrestrial plants are adapted to receive light primarily from above. This difference in morphology may have more to do with movement of the plant by its medium than with the directional availability of light. As we have seen (chapter 4), hydrodynamic forces are relatively larger on aquatic than terrestrial plants, and the fronds of marine algae, for instance, are constantly stirred by water motion. If the orientation of the frond continually changes, it would not be advantageous to confine photosynthetic cells to one surface.

The second biological consequence of refraction at the water's surface concerns the displacement of the apparent position of an image from its true position. This might be a problem for an aquatic organism attempting to capture terrestrial prey. Consider, for instance, the archer fish, *Toxotes jaculator*, which hunts by squirting water at terrestrial insects (fig. 11.27). If these jets hit the prey, the insect may fall into the water where it can be eaten by the fish. The ability to hunt terrestrial prey is of obvious advantage to the archer fish, but requires the fish to make careful adjustment for the refraction in its line of sight (Dill 1977). The excellent marksmanship of these fish is testament to their ability to cope behaviorly with refraction.

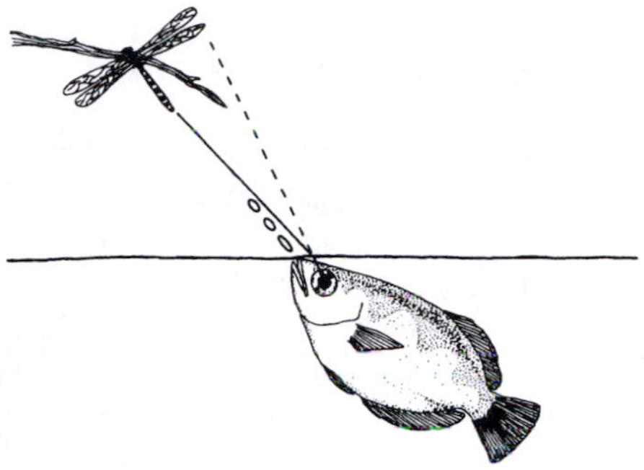

Fig. 11.27 The archer fish (*Toxotes jaculator*) is adept at spitting water at terrestrial prey despite the refraction imposed on its line of sight.

11.6 Polarized Light

We now return to the description of electromagnetic waves briefly introduced at the beginning of this chapter. If the electrons in a source of radiation vibrate along a defined axis, the electric field of the resulting light wave lies in a plane that contains that axis. For example, a vertical antenna (such as the tower used by a radio station) transmits electromagnetic waves in all directions. As we have seen (fig. 11.3), the intensity of these waves is greatest in the horizontal direction, and the electric field of horizontal waves, regardless of their direction of propagation, is always vertical. These radio waves are therefore said to be *polarized*, in this case in a vertical plane. Because radio waves from commercial towers are vertically polarized, the radio antenna on your car should be vertical to best receive the energy transmitted. If your car's antenna is horizontal, the vertically oriented electric field of a radio wave cannot act to move electrons along the antenna (just as a vertical gravitational field won't accelerate objects lying on a horizontal surface), and no signal is received.

As with radio waves, light can be polarized, but polarization occurs only under special circumstances. Consider sunlight, for instance. This light is produced by the vibration of electrons in atoms of the sun's surface. Because of thermal agitation, the axis along which electrons move is continually changing, and it is very unlikely that all atoms on the solar surface will ever have their electrons vibrating along parallel axes. As a result, light emitted by the sun consists of waves with randomly oriented electric fields. In other words, sunlight is unpolarized.

It is possible to polarize a beam of sunlight (or other unpolarized light) by removing all those waves with electric fields not parallel to a particular direction. This can be accomplished by a variety of mechanisms. That most familiar to the reader may be through the use of a piece of polaroid film such as that commonly found in sunglasses. This film is constructed from a special plastic, the molecules of which have been aligned parallel to each other. Because of this molecular orientation, the film is transparent only to waves that have electrical fields oriented

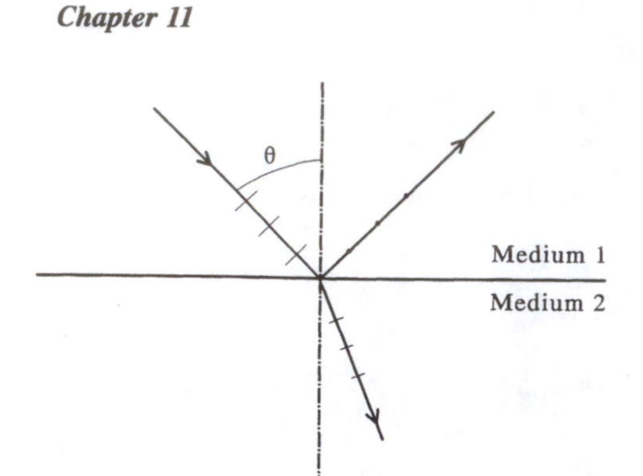

Fig. 11.28 When light strikes the interface between air and water, it can be both reflected and transmitted. The process results in some polarization of the light, however.

in a single direction. As a result, light passing through polaroid film emerges polarized because all other light has been absorbed.

There are two other mechanisms of polarizing light that can be of biological importance. First, light can become polarized when it is reflected from a surface. Consider, for instance, the situation shown in figure 11.28. A ray of unpolarized light impinges on a still water's surface at an angle of incidence, θ. Part of the light is reflected, and part refracted. It can be shown experimentally that light which has its electric field lying perpendicular to the plane of incidence is reflected more effectively than light with an electric field lying parallel to the plane of incidence. As a result, the reflected beam is enriched in light with an electric field that lies in one particular direction. In other words, the reflected light is at least partially polarized.

The fraction of light reflected depends both on the orientation of its electric field and on the angle of incidence. Experimental results can be accurately described by *Fresnel's law of reflection*:

$$\frac{\mathcal{I}_{r,\perp}}{\mathcal{I}_0} = \left(\frac{\sin\theta - \sin\theta/n_{2,1}}{\sin\theta + \sin\theta/n_{2,1}}\right)^2 \tag{11.30}$$

$$\frac{\mathcal{I}_{r,\parallel}}{\mathcal{I}_0} = \left(\frac{\tan\theta - \sin\theta/n_{2,1}}{\tan\theta + \sin\theta/n_{2,1}}\right)^2, \tag{11.31}$$

where $\mathcal{I}_{r,\perp}$ is the intensity of light reflected for which the electric field is perpendicular to the plane of incidence, and $\mathcal{I}_{r,\parallel}$ is the intensity of light reflected for which the electric field is parallel to the plane of incidence. $\mathcal{I}_0$ is the intensity of the incident light. These results are graphed in figure 11.29. Note that at one particular angle (known as *Brewster's angle*), $\mathcal{I}_{r,\parallel}$ goes to zero. At this angle, only the light with its electric field perpendicular to the plane of incidence (that is, horizontally polarized light) is reflected, and thus at Brewster's angle, the reflected beam is totally polarized.

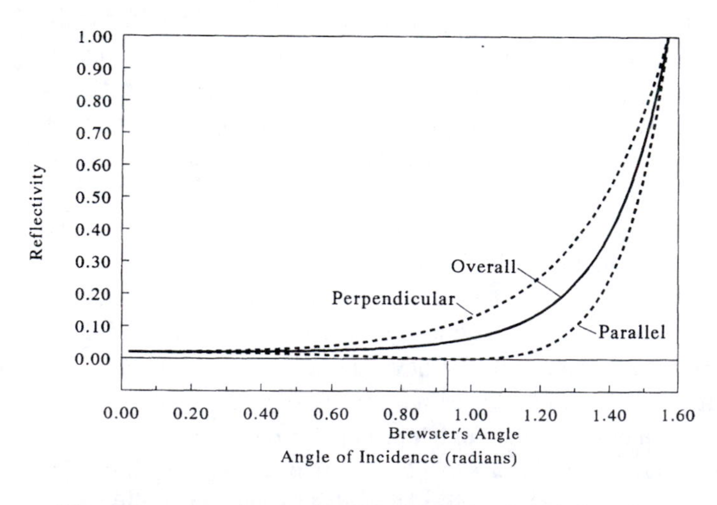

Fig. 11.29 The reflectivity of an air-water interface is different for light with its electrical field aligned perpendicular to the plane of incidence than it is for light with an electrical field parallel to the plane. At Brewster's angle (0.93 radians), none of the "parallel" light is reflected, leaving only "perpendicular" light in the polarized, reflected beam.

Now it can be shown that total polarization occurs when the reflected beam is perpendicular to the refracted beam. A consideration of the geometry of the situation thus reveals that Brewster's angle, θ_B is

$$\theta_B = \arctan n_{2,1}. \tag{11.32}$$

For the interface between air and water, $n_{2,1}$ is approximately 1.33, and Brewster's angle is about 53° (0.93 radians).

The fact that light reflected from the water's surface is at least partially polarized can be used to biological advantage. In many cases, this reflected light forms a glare that makes it difficult to see what is happening below the water's surface. If, however, the eyes of an organism are insensitive to horizontally polarized light, they are insensitive to much of this glare, and their vision of subsurface activities is augmented. This is the primary reason for polarized lenses in sunglasses. The polaroid film in the glasses is oriented such that only vertically polarized light is transmitted. As a result, the horizontally polarized glare from a water surface is largely eliminated.

A similar strategy is used by some surface-dwelling insects. For instance, the water strider *Gerris lacustris* has eyes that are sensitive only to vertically polarized light. As a result, they are relatively insensitive to glare (Trujillo-Cénoz 1972).

11.6.1 *Polarization by the Sky*

We next consider what happens when a wave of sunlight is scattered by a molecule. The oscillating electric field of the wave causes electrons in the molecule to vibrate, and the axis of this vibration is parallel to the plane that contains the electric field. In this sense, the electrons in a single atom or molecule behave just like the electrons in a radio tower, with a similar result. The scattered light emitted by the molecule is polarized, with its electric field parallel to the axis in which the electrons vibrate. Thus, each individual incident light wave produces scattered light that is polarized in a specific direction.

But, because incident sunlight is not polarized, the different waves that make up a beam each result in scattered light polarized in a different direction. One might think, therefore, that the scattered light would be no more polarized than the incident light. This is true if all the scattered light is considered together, but is not true for light scattered in a particular direction.

Consider the situation shown in figure 11.30A. Here we are looking at an atom along the direction in which incident light arrives. In this case, the electric field of the incident light is vertical, causing electrons in the atom to vibrate along a vertical axis, emitting light that is most intense in the horizontal plane. This scattered light is vertically polarized. Note that because electrons vibrate along a vertical axis, no light is emitted vertically. In this situation, an observer horizontally to the left of the atom will readily see the scattered, vertically-polarized light.

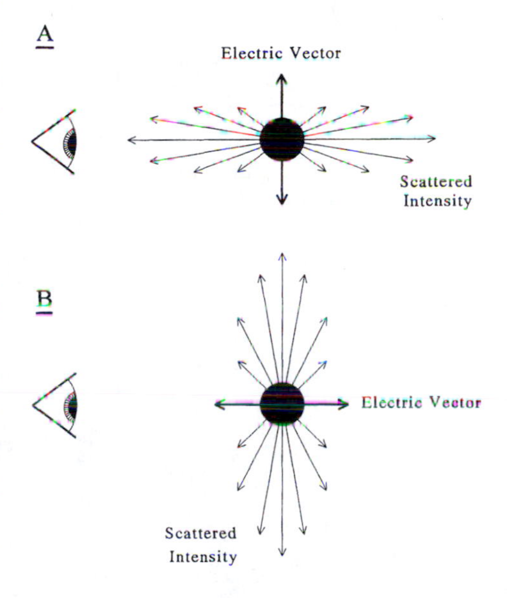

Fig. 11.30 Although a beam of light is scattered in all directions by an atom or molecule, the intensity of light you see depends on the orientation of each light wave's electrical field. As a result, light scattered from the sky, for instance, is polarized as viewed by your eye.

Now consider the situation shown in figure 11.30B. Here we are looking at the same atom from the same direction, but in this case, the electric field of the incident light striking the atom lies in a horizontal plane. Electrons vibrate as before, but along a horizontal axis in this case and polarized, scattered light is emitted. Note, however, that the intensity of this horizontally–polarized light is greatest in a vertical plane, and its intensity in the direction of the observer is zero. In other words, *in any single direction* light scattered from molecules is polarized.

Consideration of the geometry of figure 11.30 should convince you that no matter in what direction you look (except directly toward the source of incident light), the direction of polarization is perpendicular to a line connecting the source with the spot at which you look. For instance, if you look at a particular point in the sky, the light reaching your eye is polarized perpendicular to a line between the sun and the point that you are observing. This fact can be confirmed by viewing the sky through a pair of polaroid sunglasses. If the glasses are rotated around the line of sight, the sky appears to become brighter and darker as the polarizing axis of the lenses is either parallel or perpendicular to that of the sky's light.

The polarization of light by the atmosphere is used by animals as diverse as sea turtles and sea hares (*Aplysia*) as a means of orientation. Bees, for instance, have eyes that can sense the polarization of UV light and therefore can sense the direction to the sun by looking at any patch of clear sky. This allows them to orient themselves even when the sun is not visible. For an intriguing discussion of the ways in which bees use this ability in their foraging behavior, the reader is urged to consult von Frisch (1950). A general discussion of the use of polarized light by insects can be found in Waterman (1981).

Light scattered by water is polarized in the same way as that scattered by air (Waterman 1981), and it has been shown that a variety of aquatic crustacea use polarized light as a means of orientation.

Curiously enough, human beings can detect the polarization of light under certain circumstances. For example, when light is reflected from a water surface at the Brewster angle, a faint yellowish "brush" flanked by two faintly blue "clouds" can be seen. The long axis of the brush (known as *Haidinger's brush*) lies perpendicular to the direction in which the reflected light is polarized. The perception of this phenomenon requires some practice, and you may wish to consult Minnaert (1954) for a primer on naked-eye observations of polarized light.

11.7 On Being Invisible

Every few years a dectective novel is published in which the villain manages to hide the stolen jewels in some new and clever fashion. In dreaming up these schemes, authors often rely on the difference in refractive index between air and water. For example, the thief could hide the diamonds in a glass of water! Because water has a refractive index close to that of diamond, light at the diamond-water interface is minimally refracted, and the visibility of the jewels is thereby diminished.

The same trick is available to aquatic organisms. Because cytoplasm has a refractive index very close to that of water, an aquatic organism can become well nigh invisible merely by making sure that its tissues do not contain any colored substances. A wide variety of aquatic organism have taken advantage of this strategy. Jellyfish, ctenophores, salps, and many larval fishes (to name just a few) are both transparent and nonrefracting, and therefore are nearly invisible.

Transparency is not without its problems, however. A transparent body exposes all of an organism's tissue to ambient light, some of which can be harmful. For example, due to the high energy associated with its short wavelength, UV light can be destructive to living cells. In particular, DNA—the genetic material of the cell—strongly absorbs UV light and can have its structure altered by exposure to UV. Reference to figure 11.15 shows that in clear water UV light retains almost 25% of its incident intensity at a depth of 50 m, suggesting that corporeal transparency can be hazardous near the ocean surface. For instance Damkaer et al. (1980) have shown that ambient levels of ultraviolet light can pose a significant threat to the development and survival of shrimp and crab larvae. Some benthic organisms maintain the equivalent of a suntan: both corals (Siebeck 1988) and red algae (Wood 1989) produce UV-absorbing compounds to protect their tissues. Thus, while transparency has evolved in a variety of aquatic organisms, its use is largely confined to organisms that maintain a sufficient depth or enter the surface water only at night. For a review of the effects of ultraviolet light in marine systems, you are invited to consult Calkins (1982).

Transparency as a means of camouflage does not work well in air. First, even if an organism is transparent, its high refractive index relative to air causes it substantially to refract light passing through it. This pattern of refraction makes the organism visible in much the same way that a transparent glass of water is visible in air. Note that this fact would make it very difficult to construct an invisible man. Even if through some devious science fiction apparatus one could create a perfectly *transparent* man, he would be readily visible.

Second, as for marine organisms, it can be dangerous for a terrestrial organism to be transparent. Anyone who has ever suffered a sunburn can attest to the fact that a substantial amount of UV light impinges on the earth's surface.

Because of these problems, transparency as a means of camouflage has seldom evolved among terrestrial organisms. The most prevalent examples of transparency in terrestrial organisms are the wings of insects. In many cases, however, the transparent parts of the wings are formed from dead cuticle, and are therefore not bothered by UV damage. To avoid an overdose of ultraviolet light, most terrestrial organisms have light-absorbing pigments in their skin. The presence of these pigments, of course, keeps the organism from being transparent, and terrestrial organisms have evolved other schemes for blending into their visual environments. Their pigments may match the color of the substratum, for instance, and they may be shaped to mimic innocuous items in the environment. For further details about the strategy of crypsis, you may wish to consult Wickler (1968), Owen (1980), or McFall-Ngai (1990).

11.8 Optical Esthetics

The optical properties of air and water create some of the most pleasing of natural visual displays. For example, the refraction and internal reflection of sunlight by raindrops can form a rainbow, and the optics of air-borne ice crystals can form glories. Similar displays in water are much less likely because of water's high refractive index.

As tempting as it is to explore visual esthetics in the context of comparing air and water, these esthetics have little demonstrable biological importance, and thus will not be addressed here. If you are curious, however, you may wish to consult the several readable works on the subject: Minnaert (1954), Bryant and Jarmie (1974), Greenler (1980), and Boyer (1987).

11.9 Summary . . .

Light is of great biological importance because of the high energy it can impart to molecules when it is absorbed. This energy drives photosynthesis, provides warmth, and allows animals to obtain visual information regarding their surroundings.

Water attenuates light much more effectively than does air. As a consequence, plants can photosynthesize only near the surface of the oceans, and the effective range of vision is much less in water than it is in air.

The refractive index of water is larger than that of air, and is quite close to that of materials used in biological lenses. As a result, aquatic organisms typically use spherical lenses in their eyes to obtain a short focal length and thereby to project a sufficiently bright image on their retinas. A variety of optical "tricks" have evolved to cope with the high refractive index of water, including the development of lenses with spatially varying refractive indices and the use of mirrors rather than lenses.

The resolving power of an eye is limited, in part, by how small visual sensory cells can be made. The lower limit to size is set by the ability of sensory cells to act as light guides, an ability that is itself limited by the high refractive index of the aqueous medium that surrounds cells.

Light can be polarized by its interactions with both air and water, and the ability to detect polarized light is used by a variety of animals as an aid to navigation. The ability to detect light with only a certain polarization is used by water striders to decrease the glare from the water's surface.

Because the refractive index of animal tissues is similar to that of water, aquatic plants and animals can be virtually invisible. The low refractive index of air, however, insures that even transparent terrestrial organisms are readily visible.

11.10 . . . and a Warning

It would be difficult to overstate the importance of light in biology. Both as a source of energy and of information, light is intimately associated with virtually every aspect of life on earth. In response to its importance, the study of the biology of light has a long and rich history, only a brief and superficial sampling of which is presented in this chapter. Many of the processes explored here are much more complex than you might assume from their simplistic presentation (vision and photosynthesis are obvious examples), and you are urged to consult the original literature before applying the concepts of this chapter to any real-world situation.

Chapter 12

Surface Tension: The Energy of the Interface

We are now at the first of three chapters in which we explore the physics of the interface between water and air. As we will soon discover, the interface is a bizarre and fascinating place. To begin with, it is truly two-dimensional: it has neither outside nor inside. It is not contained in the water nor is it contained in the air; it is simply the place at which they meet. As such, its properties are not those of air or water alone, but of their mutual interaction, and these can be both surprising and nonintuitive.

Why should a biologist care about the properties of the air-water interface? There are many reasons. For example, terrestrial plants and animals, like all organisms, consist mostly of water. As a result, contact between the organism and its environment is likely to occur at an air-water interface. Transport of heat and material across this junction is therefore of profound importance to terrestrial plants and animals. We may also recall (chapter 2) that 70% of the earth's surface is covered by oceans and another few percent by lakes and rivers. This huge expanse of air-water interface forms a habitat for many organisms. How do the properties of the interface affect the plants and animals that live near it?

We begin this exploration with an examination of the phenomenon known as *surface tension*. This is the force that keeps water droplets spherical, and it has a variety of biological consequences. For instance, we will see how surface tension allows trees to grow to majestic heights and flies to adhere to glass; how surface tension allows some insects to breathe under water and others to walk upon it.

12.1 The Physics

12.1.1 *Surface Energy*

Consider the structure of a single water molecule—an oxygen atom bound to two hydrogen atoms (fig. 12.1). The nature of the bonds between these atoms is such that hydrogen atoms do not lie at opposite poles of the oxygen. Instead, the angle between them is only about 105°, and the hydrogens stick off to one side of the molecule. Now, oxygen is strongly electronegative (more so than hydrogen) so the oxygen atom tends to pull electrons away from the hydrogens. The shift of electrons toward the oxygen atom gives that end of the molecule a slight negative charge and leaves the hydrogens each with a slight positive charge. In other words, the water molecule is *polar*.

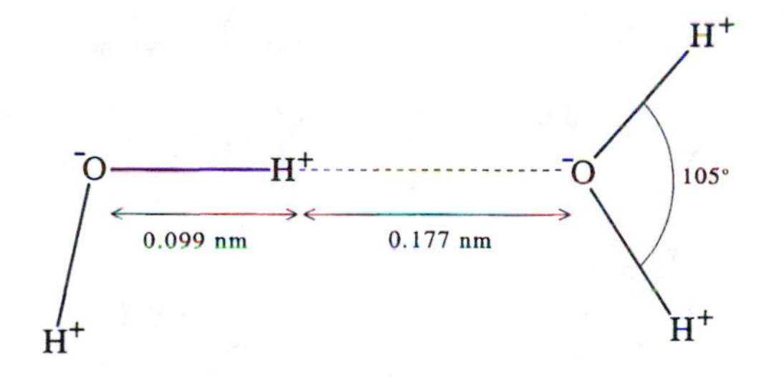

Fig. 12.1 The positive charge on the hydrogen atom of one water molecule is attracted to the negative charge on the oxygen atom of another, forming a hydrogen bond.

The separation of charges in the water molecule means that adjacent molecules can be attracted to each other. The negatively charged oxygen of one molecule is attracted to the positively charged hydrogens of another, forming a *hydrogen bond*. This bond is not as strong as the covalent bonds that hold oxygen and hydrogen together, but is nonetheless a substantial attraction. For example, it takes about 3.3×10^{-20} J to break each hydrogen bond, roughly five times the average kinetic energy of a water molecule at 20°C. In other words, at a typical biological temperature, most water molecules are firmly, but not rigidly, attached to their neighbors.

This is just another way of saying that water is a liquid. At temperatures between 0°C and 100°C the attraction of molecules to one another is, on average, sufficiently strong to keep them from flying apart as they would in a gas, but not so rigid that the material behaves as a solid.

In chapter 15 we will explore the role that hydrogen bonds play in the exchange of water vapor across the air-water interface. In this chapter, we concentrate instead on the effect that these bonds have on water molecules at the interface itself.

Consider the situation of a single water molecule in the midst of others. At any instant, the molecule is attracted on all sides by the polar nature of its neighbors. What happens if we now transport this molecule to the air-water interface? Whereas the molecule was previously attracted equally in all directions, this can no longer be the case. One side of the molecule is, by definition, now in contact with air, and, as a gas, air has much less attraction for the water molecule than does water. Thus, the portion of the water molecule that is exposed to air is subject to a small attractive force attempting to pull the molecule out of the liquid, while the remainder is subject to a large attractive force attempting to pull the molecule into the bulk of the water. As a result of this unequal attraction, a water molecule at the interface attempts to escape back into the liquid. Only by the expense of substantial work can a water molecule be dragged to the junction between air and water.

Now, each water molecule that is coerced into the interface creates a small amount of new interfacial area. If we express the work required to move a molecule to the interface relative to the new area created, we have a measure of the *surface energy* of water when in contact with air, γ,[1] expressed in J m^{-2}. Careful measurements on pure water show that it has a surface energy (at 20 °C) of 0.0728 J m^{-2}. In other words, anything one does to water that involves the creation of a new square meter of air-water interface requires the expenditure of 0.0728 J.

This is an exceptionally high surface energy for a liquid. For instance, ethanol has a surface energy of only 0.0228 J m^{-2}. In fact, due to its ability to form hydrogen bonds, water has the highest surface energy of any room-temperature liquid.

The concept of surface energy is not confined to liquids—the creation of new surface area in a solid requires energy in the same sense. To expose a molecule to air one must break the bonds that previously acted between it and other molecules in the solid, and work must therefore be done. These bonds have a wide range, from energies considerably lower than that of water to those much higher. The surface energies for a variety of solids are given in table 12.1.

Table 12.1 The surface energy of a variety of solids and liquids at 20°C.

	Surface Energy (J m^{-2})
Solids	
Polytetrafluoroethylene	0.0185
Polyethylene	0.0310
Tooth enamel	0.0380 to 0.0400
Wool	0.0425
Cellulose	0.0450
Polyglycine	0.0450 to 0.0510
Glass	0.1700
NaCl	0.3000
Liquids	
1% Gelatin	0.0083
Ethanol	0.0228
Benzene	0.0289
Phenol	0.0409
Water	0.0728

Source: Data from Weast (1977).

The surface energy of water decreases slightly with increasing temperature (fig. 12.2; table 12.2), ranging from 0.0756 J m^{-2} at 0°C to 0.0696 at 40°C, a variation of about 9% relative to the lower value. The surface energy of seawater is about 1% higher than that of fresh water (table 12.2).

Note that the values cited here are for pure water and "pure" seawater (if such a thing exists). In nature it is extremely rare to find an air-water interface that is not fouled to some extent with an ill-defined organic film. Most biological molecules (fatty acids in particular) can lower the surface energy to a fraction of that found in pure water. As a result, the surface of all but the cleanest bodies of water is likely to have a lower surface energy than that reported here. Unfortunately, the effects of surface films are quite variable and not well documented, and with but one exception we will not be able to take them into account. You are hereby warned

[1] In the general context of surface physics, one really should denote the surface tension of water in contact with air by the symbol γ_{wa} to distinguish it from the surface tension of water in contact with another medium. However, in this chapter we are only concerned with the interface between air and water, so no ambiguity should arise through the use of a streamlined symbol.

that some of the results derived in this chapter may need to be revised to take into account the presence of a surface film if they are to be applied accurately to real-world situations.

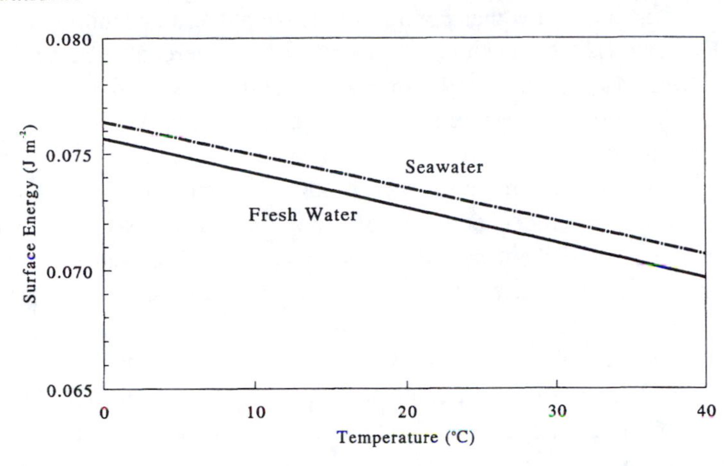

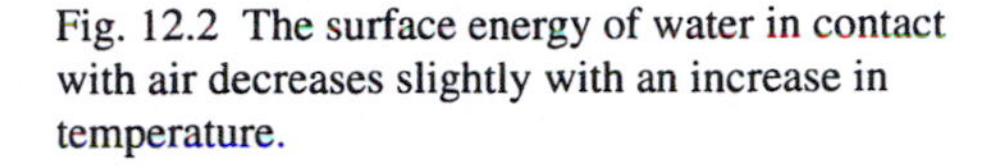

Fig. 12.2 The surface energy of water in contact with air decreases slightly with an increase in temperature.

12.1.2 *Cohesion and Adhesion*

Imagine, if you will, a column of water arranged in a fashion that allows us to pull on its ends, thereby placing the column in tension (fig. 12.3A). This is a task that can be achieved only with some ingenuity. In practice, it is often accomplished by placing water in a Z-shaped tube and spinning the tube about its center (fig. 12.3B). As the tube spins, the centrifugal force tends to pull the ends of the water column radially outward, placing the center of the column in tension. If we spin the tube fast enough, and thereby pull hard enough on the water, we can measure the force required to break (or *cavitate*) the column. This force, expressed relative to the cross-sectional area of the column, is a measure of the *tensile* or *cavitation strength* of water.

Table 12.2 Surface energy for fresh water and seawater (S = 35) as a function of temperature.

T (°C)	Surface Energy ($J\ m^{-2}$)	
	Fresh Water	Seawater
0	0.0756	0.0764
10	0.0742	0.0750
20	0.0728	0.0735
30	0.0712	0.0721
40	0.0696	0.0707

Source: Fresh water data from Weast (1977); seawater data from Sverdrup et al. (1942.)

Measured values of tensile strength vary with the material from which the tube is made, the size of the tube, and the purity of the water. Values as high as 28 MPa (280 atmospheres) have been measured for distilled water in glass tubes (Hayward 1971; Oertli 1971), a strength equivalent to that of a piece of rubber. Measurements on seawater are considerably lower, however. Smith (1991) obtained tensile strengths of 0.05 to 0.07 MPa in glass tubes, and values approaching 0 in nonwettable tubes, that is, in tubes with low surface energy, for example those coated with silicone grease.

The tensile strength of seawater can have biological consequences. For example, octopods use suckers to grasp their prey. Muscles in the sucker wall create a negative pressure in the water under the sucker, and the resulting pressure difference across the wall accounts for the strength of the sucker's adhesion (Smith 1991; Kier and Smith 1990; Denny 1988). The more tension that can be sustained by water under the sucker, the greater the adhesive strength. Seawater has a relatively low tensile strength, however, and this limits the strength with which an octopus can hold its prey (Smith 1991). We will discuss another example in which the cavitation of water can potentially limit biological function when we examine the plumbing of trees later in this chapter.

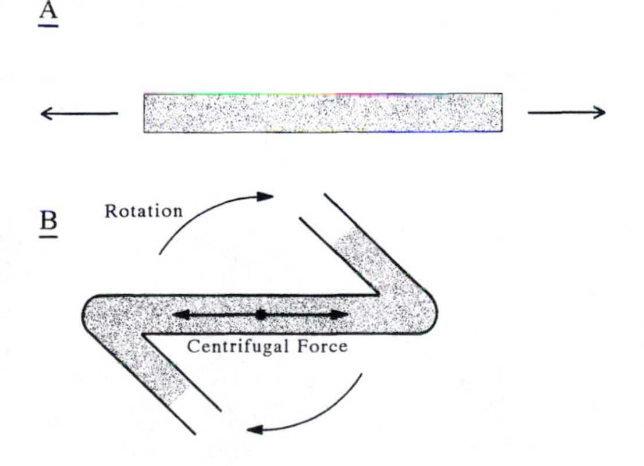

Fig. 12.3 (A) The centrifugal force in a spinning tube places water in tension. (B) The bent ends of the Z-shaped tube insure that the water column remains centered.

In at least one instance, however, an animal has made use of the cavitation of water. The snapping shrimp *Alpheus californiensis* has one enlarged claw that it snaps together as a defensive display. The abrupt closure of the claw is effected through the use of a novel triggering mechanism (Ritzman 1973). As the claw is opened, a smooth surface on the dactyl (the "finger" of the claw) is brought close to another on the propus (the "hand'). A thin layer of water remains in between. Muscles in the claw then contract, increasing the force tending to close the claw. The claw is held "cocked," however, by the tensile strength of the water sandwiched between dactyl and propus. This allows the muscles to contract isometrically and thereby to impose a large force. The water between dactyl and propus cavitates when the muscular force imposes a negative pressure of about 2.9 MPa. Upon cavitation the claw rapidly closes, producing a loud snap.

Now, in the process of breaking a column of water, two new air-water interfaces are created, one for each broken piece, and energy is expended in their formation. This *energy of cohesion*, W_c, is the energy per area (rather than the force per area) required to separate the material from itself, and, because there are two interfaces created, is equal to 2γ.

In a similar fashion, we can imagine pulling on a composite column formed half of water and half of a solid (fig. 12.4). By tugging on this column we can measure the energy required to pull the water away from the solid. This energy, the *energy of adhesion*, W_a, is the work that must be expended to create both a new interface between the solid and air and a corresponding interface between water and air.

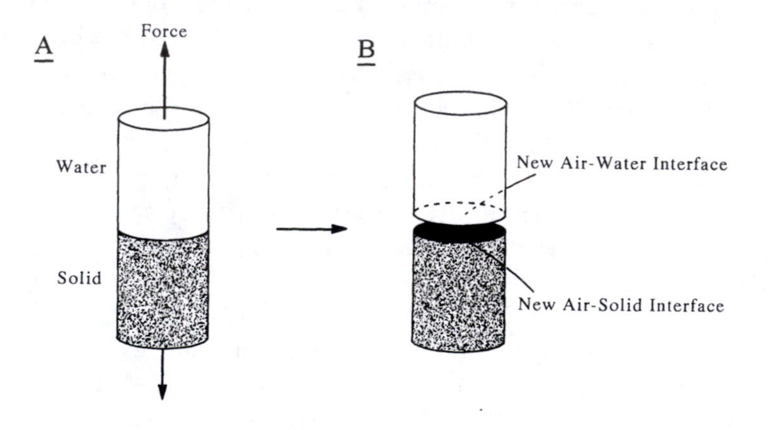

Fig. 12.4 A tensile force (A) can break the adhesive bond between water and a solid, forming two new interfaces (B).

In general, the work of cohesion is different from the work of adhesion. For instance, if water molecules are attracted more strongly to each other than to molecules of the adjacent solid, the work of adhesion is less than the work of cohesion. This usually occurs when the surface energy of the solid is less than that of water. The opposite is true if water molecules are attracted more strongly to the solid than to themselves, which happens when the surface energy of the solid is greater than that of water.

The difference between the work of adhesion and cohesion provides an explanation for the behavior of water drops on a solid surface. If the energy of cohesion is larger than that of adhesion, the molecules in the drop are preferentially drawn to themselves and the drop "beads up" on the solid surface (fig. 12.5). The drop pulls in on itself until the force of its own cohesion is just offset by the combined forces of gravity and adhesion. For small drops, the force of gravity is small compared to that of adhesion, and we may safely neglect it. In this case, the shape of the mounded drop is set by the interplay between cohesion and adhesion.

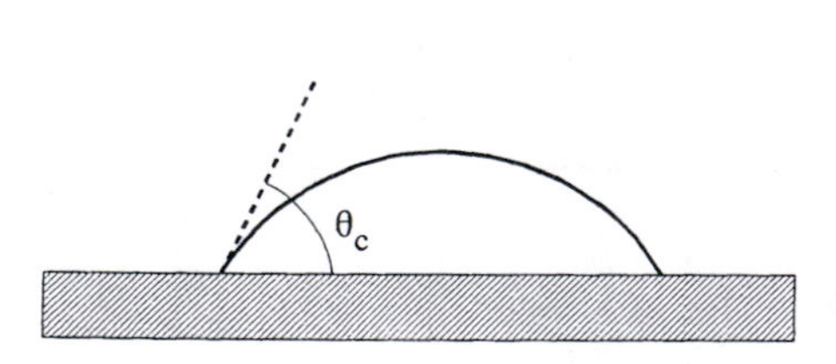

Fig. 12.5 The contact angle (θ_c) between a liquid and a solid is measured through the liquid rather than outside it.

This relationship was described independently in 1805 by Young and Laplace in terms of the *contact angle*, θ_c, with which the liquid meets the solid (see fig. 12.5).

Note that for our purposes the contact angle is always measured *in the liquid*. Young and Laplace proposed that

$$W_a = W_c \frac{1 + \cos\theta_c}{2} = \gamma(1 + \cos\theta_c). \qquad (12.1)$$

In other words, if the drop mounds up (indicating that $\theta_c > 0$), $(1 + \cos\theta_c)/2 < 1$, and the energy of adhesion is less than the energy of cohesion. The lower the work of adhesion, the larger θ_c must be. In the limit of zero work of adhesion, $\theta_c = \pi$ (180°) and $\cos\theta_c = -1$.

If θ_c is equal to 0, the drop spreads freely over the solid because its attraction to the solid is less than that to itself. In this case, the liquid is said to *wet* the solid.

The utility of eq. 12.1 is its ability to express the energy relationship between water and a solid surface—a relationship that would be difficult to measure directly—in terms of a parameter (the contact angle) that is easily observed.

12.1.3 *Surface Tension*

To this point we have discussed the air-water interface in terms of its surface energy. Why, then, is this chapter entitled "Surface Tension"? It turns out that surface tension is just another way of expressing surface energy.

In the abstract, this is easily seen by comparing the units of the two expressions. Surface energy is expressed as J m^{-2}. But a joule is one newton meter, so J m^{-2} is the same as N m^{-1}, that is, force per distance, a *tension*.[2] For example, when, by applying a force, you stretch a rubber band, a tension is created in the band. Thus, by definition, surface energy and surface tension are the same thing.

This equality can be explained on an intuitive level as well. Consider the apparatus shown in figure 12.6. A U-shaped piece of wire is outfitted with a sliding wire across its open end. This sliding wire has length ℓ, and a film of water is suspended within the perimeter formed by the U and the slider. If one pulls the sliding wire a distance Δx away from the base of the U, the area of air-water interface is expanded by $2\ell\Delta x$, a process that we know requires an expenditure of energy equal to $2\ell\Delta x\gamma$. The increase in area is twice $\ell\Delta x$ because the water film has two sides. The force required to move the slider is equal to the quotient of the work done and the distance over which the force was applied. Thus,

$$\text{force} = \frac{2\ell\Delta x\gamma}{\Delta x} = 2\ell\gamma. \qquad (12.2)$$

In other words,

$$\gamma = \frac{\text{force}}{\text{length}}, \qquad (12.3)$$

as advertised.

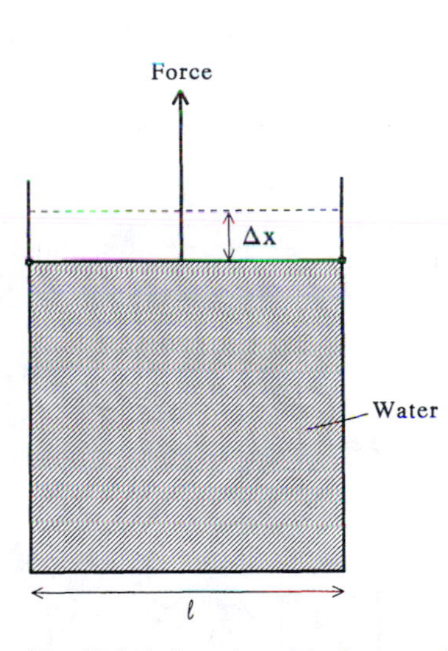

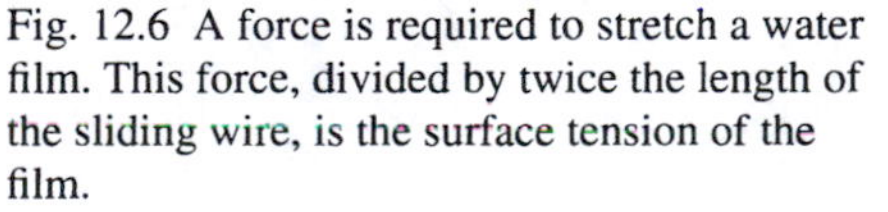
Fig. 12.6 A force is required to stretch a water film. This force, divided by twice the length of the sliding wire, is the surface tension of the film.

There are several peculiar aspects to this simple experiment. First, the force that must be applied to the slider to maintain its position on the U does not depend on how far the slider is from the base of the U. As the slider is moved, more water molecules are coerced out of the bulk liquid into the interface. In essence, the area of the water film adjusts itself so that the force exerted on the slider remains the same. This is quite different from the situation obtained by tugging on a film of rubber (or any other solid) rather than a film of water. In a solid, molecules are not free to move from the interior to the interface, and a deformation of the film

[2]Note that tension, the force required to stretch an object by a given distance, is different than tensile stress, the force per area obtained when an object is placed in tension.

requires an increased stretching of those molecules already at the surface. The greater the stretch, the greater is the force required.

We next note that there is nothing special about the particular sliding wire we have chosen. We could just as easily have made the slider from one of the arms of the U or the base. In each case we would have obtained the same answer: the force acting on a segment of the perimeter is equal to twice the product of the segment's length and the surface energy. Thus, the overall force exerted by the surface tension in the water film is equal to twice the total perimeter of the interface times γ.

And finally, we make explicit a fact that has been implicit in this experiment: *the direction of the force imposed by a surface tension lies tangent to the air-water interface*. In this case, we simply mean that the force acting on the wire perimeter acts inward in the plane of the water film. In many cases, however, the interface between water and air is curved. At any point along the curve, the tension can act only in the interface itself. In this respect, and this respect alone, surface tension behaves in a manner similar to that of a thin rubber membrane.

12.2 Capillarity

We are now in a position to apply the physics of surface tension to a problem of biological interest. How does water get to the tops of tall trees?

We begin by exploring the simple situation shown in figure 12.7. A rigid-walled tube of radius r pierces the surface of a bowl of water. The tube is open at both ends, allowing water to enter from below, and an air-water interface is created at the top of the water in the tube. The existence of this interface tells us that a surface tension must also exist. Our task is to quantify the effect of this tension on the column of water in the tube.

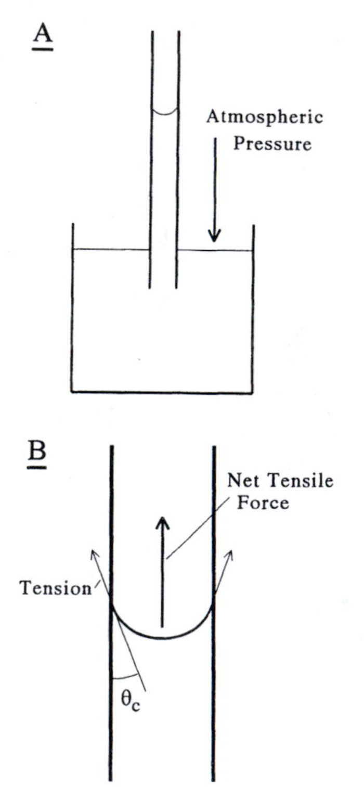

Fig. 12.7 Surface tension reduces the pressure of water in a glass tube (A), allowing atmospheric pressure to force water upward (B).

We first note that the water meets the walls of the tube with a contact angle θ_c (fig. 12.7B). Because surface tension acts tangential to the interface, we know that the tension has a vertical component, $\gamma \cos \theta_c$. If $\theta_c < 90°$ ($\pi/2$ radians) this tension pulls up on the column of water, if $\theta_c > 90°$, it pulls down.

The force applied to the water by the interface is simply the surface tension multiplied by the perimeter of the interface, which in this case is equal to the inner circumference of the tube ($2\pi r$). Thus,

$$\text{vertical force} = 2\pi r \gamma \cos \theta_c. \tag{12.4}$$

It is best if we think of this upward force in terms of the pressure it applies to the water. Dividing force by the cross-sectional area of the tube (πr^2), we see that

$$\text{capillary pressure} = -\frac{2\gamma \cos \theta_c}{r}, \tag{12.5}$$

where the negative sign signifies that the pressure acts to pull the water upward.

Now, as we saw with the slider on the U, it doesn't matter where in the tube the interface lies; the same force (and therefore, the same pressure) is always applied. Thus, the water in the tube can move up or down without changing the pressure imposed on its surface. If, for instance, the contact angle is less than 90°, the attraction of the water to the material of the tube exerts an upward force and water is drawn into the tube. This vertical motion continues until the pressure imposed by surface tension is just equal to that due to the weight of water lifted above the surrounding level. The weight of a column of water in the tube is

$$\text{weight} = \pi r^2 hg\rho_e, \tag{12.6}$$

where g is the acceleration due to gravity, ρ_e is the effective density of the water in air (approximately 1000 kg m^{-3}), and h is the height to which the column rises above the surrounding water level. Note that h is positive if the water moves up and negative if it moves down. The height h is known as the *capillary rise*, and the general process by which water is drawn into a small tube is known as *capillarity*.

The pressure applied by the water's weight is

$$\text{pressure} = -h\rho_e g. \tag{12.7}$$

Equating capillary pressure and the pressure due to weight, we find that the equilibrium level of the water in the tube is

$$h = \frac{2\gamma \cos\theta_c}{r\rho_e g}. \tag{12.8}$$

In other words, the smaller the radius of the tube, the higher the column of water will rise. Note that because the water in the tube is, in essence, hanging from the air-water interface, it is at a lower pressure than that of the surrounding air. As a result, if one were to poke a hole in the side of the tube, air would be drawn in rather than water being forced out. This is in contrast to a column of water that had been forced up from below.

If $\theta_c > 90°$, its cosine is negative and the column of water is depressed in the tube below the level of the surrounding water. In this case, the pressure of water in the column is higher than that in the air above.

12.2.1 *Water Transport in Trees*

Now, water in higher plants is carried primarily by the vessel cells of the xylem. The material of these hollow, dead tubes is effectively wetted by water, so the contact angle is approximately 0. In this case, eq. 12.8 simplifies to

$$h = \frac{2\gamma}{r\rho_e g}. \tag{12.9}$$

The diameter of these cells varies greatly among species, but a radius of 20 μm is typical (Nobel 1983). Using values of γ = 0.073 J m^{-2} and g = 9.8 m s^{-1}, we see that the capillary rise in a typical vessel would be about 0.74 m. In other words, plants with a height less than about 74 cm could deliver water to their leaves by capillary action alone.

This is fine for small plants, but we are still left with our original question: How does water get to the tops of tall trees? To answer this question we apply the same logic, but in reverse.

Consider the situation shown in figure 12.8. An unbroken column of water extends through the xylem of a tree from the root to a leaf. At its upper end, this xylem "tube" is closed by cells, but these cells are permeable to water. In other words, the top of the tube is not really sealed, it has just been subdivided into a large number of small "tubes," each with its own air-water interface. In an actual

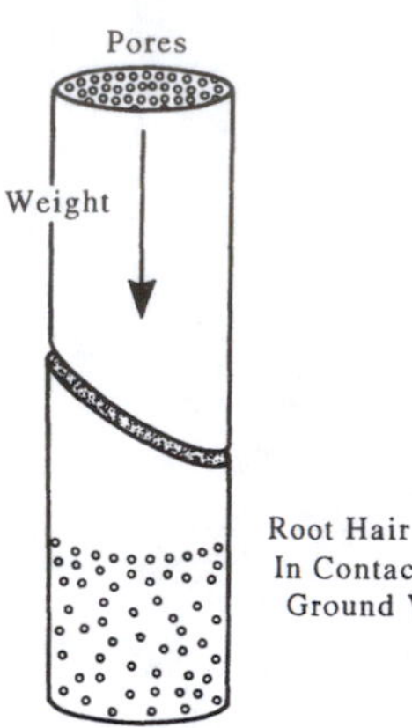

Fig. 12.8 Surface tension in the small pores of a leaf supports the weight of a column of water leading down to the roots.

leaf cell, these small tubes are formed by the interstices between cellulose fibers in the cell wall. Nobel (1983) suggests that the effective radius of these pores is typically 5 nm (5×10^{-9} m).

We know from the calculation made above that if the tree is higher than 74 cm, water would not rise by capillarity alone to fill these interstices. However, once water has been delivered by some other means to these small tubes, the fact that the xylem vessels below them are substantially wider no longer matters. Once the pores of the cell walls in the leaf have been filled, only the negative *pressure* in the water column (not its weight) can act to empty them. How high would the tree have to be before the negative pressure of water in the xylem would pull water down through the pores?

If we insert a radius of 5 nm into eq. 12.9, we find that a tree could be nearly 3 km high before the negative pressure in its xylem would pull water out of the cell wall. This is far higher than any tree on earth, and we may conclude that the height of trees is not limited by the ability of their cell walls to support a water column through surface tension.

This conclusion requires some qualification. First, the negative pressure at the top of a 3 km column of water would be about −300 MPa. This is more than ten times the measured cohesive strength of water. Thus, while the cell-wall pores of a 3-km-high tree might not empty, the column of water in the xylem would most assuredly break. The tensile strength of water would limit trees to a height of about 280 m. This is still several times higher than the highest tree ever found on earth (126.5 m), and one must look elsewhere for the operative factor that limits the height of tress (see, for instance, the discussion of drag in chapter 4).

Second, the presence of small pores in cell walls can explain only how a tall column of water is maintained, not how it got there in the first place. In the case of trees, it seems likely that the water grew into place. In other words, the vessels are first filled when the tree is short by a combination of capillarity and root pressure (a consequence of osmotic pressure in the roots). The column is then slowly pulled upward as the tree grows. As long as no air is introduced into the column, it can be maintained during growth by the capillary pressure of the pores.

Because water in the xylem is under negative pressure, any break in the "tube's" wall is quickly filled with air and any bubble that manages to form inside a vessel quickly expands. As a result, even a minor injury to a vessel can fill the entire tube with gas rather than liquid. Because a gas cannot sustain a negative pressure, the capillarity of pores cannot maintain the water column, and the vessel can no longer be used.

If the capillary rise in small pores is so great, why, one might ask, do plants bother with xylem? A stem formed of nothing but porous cell-wall material would indeed form an effective wick, and would be capable of transporting water to the top of even the tallest trees by capillarity alone. There is a severe problem, however, concerning the *rate* with which water could be transported.

Recall from chapter 5 that at a low Reynolds number, the pressure difference required to force water through a small tube at a given rate is inversely proportional to r^4:

$$\Delta p = \frac{8\mu J \ell}{\pi r^4}, \tag{12.10}$$

where J is the volume flux of water ($m^3\ s^{-1}$) and ℓ is the length of the tube.

Now, 16 million pores (r = 5 nm) could fit into the same cross-sectional area as one vessel (r = 20 μm), so the flux through each pore need be one sixteen-millionth of that in the larger pipe for the overall flux to be the same. But because the radius of a pore is only 0.00025 times that of a vessel, the factor of r^4 in the denominator of eq. 12.10 is (16 million)^2 times smaller for the smaller tube. Working through the math, we find that the pressure difference required to pump water at a given flux through a group of pores would be 16 million times that required to force the same flux through a vessel. Because the water at the bottom of a xylem tube is only slightly above ambient air pressure, this immense pressure difference could only be obtained by having a large negative pressure at the top of the tree. Even the small pores of the cell wall could not support a negative pressure 16 million times that found in extant trees. For this reason, if for no other, trees cannot act as wicks and still transport water at a practical rate.

The calculations made above show how trees and other plants use small pores to their advantage. Under certain circumstances, however, the tenacity with which water remains in small pores can be problematic. For instance, small interstices between soil particles can harbor substantial amounts of water, a commodity of substantial biological value. But the high surface tension of water and the small size of these interstices make it difficult, if not impossible, for a plant to draw the water out. Thus, as a result of surface tension, a fraction of the water in soil is effectively unavailable to plants.

12.2.2 *Insect Tracheae*

In a similar fashion, capillarity is not always a "good thing" for animals, either. Consider, for instance, the problem faced by insects. These animals rely on their tracheae and tracheoles to deliver oxygen to their muscles and to remove carbon dioxide. As we have seen (chapter 6), this system works only because the tracheae and tracheoles are filled with air. If these small tubes become filled with water, the rate at which they transport O_2 and CO_2 decreases 10,000-fold. How does the respiratory system of insects keep from filling up with water via capillarity?

The answer is likely to be that the inner surface of the respiratory system is coated with some substance that is not wetted by water. Recall from eq. 12.8, that the height to which a column of water rises in a small tube is proportional to the cosine of the contact angle. If the contact angle is greater than 90° (as it would be for a nonwettable surface), the cosine is negative and water is actively forced out of the tube. Thus, if the tracheae and tracheoles are coated with a waxy substance similar to that found on the external cuticle of many insects, water has no tendency to fill the system, and effective respiration is possible. Due to the small size of these tubes, however, it will be difficult to detect the presence of wax on their surfaces, and I know of no attempt actually to confirm this speculation.

12.2.3 *Pressure in Spherical Bubbles*

Let us now turn our attention to bubbles of air in water. In particular, the bubbles of interest are those that form the lungs of amphibians, reptiles, and mammals.

The smallest divisions of the vertebrate lung are the alveoli—small, more or less spherical air spaces surrounded by moist lung epithelium. As air is drawn into the lung, these alveoli increase in volume and their surface area is stretched. As a result, the area of the lung's air water interface is increased, a process that we know requires energy. How much work is required to expand the lung?

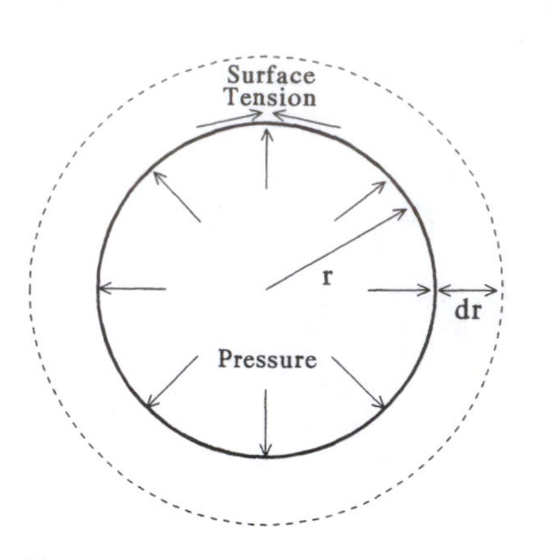

Fig. 12.9 Surface tension increases the pressure inside a bubble. The smaller the bubble's radius, the greater is the pressure.

The complete answer to this question is highly complex, but we can achieve a qualitative understanding through the examination of a simplified model. In this analysis we treat an alveolus as if it were a bubble of air suspended in water (fig. 12.9) and explore the pressure required to increase the bubble's volume.

We know that the work required to expand the surface area of the bubble by an amount dA is γdA. Now, the surface area of a spherical bubble of radius r is $4\pi r^2$, so we can replace the differential area dA with the derivative of $4\pi r^2$, which is $8\pi r dr$. Thus, the work entailed in increasing the radius of the bubble by an amount dr is

$$\text{surface work} = 8\gamma\pi r dr. \tag{12.11}$$

The energy required to effect this increase in radius is provided by an increase in pressure within the bubble. For instance, we could insert a small needle into the bubble and pump in some additional air. The pressure-volume work done in this fashion is equal to the pressure, p, times the change in volume. For a small change in radius, dr, the change in volume of a sphere is $4\pi r^2 dr$. Thus, the pressure volume work involved in inflating the sphere to its new size is

$$\text{pressure-volume work} = 4p\pi r^2 dr. \tag{12.12}$$

Equating the work done against surface tension with pressure–volume work, we find that

$$8\gamma\pi r\, dr = 4p\pi r^2\, dr \tag{12.13}$$

$$p = \frac{2\gamma}{r} \quad \text{(spherical interface).} \tag{12.14}$$

In other words, the pressure required to inflate the sphere is proportional to the surface tension and inversely proportional to the radius of the bubble.

This relationship (known as *Laplace's law*) has two biological implications. First, the smaller the bubble, the more difficult it is to inflate. Thus it takes more work to inflate a lung made of small alveoli than it does to inflate one made of large alveoli. In this respect, it would be advantageous to build a lung from large alveoli. There is a disadvantage to this strategy, however. For a given overall lung volume, a group of large alveoli has less surface area for gas exchange than does the same volume of small alveoli. In other words, there is an inherent trade-off between the efficiency of gas exchange and the work required to inflate the lung.

A more severe consequence can be understood through an examination of figure 12.10. Here two bubbles of different sizes are connected by a tube, as alveoli would be connected by branchioles in the lung. Because the smaller bubble has a higher pressure, it tends to collapse, forcing air into the larger bubble. This tendency renders an alveolar lung unstable. Unless all the alveoli are *exactly* the same size, the smaller ones collapse, and the larger ones thereby grow larger. The process continues until, in theory, only one expanded alveolus (the largest one) remains. In practical terms, the dependence of pressure on bubble radius makes it very difficult to inflate the alveoli of a lung evenly.

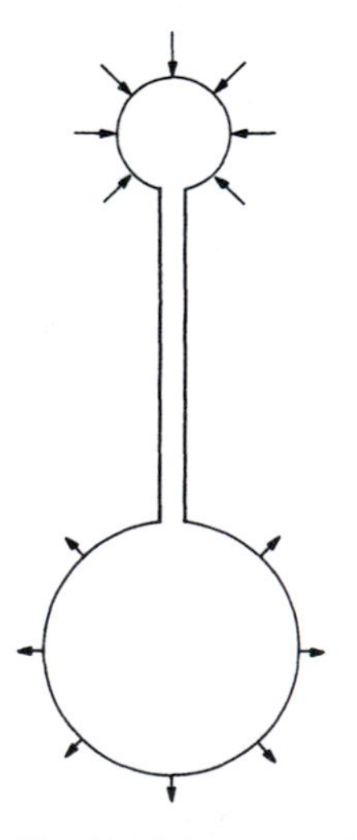

Fig. 12.10 Due to Laplace's law, a small bubble has a greater internal pressure than a large one. If the two are connected, the small will empty into the large.

That is, it would be difficult if a trick had not evolved to circumvent the problem. The fluid that coats the alveolar epithelium contains a surface-active phospholipid (a *surfactant*) that has two properties. First, it lowers the surface energy of the fluid, and thereby reduces the amount of work required to inflate the lung. Second, the molecules interact to yield a variable surface tension. The smaller the area of the alveolus, the lower the effective γ. The net result is that the surfactant tends to reduce the dependence of pressure on radius, and thereby to stabilize the lung structure. Human babies that are born prematurely often do not have sufficient surfactant in their lungs (a condition known as hyaline membrane disease), and therefore have tremendous difficulty breathing.

The importance of surface active molecules in the evolution of lungs is reviewed by Clements et al. (1970) and Hills (1988).

12.2.4 *Cylindrical Bubbles*

It is useful to extend the approach we followed above to shapes other than spheres. For instance, water is often held between two more-or-less straight sections of solid. Between hairs, for example, or between the needles on a pine tree. The resulting air-water interface is shaped like a portion of a cylinder (fig. 12.11A). What is the pressure distribution around this type of curved surface?

Consider the cylindrical bubble shown in figure 12.11B. The bubble has a radius r and a length ℓ. For simplicity, we assume that the length is much greater than the radius, so that most of the interface is in the walls of the cylinder rather than in its ends. In this case, the ends form a small part of the surface; they therefore contribute little to the pressure, and we can safely ignore them.

The pressure within the bubble is calculated as before. The work required to expand the area of the bubble's walls is γdA, where dA is the change in area. We can express dA in terms of the change in radius by assuming that the length of the cylinder does not change. In this case, $dA = 2\pi\ell dr$, and the work required to expand the bubble against surface tension is

$$\text{surface work} = 2\pi dr\ell\gamma. \tag{12.15}$$

The energy required to expand the bubble is, again, provided by the pressure within the cylinder. The pressure-volume work is equal to the product of pressure and the change in volume ($2\pi r\ell dr$), which in this case is

$$\text{pressure-volume work} = 2\pi r\ell p dr. \tag{12.16}$$

Equating the work done against surface tension with pressure-volume work, we see that

$$p = \frac{\gamma}{r} \quad \text{(cylindrical interface).} \tag{12.17}$$

In other words, the pressure inside a cylindrical bubble is proportional to the surface tension and inversely proportional to the cylinder's radius. This result is similar to that for a spherical bubble, but note that the pressure is only half that which would occur in a sphere of the same radius (eq. 12.14). The difference is due to the fact that the interface of a cylinder curves around only a single axis, whereas a spherical interface is, in a sense, curved in two ways at once.

It is difficult to imagine where one would find a complete cylindrical bubble in nature, but the results of eq. 12.17 can be applied directly to sections of bubbles, the cylindrical interfaces cited above. Consider, for example, the situation depicted in figure 12.12. A series of square rods (here viewed end-on) separate a body of water (above) from air (below). The rods are spaced a distance $2r$ apart. In the absence of surface tension, water would be free to flow between the rods. But as the water intrudes upon the air, the interface between media is stretched, a process we know

A

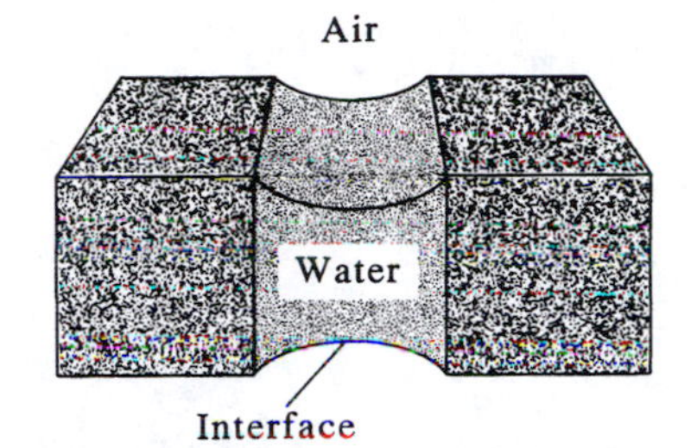

B

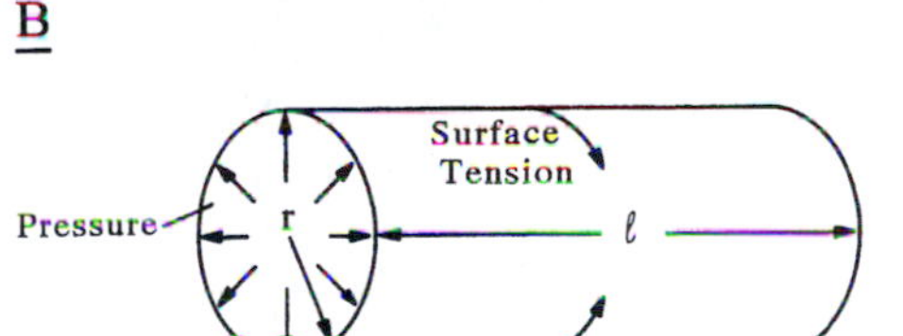

Fig. 12.11 Water held between two straight sections of solid has a cylindrical interface with air (A). The tension in such an interface increases the pressure inside a hypothetical cylindrical bubble (B).

requires the expenditure of pressure-volume energy under these circumstances. How much of a pressure difference can be sustained by surface tension before the water breaks free and flows into the air?

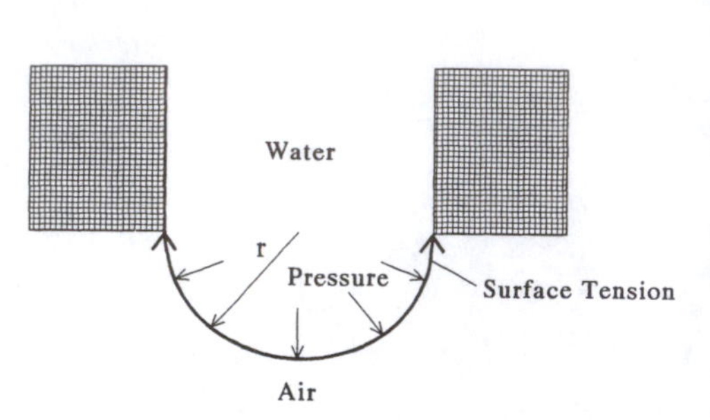

Fig. 12.12 Surface tension in the air-water interface between two solid rods acts like a "membrane," preventing the water from moving into the air. Here the rods are assumed to be nonwettable.

Let us assume that the material from which the rods are constructed is nonwettable, in which case the contact angle between the interface and the rods approaches 180° (π radians). Given this situation, the air-water interface approximates half a cylinder with a radius r, and we may suppose that the pressure difference between air and water is γ/r as predicted by eq. 12.17.

Note that in this case the curvature of the interface is the opposite of that in our earlier analysis; the water protrudes into the air rather than the air into the water. As a result, the increased pressure due to surface tension is found in the water rather than the air. You can convince yourself of this fact by thinking of the air-water interface as a stretched rubber "membrane." The curvature of this "membrane" is such that it pulls inward on the water, thereby increasing its pressure.

The net result of this analysis is that the presence of nonwetting rods can maintain a pressure difference between air and water of γ/r. If the spacing between rods is small, this pressure difference can be substantial. For instance, if $r = 0.5$ μm (a 1 μm gap between rods), a pressure difference of 1.46×10^5 Pa (1.46 atmospheres) can be maintained.

Plastron Respiration

The ability of surface tension to separate water from air is used by several bugs and beetles to breathe underwater. To see how this works, let us first consider what happens to the gas in an ordinary bubble held at a given depth below the water's surface.

Although water and air are separated at the surface of the bubble, gases are free to diffuse across the interface. As always, the direction of net transport is from the area of higher concentration to that of lower concentration. Now, because water is virtually incompressible, the concentration of oxygen dissolved in water remains the same as that in the atmosphere (2×10^4 Pa) regardless of depth. In contrast, the concentration of oxygen in the bubble varies with pressure and therefore *does* vary with depth.

Consider, for example, a bubble 1 m below the surface. At this depth, the pressure is approximately 1.1 atmospheres, so the partial pressure of oxygen in the bubble (20.9% of the overall pressure) is about 2.2×10^4 Pa. Because this is higher than the pressure of O_2 in the surrounding water, oxygen diffuses out of the bubble. The same is true for the other gases in the bubble. As a result, the bubble gradually shrinks.[3] This analysis leads us to conclude that a bubble can be maintained under water only if there is some mechanism available to maintain its internal pressure at something less than atmospheric pressure.

Therein lies a constructive role for surface tension. The abdomens of underwater insects are covered by fine hairs that are typically bent over near their distal ends to lie parallel to the cuticle. These bent ends are analogous to the rods in our example above, and they serve to maintain an air space around the abdomen when the insect is submerged. When the insect arrives at a given depth, the gas in the air space diffuses into the surrounding water until the pressure in the air space is less

[3]From Laplace's law we know that the pressure in the bubble is greater than the surrounding hydrostatic pressure due to surface tension. The smaller the bubble, the greater the surface-tension-induced pressure, and the greater the concentration difference between gases in the bubble and those dissolved in the surrounding water. Thus, the smaller the bubble, the faster it shrinks.

than atmospheric. At that point the air space and the water are at equilibrium. Any oxygen subsequently removed from the air space by the insect's respiration lowers the pressure of oxygen in the air space below atmospheric pressure, and oxygen diffuses into the air space from the surrounding water. In other words, because the surface tension acting among hairs allows air to be at a lower pressure than the water, the hairs (collectively referred to as a *plastron*) act much like a gill.

The depth to which plastron respiration is effective is limited primarily by the spacing between hairs. For a 1 μm spacing as analyzed above, we know that the maximum difference in pressure between water and air is 1.46 atmospheres. Thus, a beetle with a plastron of this sort could submerge to a depth where the overall pressure was almost 2.46 atmospheres and still keep its air space at less than atmospheric pressure. A pressure of 2.46 atmospheres is reached at a depth of approximately 14.6 m, so this would be the deepest the beetle could dive and continue to rely on its plastron.

The analysis we have performed here is only a first approximation in that it has assumed a square cross section for the plastron hairs and a contact angle of 180°. A more exacting analysis has been carried out by Crisp and Thorpe (1948) in which the hairs are assumed to be circular in section (a more likely situation) and the contact angle is allowed to vary. Their results show that the maximum difference in pressure across the plastron is

$$\Delta p = \frac{\gamma \cos(\theta_c + \phi)}{\ell/2 - r\cos\phi}, \tag{12.18}$$

where r is the radius of each hair and ℓ is the center-to-center spacing between hairs (fig. 12.13). ϕ is the angle (measured from the line connecting the centers of hairs) at which the interface meets the hair. At the maximum pressure difference,

$$\phi = \arcsin\left(\frac{2r}{\ell}\sin\theta_c\right) - \theta_c. \tag{12.19}$$

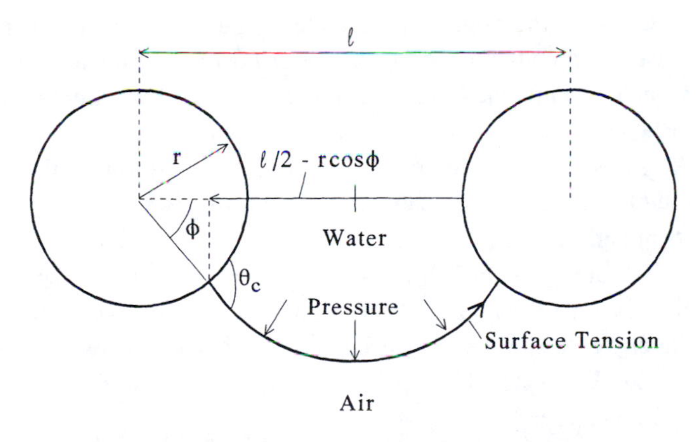

Fig. 12.13 A cross section through the plastron hairs of an aquatic insect. Surface tension prevents water from flowing between hairs, thereby maintaining an air space next to the animal's body.

The cuticle of many insects is covered with a waxy substance for which the contact angle is approximately 105° to 110°. We may assume that $r = 0.5$ μm and $\ell = 2$ μm, which yields the same 1 μm spacing between hairs that we dealt with previously. Inserting these values into eq. 12.19 we calculate that $\phi = -76°$ to $-82°$, where the negative sign indicates that the interface contacts the hairs below the line connecting their centers, as shown in figure 12.13. Using an average value of $-79°$, we calculate from eq. 12.18 that the maximum pressure difference that can be maintained is 6.9×10^4 Pa, about half the value we obtained previously.

Note that when $\theta_c < 90°$ (in other words, when the material is wettable), the contact between the interface and the hair occurs at a point such that $\phi < -90°$. In this case, the interfaces wrapping around the two sides of the hairs meet in the middle of the hair, and the plastron ceases to work. In other words, the full depth to which an animal can dive with hairs of a given dimension is available only if the material of the hairs is sufficiently nonwetting so that $\theta_c > 90°$.

Crisp and Thorpe also assumed that the depth to which an insect could dive was limited by the strength of the hairs. If the pressure difference across the air-water interface was too large, the hairs would buckle, and the airspace would collapse. A more recent study by Hinton (1976) suggests that plastron hairs are robust enough to allow insects the full depth possible given the surface tension of water.

Insects could increase their depth range by decreasing the spacing between their plastron hairs while maintaining a given hair radius (eq. 12.18). At some point, however, this strategy would become impractical. For hairs of a given diameter, the closer the spacing, the smaller the area across which gas can diffuse, and the less efficient respiratory exchange becomes. The area available for gas exchange could be maintained constant if both the spacing between hairs and the radius of the hairs themselves were decreased. In this case, the hairs would become more flexible, and at some point would buckle as suggested by Crisp and Thorpe.

Actual plastron hairs come in a bewildering variety of shapes and sizes (Hinton 1976), and you are invited to calculate for yourself the depth to which various species may dive.

Human technology has recently made use of the principle that underlies plastrons. Fabrics such as GoreTex have within them a layer of nonwettable material that is pierced by a myriad of small holes. Surface tension prevents water from moving through these holes, thereby making the fabric waterproof, but air is free to diffuse through the holes, allowing the fabric to "breathe."

12.2.5 *Gas Vesicles in Blue-Green Algae*

Consider the plight of blue-green algae. To maintain their place in the sun, these small cells must be neutrally or positively buoyant, and buoyancy is provided by membrane-bound, cylindrical gas vesicles within the cell. Some of these vesicles are extremely small, having a radius of perhaps 0.05 μm (Walsby 1972). At this size, the pressure exerted by surface tension on the gas in a cylindrical bubble (eq. 12.17) is greater than 14 atmospheres. How can the vesicle maintain its integrity?

The answer seems to lie with the surface energy of the vesicle's membrane. Walsby (1972) suggests that the outer surface of the membrane is hydrophilic, thereby reducing the difference in surface energy between the "bubble" and the surrounding water. The concomitant reduction in surface tension reduces the applied pressure.

12.3 Capillary Adhesion

The concept of eq. 12.17 can be applied to a classic biological problem. Anyone who has spent a summer afternoon sitting next to a window has noticed the amazing ability of flies to stick to glass. How can a fly adhere to a surface that is so smooth?

Even if the glass is held horizontally, a fly can attach to its underside with no apparent problem. How do they, and other small insects, do it?

Consider the situation shown in figure 12.14. A small volume of water is sandwiched between two parallel plates that are separated by a distance x. Let us suppose that water wets the plates so that the contact angle is 0. In this case, the water is more attracted to the material of the plates than it is to itself, and it attempts to spread radially across the plates. In this respect, the situation is analogous to the way in which water is drawn upward into a tube by capillarity. As a result of the water's tendency to spread, a negative pressure is imposed on the water. How large is this negative pressure?

A close approximation to the answer can be obtained if we think of the air-water interface betwen the plates as a "hemicylindrical" interface that has been wrapped around so that its ends meet. If we look at a radial cross section through the interface (fig. 12.14A), it indeed looks like the cross section of half a cylinder with a diameter of x. The curvature of the interface is such that air protrudes into the water, and the low pressure induced by surface tension occurs on the water side of the interface. Again, an appeal to the analogy of a rubber membrane should convince you that the tug of the surface tension serves to reduce the pressure in the water.

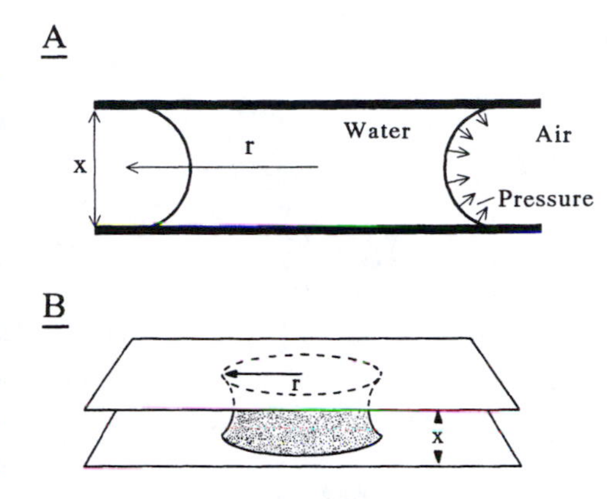

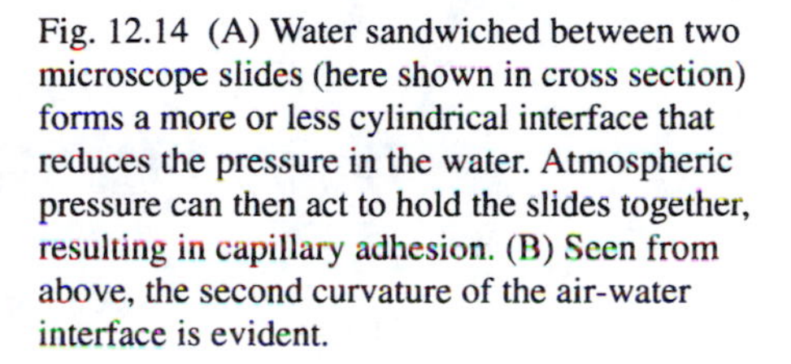

Fig. 12.14 (A) Water sandwiched between two microscope slides (here shown in cross section) forms a more or less cylindrical interface that reduces the pressure in the water. Atmospheric pressure can then act to hold the slides together, resulting in capillary adhesion. (B) Seen from above, the second curvature of the air-water interface is evident.

Now, if the diameter of the cylindrical interface is x, its radius is $x/2$ and from eq. 12.17 we can guess that the pressure in the water is $\gamma/r = 2\gamma/x$ less than that of the surrounding air. This guess would be close to the truth, but not exact. In the process of wrapping a cylindrical interface so that its ends meet, we have introduced a second curvature. This is the curvature one would see when looking at the flat surface of the plates (fig. 12.14B). Again, if we only think of this single curvature, the interface resembles a cylinder, in this case with a radius r. Because this curvature of the water is in the opposite direction from the other, the action of surface tension is to *increase* the pressure in the water by an amount γ/r. This pressure must be added to that of the wrapped interface to calculate the overall difference in pressure between the water and the surrounding air. Thus,

$$\Delta p = \gamma\left(\frac{1}{r} - \frac{2}{x}\right). \qquad (12.20)$$

For example, a drop of water placed between two microscope slides might have a radius r of 1 cm and a thickness x (set by imperfections in the slides) of 10 μm. In this case, the pressure inside the water is about 1.5×10^4 Pa less than that of the surrounding air. As a result of this difference in pressure, a force equal to $\pi r^2 \Delta p = 4.6$ N acts, tending to push the slides together. In other words, due to the action of capillarity, the slides adhere with a force of 4.6 N, and a force in excess of this would have to be applied to separate them.

This is probably the mechanism by which some insects and frogs adhere to smooth surfaces. For instance, Dixon et al. (1990) have shown that this type of capillary adhesion is sufficient to explain the force with which aphids adhere to a variety of surfaces, and is likely to be involved in the adhesion shown by some beetles (Stork 1980). Emerson and Diehl (1980) have shown that capillarity plays a major role in the adhesion of small frogs to glass. Provided that there is sufficient liquid available, either on the substratum or on the animal's foot, this type of adhesion seems likely to be common among small organisms.

It should be noted that capillarity resists only those forces that are directed perpendicular to the plates between which the water is sandwiched. A force applied parallel to the plates simply shears the fluid, as described in chapter 5. In this case, the movement of one plate relative to the other is resisted by the viscosity of the water, and the resistive force is therefore a function of how fast the fluid is sheared. Only if the layer of water is very thin is the the force resisting shear substantial. For instance, in the situation described above, if one plate is moved parallel to the other at a speed of 1 cm s^{-1}, the shear stress in the fluid is 1 Pa, and a force of only 0.0001 N would be required to move the plates at this speed. Thus, capillary adhesion can often be accompanied by a tendency to slip along the surface. In practice, slippage may be reduced by the small-scale roughness of the substratum. On the scale of a few micrometers even a glass plate can be "rough" enough to keep a fly from slipping.

To this point we have tacitly assumed that both plates in a capillary adhesive system are made of the same, wettable material. If this is not true, eq. 12.20 needs to be modified. If the substratum has a surface energy such that the contact angle with water is θ_1 while the foot of the insect has a contact angle with water of θ_2, and if r is much greater than x, then Dixon et al. (1990) purport that

$$\Delta p = \gamma \frac{\cos\theta_1 + \cos\theta_2}{x}. \tag{12.21}$$

When both the substratum and the foot are wettable, this equation simplifies to eq. 12.20, provided $r \gg x$. If, instead, the substratum is wettable and the foot of the insect is coated with wax ($\theta_c = 110°$), the difference in pressure is only about a third of that which would be obtained for a wettable foot.

You should not infer from this discussion that capillary adhesion is the *only* mechanism by which insects adhere. The fine hairs on insect legs are small enough so that they may not be prevented by surface irregularities from making very intimate contact with a substratum. If the hair can approach closely enough, van der Waals forces between cuticle and substratum can result in adhesion. It is a matter of current debate as to what fraction of the overall adhesive ability is due to capillary adhesion and what fraction to close molecular contact. It is safe to say that we still do not know *exactly* how a fly sticks to a pane of glass (Wigglesworth 1987).

12.4 Walking on Water

The surface of lakes and streams provides a unique opportunity for terrestrial organisms. An animal that can walk on water has available to it a flat substratum from which to hunt aquatic prey and a refuge on which to escape from predators. There is just this one problem: How does an animal manage to walk on water?

Surface tension, of course, provides the answer. If the animal contacts the water with a nonwettable structure of sufficient perimeter, the upward force of surface tension can support the organism's weight.

Consider, for instance, water striders, the bugs of the genera *Gerris* and *Halobates* (fig. 12.15). These insects have a mass of about 10 mg, and therefore weigh about 10^{-4} N. We may assume that their legs are made from the same nonwettable material as other insect cuticles, in which case the contact angle between the leg and the air-water interface is about 110°. Now the cosine of 110° is −0.34, indicating that, at best, only about 34% of the force due to surface tension is directed upward to support the animal's weight. The total force of surface tension is $\gamma\ell$, where ℓ is the total perimeter of leg in contact with the water. Thus,

$$10^{-4}\,\text{N} = 0.34\gamma\ell, \tag{12.22}$$

and ℓ need be only about 4 mm to support the animal, about 0.67 mm per foot. In general, water striders have a foot perimeter about 10-fold larger than this. In other words, the animal could support itself by standing on its toes alone! It is evident that walking on water poses little problem for small organisms.

In some cases, the ease with which surface tension supports a small animal can be problematic. Consider, for instance, the plight of marine crustacea such as copepods and the larvae of crabs and barnacles. The cuticle of these species is hydrophobic, a fact that poses no problem as long as the animal is entirely submerged. If, however, the organism accidentally contacts the air-water interface, surface tension strips the water away from the animal, thrusting it into the air. Once out of water, the animal may not be able to force its way back in.

The prospect of walking on water becomes less likely the larger the animal becomes. The crux of the problem is that an animal's weight increases as the cube of a linear measure of its size, whereas the force due to surface tension increases in direct proportion to a linear measure (in this case, perimeter). For example, a mouse with a mass of 100 g weighs 1 N. If the animal had feet coated with the same wax available to a water strider, it would need a perimeter of 40.3 m in contact with the water, roughly 10 m per foot. Feet of this size would clearly be impractical, and for this reason animals the size of mice do not walk on water.

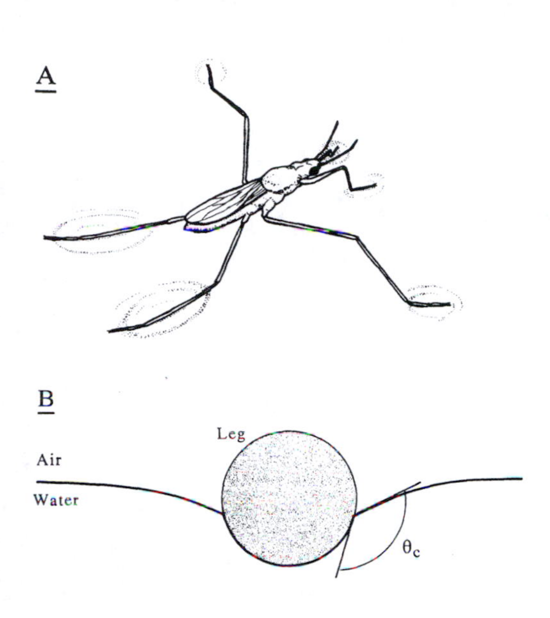

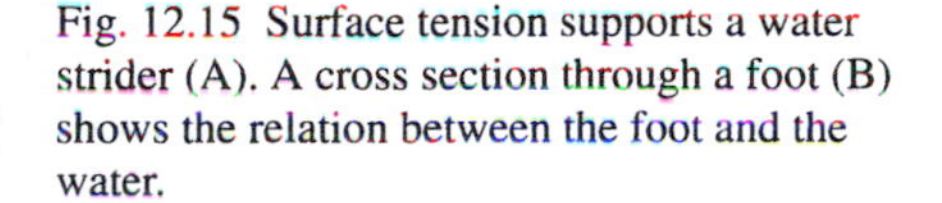

Fig. 12.15 Surface tension supports a water strider (A). A cross section through a foot (B) shows the relation between the foot and the water.

Vogel (1981) points out that a consideration of surface tension allows one to calculate how much an organism the size and shape of a human being could weigh and still walk on water. A typical pair of sandals has a perimeter of about 0.7 m. Even if these sandals are totally unwetted by water, they could support at most about 0.05 N, the equivalent of 5 grams. As Vogel (1988) aptly notes, the theological implications of this calculation are best left to the reader.

The calculations that we have made to this point assume that the organism is standing still on the water's surface. Any movement changes the balance of forces. In some cases, the result is simple to imagine. For example, if the animal attempts to jump up, its acceleration imposes an additional vertical force that must be resisted by surface tension. For small organisms like water striders, this poses no problem; they just need to have a slight surplus of perimeter in contact with the water.

Horizontal movement presents a different problem. How can the rowing motion of a water strider's legs, for instance, propel the animal forward when the legs never actually enter the water? The answer to this question is interesting and involves surface tension, but in a manner that we will leave until our discussion of capillary waves in chapter 14.

One unusual mechanism of locomotion does bear mention here. Small beetles of the genus *Stenus* often walk on water. When disturbed, these animals secrete a small drop of surfactant from their abdomen and bring it into contact with the water. As the surfactant spreads over the surface behind the beetle, it locally reduces the surface tension. The resulting excess surface tension in front of the beetle pulls the animal forward at a speed of up to 0.7 m s^{-1} (Chapman 1982).

This effect can easily be reproduced. If a sliver of soap is inserted into a notch at one end of a small stick, the stick can be propelled across the surface of a pond.

12.5 Summary . . .

The ability of water molecules to form hydrogen bonds to each other gives water an unusually high surface energy. This energy, in the guise of a surface tension, can support the water columns of tall trees and the weight of surface-striding insects. Aquatic insects take advantage of surface tension to maintain an airspace around their bodies, and all insects apparently use surface tension to keep water out of their respiratory plumbing.

The forces exerted by surface tension act only tangential to the air-water interface. If, however, the interface is curved, surface tension can change the pressure in the fluids adjacent to it. It is this change in pressure, for instance, that may explain how small insects adhere to glass.

12.6 . . . and a Warning

This chapter is intended to give the reader an appreciation for the physics of surface tension and a rudimentary understanding of its biological consequences. Many of the analyses presented here are only first approximations, however. For example, in calculating the foot perimeter required to support a water strider, we ignored the buoyancy that the organism would gain from those parts of the foot that are depressed into the fluid and any effects of the surface films commonly encountered on lakes and ponds. This, and other complicating effects are treated in standard texts on the physics of surfaces, such as Davies and Rideal (1963), Adamson (1967), and Bikerman (1970). Princen (1969) presents a particularly thorough examination of the shapes of drops at various interfaces, and of the forces acting on solid particles supported by surface tension. You are urged to consult these sources before making quantitative use of the idea of surface tension.

Chapter 13

Surface Waves

The interface between air and a large body of water is seldom still. Every slight breeze or splash sends ripples coursing across the surface of a pond, and the vast expanse of the oceans provides ample space for storm winds to raise the seas that buffet ships and crash upon the shore. In this chapter we explore the tendency for the air-water interface to form waves and the consequences that these surface oscillations have for plants and animals.

For example, we will calculate why the speed at which waves travel forms a practical upper limit to the speed at which ducks and humans can swim and a lower limit to the effective speed with which water striders move their limbs, but is unlikely to affect the speed of whirligig beetles. We will see why it is more costly for whales and dolphins to swim near the surface rather than deep in the water, and will explore the potential use of ripples as a means of communication.

13.1 Waves and Orbits

We begin by defining the basic terms applied to surface waves (fig. 13.1). In a wave, the water's surface is displaced from the still-water level by an amount we designate as η. The point of maximum upward displacement is the *wave crest* and that of maximum downward displacement is the *wave trough*. The vertical distance between crest and trough is the *wave height*, H, and the horizontal distance between two crests is the *wavelength*, λ. As the wave travels across the surface, each crest moves with a speed called the *celerity* or *phase speed*, c, and the time it takes for one crest to travel one wavelength is called the *wave period*, $\mathcal{T}$. Given these definitions, it is clear that

$$c = \frac{\lambda}{\mathcal{T}} \tag{13.1}$$

$$\mathcal{T} = \frac{\lambda}{c} \tag{13.2}$$

$$\lambda = c\mathcal{T}. \tag{13.3}$$

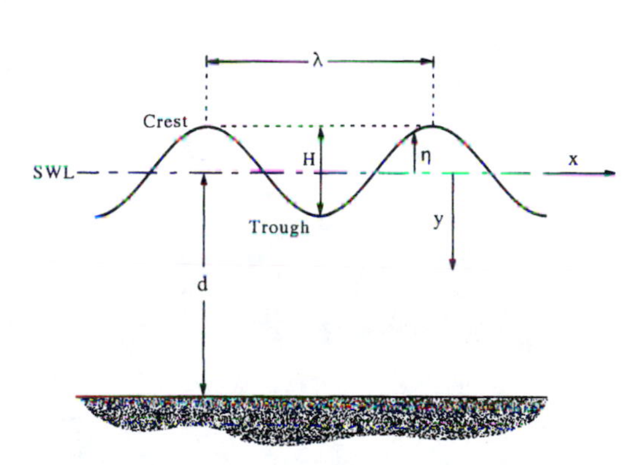

Fig. 13.1 The dimensions of a surface wave. Note that y is measured from the still water level (SWL) and is positive downward.

The vertical distance from still-water level to the substratum is the depth of the water column, d. As we will see, surface waves behave differently when d is large compared to λ than when d is small. As a rule of thumb, we refer to a water column as "deep" if $d > \lambda/2$, and "shallow" if $d < \lambda/20$. Depths in between are termed "intermediate." According to these rules, the effective depth of a water column is a function of a wave's length: $d = 1$ m is deep water for a ripple with a wavelength of 10 cm, but is shallow for an ocean wave for which $\lambda = 100$ m. The sense behind these rules will become apparent when we discuss the subsurface water motion that accompanies surface waves.

With these terms in hand, we now perform an experiment. Armed with a fishing bob, we travel to the edge of a still pond. The bob is tossed into the water a few meters from shore, where it floats motionless. We then splash the water at the shoreline to create a train of waves that radiates outward and eventually passes the bob. Let us assume for the moment that $d > \lambda/2$ so that the bob is in effectively deep water. What is the motion of the bob as the waves move by?

The most obvious motion is up and down. As the crest of a wave approaches, the bob (along with the surface water on which it floats) moves upward. After the crest has passed, the bob moves down as the subsequent trough approaches. Careful observation reveals, however, that there is also a horizontal motion of the bob. During the time the bob is above the level of still water, the bob moves forward in the direction of wave travel. While below still-water level, the bob is drawn backward, opposite the direction of wave propagation. In sum, the motion of the bob is forward and up, then forward and down, backward and down, and then backward and up, a sequence that is repeated for each wave that passes by (fig. 13.2A).

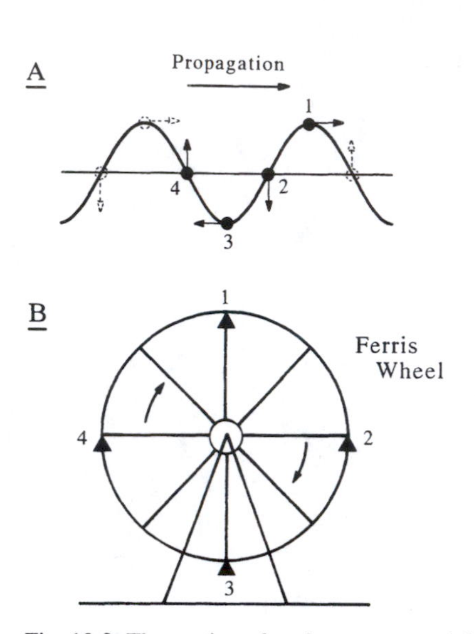

Fig. 13.2 The motion of surface water particles during the passage of a wave (A) is analogous to the motion of a car on a Ferris wheel (B). (From Denny 1988)

This is the same sequence of motions one experiences when riding a Ferris wheel (fig. 13.2B), an analogy that serves to make two points. First, while a Ferris wheel rotates, it does not progress across the fairgrounds. Similarly, the motion of the fishing bob (and therefore of the water at the surface) results in no net movement as a wave passes by. For every movement upward and forward, there is an equivalent movement downward and back, and with the passage of each wave the water at the surface ends up exactly where it started.[1] Second, during the passage of a wave, the water at the surface moves through a circular path, called an *orbit*. We note for future use that the diameter of this surface orbit is equal to the wave height, H (fig. 13.2A).

The motion of the water at a point on the surface (as revealed by our fishing bob) is in distinct contrast to the motion of the waveform. We produced the waves at the shoreline and they subsequently traveled out to the bob, so the waveform (unlike the water below it) does indeed have a net displacement that increases with time. The steady progression of the waveform is therefore distinct from the orbital motion of the water, and this distinction must be kept in mind when exploring the physics of surface waves.

Although the motion of waveform and water are distinct, they are intimately coupled, and to understand the biological consequences of surface waves we must understand the nature of this union. Before we can do so, however, we need two pieces of information.

13.2 Streamlines

We first require a means to describe the path along which water flows. Consider an infinitesimal volume of water (a so-called *fluid particle*) in the midst of a steady flow (fig. 13.3). When we first observe the particle it is located at x,y,z. At this location we can, in theory, measure the direction in which the particle is moving and the speed with which it moves. From these data we can calculate where the particle will be a short time later. We can then follow the particle to its new location, where it may be acted upon by a new set of forces. We note any change in the particle's speed and direction and again predict its next move. In this fashion, we can completely define the path that a particle takes, and thereby define one aspect of the flow.

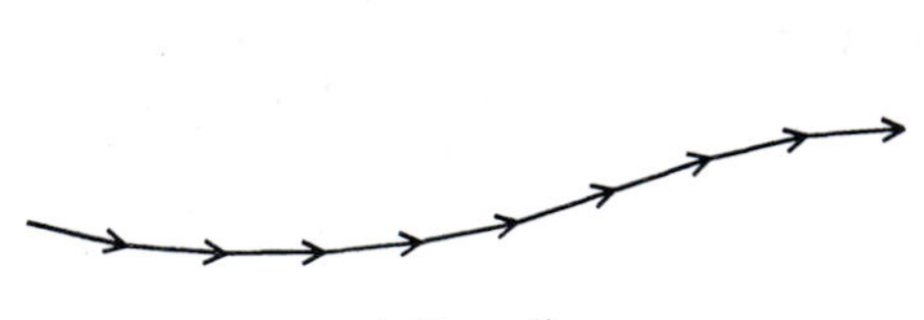

Fig. 13.3 A streamline traces the path of a fluid particle as the particle moves in time-invariant flow.

[1] Actually, this is an overstatement. There is a slight net movement of water in the direction of wave propagation, a tendency that can be predicted from wave theories such as Stokes's second-order theory. For a brief explanation of the phenomenon, see Denny (1988); or for a more complete mathematical treatment, consult Kinsman (1965) under the subject of mass transport. The net motion associated with deep-water waves is so slight, however, that its neglect in this discussion poses no threat to the validity of our conclusions.

Now, if the overall flow of the water does not vary with time (this is what we mean by the flow being *steady*), it does not matter when we make our observations. In other words, if the flow is truly steady, a fluid particle released at x,y,z at one time must follow the same path as a fluid particle released at any other time. In this case, the path we have defined is called a *streamline*.

Note that in its travel down a streamline, a fluid particle can speed up and slow down and it can change its direction. The concept of the streamline requires only that every particle that passes through point x,y,z subsequently go through the same sequence of changes in speed and direction. Note also that as a consequence of our definition, streamlines can never cross. If they did, particles at their intersection could move in either of two directions, which, by definition, is not allowed.

13.3 Bernoulli's Equation

As fluid particles move along streamlines, they have associated with them a certain amount of energy. For example, if a moving particle has mass m, it has a kinetic energy,

$$\text{kinetic energy} = \frac{mu^2}{2}, \tag{13.4}$$

where u is the particle's speed (see chapter 3).

The particle may also have some gravitational potential energy associated with its vertical position in the water column. In this case,

$$\text{potential energy} = mgy, \tag{13.5}$$

where g is the acceleration due to gravity and y is the vertical distance from some reference height.

If water particles connect areas of the fluid that are at different pressures, they can transmit energy between these areas. In the process, particles appear locally to have a third type of energy in addition to their kinetic and gravitational potential energies. This is most easily seen through an example. Consider the apparatus shown in figure 13.4. A reservoir of water is connected at its base to a horizontal pipe. The pipe contains a piston that is attached to a block of concrete. The hydrostatic pressure of the fluid in the reservoir imposes a force on the piston and thereby on the block. When this force is first applied, the piston and block are accelerated, but when the frictional force of the block moving over the substratum is equal to that applied to the piston, the acceleration ceases and the block moves at a constant speed. We examine the system once it has reached this steady-state condition.

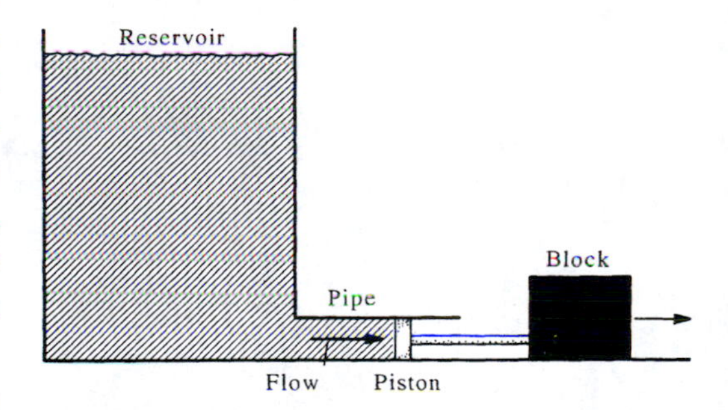

Fig. 13.4 Water in a horizontal pipe connects a region of high pressure to one of lower pressure, and can thereby do work. This type of energy is called *flow energy*. (From Denny 1988)

Now, the water in the pipe moves at a constant speed, so its kinetic energy is constant. Similarly, the fluid in the pipe does not change its vertical position, so its gravitational potential energy is constant. Nonetheless, the water in the pipe does work by moving the block against a frictional resistance. This energy (which is neither kinetic nor potential) is termed *flow energy* (Streeter and Wylie 1979; Massey 1983), and we can easily calculate its magnitude.

The work done by the water in the pipe is equal to the force it exerts times the distance, ℓ, that the block is moved. Now, the force applied to the piston is the product of hydrostatic pressure and the piston's area, pA. Thus,

$$\text{flow energy} = pA\ell = pV, \tag{13.6}$$

where V is the volume of fluid that enters the pipe. In other words, because the water in the pipe connects areas of different pressures, it is capable of doing an amount of work equal to pV. Flow energy is subtly different from kinetic and potential energy in that it is not actually possessed by a water particle. Rather, it represents work that can be done because the particle physically connects two areas of different pressure. In this example, the flow energy of fluid in the pipe exists only because of the pressure maintained by the gravitational energy of water in the reservoir, and as fluid flows into the pipe, it does so at the expense of this potential energy.

At any point in space, the overall energy of a fluid particle (termed the *available energy*) is the sum of its kinetic, flow, and potential energy:

$$\text{available energy} = \frac{mu^2}{2} + pV + mgy. \tag{13.7}$$

We now take an important step toward understanding wave motion. We assert that *in the absence of viscous effects* the available energy of a fluid particle remains constant as it travels along a streamline. In mathematical terms,

$$\frac{mu^2}{2} + pV + mgy = \text{constant}. \tag{13.8}$$

This is one form of *Bernoulli's equation*, named in honor of Jakob Bernoulli (1654–1705).

The validity of this assertion follows from the first principles of fluid dynamics, but we will not explore the derivation here. You are invited to consult standard fluid dynamics texts (e.g., Streeter and Wylie 1979; Massey 1983) on the subject of Euler's equations.

Bernoulli's equation will be most useful to us if it is slightly rearranged. First, we will find that it is more convenient to deal with the energy associated with a given volume (rather than mass) of fluid. Now, the volume of a particle is equal to the quotient of its mass and density. Dividing both sides of eq. 13.8 by volume (in the guise of m/ρ), we see that

$$\frac{\rho u^2}{2} + p + \rho gy = \text{constant}. \tag{13.9}$$

This is the form in which we will use Bernoulli's equation.

Note that in formulating this equation, we specifically require that fluid particles be free from viscous action as they flow along a streamline. In practice, this means that the fluid with which one deals must be far enough away from a solid surface so that the shear associated with the no-slip condition cannot affect the flow. In many cases this can be a severe restriction, but for the present discussion it poses no problem. Because we are dealing with waves on the surface of the water, we are likely to be far from a solid substratum, and viscous effects are minimal.

13.4 Wave Celerity

We are now in a position to begin to relate the motions of a waveform to those of the water over which the waveform moves. In particular, we use Bernoulli's equation to calculate the celerity of surface waves in deep water.

At first this may seem an unlikely proposition. The water in waves moves up and down and back and forth, whereas Bernoulli's equation (in the form we use) applies only to steady flow. We can, however, reconcile the unsteady nature of waves with the requirements of our equation by bringing the waveform to a halt in the reference frame of the laboratory. This is done in a fashion similar to that in which we treated sound waves in chapter 10. If our waves travel from left to right, we can make them stationary to our eye by moving the water from right to left with a speed equal to that of the waveform. In other words, if waves move with celerity c, they can be brought to a halt by a countercurrent moving with velocity $-c$ (fig. 13.5).

If we are careful in the production of our waves, the waveform is the same across the width of our laboratory tank. This allows us to look at one vertical "slice" of the wave (taken in a plane parallel to the direction of wave propagation) and know that this slice is representative of the entire tank. Thus, by bringing the waveform to a halt and looking at a single slice, we can reduce a complex, unsteady, three-dimensional problem to a relatively simple, two-dimensional flow.

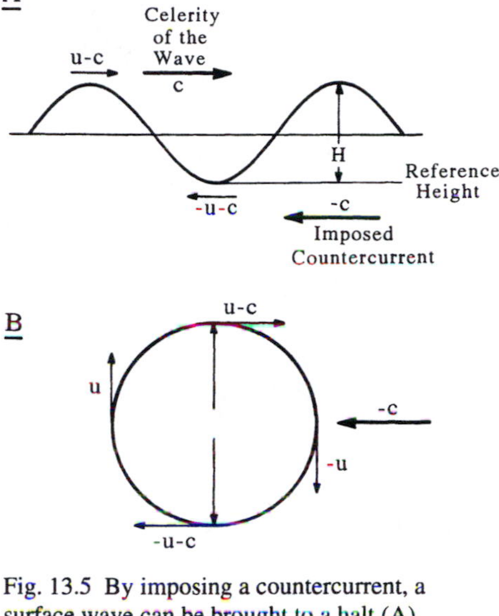

Fig. 13.5 By imposing a countercurrent, a surface wave can be brought to a halt (A), allowing us to specify the water's velocity at the crest and trough. As water flows around its orbit, its direction changes, but its speed, u, stays the same (B). At the crest, orbital velocity is in the opposite direction from the countercurrent; at the trough, orbital velocity adds to the countercurrent. (From Denny 1988)

In bringing the wave form to a halt, we have not necessarily brought the water to a standstill. In fact, we know that because of the countercurrent, the water moves to our left with an average velocity of $-c$. At the surface, this velocity is modified by the orbital motion of particles as they interact with the waveform. In particular, we know that particles at the crest of a wave move in the direction of the wave with an orbital speed that we may signify by u. Thus, the speed of surface water at the crest is $u - c$. At the trough of a wave, we know that the water moves with speed u in the direction opposite that of the waveform (in the same direction as the countercurrent), so the overall velocity in our frame of reference is $-u - c$.

We can also specify the magnitude of u. Recall from figure 13.2A that the diameter of an orbit is equal to H, the wave height, so the distance a particle travels as it moves through one complete orbit is πH. This circuit is completed once during each wave period, T, so the speed of the particle in its orbit is

$$u = \frac{\pi H}{T}. \tag{13.10}$$

We now make use of Bernoulli's equation. Because our slice of the waveform is stationary in our frame of reference, particles that move along its surface always trace the same paths. In other words, the surface of the water in our slice is a streamline, and the available energy of particles moving along it must be constant. We are thus free to pick two points on the waveform and equate the available energy between them.

For convenience, we place point 1 at the wave crest and point 2 at the adjacent wave trough. If we choose the level of the wave trough as our reference height, we find that

$$\frac{\rho u_1^2}{2} + p_1 + \rho g H = \frac{\rho u_2^2}{2} + p_2. \tag{13.11}$$

Now, both the crest and the trough are in contact with the atmosphere, and are therefore at approximately atmospheric pressure. This approximation is quite close unless the wavelength is short enough so that the curvature of the water at crest or trough is severe, in which case surface tension may affect the pressure. We will return to this possibility shortly, but for the moment we assume that the wavelength is sufficiently long so that surface tension effects can be neglected.

In this case, $p_1 = p_2$, and the pressure terms can be subtracted out. We may also divide both sides of the equation by ρ, with the result that

$$\frac{u_1^2}{2} + gH = \frac{u_2^2}{2}. \qquad (13.12)$$

We now insert the values for u_1 and u_2 calculated previously from considerations of orbital motion:

$$\frac{(\frac{\pi H}{T} - c)^2}{2} + gH = \frac{(\frac{-\pi H}{T} - c)^2}{2}. \qquad (13.13)$$

When the squared terms are expanded and like values canceled from both sides of the equation, we find that

$$c = \frac{gT}{2\pi}. \qquad (13.14)$$

In other words, the celerity of the waveform is independent of wave height and the density of water.[2] Thus, storm waves on the ocean have the same celerity as gentle swells on Lake Ontario, provided the two have the same period.

The celerity of deep-water waves does, however, depend on the acceleration due to gravity and the wave period. Waves on a lunar pond (if one existed) would move slower than those on earth due to the moon's lower gravitational acceleration. Variations in g from place to place on the earth's surface are so slight that they have a negligible effect on the speed of water waves.

Regardless of gravity, the longer the wave period, the faster the waveform moves. For example, a wave on earth with a period of one 1 s has a celerity of 1.56 m s^{-1}, whereas a wave with a period of 10 s has a celerity of 15.6 m s^{-1}.

Knowing wave celerity as a function of wave period, we can immediately calculate the deep-water wavelength. Recalling that $\lambda = cT$ (eq. 13.3), we see that

$$\lambda = \frac{gT^2}{2\pi}. \qquad (13.15)$$

A wave with a period of 1 s has a wavelength of 1.56 m. A wave with $T = 10$ s has a length of 156 m.

We may also express celerity in terms of wavelength. Recalling from eq. 2 that $T = \lambda/c$, we see that

$$c^2 = \frac{g\lambda}{2\pi} \qquad (13.16)$$

$$c = \sqrt{\frac{g\lambda}{2\pi}}. \qquad (13.17)$$

Thus, the longer the deep-water wavelength, the faster the wave form moves.

Let us return for a moment to the distinction between the movement of the waveform and that of the water. We now know that the waveform moves at a

[2]This is an overstatement. If water had the same density as air, there would be no potential energy associated with a vertical displacement of the interface between the two fluids, and waves would not propagate. If the density of water were very close to that of air (but not exactly the same), waves could be formed, but to predict their speed accurately one would have to take into account the slight change in air pressure between crest and trough. An analogous situation is found for *internal waves* in the ocean, a theoretical treatment of which is given in Kinsman (1965). However, because the density of water (either fresh or salty) is so much greater than that of air, these theoretical effects are of no importance, and the statement here is correct for all practical purposes.

speed of $g\mathcal{T}/2\pi$ whereas the water moves at a speed of $\pi H/\mathcal{T}$. The ratio of these two is

$$\frac{c}{u} = \frac{g\mathcal{T}^2}{2\pi^2 H} = \frac{\lambda}{\pi H}. \tag{13.18}$$

In other words, unless the wave height is a sizable fraction of the wave length, the celerity of the waveform is much faster than the orbital speed of the water.[3] In general, wave height is much less than wavelength, and the speed with which water particles move in the passage of a wave is therefore very small compared to the wave celerity. For example, in a deep-water ocean wave with a height of 1 m and a period of 10 s ($\lambda = 156\,m$), the wave celerity is almost fifty times the orbital speed. We will reconsider the subject of orbital velocity when we discuss breaking waves later in this chapter.

13.5 Gravity Waves vs. Capillary Waves

In the calculations made thus far, we have assumed that surface tension does not affect the pressure at the surface. Under these circumstances, the only force that acts to restore water to its still level is the one due to gravity. Water above the still level is pulled down by gravity, and water below is pushed up by buoyancy (a side effect of gravity acting on water elsewhere). Because these waves use gravity as their restoring force, they are called *gravity waves*, of which the seas and swell on the ocean surface are classic examples. The role of gravity in these waves is evident from the presence of g in the equations for celerity and wavelength.

When the wavelength is very short, however, the water's surface in a wave can have sufficient curvature so that surface tension effects can no longer be neglected. In fact, for very short wavelengths (less than about 1.7 cm) the restoring force due to surface tension outweighs that due to gravity, and waves of this length are called *capillary waves*. With what celerity do capillary waves propagate?

This question is answered through the same sort of logic we used to calculate the celerity of gravity waves. In this case, however, when we equate the available energy for two points on the waveform, we must include the effects of surface tension. This is accomplished via the term in Bernoulli's equation that relates to pressure.

Recall from chapter 12 that the pressure imposed by surface tension is $-\gamma/r$, where r is the radius of curvature of the cylindrical air-water interface. The term $1/r$ is also known as the *curvature* of the surface. Now, from calculus, we know that at a local maximum or minimum the curvature of a function is equal to the rate of change of the function's slope with respect to distance along the function.[4] In other words, at a maximum or minimum the curvature of the function $f(x)$ is simply the second derivative of $f(x)$ with respect to x, $d^2f(x)/dx^2$. This simple relationship allows us to calculate the effects of surface tension on waves.

[3]Eq. 14.30 should not be taken too seriously. When H is a sizable fraction of λ, the equations we have derived here (based on linear wave theory) no longer hold true in any exact sense. In fact, before H ever reaches a sizable fraction of λ, the wave breaks and orderly orbital motion is transformed to turbulence.

[4]The general expression for the curvature is

$$\frac{1}{r} = \frac{d^2y/dx^2}{\left[1 + (dy/dx)^2\right]^{3/2}}. \tag{13.19}$$

But at a maximum or minimum, $dy/dx = 0$, and the curvature is simply equal to d^2y/dx^2.

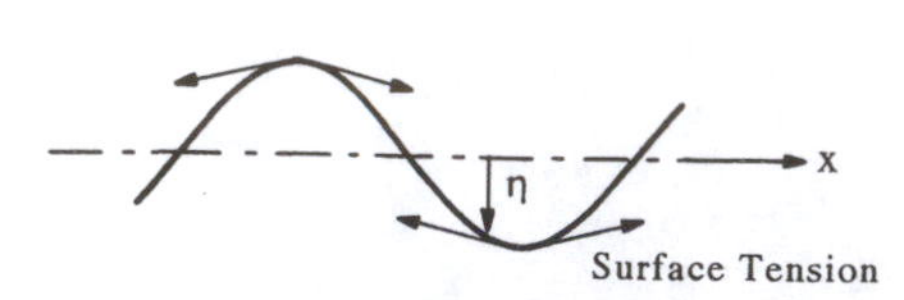

Fig. 13.6 When the wavelength is small, surface tension provides a substantial force tending to restore water to its still level.

We begin by assuming that the waves are sinusoidal. That is, the vertical displacement of the surface from the still-water level is

$$\eta = \frac{H}{2} \sin\left(\frac{2\pi x}{\lambda}\right), \tag{13.20}$$

where x is distance measured along the direction of wave propagation (fig. 13.6).

The second derivative of this surface elevation is

$$\frac{d^2\eta}{dx^2} = \frac{1}{r} = -\frac{2\pi^2 H}{\lambda^2} \sin\left(\frac{2\pi x}{\lambda}\right). \tag{13.21}$$

At the wave crest, the sine term of eq. 13.21 is 1, while at the trough it is -1. Thus,

$$\text{curvature (crest)} = \frac{1}{r} = -\frac{2\pi^2 H}{\lambda^2} \tag{13.22}$$

$$\text{curvature (trough)} = \frac{1}{r} = \frac{2\pi^2 H}{\lambda^2}, \tag{13.23}$$

and the pressures due to surface tension are

$$\text{pressure (crest)} = \frac{2\pi^2 H\gamma}{\lambda^2} \tag{13.24}$$

$$\text{pressure (trough)} = \frac{-2\pi^2 H\gamma}{\lambda^2}. \tag{13.25}$$

These pressures may then be added to atmospheric pressure, p, and inserted into Bernoulli's equation. Again placing point 1 at the crest of a wave and point 2 in the trough, we see that

$$\frac{\rho u_1^2}{2} + p + \frac{2\pi^2 H\gamma}{\lambda^2} + \rho g H = \frac{\rho u_2^2}{2} + p - \frac{2\pi^2 H\gamma}{\lambda^2}. \tag{13.26}$$

Now, in making this calculation we want to know what happens when gravitational effects are negligible compared to the effects of surface tension. We may insure ourselves of this condition by deleting from the equation the single term in which gravity appears, $\rho g H$. Thus, for pure capillary waves:

$$\frac{\rho u_1^2}{2} + p + \frac{2\pi^2 H\gamma}{\lambda^2} = \frac{\rho u_2^2}{2} + p - \frac{2\pi^2 H\gamma}{\lambda^2}. \tag{13.27}$$

We then proceed as before. Values for u at the two points are calculated on the basis of orbital motion, and these values are inserted into eq. 13.27. The squared terms are expanded and like values are canceled. When the dust settles, we find that

$$c = \sqrt{\frac{2\pi\gamma}{\rho\lambda}}. \tag{13.28}$$

The celerity of a capillary wave is proportional to the square root of surface tension and inversely proportional to the square root of the wavelength. In other words, the shorter the wavelength, the faster the wave travels.

This is in contrast to gravity waves where the longer the wavelength is, the faster the wave propagates. This difference is indicative of the counteracting mechanisms by which gravity and surface tension act in wave propagation, and leads to an intriguing situation. When both gravity and surface tension are taken into account, it can be shown (Lamb 1945) that the celerity of a wave is

$$c = \sqrt{\frac{2\pi\gamma}{\rho\lambda} + \frac{g\lambda}{2\pi}}. \qquad (13.29)$$

This expression is shown graphically in figure 13.7. There is a particular wavelength (λ_{min}) at which wave celerity is minimal (c_{min}):

$$\lambda_{min} = 2\pi\sqrt{\frac{\gamma}{\rho g}}, \qquad (13.30)$$

$$c_{min} = \sqrt{2\sqrt{\frac{g\gamma}{\rho}}}. \qquad (13.31)$$

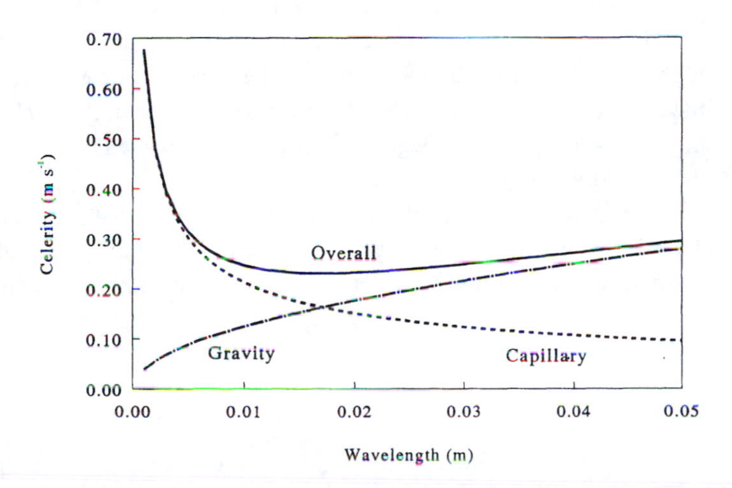

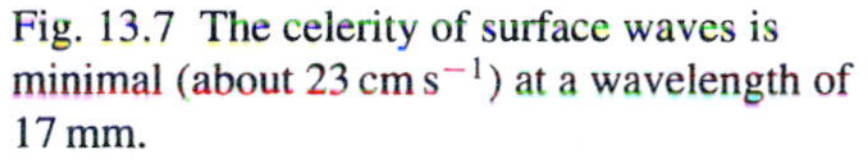
Fig. 13.7 The celerity of surface waves is minimal (about 23 cm s^{-1}) at a wavelength of 17 mm.

Using a value of 0.0728 J m^{-2} for γ, we find that the minimum wave celerity is about 0.23 m s^{-1} and it occurs at a wavelength of 17 mm. Thus, waves with lengths both longer and shorter than 17 mm travel faster than 0.23 m s^{-1}.

13.6 Hull Speed

We are now in a position to explore some of the biological consequences of surface wave propagation. We begin with an examination of the energetics of swimming at the air-water interface.

Consider the situation of figure 13.8. A duck floats on the surface of a river and propels itself upstream by paddling with its feet. For the sake of simplicity, let us assume that the duck's forward speed is just equal to the speed of the current, so that the duck is stationary relative to the riverbank and the river flows past it. To maintain its position, the duck must exert sufficient thrust to overcome the pressure and viscous drag acting on its "hull." There is, however, an additional drag in this situation—*wave drag*. To see why, we examine a streamline on the water's surface in the vicinity of the duck.

Fig. 13.8 A swimming duck creates a wave equal in length to the duck's "hull."

As with the streamlines we considered earlier, we may assume that viscous processes are negligible so that the available energy of water particles is constant as they flow along their path. At the water's surface, the pressure acting on fluid particles is constant, allowing us to simplify Bernoulli's equation to

$$\frac{u^2}{2} + gh = \text{constant.} \tag{13.32}$$

A moment's reflection on this equation reveals that u and h are inversely related: any decrease in u must be accompanied by a rise in the level of the surface if the available energy is to be kept constant. Thus, the level of the water in the duck's vicinity depends on the pattern of speed around the duck's hull.

Careful observation of the flow around the hull reveals a pattern similar to that around the sphere discussed in chapter 4. As water approaches the duck's bow, it slows down. Fluid particles then sweep along the sides of the hull, gaining speed in the process. At the duck's stern, particles slow down again to form a turbulent wake, but their average speed in the wake is not quite as slow as that reached near the bow.

The result of this pattern of velocity is a rise in water level at the bow (where speed is low), a decrease in water level along the sides of the hull (where speed is high), and another rise in the wake. In other words, the net result of flow around the duck's hull is the formation of two wave crests, with the duck sitting between them in the intervening trough. The length of this wave is approximately equal to the water-line length of the duck's hull.

Therein lies a problem. We know (from eq. 13.29) that the natural celerity of a wave is set by its wave length, $c = \sqrt{(2\pi\gamma/\rho\lambda) + (g\lambda/2\pi)}$. Thus, the wave created by the duck as it swims "wants" to move at a specific rate set by the length of the hull:

$$\text{hull speed} = \sqrt{\frac{2\pi\gamma}{\rho\ell} + \frac{g\ell}{2\pi}}, \tag{13.33}$$

where ℓ is the waterline hull length. For a duck with a hull length of 30 cm, this speed is about 0.68 m s^{-1}. If the duck tries to swim faster than its hull speed, it moves faster than the celerity of its bow wave, something it can do only by swimming uphill over the wave. As a result, the wave drag encountered by a duck rises rapidly as it approaches its hull speed (fig. 13.9), and the energetic expense of swimming near hull speed is great (Prange and Schmidt-Nielsen 1972). As a practical matter, the wave drag above hull speed is prohibitive, and the swimming speed of the bird is limited by its hull speed. Note that for a hull length of 30 cm, the wave produced is a gravity rather than a capillary wave.

We digress for a moment to examine the ratio between the swimming speed of the duck—call it u—and the speed of the gravity wave it forms:

$$\frac{\text{swimming speed}}{\text{hull speed}} = \frac{u}{\sqrt{\frac{g\lambda}{2\pi}}}. \tag{13.34}$$

The square of this ratio is

$$Fr = \frac{2\pi u^2}{g\lambda}, \tag{13.35}$$

which has a form very similar to the Froude number we dealt with in regards to the speed with which animals can walk (chapter 7). In fact, to obtain the same Froude number as before, all we need to do is divide eq. 13.35 by 2π and realize that wavelength is just another characteristic length. The conclusion of this exercise is that the maximum speed of both surface-swimming and walking animals cannot easily exceed a Froude number of 0.3 to 1, a surprising congruity between very different forms of locomotion.

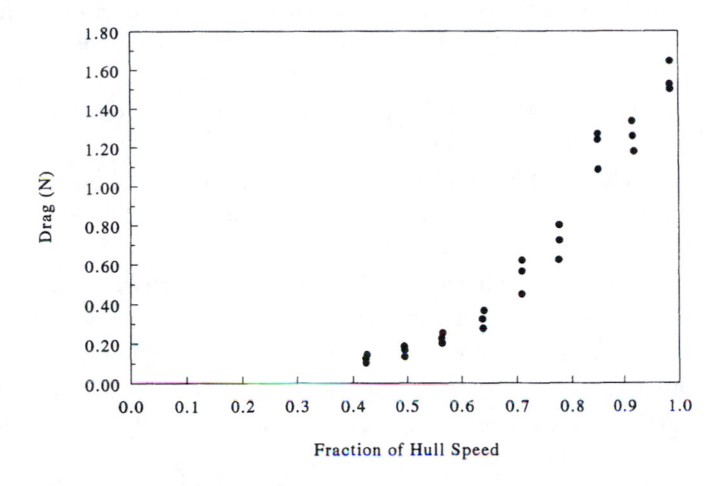

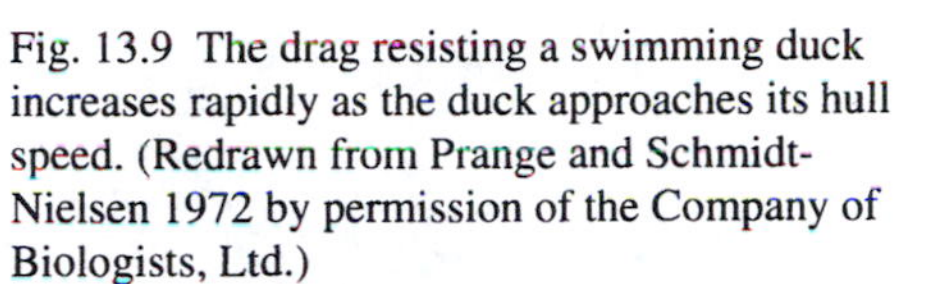
Fig. 13.9 The drag resisting a swimming duck increases rapidly as the duck approaches its hull speed. (Redrawn from Prange and Schmidt-Nielsen 1972 by permission of the Company of Biologists, Ltd.)

The limitations posed by wave drag may not pose any particular problem for the average duck, which may have little desire to swim faster than its hull speed of 0.68 m s^{-1}. After all, this is 2.3 body lengths per second, a quite respectable rate of locomotion. Note, however, that the celerity of surface gravity waves increases only with the square root of the wavelength. As a result, hull speed per body length decreases for longer animals:

$$\text{maximum body lengths per s} = \frac{\sqrt{\frac{g\lambda}{2\pi}}}{\lambda} = \sqrt{\frac{g}{2\pi\lambda}}. \qquad (13.36)$$

For example, a sea lion may have a hull length of 2 m, corresponding to a hull speed of 1.77 m s^{-1}, or only about 0.9 body lengths per second when the animal swims at the surface. This limitation may be especially stringent for large sharks and whales. For instance, a whale or shark 10 m long has a hull speed of about 4 m s^{-1}. This is a respectable absolute speed, but amounts to only 0.4 body lengths per second, which is quite slow compared to other swimmers. Thus, the surface of the water can be a restrictive environment for large animals that need (or want) to swim fast.

Although hull speed often forms a practical limit to the speed of locomotion, this limitation is not absolute, and there is at least one case in which a biological object moves faster than its hull speed. Several species of bats fish by dragging their feet through the water as they fly just above the surface. Typical flight speeds are 5 to 8 m s^{-1} (Fish et al. 1991). At these velocities, the toes (2 to 3 mm long in the direction of flow) move at twelve to forty times their hull speed. In fact, the speed of the toes is so much faster than their hull speed that the transfer of energy from the toes to the water in the form of waves is inhibited and waves are not effectively produced (Hoerner 1965). As a result, the wave drag on the toes is probably minimal. Instead, the primary source of drag is likely to be due to the spray formed in the toes' wake (Fish et al. 1991).

The problem of hull speed applies not only to those animals that move at the water's surface, but to those that swim close under it as well. For example, a dolphin swimming just under the surface of the water has the same kind of pressure distribution around its body as a duck. As a result, a bow and stern wave are formed on the surface even though the animal itself is entirely below water. Thus, a 2-m-long dolphin is subject to the same hull speed limitation calculated above for a sea lion. One would expect, then, that dolphins and other subsurface swimmers would swim at a depth where they no longer create a substantial bow wave. As a rule of thumb, this can be accomplished by submerging to a depth equal to half the hull length. Alternatively, the animal can reduce its drag by leaping into the air, as described in chapter 4, a behavior that is found in dolphins, sea lions, penguins, and flying fish.

Leaping of this sort is beyond the capability of human swimmers, but the strategy of submerging to avoid wave drag is used by competitive swimmers when doing the backstroke. From the starting position for this stroke it is easy for the swimmer to submerge, and it is common in races to see competitors remain several feet under water for nearly a full length of the pool. Swimmers who can hold their breath the longest get the most advantage of avoiding wave drag, and often win the race (provided they don't drown in the process). To prohibit a similar strategy in the breast stroke, the rules require that only one stroke at the beginning of the race can be taken underwater.

13.7 Wave Drag on Small Animals

The problem of hull speed is unlikely to apply to small organisms in the same sense as it does to whales and humans. For example, whirligig beetles (beetles of the aptly named genus *Gyrinus*) grow to a length of about 1 cm, and their hull speed (in this case set by the speed of a capillary wave) is about 25 cm s^{-1}, or twenty-five body lengths per second. This is probably plenty fast for any whirligig beetle. Smaller animals could swim at an even greater speed relative to their body size.

There are times, however, when animals *need* wave drag, and in this case small size can be problematic. Consider, for instance, the plight of water striders. We saw in chapter 12 how these bugs are supported by surface tension. We could not explain, however, how they propel themselves horizontally. The answer appears to be that when the striders' middle pair of legs is swept backward, the legs create surface waves. The drag of moving against these waves provides the "purchase" needed by the animal to skate across the water (fig. 13.10).

Fig. 13.10 A stationary water strider is amply supported by surface tension, but must rely on the production of capillary waves to get the purchase necessary for locomotion.

There are two interesting aspects of this mechanism of locomotion. First, the legs of water striders are quite slender, about 200 μm in diameter. The hull speed corresponding to this diameter is quite high, about 1.5 m s^{-1}, suggesting that the ends of the legs would have to be swept back at this rate to encounter any wave drag. For a leg that may only be 1 cm long, this amounts to a rotation rate around its base of 150 radians s^{-1}. At this rate, the leg would go through its 90° sweep in 0.01 s, which seems unlikely.

In fact, it is not the motion of the leg, per se, that is important in creating wave drag. Through its interaction with surface tension, the leg forms a dimple in the water's surface, and it is this dimple that is swept across the water as the leg moves. Although the interaction between water and dimple is dynamic and likely to be complex, we can guess that as a first approximation it is the hull speed of the dimple (rather than the leg) that sets the rate at which legs must move to create wave drag. Each dimple is about 0.5 cm across, corresponding to a hull speed of 30 cm s^{-1}, a more reasonable value.

Now, 30 cm s^{-1} is not much faster than the minimum wave celerity of 23 cm s^{-1}. What would happen if the dimples were larger than they are? Would it be possible for the legs to move at less than 23 cm s^{-1} and still provide sufficient wave drag? The answer is no. Any object that moves through the water at less than 23 cm s^{-1} simply does not make waves. Thus, no matter what size dimples (or legs) the water strider has, the ends of its legs must move faster than 23 cm s^{-1} to get a purchase on the water's surface. This does not appear to be a problem for adult striders, who readily produce waves as they skate about, but it is a problem for their young. The early instars of water striders are shaped much like the adults, and they too move about on the surface of the water. Their legs, however, are much shorter than those of adults, and apparently cannot move at 23 cm s^{-1}, because these early instars do not produce waves. Exactly how they manage to propel themselves across the water remains a mystery.

Note that the baby water strider's problem of obtaining wave drag becomes less severe if there is a film on the water that lowers its surface tension. For example, a film of phospholipids released from a rotting log might easily reduce the surface energy to 0.03 J m^{-2}. In this case, the minimum wave celerity is reduced by about 20%, to 18.5 cm s^{-1}. This speed might be within the range attainable by early instars. Of course, the reduction in surface tension reduces the ability of the strider to be supported on the water, but as we have seen (chapter 12), this is unlikely to pose a problem.

13.8 Waves in Shallow Water

To this point we have confined our exploration to waves in deep water ($d > \lambda/2$). There are, however, many cases in which waves of biological interest occur in intermediate to shallow water, and it is to these waves that we now turn our attention.

We begin by noting that waves in intermediate and shallow water often do not have the same sinusoidal shape as deep-water waves. As a wave moves into shallower water its crest becomes more peaked, the troughs become flattened, and in shallow water ($d < \lambda/20$) waves more closely resemble *solitary waves* (fig. 13.11) than they do the oscillatory waves we have dealt with so far. A solitary wave is roughly analogous to the top half of a sinusoidal wave. The entire wave form is above still-water level, and all the water in the wave moves in the direction of wave propagation. Although the similarity between real shallow water waves and solitary waves is not precise, solitary waves do provide us with a useful model for how shallow-water waves behave (Munk 1949).

We begin with a calculation of the celerity, c, of solitary waves, using much the same procedure we applied to oscillatory waves (fig. 13.11). Again we create the wave in a laboratory tank and use a countercurrent with velocity $-c$ to bring the wave form to a standstill in the reference frame of the lab. We note as before that the surface of the water is formed from streamlines, and we utilize Bernoulli's equation again by placing point 1 at the crest of the wave and point 2 a long distance in front of the wave where the water is at its still level. The difference in height between these two points is H, so, using the still-water level as our reference height, we see that

$$\frac{\rho u_1^2}{2} + p_1 + \rho g H = \frac{\rho u_2^2}{2} + p_2. \tag{13.37}$$

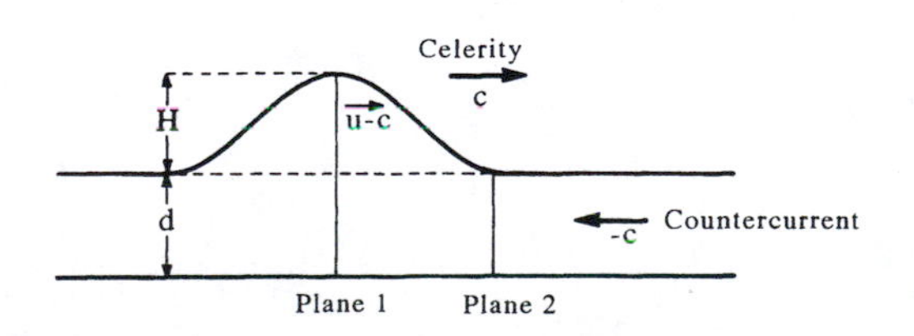

Fig. 13.11 A solitary wave can be brought to a halt by imposing a countercurrent, allowing us to specify the water velocity at the crest. From the dimensions of the system we can then calculate the celerity of the wave. (From Denny 1988)

We assume that the effects of surface tension are negligible, in which case $p_1 = p_2$, and

$$\frac{u_1^2}{2} + gH = \frac{u_2^2}{2}. \tag{13.38}$$

The water at point 2 is far upstream of the wave, and therefore moves at the same speed as the countercurrent, $-c$. Water at point 2 (the crest of the wave) is carried forward by the wave, so in our reference frame it has velocity $u - c$, where u is the speed at which the surface water would be carried if the wave were moving through still water. Inserting these values for u_1 and u_2, expanding terms and rearranging, we come to the conclusion that

$$c^2 = u^2 - 2uc + c^2 + 2gH. \tag{13.39}$$

This expression can be simplified if we assume that u is small compared to c, as it was for oscillatory waves. In this case, $u^2 \ll c$ and can safely be ignored, leaving us to conclude that

$$c = \frac{gH}{u}. \tag{13.40}$$

Now, water particles in a solitary wave do not move in closed orbits, so we cannot use orbital speed as a means for calculating u. Instead, we rely on the fact that water is virtually incompressible. The utility of this fact becomes apparent when we view the segment of wave between planes 1 and 2 as part of a pipe. The walls of the pipe are the water's surface, the bottom, and the side walls of the tank. Let us assume that the walls are a distance ℓ apart, though as we will see, the exact spacing is unimportant. A moment's thought should convince you that water which enters this "pipe" through plane 2 (fig. 13.11) must leave the pipe through plane 1. No water can move into or out of the bottom or walls, and we know because the waveform is stationary that water does not cross the surface.

Given this scenario, we can assert that water flows into the "pipe" through plane 2 at a rate of $-c\ell d$ m^3 s^{-1}. At plane 1, the depth of the water column is $d + H$. At this point the water is being carried forward by the waves, so its velocity (averaged over the water's depth) is $u - c$. This velocity times the "pipe's" area at plane 1 ($\ell[d + H]$) is the rate at which water flows out of the pipe. Thus,

$$(u - c)\ell(d + H) = -c\ell d. \tag{13.41}$$

Solving for u, we find that

$$u = c - \frac{cd}{d + H}. \tag{13.42}$$

Inserting this value for u into eq. 13.40 and rearranging, we see that

$$c = \sqrt{g(d + H)}. \tag{13.43}$$

This is an interesting result. Whereas in deep water the celerity depended only on wave period (or, equivalently, wavelength), the celerity in shallow water is independent of wave period and length and depends only on wave height and water depth. In many cases of biological interest, wave height is much less than the water's depth, and

$$c \approx \sqrt{gd}. \tag{13.44}$$

Now, if the water is shallow enough for this equation to apply, d is less than $\lambda/20$ so that the wave celerity is less than $\sqrt{g\lambda/20}$. This is only about 56% of the speed a wave of the same length would exhibit in deep water ($\sqrt{g\lambda/2\pi}$). From this calculation, we can conclude that waves slow down as they move into shallow water.

A more rigorous examination of wave celerity (linear wave theory) provides us with an expression for the celerity of waves as a continuous function of water depth:

$$c = \sqrt{\frac{g\lambda}{2\pi}}\sqrt{\tanh\frac{2\pi d}{\lambda}}. \tag{13.45}$$

This expression looks similar to the one we obtained for deep-water waves in that c is proportional to $\sqrt{g\lambda/2\pi}$. This term is multiplied, however, by a second term that is a function of water depth and wave length. In this second term, $\tanh(2\pi d/\lambda)$ is the *hyperbolic tangent* of $2\pi d/\lambda$. Now, the hyperbolic tangent of a value is defined as

$$\tanh(x) = \frac{e^x - e^{-x}}{e^x + e^{-x}}. \tag{13.46}$$

For small values of x, the hyperbolic tangent is approximately equal to x, while for large values of x, $\tanh(x)$ is approximately equal to 1 (fig. 13.12).

The effect of the term containing the hyperbolic tangent is to modify the celerity depending on the ratio of water depth to wave length. In deep water, d is greater than $\lambda/2$, so that $2\pi d/\lambda$ (the argument of the hyperbolic tangent) is greater than π. As a result, the depth-dependent expression approaches 1. In this case, $c = \sqrt{g\lambda/2\pi}$, as we originally proposed. In shallow water, the depth-dependent term approaches $\sqrt{2\pi d/\lambda}$, and $c = \sqrt{gd}$, as we suggested for a solitary wave. At intermediate depths, celerity is as shown in figure 13.13.

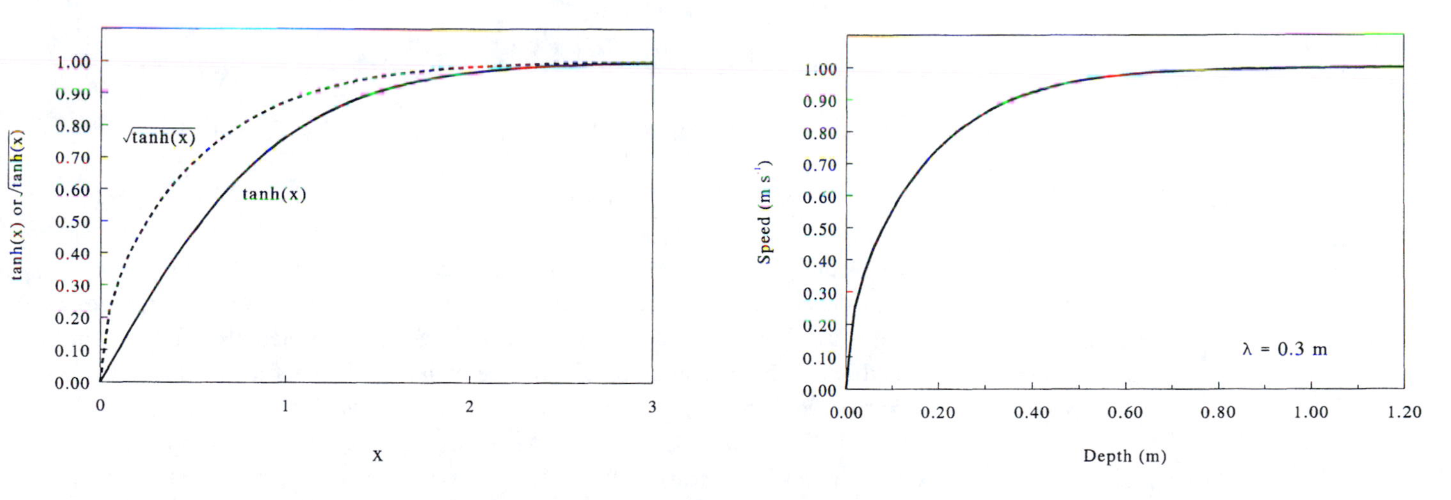

Fig. 13.12 The hyperbolic tangent asymptotes to 1 for values larger than 3.

Fig. 13.13 Hull speed is decreased in shallow water. The values shown here are for a 30-cm-long duck.

In summary, waves of a given length move slower in shallow water than in deep water, and hull speed is therefore decreased when the water is shallow. This fact has not escaped the attention of competitive human swimmers. Whereas a typical recreational swimming pool has a deep and a shallow end, a pool designed specifically for competition is uniformly deep throughout and thereby avoids the reduction in speed swimmers would otherwise encounter in the shallow end.

Note however, that the effect of water depth on hull speed is appreciable only in very shallow water. To decrease the hull speed of a 30-cm-long duck by a factor of two, the water depth would have to be less than 2.6 cm. At this depth, the duck could probably waddle through the water faster than it could swim.

A more thorough discussion of the theory that leads to eq. 13.45 can be found in Denny (1988).

13.9 Breaking Waves

Before leaving the topic of waves in shallow water, we will make one more use of eq. 13.43. When waves move into shallow water, wave crests become more peaked and the water at the crest moves faster than it does in deep water. When waves reach a depth where $d \approx H$, the water at the crest actually moves slightly faster than the speed of the waveform, and the wave *breaks*. Depending on the slope of the bottom, the crest may arch over the front of the wave to form the type of plunging breaker one sees in surfing magazines, or the crest may simply spill down the front of the wave (fig. 13.14). In either case, because at the point of breaking water at the crest moves with a speed approximately equal to the wave celerity, eq. 13.43 allows us to estimate the velocity of water in the crest. This velocity can impose itself on the benthic organisms of the shore, and is therefore of considerable biological importance.

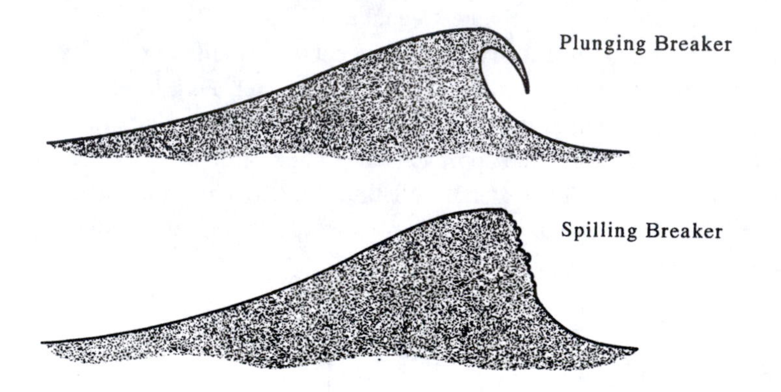

Fig. 13.14 Waves break when their height is approximately equal to the water's depth.

If a wave breaks when $d = H$, the water at the crest moves at velocity

$$\text{crest velocity} = \sqrt{2gH}. \qquad (13.47)$$

For example, if the wave is 1 m high at breaking, the velocity in the crest is approximately 4.5 m s^{-1}. This is a substantial water velocity (almost 10 mph), and it is capable of imposing large lift and drag forces on plants and animals in its path. For example, an upright human being has a drag coefficient of about 1.2 (Hoerner 1965) and a frontal area of about 1 m^2. At a water velocity of 4.5 m s^{-1}, a drag of 1.35×10^4 N (roughly the weight of 1.3 tons) would be imposed. This is one reason why it is so dangerous to be shipwrecked on a wave-swept coast. In the process of swimming ashore, one is likely to be crushed.

It is useful to compare the crest velocity at breaking to the orbital velocity typical of waves in deep water. A typical ocean swell has a period of about 10 s. Thus, a wave 1 m high in deep water has an orbital velocity of about 0.3 m s^{-1} (eq. 13.10). This is less than a tenth of the velocity that would be associated with the same height of wave at the point of its breaking. In other words, while the celerity of a wave form slows down as it enters shallow water, the maximal velocity of the water in the wave speeds up.

The high velocities associated with breaking waves play an important role in determining the survivorship of plants and animals on coral reefs and wave-swept shores. These effects, in turn, are a major factor in the ecology of wave-swept communities. A discussion of the intricacies of these effects is beyond the purview of this text, but the interested reader is urged to consult Denny (1988, 1991) and Denny and Gaines (1990).

13.10 Transmission of Information

We now turn our attention to the use of surface waves as a means of transmitting information. In this respect, we treat gravity and capillary waves in much the same fashion as we dealt with sound and light.

We begin by noting that a surface wave represents a source of energy. For example, if you drop a stone into a pond, the stone does work as it displaces water. Some of this energy is incorporated into surface waves that radiate across the pond. When these waves reach the opposite side, the energy they contain can be used to do work by vibrating the legs of a water strider, for example. This vibration provides the strider with information from which it can reasonably conclude that a stone has been dropped into its pond. In fact, water striders use surface waves as a means of communication among themselves; individuals of the species *Gerris remigis* can tell from the pattern of waves whether the strider producing them is male or female (Wilcox 1979). What are the characteristics of energy transmission by surface waves that allow this kind of information transfer?

It can be shown (Kinsman 1965) that gravity waves contain an energy proportional to the square of their height:

$$\text{energy per area} = \frac{1}{8}\rho g H^2. \tag{13.48}$$

Note that this is energy per still-water surface area, and it assumes that the entire area is filled with a train of waves of uniform height.

The energy of capillary waves is (Lamb 1945)

$$\text{energy per area} = \frac{\pi^2 \gamma H^2}{4\lambda^2}. \tag{13.49}$$

The fact that the energy of both gravity and capillary waves is proportional to the square of wave amplitude is analogous to the energy of both light and sound waves, and is in fact a general property of wave energy.

As with light and sound energy, it is not so much the energy of a water wave itself that is important, but the manner in which this energy can be transported. For example, when we dealt with the ability of an organism to detect sound, we dealt with the *intensity* of the sound wave, a measure of the rate at which sound energy is delivered to the organism's sensory area. Similarly, when we explored the effects of light, the currency with which we dealt was light intensity, the rate at which light energy arrives at a surface. Thus, to examine the role that water waves can play in the transmission of information, we need to explore the rate at which wave energy is transported.

In doing so, we encounter a curious phenomenon that is best introduced through an experiment. We return to our ever-handy pond and toss a stone into the water, thereby creating a distinct packet of waves that radiates outward in a circular pattern. If one carefully observes the leading wave crest in this packet, one finds that it gradually diminishes in height and eventually disappears! At the same time, if one observes the trailing edge of the wave train, one sees that a new wave crest gradually rises into existence. In other words, as a wave train moves across the surface, the number of waves in the packet stays the same, but they are continually trading places, the leading waves dying out to reappear at the rear. As a result, the packet of waves moves across the pond slower than the celerity of the waves from which it is formed. The rate at which the packet moves is known as the *group velocity*, c_g, and is of importance here because this is the speed at which wave energy is transmitted.

Consider an example. By inserting a board through the surface of our pond and oscillating the board horizontally, we produce a train of ten waves, each with a period of 1 s and a height of 1 cm. The energy per area within this wave train is thus $(1/8)\rho g H^2$, or about 0.12 J m^{-2}. The wave train travels across the pond and impinges on the opposite shore. What is the rate at which energy is delivered to a 1 m length of shoreline?

If energy were carried at the deep-water wave celerity, the rate of energy delivery would be

$$\text{rate of energy transport} = \frac{1}{8}\rho g H^2 c, \tag{13.50}$$

or about 0.19 W m^{-1}. In fact, though, the rate of energy delivery is

$$\text{rate of energy transport} = \frac{1}{8}\rho g H^2 c_g, \tag{13.51}$$

where we know that $c_g < c$.

The fact that the group velocity is less than the wave celerity also affects how long it takes the wave train to cross the pond. If the pond is 100 m wide, one might naively expect a wave to take $100/c$ s to travel across. In reality, the travel time is $100/c_g$, which is longer. Thus, group velocity affects both the detectability and the lag time of information reaching a water strider.

13.11 Group Velocity

What, then, is the group velocity for gravity waves? The answer requires a somewhat protracted explanation and some mathematical legerdemain. This process is instructive as to how mathematics can provide the answer to a nonintuitive phenomenon, and you are urged to follow the argument as it unfolds. The mathematical faint of heart can, however, skip to the punchline in eqs. 13.62 and 13.67 without missing any critical information.

We begin with an exploration of the way in which surface waves interact. For example, we can produce on our pond a train of gravity waves with a period of 1 s and therefore a wavelength of 1.56 m. Immediately thereafter we produce a train of waves with a period of 1.1 s, corresponding to a wavelength of 1.72 m. Having a longer period, the waves in this second train travel faster than those in the first train (eq. 13.17), and soon overtake them.

As the two wave trains mix (fig. 13.15A), their amplitudes are added together (fig. 13.15B). Wherever the crest from one wave train coincides with the crest from the other, the result is a crest of increased height. Similarly, wherever two troughs coincide, a new trough is produced that is farther below still-water level.

Where the crest from one train coincides with the trough from the other train, the two cancel out. The net result is a sinusoidal modulation of the heights of waves (fig. 13.15B), an "envelope" that defines the local wave height.

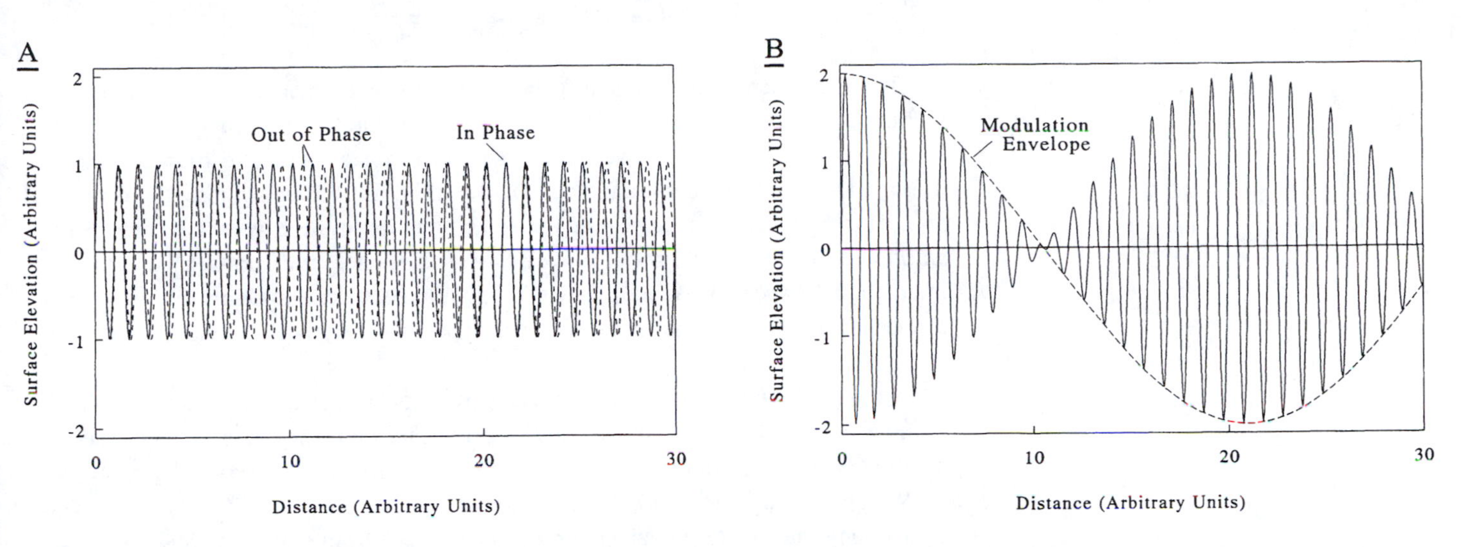

Fig. 13.15 Waves of different length travel at different speeds. As a result, the areas where they are in and out of phase (A) move in the direction of propagation. The interference of waves of different lengths produces a modulated wave form (B).

Let us now concentrate on the form of this modulation envelope. As the second wave train moves through the first, the modulation envelope also moves, in this case in the direction of wave travel. The speed with which the envelope moves is, however, different from that of the second wave train. Instead, the envelope of modulation (an abstract wave form) moves at a speed that is determined by its own wavelength and period. A consideration of the way in which waves interact (Kinsman 1965) shows that the wavelength of the modulation envelope is

$$\lambda_{mod} = \frac{2\langle\lambda\rangle^2}{\Delta\lambda}, \tag{13.52}$$

where $\langle\lambda\rangle$ is the average wavelength of the two wave trains and $\Delta\lambda$ is the difference in wavelength between the two interacting wave packets. The period of the modulation envelope is

$$\mathcal{T}_{mod} = \frac{2\langle\mathcal{T}\rangle^2}{\Delta\mathcal{T}}, \tag{13.53}$$

where $\langle\mathcal{T}\rangle$ is the average period of the waves in the wave trains and $\Delta\mathcal{T}$ is the difference in period between the two trains. From eq. 13.3 we know that the speed at which the envelope moves is simply its wavelength divided by its period: $\lambda_{mod}/\mathcal{T}_{mod}$.

We now allow the mathematics to guide us through a nonintuitive concept. What happens if we allow the period of our second wave train to become closer and closer to that of the first? It will, of course, take longer for the second train to catch up to the first, but we won't let this bother us. Let us only consider what happens after the two trains have combined.

As $\Delta\mathcal{T}$ becomes small, the period of the modulation envelope becomes large (eq. 13.53). However, when $\Delta\mathcal{T}$ is small, $\Delta\lambda$ is also small, and the wavelength of the envelope also becomes large (eq. 13.52). Now, it is the ratio of wavelength to period that determines the wave celerity. Thus, because both wavelength and period become large when $\Delta\mathcal{T}$ is small, the speed of the modulation envelope approaches a definite, finite value as $\Delta\mathcal{T}$ (and consequently, $\Delta\lambda$) approach zero.

Thus, in the limit as $\Delta\mathcal{T} \to 0$ and $\Delta\lambda \to 0$, we can calculate the speed of the wave of modulation that results *when a wave train interacts with itself*

$$c_{s,mod} = \lim_{\Delta\lambda\to 0, \Delta\mathcal{T}\to 0} \frac{\lambda_{mod}}{\mathcal{T}_{mod}}. \tag{13.54}$$

The importance of this calculation becomes apparent when one realizes that it is this abstract modulation envelope that is responsible for the strange behavior of a wave train as it travels across a pond. In essence, the wave train travels through its own modulation envelope. To see this, flip through figure 13.16 on pp. 291 to 319. Waves disappear as they reach a point in the envelope where the amplitude goes to zero. As waves move out of such a zone, they reappear. Thus, the interaction between a wave train and its modulation envelope explains the strange propensity for waves at the leading edge of a packet to disappear, only to reappear at the rear.

In turn, it is the tendency for leading waves to disappear that results in the group velocity being less than the wave celerity. Thus, it is the speed at which the modulation envelope moves that sets the group velocity. In other words, $c_{s,mod} = c_g$. With this thought in mind, we return to our calculation.

Inserting into eq. 13.54 values for λ_{mod} and $\mathcal{T}_{mod}$ from eqs. 13.52 and 13.53, and allowing $\Delta\lambda$ and $\Delta\mathcal{T}$ to approach 0, we find that

$$c_g = \frac{\lambda^2}{\mathcal{T}^2}\frac{d\mathcal{T}}{d\lambda}. \tag{13.55}$$

But from eq. 13.3, we know that $\lambda^2/\mathcal{T}^2 = c^2$, where c is the celerity of the waves in the wave train. Thus,

$$c_g = c^2\frac{d\mathcal{T}}{d\lambda}. \tag{13.56}$$

From eq. 13.15 we know that $\lambda = g\mathcal{T}^2/2\pi$, so that

$$\mathcal{T} = \sqrt{\frac{2\pi\lambda}{g}}, \tag{13.57}$$

and

$$\frac{d\mathcal{T}}{d\lambda} = \frac{1}{2}\left(\frac{2\pi\lambda}{g}\right)^{-1/2}\frac{2\pi}{g} \tag{13.58}$$

$$= \frac{1}{2}\left(\frac{g}{2\pi\lambda}\right)^{1/2}\left(\frac{4\pi^2}{g^2}\right)^{1/2} \tag{13.59}$$

$$= \frac{1}{2}\left(\frac{2\pi}{g\lambda}\right)^{1/2} \tag{13.60}$$

$$= 1/(2c). \tag{13.61}$$

Inserting this value in eq. 13.56, we see, at last, that

$$c_g = \frac{c}{2}. \tag{13.62}$$

In other words, the group velocity of gravity waves in deep water is equal to half the wave celerity.[5]

[5]In shallow water, provided $H \ll d$, $\mathcal{T} = \lambda/\sqrt{gd}$. Inserting this value into eq. 13.59 and working through the math, we find that in shallow water, $c_g = c$. The consequences of this difference in the

This immediately lets us interpret the conclusions reached earlier. The rate at which energy is transmitted by gravity waves is only half what one might expect. Thus, the "signal" produced by an organism that creates gravity waves has only half the intensity it would if energy were conveyed at the wave celerity.

Furthermore, because the group velocity is only half that of the wave celerity, the wave signal produced by a swimming animal (a duck, for instance) lags behind the animal. This may have biological importance. For example, we have seen that a water strider can use surface waves as a means of gathering information about its surroundings. Such a wave signal may indeed provide valuable information, but it cannot warn the strider that a duck is approaching. The duck's wave signal (at least the portion due to gravity waves) arrives at the bug only after the duck itself has arrived and the strider has been eaten.

13.12 Group Velocity—Capillary Waves

To this point, we have dealt only with the group behavior of gravity waves. Do capillary waves behave differently?

We may quickly obtain an answer through experimentation. If in a pool of water we create a train of capillary waves, we find that wave crests again disappear and reappear, but in the opposite direction we found for gravity waves. Capillary waves disappear from the rear of the wave packet and reemerge at its leading edge. As a result, *the group velocity of capillary waves is greater than the wave celerity!* Just how much greater can be calculated using the same logic previously employed for gravity waves.

From eq. 13.28 we know that

$$c^2 = \frac{2\pi\gamma}{\rho\lambda}. \tag{13.63}$$

But $c^2 = \lambda^2/\mathcal{T}^2$, from which we can conclude that

$$\mathcal{T} = \left(\frac{\rho}{2\pi\gamma}\right)^{1/2}\lambda^{3/2}. \tag{13.64}$$

Taking the derivative of $\mathcal{T}$ with respect to λ, we see that

$$\frac{d\mathcal{T}}{d\lambda} = \frac{3}{2}\left(\frac{\rho\lambda}{2\pi\gamma}\right)^{1/2} \tag{13.65}$$

$$= \frac{3}{2c}. \tag{13.66}$$

Inserting this value for $d\mathcal{T}/d\lambda$ into eq. 13.56, we arrive at our final answer:

$$c_g = \frac{3c}{2}. \tag{13.67}$$

That is, the group velocity of capillary waves is half again as great as the wave celerity.

This is a distinctly peculiar conclusion. For instance, if capillary waves are produced as an animal swims, the energy signal of these waves travels half again

mode of energy transport will not be explored here, but you may want to consult Denny (1988) on the subject of wave shoaling.

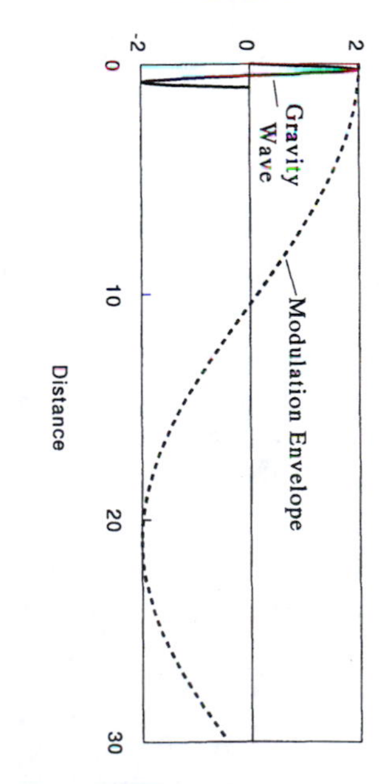

Fig. 13.16 A gravity wave in deep water moves with twice the speed of its modulation envelope. To visualize this effect, flip through the next few pages. Note that when the gravity wave has traveled the full length of the graph, the modulation envelope has traveled only half the length. Notice too how the gravity wave disappears and reappears as it moves through the modulation envelope.

as fast as the animal itself! This can be of biological importance. For example, as whirligig beetles swim, they produce capillary waves that move off ahead (fig. 13.17). These leading waves may serve to warn prey of the beetle's approach, but they may also be used by the beetle itself. Because capillary waves move faster than the animal, the beetle can use its own waves in a form of echolocation. Waves reflected by objects in front of the beetle return to the beetle and can presumably be sensed (Tucker 1969). The jerky, intermittent pace with which these beetles swim may serve to send out discreet pulses of capillary waves in much the same way that bats and dolphins send out pulses of sound.

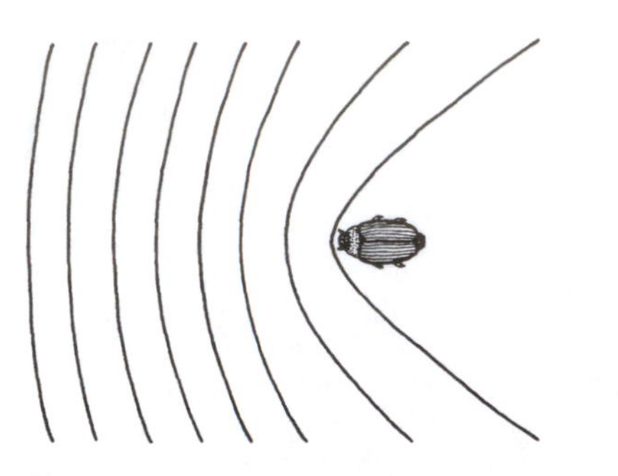

Fig. 13.17 A whirligig beetle produces capillary waves with a group velocity greater than that of the beetle itself. These waves can be used in a form of echolocation.

Capillary waves may also serve to signal the presence of stationary objects in a moving stream. For instance, a fishing line or reed held stationary through the surface of a stream produces upstream of itself a series of capillary waves. Aside from providing the knowledgable observer with visual evidence that the group velocity of capillary waves is indeed faster than the wave celerity, the presence of these waves may serve as a warning to whirligig beetles and water striders that there is an obstruction ahead.

13.13 Group Velocity—A General Formula

If gravity waves in deep water have a group velocity half that of the wave celerity and capillary waves a c_g half again larger than c, what happens with waves that are influenced by both gravity and surface tension? As one might expect, the group velocity of these intermediate waves is itself intermediate. Lamb (1945) has shown that

$$c_g = c\left(1 - \frac{1}{2}\frac{\lambda^2 - \lambda_{min}^2}{\lambda^2 + \lambda_{min}^2}\right), \tag{13.68}$$

where λ_{min} (eq. 13.30) is the wavelength at minimum wave celerity (i.e., 17 mm). A brief inspection of this equation shows that when $\lambda \ll \lambda_{min}$, c_g approaches $3c/2$ as we have calculated, and when $\lambda \gg \lambda_{min}$, c_g approaches $c/2$ as it should. When $\lambda = \lambda_{min}$, $c_g = c$. Eq. 13.68 is graphed in figure 13.18.

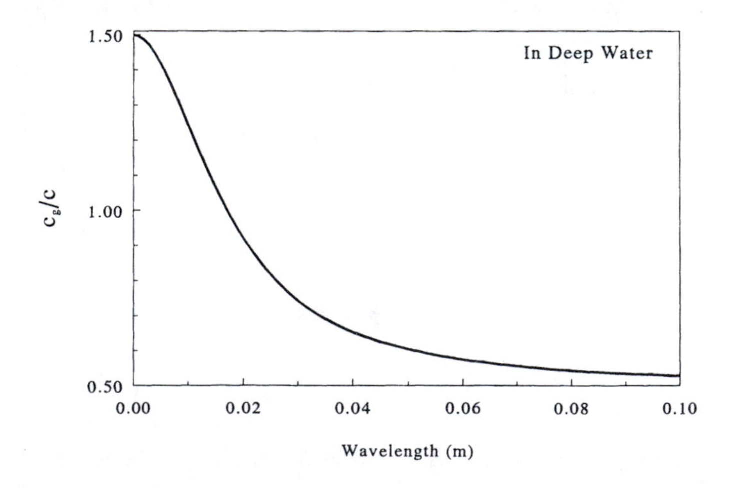

Fig. 13.18 Group velocity in deep water varies with wavelength, decreasing from $3/2$ for capillary waves to $1/2$ for gravity waves.

13.14 Beneath the Surface

Wave motion at the surface of the water is accompanied by water motions well below the surface. In deep water, fluid particles at depth move in circular orbits similar to those of particles at the surface, the only difference being that the diameter of the orbits is smaller:

$$\text{orbital diameter} = He^{-\frac{2\pi y}{\lambda}}, \tag{13.69}$$

where y is the depth at which the center of the orbit lies below the still water level (fig. 13.19). For example, when $y = \lambda/2$, the diameter of the orbit is only about 4% of that at the surface. In many cases, this amounts to negligible motion of the water, and we take it as a practical rule of thumb that water motion associated with surface waves does not extend to a depth greater than $\lambda/2$.

It is this rule that led us to the definition of "deep water" previously proposed. If the water at a depth of $\lambda/2$ is not moved by waves, the presence of a solid substratum at this depth cannot affect the wave motion. Thus, when $d > \lambda/2$, surface waves are affected to a negligible extent by the bottom, and waves behave as if the water column were infinitely deep.

The attenuation of wave-induced motion with depth is also the reason for supposing that a dolphin swimming at a depth greater than half its body length does not produce an appreciable bow wave. If surface waves do not disturb the water at $y > \lambda/2$, disturbances in the water at this depth should not result in surface waves.

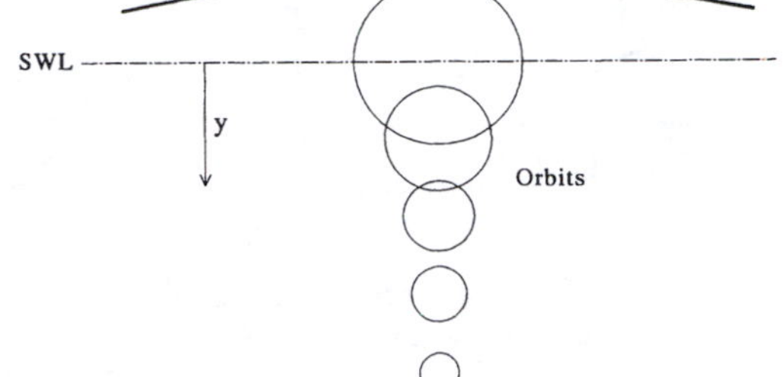

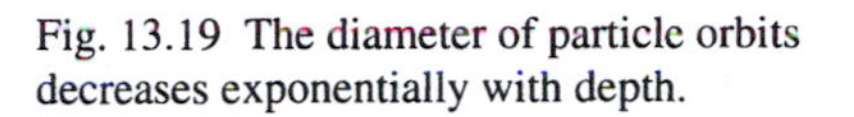

Fig. 13.19 The diameter of particle orbits decreases exponentially with depth.

13.15 Waves and Swim Bladders

Associated with a surface waveform is a change in hydrostatic pressure in the water beneath the wave. Consider, for instance, a wave with the form,

$$\eta = \frac{H}{2}\cos\left(\frac{2\pi x}{\lambda} - \frac{2\pi t}{T}\right), \tag{13.70}$$

where, as usual, x is horizontal distance and η is a measure of how far the water's surface is displaced from the still-water level. The form of this equation is quite similar to that we introduced when dealing with sound waves in chapter 10, and you may wish to refer back to that discussion.

Considerations of linear wave theory (Denny 1988) show that the pressure acting beneath a wave in deep water is

$$p = \rho g y + \frac{1}{2}\rho g H e^{-\frac{2\pi y}{\lambda}}\cos\left(\frac{2\pi x}{\lambda} - \frac{2\pi t}{T}\right). \tag{13.71}$$

This result is shown in figure 13.20 along with the surface elevation. The message of this figure is that the pressure beneath a wave is in phase with the surface elevation. Pressure is highest under a wave crest (due to the mass of water stacked on top of the water column) and is lowest under a wave trough where water has been "removed" from the water column. The magnitude of the change in pressure decreases with depth in the same fashion that the orbital diameter decreases with depth.

This pressure distribution raises an interesting question regarding the action of swim bladders. Recall from chapter 4 that many fish contain within themselves a small air space that allows the fish to adjust its effective density to that of the surrounding water. We noted in our analysis of swim bladders that they are inherently unstable. If the fish is neutrally buoyant at one depth, any upward motion reduces the pressure acting on the swim bladder, and the bladder expands. This

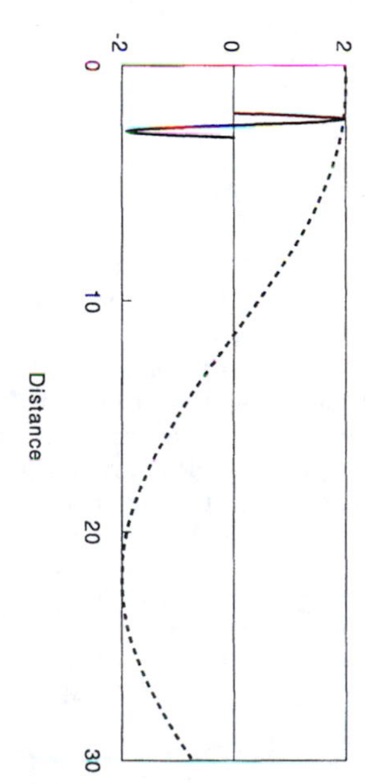

expansion reduces the fish's effective density, rendering the fish positively buoyant and thereby causing it to move up. The swim bladder then expands further, making the fish more buoyant still, the net result being a fish that rockets to the surface. Just the reverse happens if the initial displacement is downward.

Now, fish in the upper reaches of the ocean are commonly subjected to the water motions accompanying surface waves. In other words, fish at any depth less than $\lambda/2$ move in circular orbits. One might suppose that this orbital motion would trigger the sort of runaway buoyancy problems described above. There is a potential catch, however. When the fish is at the top of its orbit, and therefore displaced maximally upward, it is under the crest of the wave where the pressure is highest. This increase in pressure compresses the swim bladder, causing the fish to sink. At the bottom of its orbit, where the fish is displaced maximally downward, the pressure is at its lowest, and the consequent expansion of the swim bladder should cause the fish to rise. In other words, the variation in hydrostatic pressure under a wave should act to reduce (and perhaps to stabilize) the vertical motion of the fish.

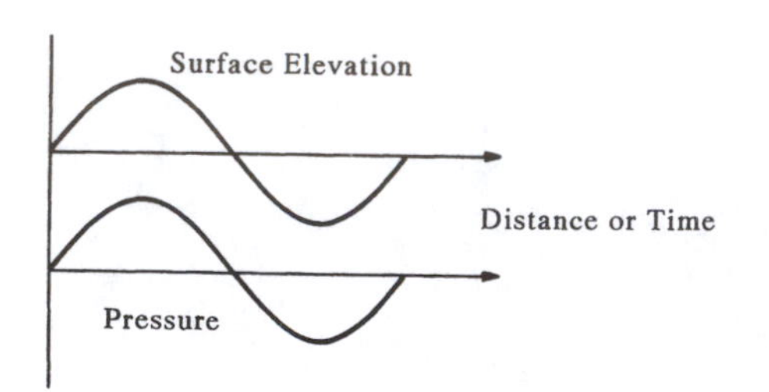

Fig. 13.20 Pressure beneath a wave is in phase with the surface elevation.

This proposition can be tested in a mathematical model. The details of the model need not be explored here, but its general outline is as follows. As usual, we choose a sphere for our model organism. The tissue of this sphere has a typical density of 1080 kg m^{-3}. Within the spherical organism is a swim bladder filled with a gas that is free to expand or compress in response to the local hydrostatic pressure. The volume of this bladder is adjusted so that the "fish" is neutrally buoyant at a depth y below still-water level.

The model fish is then subjected to wave motion. Initially the model moves with the surrounding water along an orbital path. After a small increment of time, however, the new pressure acting on the model is calculated and the resulting change in the volume of the swim bladder is made. At this point the fish is no longer neutrally buoyant, and the net force acting on it can be calculated. This force results in an acceleration of the fish relative to the surrounding water. When multiplied by the increment of time, this acceleration yields a change in velocity of the fish relative to the surrounding water. Knowledge of this velocity allows one to calculate where the fish will be at the end of that temporal increment. The process is then repeated for the next increment. The position of the fish is recorded after each increment, leading to results such as the one shown in figure 13.21A.

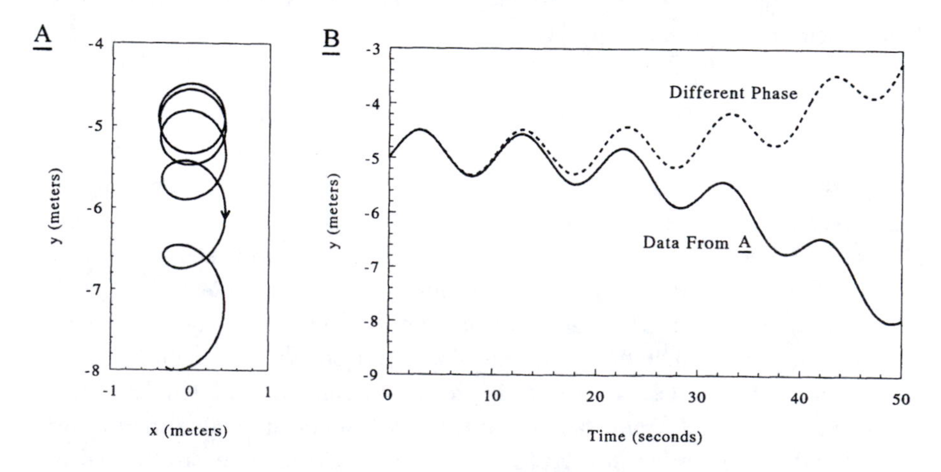

Fig. 13.21 The buoyancy system of a fish with a swim bladder is unstable in the presence of surface waves. For example, a fish initially neutrally buoyant at a depth of 5 m becomes negatively buoyant as waves pass (A). If the fish is initially stable at a different phase of the wave, its subsequent instability can cause it to move upward instead of down (B).

In this case, the fish starts at a depth of 5 m below a sea surface that sports waves 1 m high with a period of 10 s. The orbital diameter for water particles at this depth is about 0.82 m, and the first few orbits by the fish show a vertical displacement of very nearly this extent. Thus, the presence of a swim bladder does not substantially damp the vertical motion of a fish.

Furthermore, after a few orbits, our model fish becomes unstable and spirals downward. If the model is started at a different phase of the wave, a similar result is obtained, but the fish spirals up to the surface (fig. 13.21B). In other words, a buoyancy system that relies on a swim bladder is just as unstable under waves as it is in still water. Once again we conclude that a swim bladder, although a cost-effective mechanism for achieving neutral buoyancy, must be accompanied by other mechanisms for the short-term adjustment of the fish's position.

In shallow water, the pressure signal from surface waves does not diminish with depth. Thus, as a wave passes overhead, the pressure at any point in the water column is increased or decreased by an amount equal to $\rho g \eta$, where η is the local deviation of the surface from still-water level. This difference in the pattern of subsurface pressure does not affect our conclusion regarding the stability of buoyancy in fish with swim bladders; they are just as unstable if the water is shallow.

13.16 Summary . . .

Gravity and surface tension can act as restoring forces, allowing waves to propagate on an air-water interface. The speed with which these waves move can set a practical upper limit to the speed with which large animals can swim near the surface. The shallower the water, the slower the speed limit. The interplay between gravity and surface tension leads to a minimum wave celerity of 23 cm s^{-1}. As a result, animals such as water striders that rely on wave drag as a means to grip the water must move their appendages faster than 23 cm s^{-1}.

The process of wave breaking can lead to rapid water velocities, and consequently to the imposition of large hydrodynamic forces on benthic organisms. These forces may be of considerable ecological and evolutionary importance.

Water waves, like sound and light waves, can transmit information, but are subject to the limitations of group velocity. Because the group velocity of capillary waves is higher than the wave celerity, these waves are particularly useful as a means of information transmission.

13.17 . . . and a Warning

The subject of surface waves is both broad and complex. The discussion in this chapter provides only the briefest of introductions, and the conclusions reached should not be used blithely. Fortunately, there are several excellent texts on wave mechanics to which you can turn. Bascom (1980) presents a readable overview of the mechanics of ocean waves and their interaction with the shore, and Kinsman (1965) is an excellent technical reference for many aspects of ocean waves. Denny (1988) discusses wave mechanics in the context of their interaction with benthic organisms. And finally, Lamb (1945) is an invaluable reference on the myriad quirky aspects of all sorts of surface waves. Although one must be mathematically adventurous to submerge oneself in Lamb (1945), the insights that result are often worth the effort.

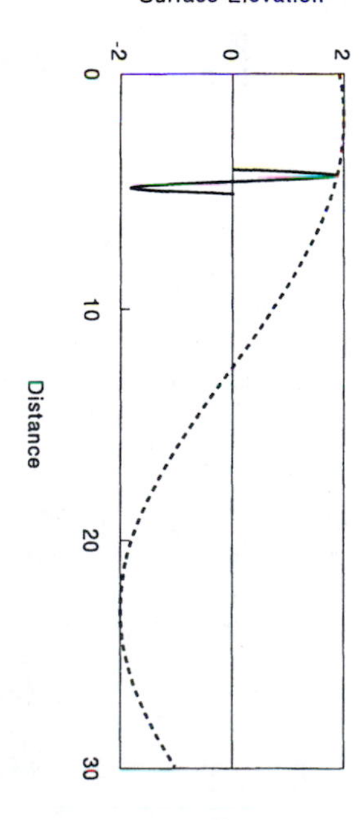

Chapter 14

Evaporation: Drying Out and Keeping Cool

This is the final chapter in which we explore the interface between air and water. Here we examine *evaporation*, the process by which liquid water becomes a gas. As we will see, evaporation can be both the bain of terrestrial existence and a key mechanism through which such existence is possible. For example, we will demonstrate why most plants lose water at a prodigious rate to gain the carbon dioxide necessary for photosynthesis and why aerial respiration is a losing proposition for bacteria. But we will also see that both plants and animals have evolved ingenious tricks to reduce the rate at which they exchange water for carbon dioxide and oxygen. We will explore the role that latent heat plays in extending the size to which animals can grow before they overheat, and show that evaporation has kept the temperature of the ocean surface virtually constant through geological time.

14.1 The Physics

Everday experience teaches us that water in contact with air gradually evaporates: wet clothes hung on the line eventually dry and water in a fishbowl must periodically be replaced. The phenomenon is so commonplace that one is likely to accept it as a proper characteristic of the terrestrial environment without giving due consideration to the physics of the matter. Why and how does water evaporate? This question is particularly perplexing in light of chapter 12, in which we spent considerable time exploring the consequences of water's adhesion to itself. If the bonds between molecules are strong enough to create a surface tension, how do water molecules ever escape the clutches of their neighbors and fly off into the air?

To understand how water evaporates we return to the concept of heat, as introduced in chapters 3 and 8. Recall that the temperature of a material is determined by the average kinetic energy of its molecules:

$$\text{average kinetic energy} = \frac{3kT}{2}, \tag{14.1}$$

where k is the Boltzmann constant (1.38×10^{-23} J K^{-1}). But, as we saw in chapter 6, averages can be deceiving. If a material has a temperature of 290 K, for instance, it does *not* follow that *every* molecule in that material has a kinetic energy of 6×10^{-21} J. All we know is that the *average* kinetic energy of molecules is 6×10^{-21} J, and, given the stochastic nature of molecular motions, we can guess that by chance some molecules travel slower than the average and some considerably faster.

In a gas, the probability that a molecule moves with a particular speed, u, is

$$P(u) = 4\pi \left(\frac{m}{2\pi kT}\right)^{3/2} u^2 e^{-\frac{mu^2}{2kT}}, \tag{14.2}$$

where m is the mass of each molecule and T is absolute temperature (Reif 1965). This relationship, known as the *Maxwell distribution of speeds*, is shown in figure 14.1. At any given temperature, there is a small probability that the speed of a molecule is near zero, a relatively high probability that the speed is near the average,[1] and a decreasing, but finite, probability that speed is much greater than average. Note that as the temperature rises, the distribution of molecular speeds spreads out. Because molecules cannot have less than zero speed,[2] there are two important consequences of this increase in the variance of u. First, the average speed must increase. This is simply a restatement of the relationship between temperature and average kinetic energy. Second, the higher the temperature, the larger the probability that an individual molecule will have a speed much in excess of the average. It is this second effect that explains evaporation.

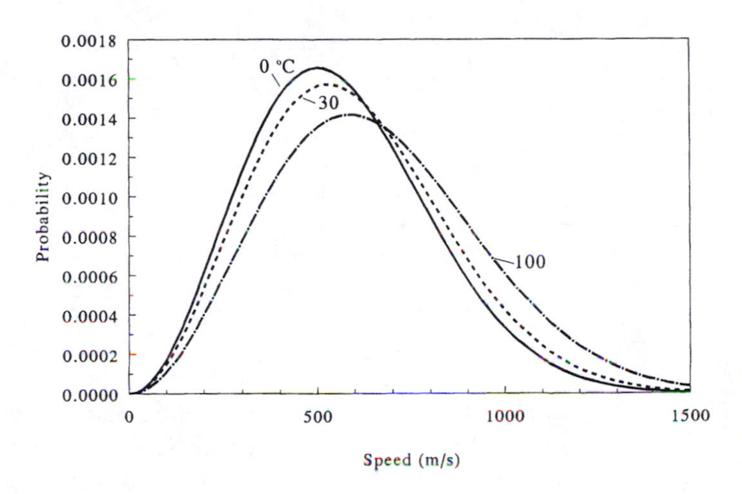

Fig. 14.1 Maxwell's distribution of molecular speeds becomes broader as the temperature rises, increasing the probability that a water molecule has a speed high enough to escape into the air.

Molecules in water have a distribution of speeds similar to the one described by the Maxwell distribution for gases. At any given temperature there is a fraction of water molecules that has a speed well in excess of the average, and a few of these molecules have kinetic energies sufficient even to break the bonds holding them in the liquid. If by chance these energetic molecules arrive at the air-water interface, they can escape from the liquid into the gas as water vapor. This, then, is the process of evaporation.

We may estimate the speed required to escape from the liquid phase by noting that at 100°C the *average speed* of molecules is sufficient to allow escape. In other words, at 100°C the water boils. At this temperature, the average speed of water molecules is 719 m s^{-1}. At any temperature below 100°C, only those molecules with speeds in excess of approximately 719 m s^{-1} can escape, and the lower the temperature, the smaller the fraction these form (fig. 14.1). As a result, evaporation is slower at low temperatures.

There are two other aspects of evaporation that deserve closer scrutiny. First, the exchange of molecules between liquid and gas can occur in both directions. Once freed from their liquid bonds, water molecules in air (water vapor) are transported by diffusion and convection. In the process, they may come into contact with the water's surface and can be "captured" by the water. The result is a constant interchange of water molecules across the air-water interface.

In a closed system (a stoppered bottle partially full of water, for example), this interchange eventually leads to a dynamic equilibrium between water molecules in the gaseous and the liquid phases. This equilibrium occurs when the concentration

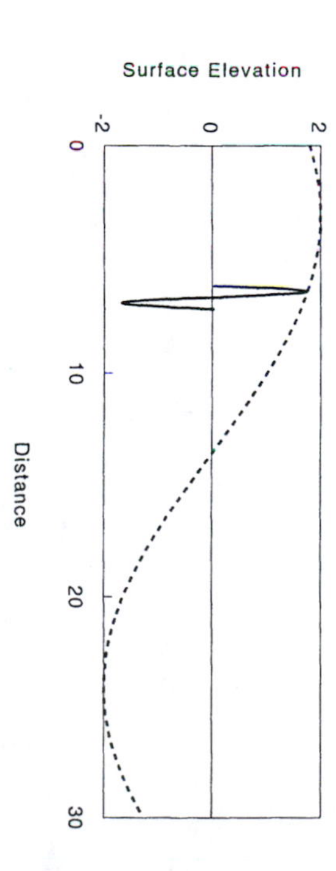

[1]Because the Maxwell distribution is skewed, with a long tail extending out to high speeds, the average velocity is somewhat greater than the *modal* velocity, the velocity with the greatest probability.

[2]A molecule can have a negative *velocity* by moving in the negative direction along some axis. The concept of *speed*, however, does not incorporate direction, and consequently a speed of zero is as slow as anything can go (see chapter 3).

of water vapor is such that the number of molecules recaptured by the water is just sufficient to offset those evaporating into the air. This *saturation vapor concentration*, C_s, is a function of temperature (fig. 14.2; table 14.1), the higher the temperature, the greater the tendency for water molecules to evaporate, and the higher the saturation concentration of water vapor.

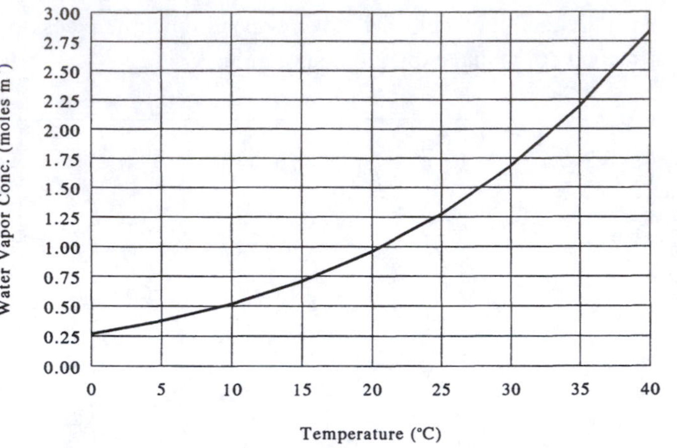

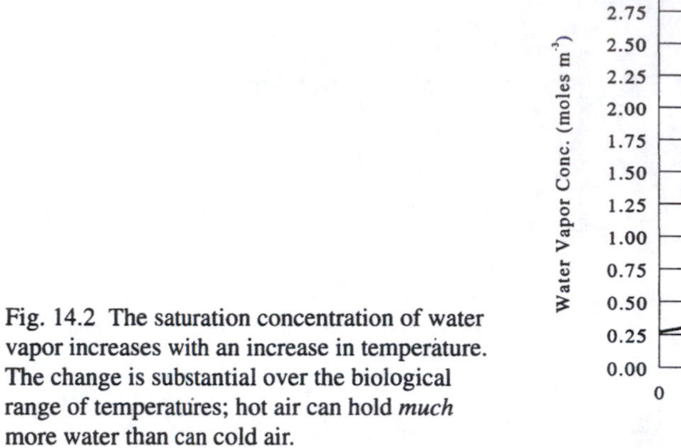

Fig. 14.2 The saturation concentration of water vapor increases with an increase in temperature. The change is substantial over the biological range of temperatures; hot air can hold *much* more water than can cold air.

The change in saturation concentration is dramatic across the biological temperature range (fig. 14.2). At 0°C, air is saturated with water vapor when the vapor has a concentration of only 0.269 moles per cubic meter, a mere 0.6% of the molecules present in the gas. At 40°C, saturated air contains 2.834 moles of water vapor per cubic meter, 7.3% of the molecules in the gas, 10.5 times the concentration at 0°C. The dependence of C_s on temperature has important biological consequences, as we will see.

Despite the ubiquitous presence of air-water interfaces in the environment, water vapor is seldom at saturation concentration in the atmosphere, and it is useful to have a means of quantifying the moisture content of ambient air. A variety of indices have been devised (see, for instance, Gates 1980; Nobel 1983; Pearcy et al. 1989). Here we use the concept of *relative humidity*, $\mathcal{H}_r$:

$$\mathcal{H}_r = \frac{C_\infty}{C_s}, \tag{14.3}$$

where C_∞ is the ambient concentration of water vapor. $\mathcal{H}_r$ is often multiplied by one hundred and expressed as a percentage of the saturated concentration, but in this chapter we deal with it simply as a fraction.

Note that for a fixed concentration of water vapor, relative humidity varies with temperature because the saturation concentration varies with temperature. For example, a concentration of 0.27 mol m^{-3} corresponds to a relative humidity of 1.00 at 0°C but only 0.095 at 40°C. Conversely, for a fixed relative humidity, the concentration of water vapor increases with temperature in proportion to the curve shown in figure 14.2.

We now consider another consequence of the process by which water molecules escape into air. We noted above that at biological temperatures, only those molecules that have kinetic energies well above the average are capable of evaporating. As a result, when they leave they take with them a biased sample of the liquid's kinetic energy, and the average kinetic energy of the remaining molecules decreases. In other words, because only the "hottest" molecules evaporate, the water they leave behind is cooled.

This effect is quantified as the *latent heat of vaporization*, Q_l, the amount by which the heat content of liquid water is decreased by the evaporation of a given mass of water vapor. The latent heat of vaporization of water is very high, about 2.5×10^6 J kg^{-1}, expressive of the fact that the bonds between water molecules are strong and only the very "hottest" molecules can escape as vapor. The latent heat of vaporization varies only slightly across the biological range of temperatures, from 2.513×10^6 J kg^{-1} at 0°C to 2.394×10^6 J kg^{-1} at 40°C (table 14.1).

To give latent heat some tangibility, consider an example. A cubic meter of water at approximately mammalian body temperature (38°C) is insulated from the surrounding air except for a small surface area across which water evaporates. Evaporation is allowed to continue until 6% of the liquid's mass (60 kg) has turned to vapor. In the process, 1.5×10^8 J of heat are removed from the remaining 940 kg of water. Recalling that water has a specific heat of about 4.2×10^3 J kg^{-1} K^{-1} (table 8.2), we find that the temperature of the remaining water is 38° lower than before. Thus, the evaporation of only 6% of an insulated mass of water is sufficient to lower its temperature from that at which mammalian tissues normally operate to that at which water freezes.

Lastly, we note that in regards to diffusion, water vapor in air behaves the same as any other gas. Because of its relatively low molecular weight (0.018 kg per mole, versus 0.032 kg for O_2 and 0.044 kg for CO_2), water vapor has a higher average velocity than oxygen or carbon dioxide, and therefore has a higher diffusion coefficient (table 14.1).

Table 14.1 Properties relating to the evaporation of water into air.

Property	Temperature (°C)				
	0	10	20	30	40
Saturation vapor concentration, C_s (mol m^{-3})	0.269	0.521	0.959	1.684	2.834
Saturation molar fraction, ξ_s	0.0060	0.01211	0.02306	0.04188	0.07282
Saturation specific humidity, $\mathcal{H}_{s,s}$	0.0038	0.0076	0.0145	0.0265	0.0467
Latent heat of vaporization, Q_l (J kg^{-1} $\times 10^6$)	2.513	2.489	2.465	2.442	2.394
Diffusion coefficient of water vapor, $\mathcal{D}_{H_2O}$ (m^2 s^{-1} $\times 10^{-6}$)	20.9	22.5	24.2	26.0	27.7
Diffusion coefficient of CO_2, $\mathcal{D}_{CO_2}$ (m^2 s^{-1} $\times 10^{-6}$)	13.9	14.9	16.0	17.0	18.1

Sources: Data from Marrero and Mason (1972) and Weast (1977).

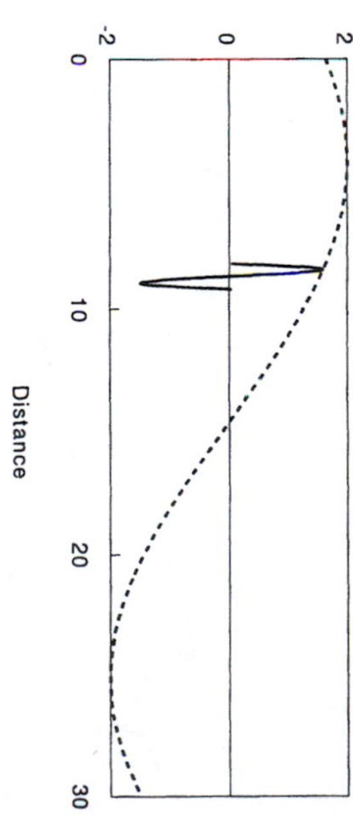

14.2 Evaporation from Leaves

We are now in a position to explore the biological consequences of evaporation. We begin with an examination of the process by which terrestrial plants obtain the carbon dioxide they need for photosynthesis.

A schematic cross section through a typical leaf is shown in figure 14.3. Most of the leaf's surface is formed by a tough epithelium that is often coated by a waxy cuticle. This epithelial layer is relatively impermeable to both water and carbon

dioxide and acts as a barrier to the exchange of these substances between leaf and atmosphere. However, at intervals along the leaf, the epithelium is pierced by small pores called *stomata* that connect the surrounding air to the spaces among the spongy mesophyll cells of the leaf's interior. It is through these pores that carbon dioxide must pass as it is delivered to the leaf's photosynthetically-active interior cells, and it is through these pores that water vapor escapes to the atmosphere.

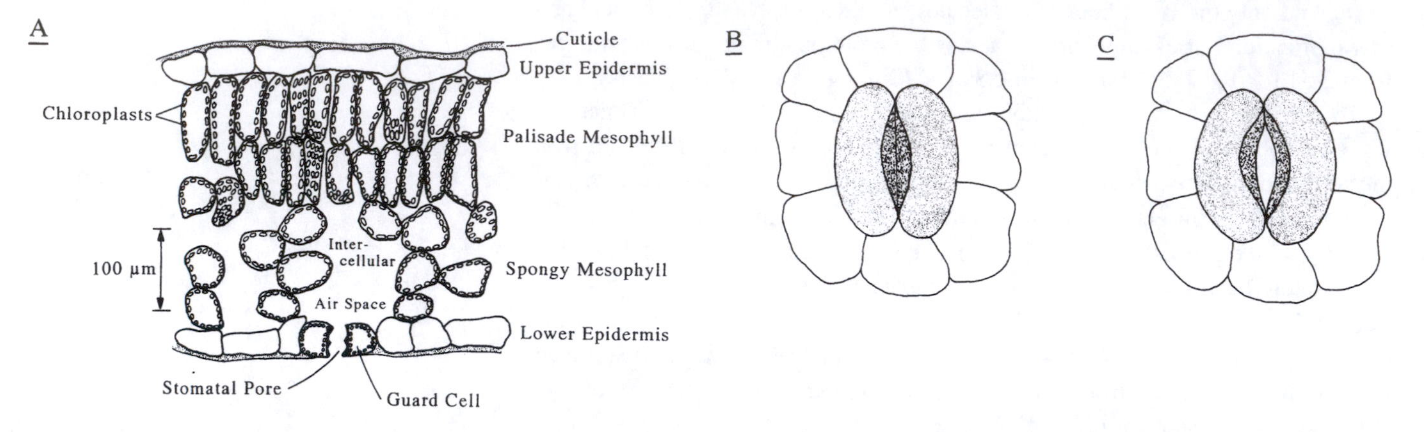

Fig. 14.3 Carbon dioxide is absorbed by the mesophyll cells of a leaf, but must pass through the stomata to reach these cells (A). When there is insufficient sunlight or when water stores are low, the stomatal openings are closed by the guard cells (B). When the plant is actively photosynthesizing, the stomata are open (C). [(A) is from *Biophysical Plant Physiology and Ecology* by Park S. Nobel. Copyright ©1983 by W. H. Freeman and Company. Reprinted by permission]

This arrangement of an impermeable, but porous, layer separating living cells from the atmosphere is reminiscent of the eggshells we considered in Chapter 6, and the same physics apply. To analyze the situation, we assume that the outer surface of the leaf is in contact with well-mixed air so that the bulk atmospheric concentration of gases is maintained at the outside end of each stomatal pore. Within the intercellular air spaces of the leaf, we assume that the concentration of carbon dioxide is zero (in effect assuming that the mesophyll cells immediately bind any CO_2 molecules that come along) and that water vapor is present at saturating concentration. The leaf and the surrounding air are at the same temperature. Each stomatal pore has a length ℓ and a cross-sectional area A.

Given this simple system, we ask the following question: What are the relative rates of water loss and carbon dioxide gain? To explore this trade-off, we proceed as we did in examining transport across eggshells. The flux of carbon dioxide (in moles per second) into the leaf through one pore is calculated from Fick's equation (eq. 6.28):

$$CO_2 \text{ flux} = \mathcal{D}_{CO_2} A \frac{\Delta C_{CO_2}}{\ell}, \tag{14.4}$$

where ΔC_{CO_2} is the difference in carbon dioxide concentration across the pore. Given our assumptions, ΔC_{CO_2} is equal to the ambient atmospheric concentration of CO_2, which at 20°C is 0.0137 moles m^{-3}. The diffusion coefficient of CO_2 is about 1.5×10^{-5} m^2 s^{-1} in air, increasing with increasing temperature, as shown in table 14.1.

The flux of water out of the leaf through one pore is

$$\text{water vapor flux} = \mathcal{D}_{H_2O} A \frac{\Delta C_{H_2O}}{\ell}, \tag{14.5}$$

where ΔC_{H_2O} is the difference in water vapor concentration across the pore. Given our assumptions,

$$\Delta C_{H_2O} = C_s(T) - \mathcal{H}_r C_s(T), \tag{14.6}$$

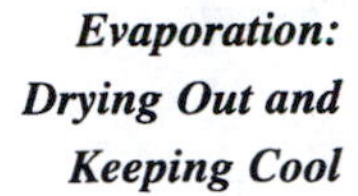

where $C_s(T)$ is the saturation vapor concentration at temperature T, and $\mathcal{H}_r$ is the relative humidity of the ambient air outside the leaf. The concentration gradient of water vapor thus depends on both temperature and relative humidity.

We may calculate the rate at which CO_2 is gained relative to the rate at which water is lost by dividing eq. 14.4 by eq. 14.6, with the result that

$$\frac{\text{rate of water loss}}{\text{rate of CO}_2\text{ gain}} = \frac{\mathcal{D}_{H_2O}}{\mathcal{D}_{CO_2}} \frac{C_s(T)(1-\mathcal{H}_r)}{C_{CO_2}(T)}. \tag{14.7}$$

This relationship is shown in figure 14.4 as a function of temperature for a variety of relative humidities. Because the diffusion coefficient of CO_2 is only about 61%

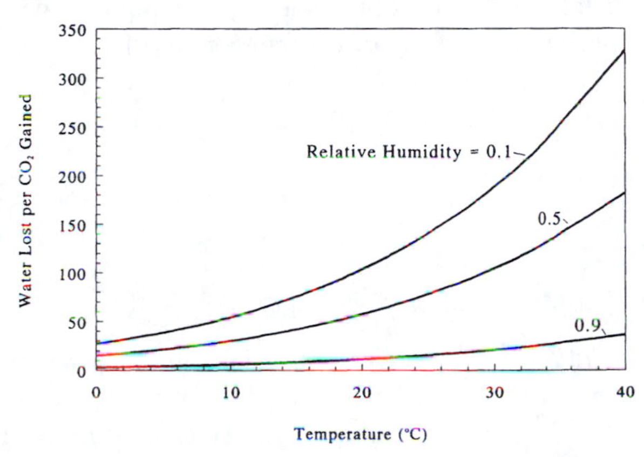

Fig. 14.4 Unless the relative humidity is very high, plants lose many molecules of water for each molecule of carbon dioxide they absorb.

that of water, and because the concentration gradient for CO_2 is very much smaller than that for water, much more water is lost than carbon dioxide is gained. For example, at 20°C and a relative humidity of 0.5, more than 50 moles of water are lost for each mole of CO_2 taken in. Only when the atmosphere surrounding the leaf is virtually saturated with water vapor can the plant gain the same number of moles of carbon as it loses of water.

This losing trade-off between water and carbon dioxide represents one of the most severe physical constraints on terrestrial plants. In order to gain the carbon necessary for growth and reproduction, plants are liable to expend exorbitant amounts of water. Although the process of loss can itself be of some utility—water moving from the roots to the leaves transports nutrients to the photosynthetically active cells—the magnitude of the loss can be problematic. In response, plants have evolved two strategies to minimize the water lost per carbon dioxide gained.

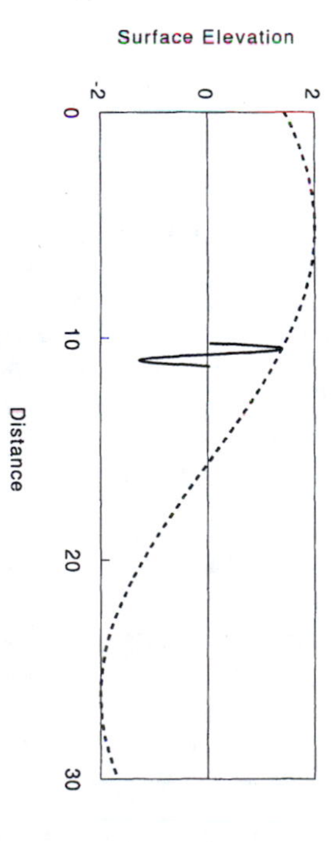

The first, and more widely spread, stratagem is for the plant to make sure that water is lost only when photosynthesis is taking place. To this end, the entrance to each stomatal pore is encircled by a pair of guard cells (fig. 14.3B,C). When the plant has sufficient stores of water and sunlight is available for photosynthesis, these cells can maintain a high internal turgor pressure. As a result of this pressure, the cells bend, uncovering the opening to the stomatal pore. If the plant's supply of water runs low, however, the guard cells can no longer maintain their turgor and they deflate, closing each pore's aperture and preventing further gas exchange. Thus, in times of water scarcity, the plant decreases its evaporative water loss. Similarly, at night when photosynthesis ceases due to a lack of light, the guard cells close the stomatal pores. Although the guard cells help the plant to avoid unnecessary loss of water, they do nothing to change the unfavorable rate at which carbon dioxide is traded for water.

Some plants native to arid environments (such as cacti and agaves) have evolved an elegant method to reduce their trade deficit: they close their stomatal pores during the day and open them to absorb carbon dioxide at night. The beauty of this stratagem lies with two aspects of the physics of water vapor. First, although the relative humidity may vary through the day due to the change in air temperature, the vapor concentration in desert air stays relatively constant. Second, in the cool of the night, the saturation concentration of water vapor in the intercellular spaces of the leaf is lower than in the day (fig. 14.2). Thus, if plants open their stomata at night, the rate of water loss can be decreased.

The magnitude of this effect can be calculated by restating eq. 14.7 to account for the constant concentration of water vapor in desert air. If we assume that ambient air has a vapor concentration of 0.17 mol m^{-3} (equivalent to a relative humidity of 0.1 at 30°C):

$$\frac{\text{water lost}}{CO_2 \text{ gained}} = \frac{\mathcal{D}_{H_2O}(T)}{\mathcal{D}_{CO_2}(T)} \frac{C_s(T) - 0.17}{C_{CO_2}(T)}, \tag{14.8}$$

where T in this case refers to the nighttime temperature at which the plant absorbs carbon dioxide. This relationship is shown in figure 14.5 as a function of temperature. The lower the nighttime temperature, the fewer moles of water lost for each mole of carbon dioxide gained. Only if the nighttime temperature is higher than that in the day (an unlikely situation given the high daytime temperatures in the desert, here assumed to be 30°C) does the plant lose more water by absorbing CO_2 at night.

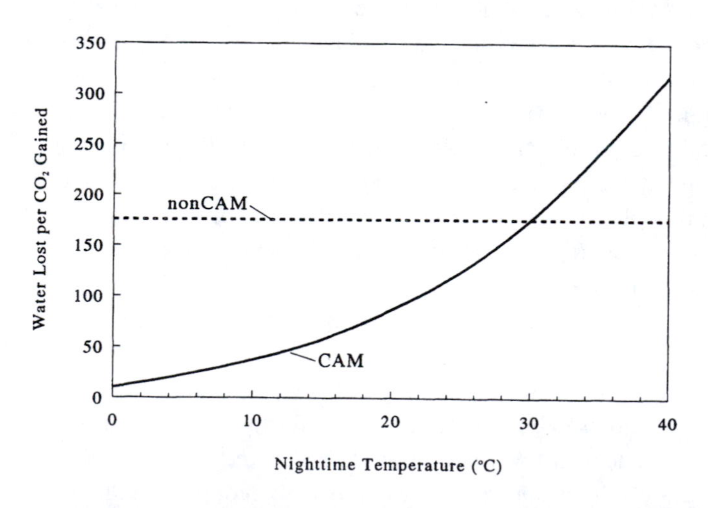

Fig. 14.5 By absorbing carbon dioxide in the cool of the night, CAM plants lose less water than do non-CAM plants. The line drawn here for a non-CAM plant assumes that the plant photosynthesizes at a daytime temperature of 30°C.

There is a problem, however. At night, there is no sunlight available to drive the conversion of CO_2 to glucose. To circumvent this problem, cacti and agaves process CO_2 into organic acids at night. During the day, when the stomatal pores are closed, these acids are hydrolyzed and the resulting carbon dioxide used by photosynthesis. This strategy is known as *crassulacean acid metabolism* (CAM for short).

Note that animals are not faced with as strict a respiratory trade-off. Eq. 14.7

can be restated in terms of oxygen transport, with the result that

$$\frac{\text{rate of water loss}}{\text{rate of O}_2\text{ gain}} = \frac{\mathcal{D}_{H_2O}}{\mathcal{D}_{O_2}} \frac{C_s(T)(1 - \mathcal{H}_r)}{C_{O_2}(T)}. \tag{14.9}$$

This relationship is shown in figure 14.6. Because oxygen is so concentrated in air (635 times the concentration of CO_2 and 3 to 35 times that of saturated water vapor), it can be gained diffusively with a relatively small concomitant water loss.

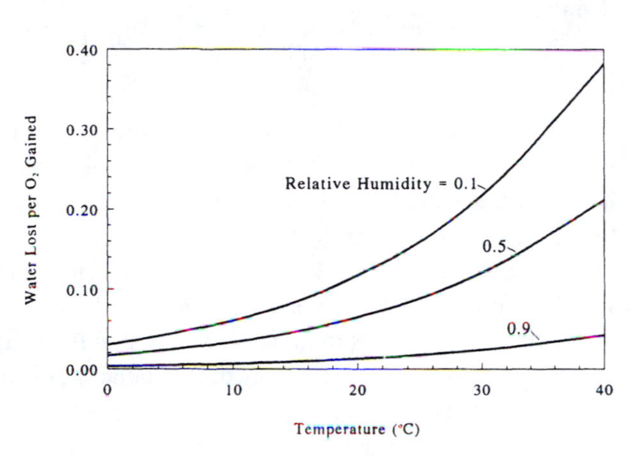

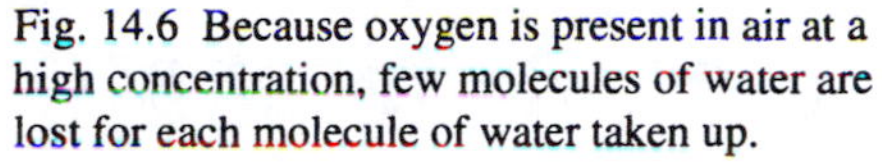
Fig. 14.6 Because oxygen is present in air at a high concentration, few molecules of water are lost for each molecule of water taken up.

14.3 Drying Up

Throughout this text we have explored the mechanics of bacteria, blue-green algae, and other small organisms in air, with the conclusion that these aerial plankton could possibly "swim" through air, have no problem shedding their metabolic heat, and could easily obtain sufficient oxygen or carbon dioxide to meet metabolic needs. On the basis of these criteria, an aerial existence seems plush. In each case, however, we have tempered these conclusions with the thought that aerial life may be difficult because of the tendency to desiccate rapidly. We are now in a position to calculate the rate at which organisms dry out in air.

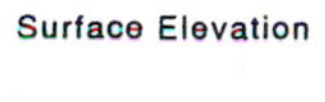

14.3.1 *Diffusive Desiccation*

As usual, we simplify matters by assuming that our organism is a sphere. We suppose that its outer surface is freely permeable to water, in which case the air directly in contact with the sphere is assumed to be saturated with water vapor. In addition, we assume that water is transported away from the organism by diffusion alone. As a consequence, our results underestimate actual rates of desiccation from a freely permeable sphere, which in most circumstances would be augmented by convection.

Given these assumptions, we can treat the flow of water away from an organism in much the same fashion as we treated the flow of oxygen or carbon dioxide in chapter 6 or the flow of heat in chapter 8. By analogy to Newton's law of cooling, we can define a *mass transport coefficient*, h_m:

$$h_m = \frac{J_w}{A \Delta C}, \tag{14.10}$$

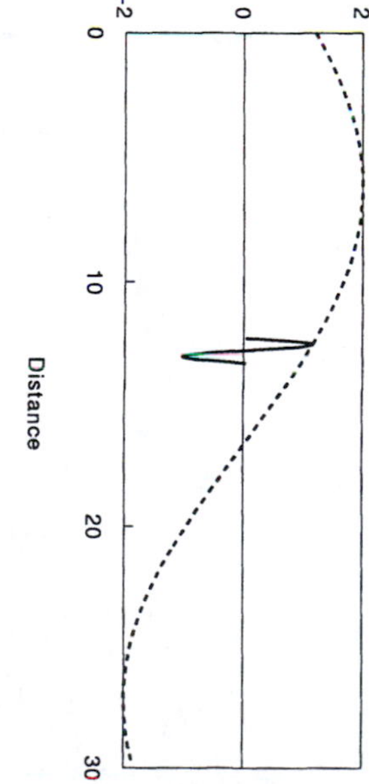

where J_w is the net flux of water away from the sphere (in mol s^{-1}), A is the area over which exchange occurs, and ΔC is a measure of the difference in the concentration of water vapor between the sphere's surface and the surrounding air (expressed as moles m^{-3}). If evaporation were completely analogous to the diffusion of heat or of molecules within a single phase (either liquid or gas), ΔC would equal $C_s - C_\infty$. The transport of molecules between the liquid and gaseous phases introduces a complication, however. As molecules evaporate, they add to the volume of gas in a system, and this addition results in a bulk transport of the air adjacent to the liquid-gas interface. This effect can be taken into account by defining ΔC as

$$\Delta C = \frac{C_s(T_b) - C_\infty(T_\infty)}{1 - \xi_s}, \tag{14.11}$$

where $C_s(T_b)$ is the saturation vapor concentration at the temperature of the sphere, $C_\infty(T_\infty)$ is the vapor concentration in the ambient air at ambient temperature, and ξ_s is the water-vapor fraction of the gas molecules (the *molar fraction*) in the saturated air adjacent to the interface (table 14.2). For a derivation of this expression, see Bird et al. (1960). At the temperatures encountered by organisms, $1 - \xi_s$ varies from 0.927 to 0.994 (not much different from 1), so minimal error is incurred if one prefers to think of ΔC as simply the difference in water vapor concentration between the air immediately adjacent to the air-water interface and the air some large distance away.

A restatement of eq. 14.11 will be useful. As long as the sphere and the surrounding air are at the same temperature, the water vapor concentration in the bulk of the air (C_∞) can be expressed as $\mathcal{H}_r C_s$. Thus,

$$\Delta C = \frac{C_s(1 - \mathcal{H}_r)}{1 - \xi_s} \quad \text{(same temperature)}. \tag{14.12}$$

As with heat transfer coefficients, mass transfer coefficients depend on the shape of the organism and the nature of the flow around it. However, for simple shapes such as our sphere, h_m is well defined. The mass transport coefficient for a sphere when transport is by diffusion alone is (Bird et al. 1960)

$$h_m = \frac{2\mathcal{D}_{H_2O}}{d}, \tag{14.13}$$

where d is the diameter of the sphere. Note the similarity of this coefficient to that for heat (eq. 8.14).

Inserting this expression for h_m in eq. 14.10, noting that the surface area of the sphere is πd^2, and rearranging, we see that the net flux of water away from the sphere (in moles per second) is

$$J_w = 2\pi d\mathcal{D}_{H_2O}\Delta C. \tag{14.14}$$

This is very similar to the conclusion we reached regarding the diffusive transport of molecules from a sphere in either water or air (eq. 6.40), with the exception of the manner in which we have defined ΔC (eq. 14.11).

If this flux (J_w) is multiplied by the molecular weight of water ($\mathcal{M}$), the result is an expression for the rate at which mass evaporates from the sphere (in kg s^{-1}):

$$\text{rate of mass loss} = 2\pi\mathcal{M}d\mathcal{D}_{H_2O}\Delta C. \tag{14.15}$$

It is useful to relate this rate to the mass of the sphere itself. Noting that the mass of the sphere is $\rho_b \pi d^3/6$, we see that

$$\frac{\text{rate of mass loss}}{\text{total mass}} = \frac{12\mathcal{D}_{H_2O}\mathcal{M}\Delta C}{d^2 \rho_b}. \tag{14.16}$$

This expression is shown in figure 14.7 for a range of relative humidities. It is clear that small organisms are likely to desiccate rapidly in air. For example, at 20°C in dry air, a sphere 10 μm in diameter loses water at a rate equal to 47 times its mass every second. A sphere the size of a bacterium (1 μm in diameter) evaporates water at a rate equal to about 4700 times its weight each second! Unless the relative humidity is virtually 1.0, it is difficult for an aerial bacterium to maintain a favorable water balance.

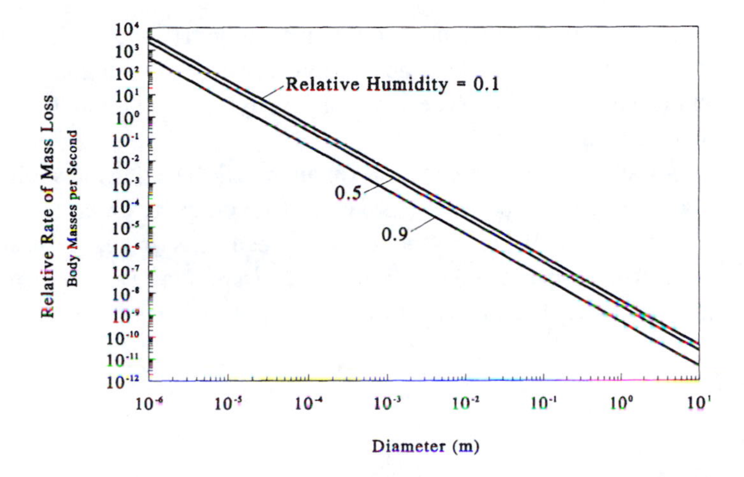

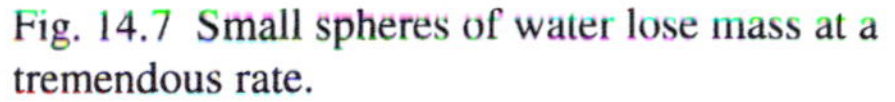
Fig. 14.7 Small spheres of water lose mass at a tremendous rate.

There is a major caveat regarding these calculations. We have assumed that the entire surface of the organism is freely permeable to water, and we thereby may be overestimating the rate of water loss. It is possible, for instance, that small aerial organisms could coat their bodies with an impermeable substance, thereby reducing evaporative losses. There are limits to this strategy, however. Aerobic organisms must exchange oxygen across their surfaces, and photosynthetic plants must exchange carbon dioxide. Thus, the reduction of water loss necessarily entails a reduction in gas exchange.

In this respect, small organisms face the same trade-offs as large organisms, and must be similarly vigilant in their water conservation. The primary difference between small and large organisms is the incredible speed with which a lapse in vigilance results in the complete desiccation of a small plant or animal. This swift retribution imposed on small organisms may have been a potent selective factor favoring large body size in terrestrial organisms.

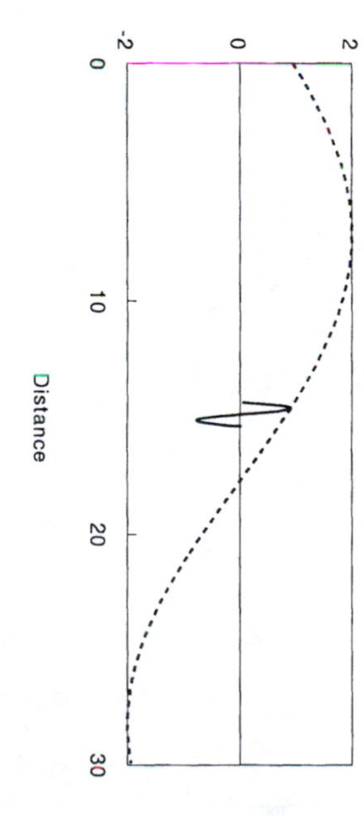

14.4 Evaporative Cooling

Having calculated the rate at which water mass is lost from a sphere, we are now in a position to return to the theme of chapter 8 and explore the rate at which heat can be dissipated by evaporation. Recall that large terrestrial organisms, especially mammals and birds, have a problem getting rid of the heat produced by their metabolism. Our calculations of heat transfer showed that in the absence of evaporation, animals the size of rhinoceroses and elephants would be likely to overheat even when standing in a strong breeze. Can evaporative heat loss solve this problem?

We begin by examining the rate at which heat is dissipated if water vapor is transported by diffusion alone. From eq. 14.15 we know that $2\pi\mathcal{M}d\mathcal{D}_{H_2O}\Delta C$ kg of water are lost from a sphere each second. Multiplying this flux by the latent heat of vaporization Q_l, we find that the rate at which heat is dissipated is

$$\text{rate of heat dissipation} = 2\pi Q_l \mathcal{M} d\mathcal{D}_{H_2O}\Delta C. \tag{14.17}$$

From eq. 8.17, we know that the metabolic heat produced by a spherical organism is

$$\text{rate of heat production} = M\left(\frac{\rho_b \pi d^3}{6}\right)^{\alpha}, \tag{14.18}$$

where M is the coefficient of resting metabolic rate and α is the allometric coefficient of metabolism. M varies from one animal group to the next and is highest in mammals and birds. The exponent α is typically close to $3/4$, a value that we assume in our calculations here.

As before, we calculate the maximum practical resting metabolic rate by equating the rate at which heat is produced to the rate at which it is dissipated, and solving for M. If the metabolic rate exceeds this value, the heat produced exceeds that dissipated and the organism heats up. At a steady body temperature, the maximum metabolic rate set by evaporative heat loss is

$$M_{max} = \frac{10.2 Q_l \mathcal{M}\mathcal{D}_{H_2O}\Delta C}{\rho_b^{3/4} d^{5/4}}. \tag{14.19}$$

The form of this equation is very similar to that for M_{max} when heat is lost by conduction alone:

$$M_{max} = \frac{10.2\mathcal{K}\Delta T}{\rho_b^{3/4} d^{5/4}}. \tag{14.20}$$

The ratio of eq. 14.19 to eq. 14.20 is a measure of the relative rates at which animals may metabolize when they lose heat by evaporation or by simple conduction:

$$\text{ratio of metabolic rates} = \frac{Q_l \mathcal{M}\mathcal{D}_{H_2O}\Delta C}{\mathcal{K}\Delta T}. \tag{14.21}$$

This relationship is shown in figure 14.8 as a function of air temperature for a fixed body temperature of 37°C and a fixed relative humidity of 0.5. Only at low temperatures is the rate of heat loss by conduction near that of evaporation, and at these low temperatures animals are less likely to have problems with overheating. At high air temperatures, where overheating is a problem, evaporative cooling is much more effective than conduction. In figure 14.9 the relationship of eq. 14.21 is shown as a function of relative humidity for a fixed air temperature of 20°C and a fixed body temperature of 37°C. Note that even at $\mathcal{H}_r = 1$ the body loses more heat by evaporation than by conduction. This is due to the fact that air adjacent to the body is saturated at 37°C while the surrounding air is assumed to be at 20°C. C_s is lower at 20°C than at 37°C.

These results show that evaporative cooling can be an effective mechanism by which large mammals and birds can cope with high metabolic rates. Note, however, that the rates of evaporative heat loss calculated here assume that the entire surface of the organism is freely permeable to water. If this is not so, the rate of evaporative cooling is decreased. Therein lies one of the primary mechanisms by which large terrestrial animals regulate their temperature. In times of heat stress, many animals actively facilitate the transport of water to the surface of the body by sweating. The evaporation of this sweat then cools the organism. When evaporative cooling is undesirable (when the air is cold, for instance), the animal stops sweating and the rate of evaporation is thereby decreased.

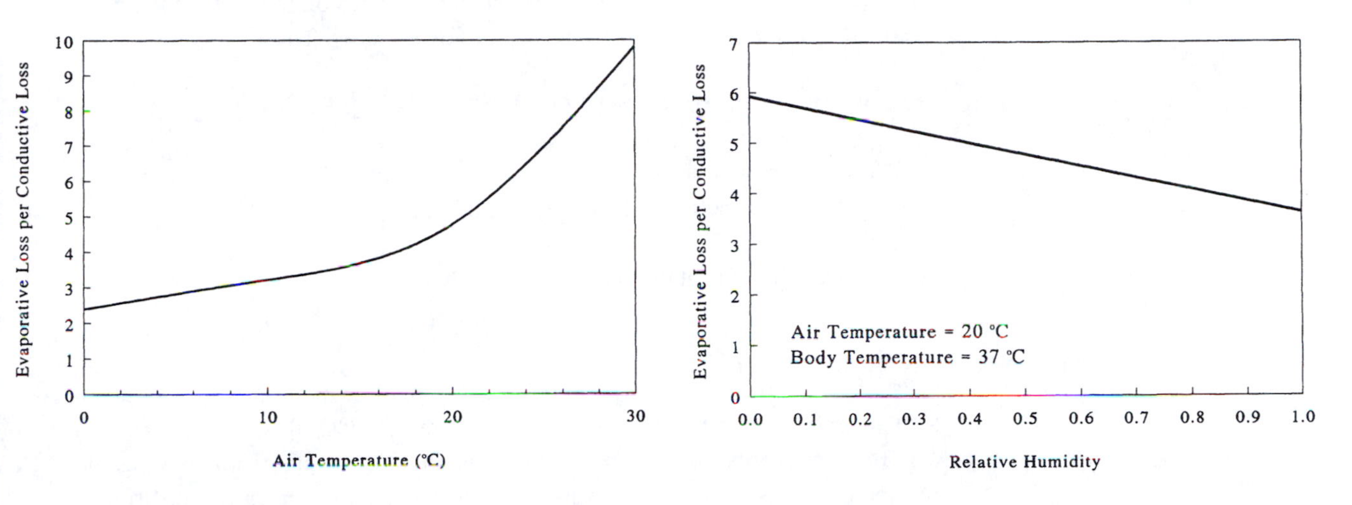

Fig. 14.8 A spherical organism with a body temperature of 37°C loses heat much more effectively by evaporation than by conduction.

Fig. 14.9 Even at a relative humidity of 1.0, heat is lost more effectively by evaporation than by conduction because the body is warmer than the air.

Those animals that do not sweat often regulate their rate of evaporative heat loss by panting. When evaporation is advantageous, panting motions force air rapidly over the wet surfaces of the lungs and nasal lining and the rate of evaporation is thereby increased. For a more thorough exploration of the physiology of temperature regulation, you are invited to consult Schmidt-Nielsen (1979).

14.4.1 *Thermal Cost of Respiratory Evaporation*

In chapter 8 we calculated the rate at which heat is inevitably lost in the process of breathing. At that time we only considered the thermal cost of heating air to the same temperature as the body. However, when air is drawn into the lungs or tracheae, water evaporates until the gas is saturated, and heat is thereby lost due to the latent heat of vaporization. How large a drain on the thermal reserves of an organism is this heat loss?

We pick up the calculation from chapter 8 by recalling that for each liter of oxygen absorbed by an insect or mammal, approximately 0.0192 m^3 of air must be brought into the lungs or tracheae.[3] In the process, this volume becomes saturated with water vapor. Thus,

$$\frac{\text{moles of water vapor}}{\text{liter } O_2} = 0.0192\,[C_s(T_b) - C_\infty(T_\infty)], \qquad (14.22)$$

where $C_s(T_b)$ is the saturation water vapor concentration at body temperature, and $C_\infty(T_\infty)$ is the water vapor concentration in the surrounding air at ambient

[3] Here we have assumed that the animal absorbs only 25% of the oxygen present in the lungs. Birds may be more efficient at absorbing O_2, and therefore may require only 0.0144 m^3 of air per liter of oxygen.

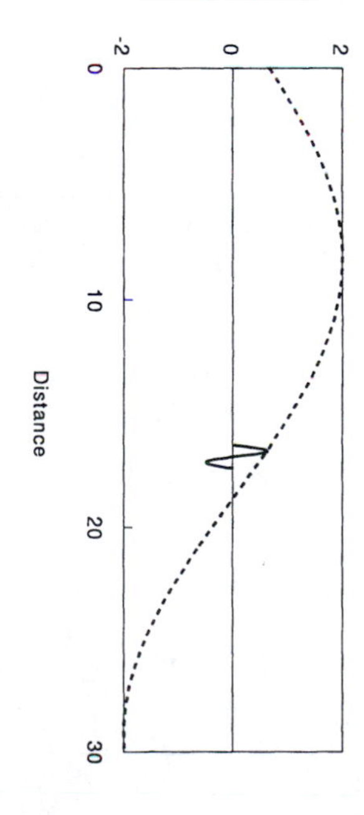

temperature. The heat dissipated by the evaporation of this water is

$$\frac{\text{heat dissipated}}{\text{liter O}_2} = 0.0192\, Q_l \mathcal{M}\, [C_s(T_b) - C_\infty(T_\infty)], \qquad (14.23)$$

where Q_l is the latent heat of vaporization (about 2.5×10^6 J kg^{-1} k^{-1}), and $\mathcal{M}$ is the molecular weight of water vapor (0.018 kg).

Now, the rate at which oxygen is consumed depends on the metabolic rate of the organism, which we know from chapter 8 is $M'm^{3/4}$. Thus,

$$\text{rate of heat dissipation} = 0.0192\, Q_l \mathcal{M}\, [C_s(T_b) - C_\infty(T_\infty)]\, M'm^{3/4}. \qquad (14.24)$$

This rate can then be compared to the resting metabolic rate of the organism, $Mm^{3/4}$. Noting as before (Chapter 8) that $M'/M = 4.98 \times 10^{-5}$, we see that

$$\text{fraction of heat dissipated by evaporation} = 1.7 \times 10^{-8} Q_l\, [C_s(T_b) - C_\infty(T_\infty)]. \qquad (14.25)$$

This relationship is shown in figure 14.10. At a relative humidity of 0.5, with an air temperature of 0°C and a body temperature of 37°C, this fraction is about 0.1. In other words, under arctic conditions about 10% of an animal's resting metabolic rate is "wasted" in evaporating water from the respiratory surfaces. This is about ten times the fraction of the resting metabolic rate needed to heat the air alone (see chapter 8). The thermal cost of respiratory evaporation is lower if the relative humidity is very high, or if the ambient air temperature is near that of the organism.

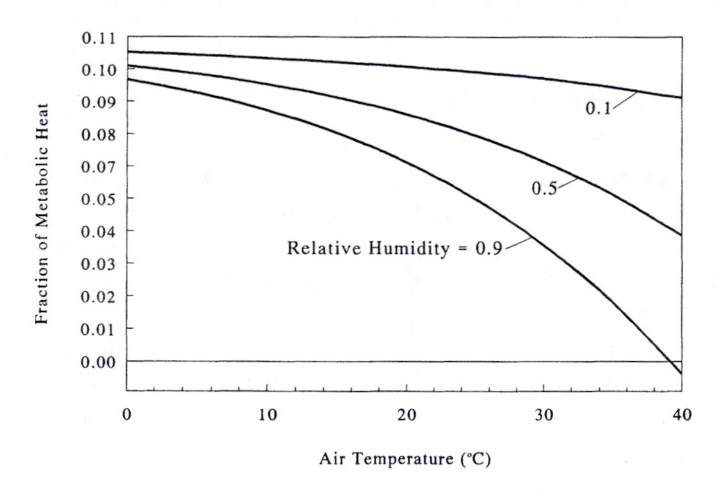

Fig. 14.10 At low temperatures or low humidity, a substantial portion of an animal's metabolic heat can be "wasted" in evaporation of water from the lungs.

Birds have a higher body temperature than mammals (41°C typically), and therefore might be thought to have a higher thermal cost of respiration than mammals. However, the increased cost due to body temperature is likely to be more than offset by the increased efficiency with which birds extract oxygen from the air in their lungs.

14.4.2 *The Advantage of a Cold Nose*

The calculations made above make a "worst case" prediction in that we assumed that air is exhaled at body temperature. This need not be the case. For instance, as dry desert air is inhaled through the nose of a kangaroo rat, evaporation cools the nasal epithelia to a temperature several degrees below that of the ambient air. The epithelium is still cold when the saturated air is released, and it cools the exhaled air. As the breath cools, its saturation vapor concentration decreases (fig. 14.2) and water condenses onto the nasal epithelium. Thus, some of the water lost by evaporation in the lungs is reclaimed by condensation in the nose. This cycle is repeated with the next breath. Schmidt-Nielsen (1972b) reports that when the ambient air is at 15°C, this trick allows kangaroo rats to decrease by 83% the rate at which they lose water in respiration, with a concomitant saving in heat. Savings are smaller at higher ambient temperatures.

Although the kangaroo rat is exceptional in its ability to lower the temperature of exhaled air, the mechanism it uses is not atypical. Many mammals and birds use a nasal countercurrent heat exchanger to reduce the respiratory loss of water. For a full description of both countercurrent heat exchangers and their physiological consequences, you should consult Schmidt-Nielsen (1972b) and the references cited therein.

14.4.3 *The Role of Convection*

To this point we have only considered the rate of evaporation through simple diffusion. As we have seen in chapters 7 and 8, the rates of transport can be drastically increased in the presence of convection. How fast would a spherical organism lose water if exposed to a breeze?

Bird et al. (1960) suggest that the coefficient of mass transfer for a sphere in the presence of convection is

$$h_m = \frac{2.0\mathcal{D}_{H_2O}}{d} + \frac{0.60Re^{1/2}Sc^{1/3}\mathcal{D}_{H_2O}}{d}, \tag{14.26}$$

where the Reynolds number is calculated using the diameter of the sphere as the characteristic length. This expression assumes that Re is low enough so that the boundary layer around the sphere is laminar. Sc is the Schmidt number, a dimensionless number we first encountered in chapter 7. The Schmidt number is equal to the ratio of the diffusivity of momentum (ν, the kinematic viscosity) to the diffusivity of the molecules in question, in this case $\mathcal{D}_{H_2O}$:

$$Sc = \frac{\nu}{\mathcal{D}_{H_2O}}. \tag{14.27}$$

Note that eq. 14.26 is very similar to the expression for the heat transfer coefficient in forced convection (eq. 8.16), the Prandtl number simply being replaced by the Schmidt number. Once again, the conceptual similarity among transport processes is apparent.

Using this expression for h_m, we can calculate the maximum metabolic rate that can just be offset by evaporative cooling,

$$M_{max} = \frac{3.06Q_l\mathcal{M}\mathcal{D}_{H_2O}^{2/3}u^{1/2}\Delta C\rho_f^{1/6}}{d^{3/4}\rho_b^{3/4}\mu^{1/6}}. \tag{14.28}$$

As expected, this equation is similar to that for the maximum metabolic rate set

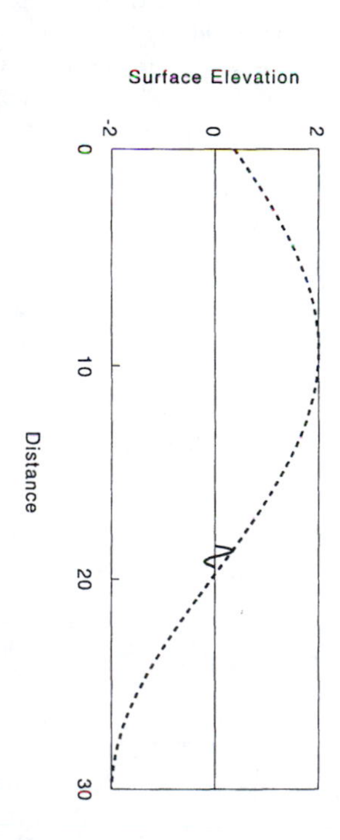

by forced convective heat loss (chapter 8):

$$M_{max} = \frac{3.06 Q_s^{1/3} \mathcal{K}^{2/3} u^{1/2} \Delta T \rho_f^{1/2}}{d^{3/4} \rho_b^{3/4} \mu^{1/6}}, \qquad (14.29)$$

where Q_s is the specific heat of water and $\mathcal{K}$ is the thermal conductivity of water.

By dividing eq. 14.28 by eq. 14.29, we can compare the two maximum metabolic rates:

$$\text{ratio of metabolic rates} = \frac{Q_l \mathcal{M} \mathcal{D}_{H_2O}^{2/3} \Delta C}{Q_s^{1/3} \mathcal{K}^{2/3} \rho_f^{1/3} \Delta T}. \qquad (14.30)$$

This relationship is shown in figure 14.11 as a function of air temperature. In this example we assume that the organism has a body temperature of 37 °C and is immersed in air with a relative humidity of 0.5. Even at very low temperatures, evaporative heat loss is much larger than convective heat loss, allowing an increase in metabolic rate for a given body temperature. Note that the result for convection is very similar to that for heat transfer by diffusion alone. The *ratio* between the metabolic rate possible with and without evaporative cooling is nearly the same regardless of the precise mechanism of transport. The absolute metabolic rates are of course much higher in the presence of convection.

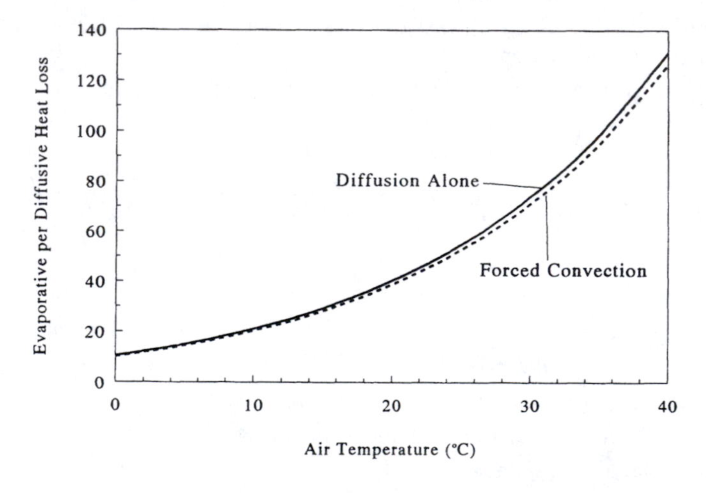

Fig. 14.11 The ratio of evaporative to diffusive heat loss is surprisingly independent of whether an organism is in still or moving air. Note that the absolute rate of heat loss (rather than its ratio) is much higher in moving air.

Now, if we examine the *absolute* rate of metabolism allowed by evaporative heat loss alone (eq. 14.28) we note that it depends on the square root of wind speed (fig. 14.12). In other words, a slow breeze drastically increases the equilibrium rate at which an animal can metabolize compared to that possible in still air. Increasing u increases M_{max}, but the increase is smaller the larger the wind speed.

Viewed from a different perspective, this relationship also means that to maintain a given body temperature, an organism must increase its metabolic rate in the presence of wind. This is why one begins to shiver after stepping out of a swimming pool or shower into a slight breeze.

14.5 Ocean Surface Temperature

Plants and animals are not the only objects in the environment affected by water's large latent heat of vaporization. For example, we noted in chapter 2 that the surface temperature of lakes and oceans seldom exceeds 28°C to 30°C owing to the effects of evaporation (Vermeij 1978). It is now time to examine this assertion in greater detail. The following discussion is based on Newell (1979).

As the sun shines on the surface of a body of water, radiant energy is absorbed by the liquid, tending to heat it. In the tropics, the radiation from the sun averaged over the year is 412 W m^{-2} (List 1958). Of this amount, roughly 10% is absorbed by the atmosphere before it reaches the water's surface, and about 8% of the remainder is reflected rather than being absorbed. The net heat influx to the surface water is thus about 341 W m^{-2}. The effect of this heat influx is to raise the temperature of the water to a point where the water loses heat at the same rate at which it is gained. There are three primary methods by which heat may be lost.

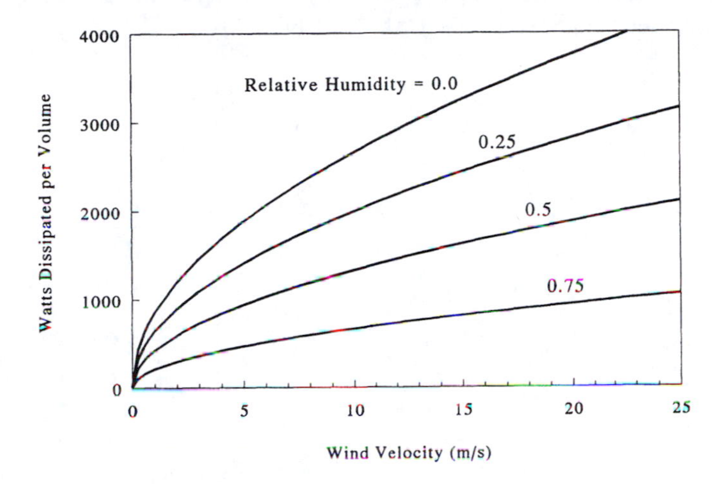

Fig. 14.12 The rate at which heat is lost by evaporation increases with the square root of wind speed.

First, heat can be lost by evaporation. A simple approximation of the evaporative flux density of heat, $\mathcal{J}_{Q,e}$, is provided by Budyko (1974):

$$\mathcal{J}_{Q,e} = 6080(\mathcal{H}_{s,s} - \mathcal{H}_{s,\infty})u, \tag{14.31}$$

where $\mathcal{H}_{s,s}$ is the saturation specific humidity of air above the water's surface at the temperature of the water, and $\mathcal{H}_{s,\infty}$ is the specific humidity of the ambient air above the water's surface.

Specific humidity is different from relative humidity; it is the density of water vapor expressed as a fraction of the total density of the moist air:

$$\mathcal{H}_s = \frac{\rho_{wv}}{\rho_a}. \tag{14.32}$$

Specific humidity rises with temperature (table 14.1; fig. 14.13).

The leading coefficient of 6080 in eq. 14.31 has units of J m^{-3}, so that when u is expressed in m s^{-1}, $\mathcal{J}_{Q,e}$ has the units of W m^{-2}.

This expression is based on a combination of theory and empirical measurements, and is therefore difficult to compare directly to the simple models we have dealt with previously. Note, however, that the rate of evaporative heat loss increases with a decrease in humidity and increases with an increase in wind speed, as we would expect. $\mathcal{J}_{Q,e}$ increases directly with wind speed rather than with its square root, in contrast with our results for a sphere.

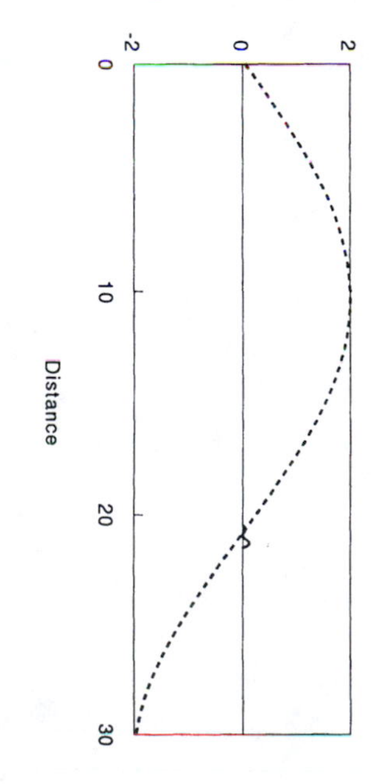

Heat can also be lost as infrared radiation is emitted from the surface. This flux density, $\mathcal{J}_{Q,b}$, is

$$\mathcal{J}_{Q,b} = 0.94\sigma T_s^4 \,(0.56 - 0.065\sqrt{1000 \times H_{s,\infty}}), \qquad (14.33)$$

where σ is the *Stefan-Boltzmann constant*, 5.6696×10^{-8} W m^{-2} K^{-4}, and T_s is the surface temperature of the water. The rate at which heat is radiated from the water's surface depends on the difference in temperature between the water and the object toward which it is radiating, in this case, the adjacent air. The effective temperature of this air is affected by its moisture content, hence the presence of the term containing the air's specific humidity.

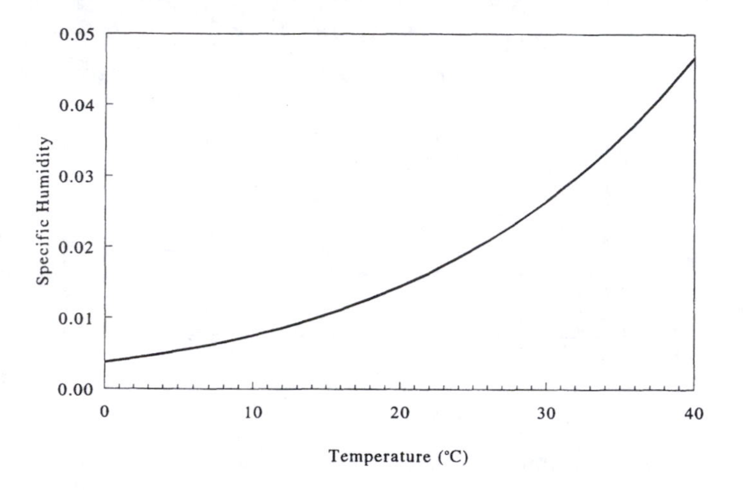

Fig. 14.13 The saturation specific humidity of air increases with an increase in temperature.

Finally, heat can be lost from the surface by convection. For simplicity, we assume that no heat is conducted to water below the surface, so that any heat lost by conduction is lost to the air. When wind is present, this heat loss is primarily in the form of a forced convective flux density, $\mathcal{J}_{Q,s}$:

$$\mathcal{J}_{Q,s} = 2.51(T_s - T_\infty)u, \qquad (14.34)$$

where T_∞ is the temperature of the air. The coefficient of 2.51 has units J m^{-3} K^{-1}, so that when u is expressed in m s^{-1}, $\mathcal{J}_{Q,s}$ has units of W m^{-2}. Again this expression is based largely on empirical measurements, and therefore bears only a qualitative resemblance to the the models we have dealt with here.

We now proceed by assuming that the air temperature remains at a typical tropical value, 27°C, regardless of the temperature of the sea surface. We fix the specific humidity at 0.7, a typical value, and assume that the wind blows at a steady 3 m s^{-1}, again a value typical of tropical seas.

Given these assumptions, we may calculate individual flux densities of heat and the total flux ($\mathcal{J}_{Q,e} + \mathcal{J}_{Q,b} + \mathcal{J}_{Q,s}$), each as a function of sea surface temperature. These values are shown in figure 14.14. As the sea surface temperature rises, the rate of heat dissipation rises rapidly, primarily due to evaporative heat loss. At a temperature of approximately 30°C, the total rate of heat loss equals the solar heat influx. In other words, any tendency for solar irradiation to raise the temperature of the ocean surface above 30°C only results in an increased rate of evaporation, and the surface temperature does not change.

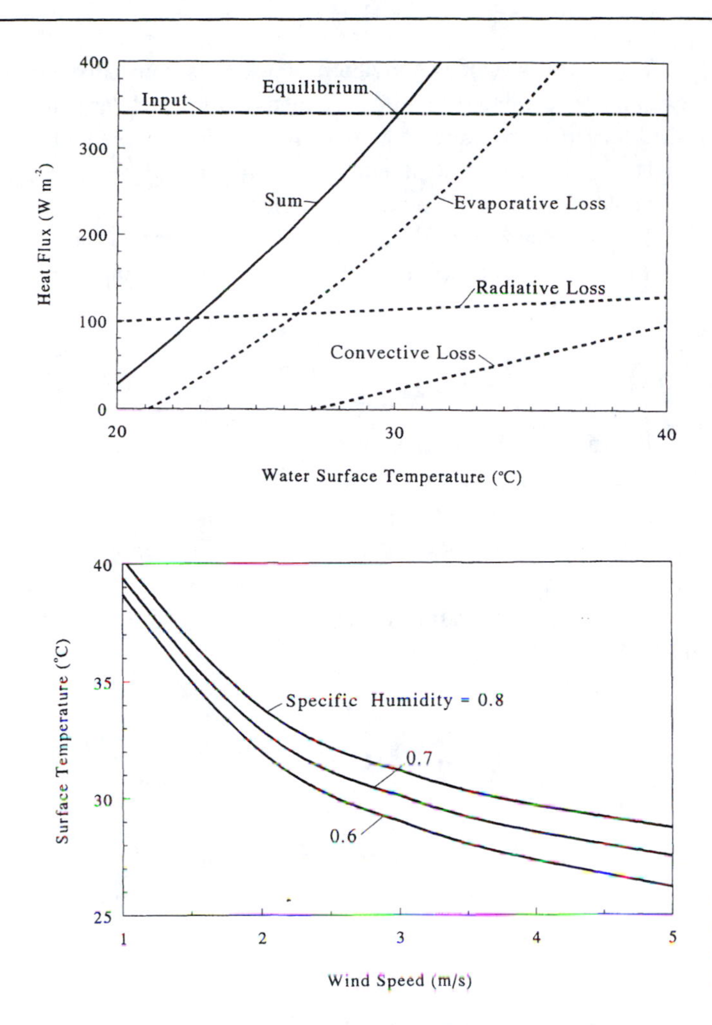

Fig. 14.14 The temperature of ocean surface water reaches an equilibrium when the sum of various forms of heat loss equals the influx of heat from solar radiation. (Redrawn from Newell 1979 by permission of Sigma Xi, the Scientific Research Society)

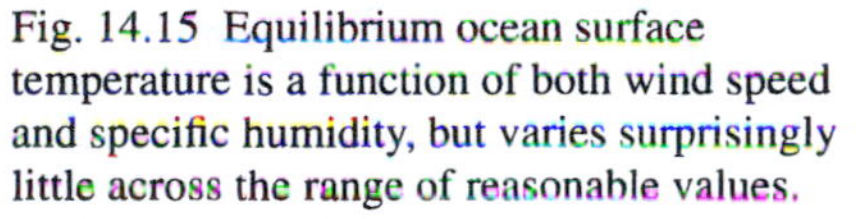

Fig. 14.15 Equilibrium ocean surface temperature is a function of both wind speed and specific humidity, but varies surprisingly little across the range of reasonable values.

Because conductive heat loss is such a small fraction of the overall heat loss, these calculations are not particularly sensitive to our choice of air temperature. The calculated sea surface temperature is, however, sensitive to the average wind speed and to the specific humidity of the air (fig. 14.15). For example, if the wind drops to 1 m s^{-1} and the specific humidity rises to 0.8, the equlibrium sea surface temperature is predicted to be 31.2°C. Conversely, if the average wind blows at 5 m s^{-1} and $\mathcal{H}_s$ is only 0.6, the sea surface temperature can fall as low as 26.2°C. However, even these relatively drastic changes in the average wind and humidity do not substantially change the predicted ocean temperature, and the assertion made in chapter 2 is likely to be valid: due to water's high latent heat of vaporization, the ocean surface temperature is unlikely to have ever risen much above 30°C. Thus, the physics of evaporation serves to set one of the primary characteristics of the biosphere.

14.6 Summary . . .

Because the concentration gradient of water vapor around leaves is large and the diffusion coefficient of water vapor relatively high, plants rapidly lose water in the process of taking up carbon dioxide. CAM plants circumvent this problem by absorbing CO_2 at night when the ambient air temperature is low and the relative humidity is, as a consequence, high.

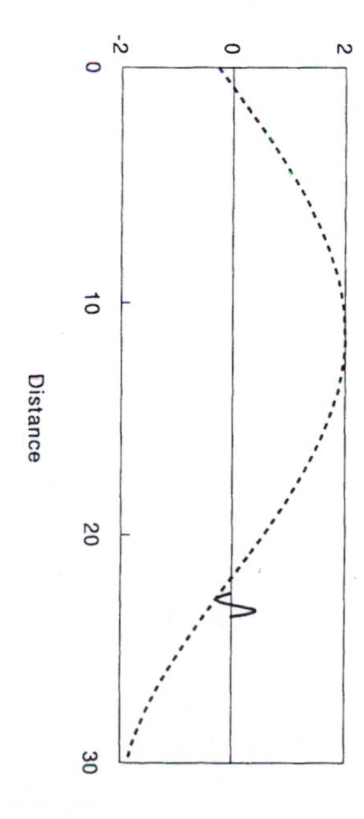

The latent heat of vaporization of water is unusually large, and, as a result, evaporative cooling is an effective method by which large terrestrial animals can shed their excess heat of metabolism. Heat lost to evaporation in the process of breathing can be a substantial thermal drain to small animals, but the maintenance of a cold nose can ameliorate the problem.

The maximum temperature of ocean surface water is constrained by evaporative cooling, with the consequence that ocean surface temperature has been virtually constant through geological time.

14.7 . . . and a Warning

The material in this chapter is presented as an exploration of the physics of water evaporating in a biological context. In the process we have touched upon the physiology of water loss by plants and the evaporative cooling of animals, but this discussion should not be construed as a thorough examination of these subjects. It is hoped, rather, that you will view the present exposition as an invitation to pursue the subject of evaporation in texts such as Nobel (1983) and Schmidt-Nielsen (1979).

Chapter 15

A Thought at the End

Albert Einstein once remarked, "It is the theory which decides what we can observe." The comment was made to Werner Heisenberg, who interpreted it in a literal sense, and used it as a guidepost in the elucidation of the uncertainty principle of quantum mechanics (Heisenberg 1971). But taken somewhat less literally, Einstein's remark applies to all of science. Much of what we see is determined by the theories to which we have previously been exposed.

There has been theory aplenty in the last fourteen chapters, but the value of its exposure lies not in the conclusions drawn here, but instead in the possibility that it will allow you to see something new and interesting in the world around you. I hope that someday you will sit and watch a tree sway in the wind, a beetle crawl, or a ripple dance on a pond, and some piece of the great puzzle will fall into place in your mind.

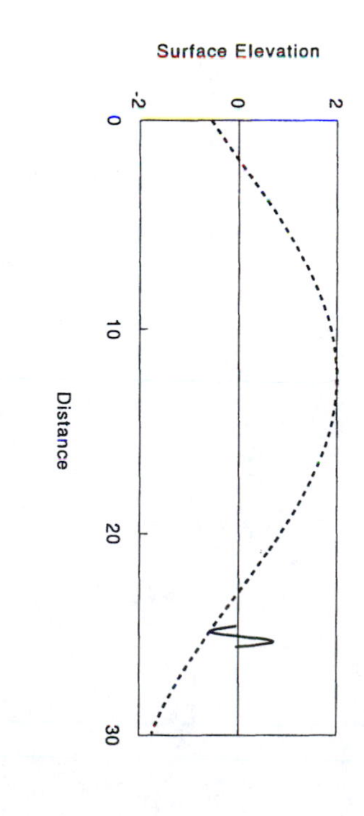

Literature Cited

Adamson, A. W. 1967. *Physical Chemistry of Surfaces*. 2d ed. Wiley Interscience, New York.

Alexander, R. McN. 1966. Physical aspects of swimbladder function. *Biol. Rev.* 41:141–176.

Alexander, R. McN. 1968. *Animal Mechanics.* University of Washington Press, Seattle.

Alexander, R. McN. 1971. *Size and Shape.* Edward Arnold, London.

Alexander, R. McN. 1982. *Locomotion of Animals.* Chapman and Hall, New York.

Alexander, R. McN. 1983. *Animal Mechanics.* 2d ed. Blackwell Scientific, London.

Alexander, R. McN. 1989. *Dynamics of Dinosaurs and Other Extinct Giants.* Columbia University Press, New York.

Alexander, R. McN. 1990. Size, speed and buoyancy adaptations in aquatic animals. *Amer. Zool.* 30:189–196.

Amsler, C. D., and M. Neushul. 1989. Chemotactic effects of nutrients on the spores of the kelps *Macrocystis pyrifera* and *Pterogophera californica. Mar. Biol.* 102:557–564.

Amsler, C. D., and R. B. Searles. 1980. Vertical distribution of seaweed spores in a water column offshore of North Carolina. *J. Phycol.* 16:617–619.

Armstrong, W. 1979. Aeration in higher plants. *Adv. in Bot. Res.* 7:225–332.

Atkins, P. W. 1984. *The Second Law.* Scientific American Library, New York.

Au, D., and D. Weihs. 1980. At high speeds dolphins save energy by leaping. *Nature* 244:548–550.

Augspurger, C. K., and S. E. Franson. 1987. Wind dispersal of artificial fruits varying in mass, area, and morphology. *Ecology* 69:27–42.

Autrum, H., ed. 1979. Comparative physiology and evolution of vision in invertebrates. A: Invertebrate photoreceptors. In *Handbook of Sensory Physiology*, vol. 7/6A. Springer-Verlag, New York.

Autrum, H., ed. 1981. Comparative physiology and evolution of vision in invertebrates. B: Invertebrate visual centers and behavior. In *Handbook of Sensory Physiology*, vol. 7/6B. Springer-Verlag, New York.

Bagnold, R. A. 1942. *The Physics of Blown Sand and Desert Dunes.* Wieliam Morrow, New York.

Bascom, W. 1980. *Waves and Beaches.* 2d ed. Anchor Press/Doubleday, New York.

Bearman, G., ed. 1989. *Seawater: Its Composition, Properties and Behaviour.* Pergamon Press, Oxford.

Berg, H. 1983. *Random Walks in Biology.* Princeton University Press, Princeton, N.J.

Berg, H. C., and D. A. Brown. 1972. Chemotaxis in *Escherichia coli* analysed by three-dimensional tracking. *Nature* 239:500–504.

Berg, H. C., and E. M. Purcell. 1977. Physics of chemoreception. *Biophys. J.* 20:193–219.

Bikerman, J. J. 1970. *Physical Surfaces.* Academic Press, New York.

Bird, R. B., W. E. Stewart, and E. N. Lightfoot. 1960. *Transport Phenomena.* John Wiley, New York.

Block, B. A. 1991. Endothermy in fish: Thermogenesis, ecology and evolution. In P. W. Hochochka and T. M. Mommsen, eds., *Biochemistry and Molecular Biology of Fishes*, vol. 1, *Phylogenetic and Biochemical Perspectives*, pp. 269–311. Elsevier, Amsterdam.

Bond, C. 1979. *Biology of Fishes.* W. B. Saunders, Philadelphia.

Bowden, K. F. 1964. Turbulence. *Oceanogr. Mar. Biol.* 2:11–30.

Boyer, C. B. 1987. *The Rainbow: From Myth to Mathematics.* Princeton University Press, Princeton, N.J.

Brett, J. R. 1965. The relation of size to rate of oxygen consumption and sustained swimming speed of sockeye salmon (*Oncorhychus nerka*). *J. Fish. Res. Bd.* (Canada) 22:1491–1497.

Bryant, H. C., and N. Jarmie. 1974. The glory. *Sci. Amer.* 231(7):60–73.

Budyko, M. I. 1974. *Climate and Life.* Academic Press, New York.

Bullock, T. H., and R. B. Cowles. 1952. Physiology of an infrared receptor: The facial pits of pit vipers. *Science* 115:541–543.

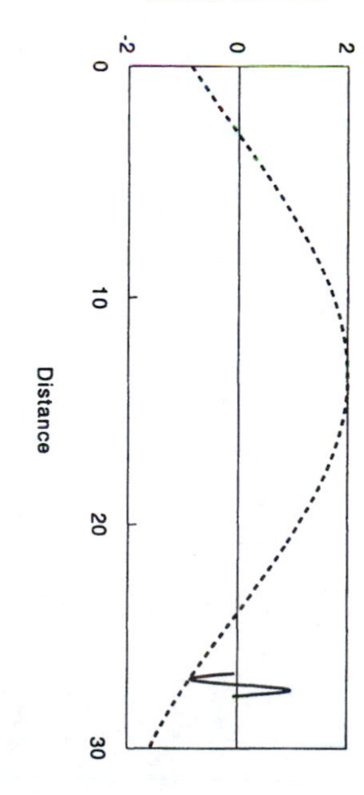

Literature Cited

Bullock, T. H., and W. Heiligenberg. 1986. *Electroreception.* Wiley Interscience, New York.

Bullock, T. H., and G. A. Horridge. 1965. *Structure and Function in the Nervous Systems of Invertebrates.* 2 vols. W. H. Freeman, San Francisco.

Burkhardt, D. 1989. UV vision: A bird's eye view of feathers. *J. Comp. Physiol.* A 164:787–796.

Burkhardt, D., and E. Maier. 1989. The spectral sensitivity of a passerine bird is highest in the UV. *Naturwissenschaften* 76:82–83.

Calkins, J., ed. 1982. *The Role of Solar Ultraviolet Radiation in Marine Ecosystems.* Plenum Press, New York.

Campbell, G. S. 1977. *An Introduction to Environmental Biophysics.* Springer-Verlag, New York.

Carey, F. G., and J. M. Teal. 1966. Heat conservation in tuna fish muscle. *Proc. Natl. Acad. Sci.* (USA) 56:1464–1469.

Carey, F. G., and J. M. Teal. 1969. Mako and porbeagle: Warm-bodied sharks. *Comp. Biochem. Physiol.* 28:199–204.

Carey, F. G., J. M. Teal, J. W. Kanwisher, and K. D. Lawson. 1971. Warm-bodied fish. *Amer. Zool.* 11:137–145.

Carslaw, H. S., and J. C. Jaeger. 1959. *Conduction of Heat in Solids.* 2d. ed. Oxford University Press, New York.

Chapman, R. F. 1982. *The Insects: Structure and Function.* 2d ed. Harvard University Press, Cambridge, Mass.

Clarke, M. R. 1979. The head of the sperm whale. *Sci. Amer.* 240(1):128–141.

Clay, C. S., and H. Medwin. 1977. *Acoustical Oceanography.* John Wiley, New York.

Clements, J. A., J. Nellenbogen, and H. J. Trahan. 1970. Pulmonary surfactant and the evolution of the lung. *Science* 169:603–604.

Cloud, P. 1988. *Oasis in Space.* W. W. Norton, New York.

Crank, J. 1975. *The Mathematics of Diffusion.* 2d ed. Oxford University Press, New York.

Crenshaw, H. C. 1990. Helical orientation—a novel mechanism for the orientation of microorganisms. *Lecture Notes in Biomath.* 89:361–386.

Crisp, D. J., and W. H. Thorpe. 1948. The water-protecting properties of insect hairs. *Discussions of the Faraday Society* 3:210–220.

Dacey, J. W. H. 1981. Pressurized ventilation in the yellow water lily. *Ecology* 62:1137–1147.

Damkaer, D. M., D. B. Day, G. A. Heron, and E. F. Prentice. 1980. Effects of UV-B radiation on near-surface zooplankton of Puget Sound. *Oecologia* (Berlin) 44:149–158.

Daniel, T. L. 1982. The role of added mass in impulsive locomotion with special reference to medusae. Ph.D. thesis, Duke University, Durham, N.C.

Daniel, T. L. 1984. Unsteady aspects of aquatic locomotion. *Amer. Zool.* 24:121–134.

Davies, J. T., and E. K. Rideal. 1963. *Interfacial Phenomena.* Academic Press, New York.

Davson, H. 1972. *The Physiology of the Eye.* 3d ed. Academic Press, New York.

Dejours, P. 1975. *Principles of Comparative Respiratory Physiology.* North-Holland Publishing Co., Amsterdam.

DeMont, M. E., and J. M. Gosline. 1988. Mechanics of jet propulsion in the hydromedusan jellyfish, *Polyorchis penicillatus.* III: A natural resonating bell; the presence and importance of a resonant phenomenon in the locomotor structure. *J. Exp. Biol.* 134:347–361.

Denny, M. W. 1976. The physical properties of spider's silk and their role in the design of orb-webs. *J. Exp. Biol.* 65:483–506.

Denny, M. W. 1987. Lift as a mechanism of patch initiation in mussel beds. *J. Exp. Mar. Biol. Ecol.* 113:231–245.

Denny, M. W. 1988. *Biology and the Mechanics of the Wave-Swept Environment.* Princeton University Press, Princeton, N.J.

Denny, M. W. 1991. Biology, natural selection, and the prediction of maximal wave-induced forces. *S. Afr. J. Mar. Sci.* 10:353–363.

Denny, M. W., and S. D. Gaines. 1990. On the prediction of maximal intertidal wave force. *Limnol. Oceanogr.* 35:1–15.

Denny, M. W., T. L. Daniel, and M.A.R. Koehl. 1985. Mechanical limits to size in wave-swept organisms. *Ecol. Monogr.* 55:69–102.

Denny, M. W., V. Brown, E. Carrington, G. Kramer, and A. Miller. 1989. Fracture mechanics and the survival of wave-swept macroalgae. *J. Exp. Mar. Biol. Ecol.* 127:211–228.

Diamond, J. 1989. How cats survive falls from New York skyscrapers. *Natural History* (August), pp. 21–26.

Dill, L. M. 1977. Refraction and the spitting behavior of the archer fish (*Toxotes chatareus*). *Behav. Ecol. Sociobiol.* 2:169–184.

Dixon, A.F.G., P. C. Croghan, and R. P. Gowing. 1990. The mechanism by which aphids adhere to smooth surfaces. *J. Exp. Biol.* 152:243–253.

Emerson, S., and D. Diehl. 1980. Toe-pad morphology and mechanisms of sticking in frogs. *Biol. J. Linn. Soc.* 13:199–216.

Ewing, A. W. 1989. *Arthropod Bioacoustics.* Comstock, Ithaca, N.Y.

Farlow, J. O., C. V. Thompson, and D. E. Rosner. 1976. Plates of the dinosaur *Stegasaurus*: Forced convection heat loss fins? *Science* 192:1123–1125.

Fay, R. R., and A. N. Popper. 1974. Acoustic stimulation of the ear of the goldfish (*Carassius auratus*). *J. Exp. Biol.* 61:243–260.

Feinsinger, P., R. K. Colwell, J. Terborgh, and S. B. Chapin. 1979. Elevation and the morphology, flight energetics, and foraging ecology of tropical hummingbirds. *Amer. Nat.* 113:481–497.

Fessard, A., ed. 1974. Electroreceptors and other specialized receptors in lower vertebrates. In *Handbook of Sensory Physiology*, vol. 3/3. Springer-Verlag, New York.

Feynman, R. P., R. B. Leighton, and M. Sands. 1963. *The Feynman Lectures in Physics*, vol. 1. Addison-Wesley, Reading, Mass.

Fish, F. E. 1990. Wing design and scaling of flying fish with regard to flight performance. *J. Zool.* (Lond.) 221:391–403.

Fish, F. E., B. R. Blood, and B. D. Clark. 1991. Hydrodynamics of the feet of fish-catching bats: Influence of the water surface drag and morphological design. *J. Exp. Zool.* 258:164–173.

Fuchs, N. A. 1964. *The Mechanics of Aerosols.* Pergamon Press, Oxford.

Gamow, R. I., and J. F. Harris. 1973. The infrared receptors of snakes. *Sci. Amer.* 228(5):94–101.

Gates, D. M. 1980. *Biophysical Ecology.* Springer-Verlag, New York.

Gosline, J. M., and R. E. Shadwick. 1983. The role of elastic energy storage mechanisms in swimming: An analysis of mantle elasticity in escape jetting in the squid, *Loligo opalescens*. *Can. J. Zool.* 61:1421–1431.

Grace, J. 1977. *Plant Response to Wind.* Academic Press, New York.

Grant, W. D., and O. S. Madsen. 1986. The continental-shelf bottom boundary layer. *Ann. Rev. Fluid Mech.* 18:265–305.

Greenhill, A. G. 1881. Determination of the greatest height consistent with stability that a vertical pole or mast can be made, and of the greatest height to which a tree of given proportions can grow. *Proc. Cambridge Philos. Soc.* 4:65–73.

Greenler, R. 1980. *Rainbows, Halos, and Glories.* Cambridge University Press, New York.

Griffin, D. R. 1986. *Listening in the Dark.* Comstock, Ithaca, N.Y.

Gross, M. G. 1990. *Oceanography: A View of the Earth.* 5th ed. Prentice Hall, Englewood Cliffs, N.J.

Gulick, W. L., G. A. Gascheider, and R. O. Frisina. 1989. *Hearing.* Oxford University Press, Oxford.

Haldane, J.B.S. 1985. *On Being the Right Size and Other Essays.* Oxford University Press, Oxford.

Hamilton, W. J., and M. K. Seely. 1976. Fog basking by the Namib Desert beetle, *Onymacris unguicularis*. *Nature* 262:284–285.

Happel, J., and H. Brenner. 1983. *Low Reynolds Number Hydrodynamics.* Martinus Nijhof, Amsterdam.

Harris, J. F., and R. I. Gamow. 1971. Snake infrared receptors: Thermal or photochemical mechanism? *Science* 172:1252–1253.

Hawkins, A. D., and A. A. Myrberg. 1983. Hearing and sound communication under water. In B. Lewis, ed., *Bioacoustics: A Comparative Approach*, pp. 347–405. Academic Press, New York.

Hayward, A.T.J. 1971. Negative pressure in liquids: Can it be harnessed to serve man? *Amer. Sci.* 59:434–443.

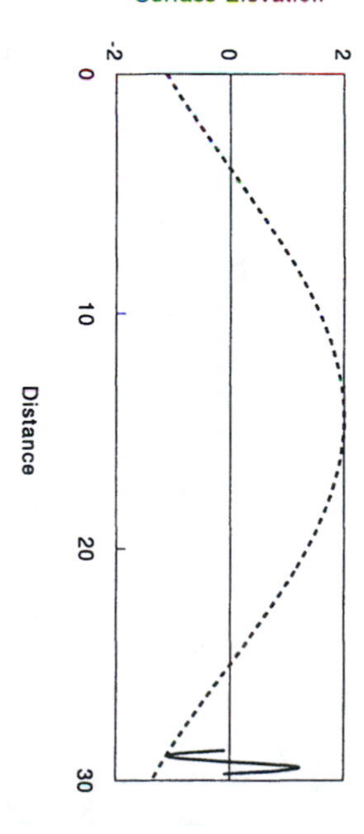

Literature Cited

Heinrich, B. 1979. *Bumble-Bee Economics.* Harvard University Press, Cambridge, Mass.

Heisenberg, W. 1971. *Physics and Beyond.* Harper, New York.

Hemmingsen, A. M. 1950. The relation of standard (basal) energy metabolism to total fresh weight of living organisms. *Reports Steno Mem. Hosp. & Nordisk Insulinlab.* 4:7–58.

Hills, B. A. 1988. *The Biology of Surfactant.* Cambridge University Press, New York.

Hinton, H. E. 1976. Plastron respiration in bugs and beetles. *J. Insect Physiol.* 22:1529–1550.

Hoerner, S. F. 1965. *Fluid-Dynamic Drag.* Hoerner Fluid Dynamics, Bricktown, N.J.

Holland, H. D. 1984. *The Chemical Evolution of the Atmosphere and Oceans.* Princeton University Press, Princeton, N.J.

Horridge, G. A., ed. 1975. *The Compound Eye and Vision of Insects.* Clarendon Press, Oxford.

Hughes, D. E., and J.W.T. Wimpenny. 1969. Oxygen metabolism by microorganisms. In A. H. Rose and J. F. Wilkinson, eds., *Advances in Microbial Physiology,* vol. 3, pp. 197–231. Academic Press, New York.

Hughes, D. J., and R.N.P.G. Hughes. 1986. Metabolic implications of modularity: Studies on the respiration and growth of *Electra pilosa. Phil. Trans. Roy. Soc.* (Lond.) B 313:23–29.

Humphrey, J.A.C. 1987. Fluid mechanic constraints on spider ballooning. *Oecologia* (Berlin) 73:469–477.

Hutchinson, G. E. 1957. *A Treatise on Limnology.* John Wiley, New York.

Irving, L. 1969. Temperature regulation in marine mammals. In H. T. Andersen, ed., *The Biology of Marine Mammals,* pp. 147–174. Academic Press, New York.

Jenkins, F. A., and H. E. White. 1957. *Fundamentals of Optics.* 3d ed. McGraw-Hill, New York.

Jerlov, N. G. 1976. *Marine Optics.* Elsevier, Amsterdam.

Kalmijn, A. J. 1971. The electric sense of sharks and rays. *J. Exp. Biol.* 55:371–383.

Kalmijn, A. J. 1974. The detection of electric fields from manmade and animal sources other than electric organs. In A. Fessard, ed., *Handbook of Sensory Physiology,* vol. 3/3, pp. 147–200. Springer-Verlag, New York.

Kalmijn, A. J. 1984. Theory of electromagnetic orientation: A further analysis. In L. Bolis, R. D. Keynes, and S.H.P. Maddrell, eds., *Comparative Physiology of Sensory Systems,* pp. 525–560. Cambridge University Press, New York.

Kerfoot, O. 1968. Mist precipitation in vegetation. *Forestry Abst.* 29:8–20.

Khurana, A. 1988. Numerical simulations reveal fluid flows near solid boundaries. *Physics Today* 41(5):17–19.

Kier, W. M., and A. M. Smith. 1990. The morphology and mechanics of octopus suckers. *Biol. Bull.* 178:126–136.

King, D., and O. L. Loucks. 1978. The theory of tree bole and branch form. *Rad. and Environ. Biophys.* 15:141–165.

Kinsler, L. E., and A. R. Frey. 1962. *Fundamentals of Acoustics.* 2d ed. John Wiley, New York.

Kinsman, B. 1965. *Wind Waves.* Prentice Hall, Englewood Cliffs, N.J.

Knudsen, E. I. 1975. Spatial aspects of the electric fields generated by weakly electric fish. *J. Comp. Physiol.* 99:103–118.

Knudsen, E. I. 1981. The hearing of the barn owl. *Sci. Amer.* 245(6):113–125

Knutson, R. M. 1974. Heat production and temperature regulation in Eastern skunk cabbage. *Science* 186:746–747.

Kober, R. 1986. Echoes of fluttering insects. In P. E. Nachtigall and P.W.B Moore, eds., *Animal Sonar,* pp. 477–487. Plenum Press, New York.

Koehl, M.A.R. 1977. Mechanical diversity of connective tissue of the body wall of sea anemones. *J. Exp. Biol.* 69:107–126.

Koehl, M.A.R., and J. R. Strickler. 1981. Copepod feeding currents: Food capture at low Reynolds number. *Limnol. Oceanogr.* 26:1062–1073.

Konishi, M. 1975. How the owl tracks its prey. *Amer. Sci.* 61:414–424.

LaBarbera, M. 1984. Feeding currents and particle capture mechanisms in suspension feeding organisms. *Amer. Zool.* 24:71–84.

LaBarbera, M. 1990. Principles of design of fluid transport systems in zoology. *Science* 249:992–1000.

Lamb, H. 1945. *Hydrodynamics.* Dover, New York.

Land, M. F. 1965. Image formation by a concave reflector in the eye of a scallop, *Pecten maximus. J. Physiol.* 179:138–153.

Land, M. F. 1978. Animal eyes with mirror optics. *Sci. Amer.* 239(6):126–134.

Land, M. F. 1981. Optics and vision in invertebrates. In H. Autrum, ed., *Comparative physiology and evolution of vision in invertebrates. B: Invertebrate visual centers and behavior* I. *Handbook of Sensory Physiology*, vol. 7/6B, pp. 471–594. Springer-Verlag, New York.

Langbauer, W. R., Jr., K. B. Payne, R. A. Chairif, L. Rapaport, and F. Osborn. 1991. African elephants respond to distant playbacks of low-frequency conspecific calls. *J. Exp. Biol.* 157:35–46.

Lasiewski, R. C. 1963. Oxygen consumption of torpid, resting, active, and flying hummingbirds. *Physiol. Zool.* 36:122–140.

Laties, G. G. 1982. The cyanide resistant alternative path in higher plants. *Ann. Rev. Plant. Physiol.* 33:519–555.

Lazier, J.R.N., and K. H. Mann. 1989. Turbulence and the diffusive layers around small organisms. *Deep-Sea Res.* 36:1721–1733.

Lewis, B. 1983. *Bioacoustics: A Comparative Approach.* Academic Press, New York.

Lillywhite, H. B. 1987. Circulatory adaptations of snakes to gravity. *Amer. Zool.* 27:81–95.

Lillywhite, H. B. 1988. Snakes, blood circulation and gravity. *Sci. Amer.* 259(6):92–98.

Lissman, H. W., and K. E. Machin. 1958. The mechanism of object location in *Gymarchus niloticus* and similar fish. *J. Exp. Biol.* 35:451–486.

List, R. J. 1958. Smithsonian meteorological tables. 6th ed. Smithsonian Misc. Collections 114 (entire volume).

Louw, G., and M. Seely. 1982. *Ecology of Desert Organisms.* Longman Group, Ltd., Essex, U.K.

McFall-Ngai, M. J. 1990. Crypsis in the pelagic environment. *Amer. Zool.* 30:175–188.

McFarlan, D. 1990. *The Guinness Book of World Records.* Bantam Books, New York.

Mann, K. H., and J.R.N. Lazier. 1991. *Dynamics of Marine Ecosystems.* Blackwell Scientific, New York.

Marrero, T. R., and E. A. Mason. 1972. Gaseous diffusion coefficients. *J. Physical and Chem. Ref. Data* 1:3–118.

Massey, B. S. 1983. *Mechanics of Fluids.* 5th ed. Van Nostrand-Reinhold, New York.

Meeuse, B.J.D. 1966. The voodoo lily. *Sci. Amer.* 215(7):80–88.

Meeuse, B.J.D. 1975. Thermogenic respiration in aroids. *Ann. Rev. Plant. Physiol.* 26:117–126.

Meglitsch, P. A. 1972. *Invertebrate Zoology.* 2d ed. Oxford University Press, New York.

Mehrbach, C., C.H. Culberson, J. E. Hawley, and R. M. Pytkowicz. 1973. Measurement of the apparent dissociation constants of carbonic acid in seawater at atmospheric pressure. *Limnol. Oceanogr.* 18:897–907.

Middleton, G. V., and J. B. Southard. 1984. *Mechanics of Sediment Movement.* Lecture notes for short course no. 3, Society of Experimental Paleontologists and Mineralogists, Tulsa, Oklahoma.

Miller, R. L. 1982. Sperm chemotaxis in ascideans. *Amer Zool.* 22:827–840.

Miller, R. L. 1985a. Sperm chemotaxis in echinodermata: Asteroidea, Holothuroidea, Ophiuroidea. *J. Exp. Zool.* 234:383–414.

Miller, R. L. 1985b. Sperm chemo-orientation in the metazoa. In C. B. Metz and A. Monroy, eds., *The Biology of Fertilization*, vol. 2, pp. 275–337. Academic Press, New York.

Minnaert, M. 1954. *The Nature of Light and Colour in the Open Air.* Dover, New York.

Monteith, J. L. 1973. *Principles of Environmental Physics.* Edward Arnold, London.

Morse, P. M., and K. U. Ingard. 1968. *Theoretical Acoustics.* Princeton University Press, Princeton, N.J.

Munk, W. H. 1949. The solitary wave theory and its application to surf problems. *Annals N.Y. Acad. Sci.* 51:376–424.

Murray, C. D. 1926. The physiological principle of minimum work applied to the angle of branching of arteries. *J. Gen. Physiol.* 9:835–841.

Muschenheim, D. K. 1987. The dynamics of near-bed seston flux and suspension-feeding benthos. *J. Mar. Res.* 45:473–496.

Nachtigall, P. E., and P.W.B. Moore, eds. 1986. *Animal Sonar: Processes and Perfor-*

mance. Proc. of a NATO Advanced Study Institute conference on animal sonar systems, September 10–19, 1986, Helsingor, Denmark. Plenum Press, New York.
Nagy, K. A., D. K. Odell, and R. S. Seymour. 1972. Temperature regulation by the inflorescence of Philodendron. *Science* 178:1195–1197.
Newell, R. E. 1979. Climate and the ocean. *Amer. Sci.* 67:405–416.
Niklas, K. J. 1982a. Simulated and empiric wind pollination patterns of conifer ovulate cones. *Proc. Natl. Acad. Sci.* (USA) 79:510–514.
Niklas, K. J. 1982b. Pollination and airflow patterns around conifer cones. *Science* 217:442–444.
Niklas, K. J. 1987. Aerodynamics of wind pollination. *Sci. Amer.* 257(1):90–95.
Nobel, P. S. 1983. *Biophysical Plant Physiology and Ecology.* W. H. Freeman, New York.
Nowell, A.R.M., and P. A. Jumars. 1984. Flow environment of aquatic benthos. *Ann. Rev. Ecol. Syst.* 15:303–328.
Oertli, J. J. 1971. The stability of water under tension in the xylem. *Zeitschr. für Pflanzenphysiol.* 65:195–209.
Okubo, A. 1987. Fantastic voyage into the deep: Marine biofluid mechanics. In E. Teromoto and M. Yamaguti, eds., *Mathematical Topics in Population Biology, Morphogenisis, and Neuroscience*, pp. 32–47. Springer-Verlag, New York.
Owen, D. 1980. *Camouflage and Mimicry.* University of Chicago Press, Chicago.
Payne, R., and D. Webb. 1971. Orientation by means of long-range acoustic signaling in baleen whales. *Ann. N.Y. Acad. Sci.* 188:110–141.
Pearcy, R. W., J. Ehleringer, H. A. Mooney, and P. W. Rundel, eds. 1989. *Plant Physiological Ecology.* Chapman and Hall, New York.
Pich, J. 1966. Theory of aerosol filtration by fibrous and membrane filters. In C. N. Davies, ed., *Aerosol Science*, pp. 223–285. Academic Press, New York.
Pickard, W. F. 1974. Transition regime diffusion and the structure of the insect tracheolar system. J. *Insect Physiol.* 20:947–956.
Poisson, A. 1980. Conductivity/salinity/temperature relationship of diluted and concentrated seawater. *IEEE J. Oceanic Engineering.* OE-5(1):41–50.
Popper, A. N. 1980. Sound emission and detection by dolphinids. In L. M. Herman, ed., *Cetacean Behavior*, pp. 1–52. Wiley Interscience, New York.
Prange, H. D., and K. Schmidt-Nielsen. 1972. The metabolic cost of swimming in ducks. *J. Exp. Biol.* 53:763–777.
Princen, H. M. 1969. The equilibrium shapes of interfaces, drops, and bubbles: Rigid and deformable particles at interfaces. *Surf. and Colloid Sci.* 2:1–84.
Prothro, J. W. 1979. Maximal oxygen consumption in various animals and plants. *Comp. Biochem. Physiol.* A 64:463–466.
Pumphry, R. J. 1961. Concerning vision. In J. A. Ramsay and V. B. Wigglesworth, eds., *The Cell and the Organism*, pp. 193–208. Cambridge University Press, Cambridge, U.K.
Rahn, H., and C. V. Paganelli. 1979. How bird eggs breathe. *Sci. Amer.* 240(2):46–55.
Raskin, I., and H. Kende. 1985. Mechanisms of aeration in rice. *Science* 228:327–329.
Reif, F. 1965. *Fundamentals of Statistical and Thermal Physics.* McGraw-Hill, New York.
Resnick, R., and D. Halliday. 1966. *Physics.* John Wiley, New York.
Reynolds, C. S., and A. E. Walsby. 1975. Water blooms. *Biol. Rev.* 50:437–481.
Riley, G. A., H. Stommel, and D. F. Bumpus. 1949. Quantitative ecology of the plankton of the Western North Atlantic. *Bull. Bingham Oceanographic Collection*, vol. 12, art. 3.
Ritzman, R. 1973. Snapping behavior of the shrimp *Alpheus californiensis. Science* 181:459–460.
Rubenstein, D. L., and M.A.R. Koehl. 1977. The mechanism of filter feeding: Some theoretical considerations. *Amer. Nat.* 111:981–994.
Sand, O., and A. D. Hawkins. 1973. Acoustic properties of the cod swimbladder. *J. Exp. Biol.* 58:797–820.
Scheich, H., G. Langner, C. Tidermann, R. Coles, and A. Guppy. 1986. Electroreception and electrolocation in platypus. *Nature* 319:401–402.
Scheidegger, A. E. 1971. *The Physics of Flow through Porous Media.* 3d ed. University of Toronto Press, Toronto.
Schey, H. M. 1973. *Div, Grad, Curl and All That.* W. W. Norton, New York.
Schlee, S. 1973. *The Edge of an Unfamiliar World.* E. P Dutton, New York.

Schlichting, H. 1979. *Boundary-Layer Theory.* 7th ed. McGraw-Hill, New York.

Schmidt-Nielsen, K. 1972a. Locomotion: Energy cost of swimming, flying and running. *Science* 177:222–226.

Schmidt-Nielsen, K. 1972b. *How Animals Work.* Cambridge University Press, New York.

Schmidt-Nielsen, K. 1979. *Animal Physiology.* 2d ed. Cambridge University Press, New York.

Schmidt-Nielsen K. 1984. *Scaling: Why Animal Size Is So Important.* Cambridge University Press, New York.

Schnitzler, H.-U., D. Menne, R. Kober, and K. Heblich. 1983. The acoustical image of fluttering insects in echolocating bats. In F. Haber and H. Markl, eds., *Neuroethology and Behavioral Physiology*, pp. 235–250. Springer-Verlag, New York.

Schopf, J. W. 1978. The evolution of the earliest cells. *Sci. Amer.* 239(9):110–138.

Schopf, T.J.M. 1980. *Paleoceanography.* Harvard University Press, Cambridge, Mass.

Schusterman, R. J. 1972. Visual acuity in pinnipeds. In H. E. Winn and B. L. Olla, eds., *Behavior of Marine Animals*, vol. 2, *Vertebrates*, pp. 469–492. Plenum Press, New York.

Seely, M. K., and W. J. Hamilton. 1976. Fog catchment sand trenches constructed by Tenebrionid beetles, *Lepidochora*, from the Namib Desert. *Science* 193(4252):484–486.

Shaw, E.A.G. 1974. The external ear. In W. D. Keidel and W. D Neff, eds., *Handbook of Sensory Physiology*, vol. 5/1, pp. 455–490. Springer-Verlag, New York.

Sherman, T. F. 1981. On connecting large vessels to small: The meaning of Murray's Law. *J. Gen. Physiol.* 78:431–453.

Shifrin, K. S. 1988. *Physical Optics of Ocean Water.* American Institute of Physics, New York.

Shimozawa, T., and M. Kanou. 1984. The aerodynamics and sensory physiology of range fractionation in the cercal filiform sensilla of the cricket *Gryllus bimaculatus*. *J. Comp. Physiol.* A 155:495–505.

Siebeck, O. 1988. Experimental investigation of UV tolerance in hermatypic corals (Scleractinia). *Mar. Ecol. Progr. Ser.* 43:95–103.

Simmons, J. A., and A. D. Grinnell. 1986. The performance of echolocation: Acoustic images perceived by echolocating bats. In P. E. Nachtigall and P.W.B. Moore, eds., *Animal Sonar*, pp. 353–385. Plenum Press, New York.

Smayda, T. J. 1970. The suspension and sinking of phytoplankton in the sea. *Oceanogr. Mar. Biol. Ann. Rev.* 8:353–414.

Smith, A. M. 1991. Negative pressure generated by octopus suckers: A study of the tensile strength of water in nature. *J. Exp. Biol.* 157:257–271.

Snyder, A. W. 1975a. Photoreceptor optics: Theoretical principles. In A. W. Snyder and R. Menzel, eds., *Photoreceptor Optics*, pp. 38–55. Springer-Verlag, New York.

Snyder, A. W. 1975b. Optical properties of invertebrate photoreceptors. In G. A. Horridge, ed., *The Compound Eye and Vision of Insects*, pp. 179–235. Clarendon Press, Oxford.

Stefan, J. 1874. *Sitzb. Akad. Wiss. Wien* (Mathem-naturwiss. Kl) 69:713. As cited in W. L. Wake, *Adhesion and the Formulation of Adhesives*, 2d ed., 1982. Applied Science Publishers, London.

Stork, N. E. 1980. Experimental analysis of adhesion of *Chrysolina polita* (Chrysomelidae:Coleoptera) on a variety of surfaces. *J. Exp. Biol.* 88:91–107.

Størmer, L. 1977. Arthropod invasion of land during late Silurian and Devonian times. *Science* 197:1362–1364.

Streeter, V. L. and E. B. Wylie. 1979. *Fluid Mechanics.* 7th ed. McGraw-Hill, New York.

Suga, N. 1990. Biosonar and neural computation in bats. *Sci. Amer.* 262(6):60–68.

Suthers, R. A. 1965. Acoustic orientation by fish-catching bats. *J. Exp. Zool.* 158:319–348.

Sverdrup, H. W., M. W. Johnson, and R. H. Fleming. 1942. *The Oceans.* Prentice Hall, Englewood Cliffs, N.J.

Tanford, C. 1961. *Physical Chemistry of Macromolecules.* John Wiley, New York.

Tavolga, W. N. 1971. Sound production and detection. In W. Hoar and D. Randall, eds., *Fish Physiology*, pp. 135–205. Academic Press, New York.

Timoshenko, S. P., and J. M. Gere. 1972. *Mechanics of Materials.* Van Nostrand, New York.

Trujillo-Cénoz, O. 1972. The structural organization of the compound eye in insects. In M.G.F. Fuortes, ed., *Physiology of Photoreceptor Organs*, pp. 5–62. Springer-Verlag, New York.

Literature Cited

Tucker, V. A. 1969. Wave making by whirligig beetles (*Gyrinidae*). *Science* 166:897–898.

UNESCO. 1983. Algorithms for computation of fundamental properties of seawater. UNESCO Technical Papers in Marine Science, no. 44. UNESCO, Paris.

UNESCO. 1987. International oceanographic tables. UNESCO Technical Papers in Marine Science, no. 40. UNESCO, Paris.

van Bergeijk, W. A. 1967. The evolution of vertebrate hearing. In W. D. Neff, ed., *Contributions to Sensory Physiology*, vol. 2, pp. 1–41. Academic Press, New York.

van der Pijl, L. 1982. *Principles of Dispersal in Higher Plants.* Springer-Verlag, Berlin.

Vermeij, G. J. 1978. *Biogeography and Adaptation: Patterns of Marine Life.* Harvard University Press, Cambridge, Mass.

Vogel, S. 1970. Convective cooling at low airspeeds and the shape of broad leaves. *J. Exp. Bot.* 21:91–101.

Vogel, S. 1981. *Life in Moving Fluids.* Willard Grant Press, Boston.

Vogel, S. 1983. How much air passes through a silkmoth's antenna? *J. Insect Physiol.* 29:597–602.

Vogel, S. 1987. Flow-assisted mantle cavity refilling in jetting squid. *Biol. Bull.* 172:61–68.

Vogel, S. 1988. *Life's Devices.* Princeton University Press, Princeton, N.J.

Vogel, S., and C. Loudon. 1985. Fluid mechanics of the thallus of an intertidal red alga, *Halosaccion glandiforme. Biol. Bull.* 168:161–174.

von Frisch, K. 1950. *Bees: Their Vision, Chemical Senses, and Language.* Cornell University Press, Ithaca, N.Y.

Wachmann, E., and W. D. Schroer. 1975. Zur Morphologie des Dorsal- und Ventralauges des Taumelkäfers *Gyrinus substriatus* (Steph.) (Coleoptera, Gyrinidae). *Zoomorphologie* 82:43–61.

Wainwright, S. A., W. D. Biggs, J. D. Currey, and J. M. Gosline. 1976. *Mechanical Design in Organisms.* Edward Arnold, London.

Walker, J.C.G., C. Kleinn, M. Schidlowski, J. W. Schopf, D. J. Stevensen, and M. R. Walter. 1983. Environmental evolution of the Archean–early Proterozoic earth. In J. W. Schopf, ed., *The Earth's Earliest Biosphere: Its Origin and Evolution*, pp. 260–290. Princeton University Press, Princeton, N.J.

Walls, G. 1942. *The Vertebrate Eye and Its Adaptive Radiation.* Cranbrook Institute of Science, Bloomfield Hills, Michigan.

Walsby, A. E. 1972. Gas-filled structures providing buoyancy in photosynthetic organisms. In Soc. Exp. Biol. Symp., no. 26, *The Effects of Pressure on Organisms*, pp. 233–250. The Company of Biologists, Ltd., Cambridge, U.K.

Walters, V., and H. Fierstine. 1964. Measurements of swimming speeds of yellowfin tuna and wahoo. *Nature* 202:208–209.

Warren, J. V. 1974. The physiology of the giraffe. *Sci. Amer.* 231(5):96–105.

Washburn, J. O., and L. Washburn. 1984. Active aerial dispersal of minute wingless arthropods: Exploitation of boundary-layer velocity gradients. *Science* 223:1088–1089.

Waterman, T. H. 1981. Polarization sensitivity. In H. Autrum, ed., *Comparative physiology and evolution of vision in invertebrates.* B: Invertebrate visual centers and behavior I. In *Handbook of Sensory Physiology*, vol. 7/6B, pp. 283–469. Springer-Verlag, New York.

Weast, R. C., ed. 1977. *CRC Handbook of Chemistry and Physics.* CRC Press, Cleveland, Ohio.

Webb, P. W. 1975. Hydrodynamics and energetics of fish propulsion. *Bull. Fish. Res. Bd.* (Canada) 190.

Weis-Fogh, T. 1964. Diffusion in insect wing muscle, the most active tissue known. *J. Exp. Biol.* 41:229–256.

Weiss, R. F. 1970. The solubility of nitrogen, oxygen and argon in water and seawater. *Deep-Sea Res.* 17:721–735.

Weiss, R. F. 1974. Carbon dioxide in water and seawater: The solubility of a non-ideal gas. *Mar. Chem.* 2:203–215.

Wever, E. G. 1974. The evolution of vertebrate hearing. In W. D. Keidel and W. D. Neff, eds., *Contributions to Sensory Physiology*, vol. 5, pp. 423–454. Springer-Verlag, New York.

Wickler, W. 1968. *Mimicry in Plants and Animals.* McGraw-Hill, New York.

Wigglesworth, V. B. 1987. How does a fly cling to the under surface of a glass sheet? *J. Exp. Biol.* 129:373–376.

Wilcox, R. S. 1979. Sex discrimination in *Gerris remigis:* Role of surface wave signal. *Science* 206:1325–1327.

Wood, W. F. 1989. Photoadaptive responses of the tropical red alga *Eucheuma striatum* Schmitz (Gigartinales) to ultraviolet radiation. *Aq. Bot.* 33:41–51.

Wu, T. Y. 1977. Introduction to the scaling of aquatic organisms. In T. J. Pedley, ed., *Scale Effects in Animal Locomotion*, pp. 203–232. Academic Press, New York.

Yen, J., and N. T. Nicoll. 1990. Setal array on the first antenna of a carnivorous marine copepod *Euchaeta norvegica. J. Crust. Biol.* 10:218–224.

Yorke, R. 1981. *Electrical Circuit Theory.* Pergamon Press, New York.

Yost, W. A., and D. W. Nielson. 1977. *Fundamentals of Hearing.* Holt, Rhinehart and Winston, New York.

Young, J. Z. 1964. *A Model of the Brain.* Oxford University Press, New York.

Author Index

Author Index

Subject Index

"Seawater, see water, sea"—a poem from the *Handbook of Chemistry and Physics*

Italicized page numbers denote tables.

Subject Index

Subject Index

Subject Index

Subject Index